Contemporary Mathematicians

Gian-Carlo Rota
Editor

Left to right: I.M. Gelfand, R.D. MacPherson, R. Bott, and D. Kazdhan

W. Odze and R. Bott, 1942

Raoul Bott
Collected Papers

Volume 3
Foliations

Robert D. MacPherson
Editor

1995

Birkhäuser
Boston • Basel • Berlin

Robert D. MacPherson
Department of Mathematics
Massachusetts Institute of Technology
Cambridge, MA 02139

Library of Congress Cataloging In-Publication Data

Bott, Raoul, 1924- [Works, 1993]
Collected papers of Raoul Bott / Robert D. MacPherson, editor.
p. cm. -- (Contemporary mathematicians)
Includes bibliographical references.
Contents: v. 1. Topology and Lie groups -- v. 2. Differential operators -- v. 3. Foliations -- v. 4. Mathematics related to physics.
ISBN 978-0-8176-3701-9 (U.S. : acid-free : set). -- ISBN 978-0-8176-3701-9 (Switzerland : acid-free : set)
1. Topology. 2. Mathematical physics. I. MacPherson, Robert, 1944- . II. Title. III. Series.
QA3.B68 1993 93-36938
514--dc20 CIP

Printed on acid-free paper

Birkhäuser

ISBN 978-0-8176-3701-9
ISBN 978-0-8176-3613-5 (Volume 1)
ISBN 978-0-8176-3646-3 (Volume 2)
ISBN 978-0-8176-3647-0 (Volume 3)
ISBN 978-0-8176-3648-7 (Volume 4)

978-0-8176-3701-9 (set)
978-0-8176-3613-5 (Volume 1)
978-0-8176-3646-3 (Volume 2)
978-0-8176-3647-0 (Volume 3)
978-0-8176-3648-7 (Volume 4)

9 8 7 6 5 4 3 2 1

Contents — Volume 3

(Bracketed numbers are from the Bibliography)

The Papers of Raoul Bott

Bibliography of Raoul Bott

[1] (with R. J. Duffin) Impedance Synthesis Without Use of Transformers, J. Appl. Phys., **20**(1949), 816.

[2] Two New Combinatorial Invariants for Polyhedra, Portugaliae math., **11**(1952), 35–40.

[3] On the Third Symmetric Potency of S_1, Fund. Math., **39**(1952), 264–268.

[4] Symmetric Solutions to Majority Games — d'après E. Specker, in: *Contributions to the Theory of Games, Volume II*, Princeton University Press, (1953), 319–323.

[5] On Symmetric Products and the Steenrod Squares, Ann. of Math., **57**(1953), 579–590.

[6] (with R. J. Duffin) On the Algebra of Networks, Trans. Amer. Math. Soc., **74**(1953), 99–109.

[7] (with H. Samelson) On the Pontryagin Product in Spaces of Paths, Comment. Math. Helv., **27**(1953), 320–337.

[8] On Torsion in Lie Groups, Proc. NAS, **40**(1954), 586–588.

[9] Nondegenerate Critical Manifolds, Ann. of Math., **60**(1954), 248–261.

[10] (with J. P. Mayberry) Matrices and Trees, in: *Economic Activity Analysis*, John Wiley & Sons, (1954), 391–400.

[11] On Manifolds all of Whose Geodesics are Closed, Ann. of Math., **60**, (1954), 375–382.

[12] (with H. Samelson) The Cohomology Ring of G/T, Proc. NAS, **41**, (1955), 490–493.

[13] An Application of the Morse Theory to the Topology of Lie-Groups, Bull. Soc. Math. France, **84**, (1956), 251–281.

[14] On the Interation of Closed Geodesics and the Sturm Intersection Theory, Comm. Pure Appl. Math., **IX**, (1956), 171–206.

[15] Homogeneous Vector Bundles, Ann. of Math., **66**, (1957), 933–935.

[16] The Stable Homotopy of Classical Groups, Proc. NAS, **43**, (1957), 933–935.

[17] (with H. Samelson) Applications of Morse Theory to Symmetric Spaces, Proceedings from Symposium Internacional de Topologia Algebraica, (1958), 282–284.

[18] (with J. Milnor) On the Parallelizability of the Spheres, Bull. Amer. Math. Soc., **64**, (1958), 87–89.

[19] An Application of the Morse Theory to the Topology of Lie Groups, Proceedings of the International Congress of Mathematics, (1958), 14–21.

[21] Morse Theory and its Application to Homotopy Theory, Bonn, (1958), 1–42 [previously unpublished].

[22] (with H. Samelson) Applications of the Theory of Morse to Symmetric Spaces, Amer. Jour. of Math., **80**, (1958), 964–1029.

[23] On a Theorem of Lefschetz, Mich. Math. Jour., **6**, (1959), 211–216.

[24] The Stable Homotopy of the Classical Groups, Ann. of Math., **70**, (1959), 313–337.

[25] Quelques Remarques sur les Théorèmes de Periodicité, Bull. Soc. Math. France, **87**, (1959), 293–310.

[26] A Note on the Samelson Product in the Classical Groups, Comm. Math. Helv., **34**, (1960), 249–256.

[27] The Space of Loops on a Lie Group, Mich. Math. Jour., **5**, (1958), 35–61.

[28] Vector Fields on Spheres and Allied Problems, L'Enseignement Mathématique, **VII**, (1961), 125–138.

[29] A Report on the Unitary Group, Diff. Geom., **3**, (1961), 1–6.

[30] (with H. Samelson) Correction to "Applications of the Theory of Morse to Symmetric Spaces," Amer. Jour. Math., **83**, (1961), 207–208.

[31] A Note on the KO-Theory of Sphere-Bundles, Bull. Amer. Math. Soc., **68**, (1962), 395–400.

[32] (with M. Berger) Sur les Variétés à Courbure Strictement Positive, Topology, **1**, (1962), 301–311.

[33] (with M. Atiyah and A. Shapiro) On Clifford Modules, Topology, **3**, (1964), 3–38.

[34] (with M. Atiyah) The Index Problem for Manifolds with Boundary, Proceedings of a Symposium on Differential Analysis, Tata Institute, (1964), 175–184.

[35] (with M. Atiyah) On the Periodicity Theorem for Complex Vector Bundles, Acta Mathematica, **112**, (1964), 229–247.

[36] (with M. Atiyah) Notes on the Lefschetz Fixed Point Theorem for Elliptic Complexes, Harvard University, (Fall, 1964), [previously unpublished].

[37] The Index Theorem for Homogeneous Differential Operators, in: *Differential and Combinatorial Topology: A Symposium in honor of Marston Morse*, Princeton University, (1964) 167–186.

[38] (with S. Chern), Hermitian Vector Bundles and the Equidistribution of the Zeroes of their Holomorphic Sections, Acta Mathematics, **114**, (1964), 71–112.

[39] A Fixed Point Theorem for Elliptic Complexes, Proceedings of the U.S.-Japan Seminar in Differential Geometry, Kyoto, (1965), 6–12.

[40] (with M. Atiyah) A Lefschetz Fixed Point Formula for Elliptic Complexes: II, Ann. of Math., **88**, (1968), 460–515.

[41] Vector Fields and Characteristic Numbers, Mich. Math. Jour., **14**, (1967), 231–244.

[42] (with M. Atiyah) A Lefschetz Fixed Point Formula for Elliptic Complexes: I, Ann. of Math., **86**, (1967), 374–407.

[43] A Residue Formula for Holomorphic Vector-Fields, Jour. of Diff. Geom., **1**, (1967), 311–330.

[44] (with M. Atiyah) A Lefschetz Fixed Point Formula for Elliptic Complexes: II, Ann. of Math., **88**, (1968), 451–491.

[45] (with J. Mather) Topics in Topology and Differential Geometry, in Battelle Recontres, 1967 Lectures in Mathematics and Physics, W. A. Benjamin, Inc., (1968), 460–515.

[46] Lectures on $K(X)$, W. A. Benjamin, Inc., (1969).

[47] (with M. Atiyah and L. Gårding) Lacunas for Hyperbolic Differential Operators with Constant Coefficients I, Acta Mathematica, **124**, (1970), 109–189.

[48] (with S. Chern) Some Formulas Related to Complex Transgression, in: *Essays on Topology and Related Topics: Memoires dédiés à Georges de Rham*, Springer–Verlag, (1970), 48–57.

[49] (with P. Baum) On the Zeroes of Meromorphic Vector-Fields, in: Essays on Topology and Related Topics: Memoires dédiés à Georges de Rham, Springer–Verlag, (1970), 29–47.

[50] The Periodicity Theorem for the Classical Groups and Some of its Applications, Adv. in Math., **4**, (1970), 353–411.

[51] On a Topological Obstruction to Integrability, in: *Global Analysis: Proceedings of Symposia in Pure Mathematics*, **XVI**, (1970), 127–131.

[52] On a Topological Obstructions to Integrability, Actes Congrès Intern. Math., **I**, (1970), 27–36.

[53] (with J. Heitsch) A Remark on the Integral Cohomology of $B\Gamma_q$, Topology, **11**, (1972), 141–146.

[54] On the Lefschetz Formula and Exotic Characteristic Classes, Symposia Mathematics, **10**, (1972), 95–105.

[55] (with P. Baum) Singularities of Holomorphic Foliations, J. Diff. Geom., **7**, (1972), 279–342.

[56] (with A. Haeflinger) On Characteristic Classes of Γ-Foliations, Bull. Amer. Math. Soc., **78**, (1972), 1039–1044.

[57] Lectures on Characteristic Classes and Foliations, in: *Lectures on Algebraic and Differential Topology*, Lec. Notes in Math., **279**, Springer–Verlag, (1972), 1–94.

[58] (with M. Atiyah and V. Patodi) Errata to the paper On the Heat Equation and the Index Theorem, [previously unpublished].

[59] Some Remarks on Continuous Cohomology, Proceedings from the International Conference on Manifolds and Related Topics in Topology, Tokyo 1973, Univeristy of Tokyo Press, (1975), 161–170.

[60] On the Chern–Weil Homomorphism and the Continuous Cohomology of Lie-Groups, Adv. in Math., **11**, (1973), 289–303.

[61] (notes by M. Mostow and J. Perchik) Notes on Gel'fand Fuks Cohomology and Characteristic Classes, Proceedings of the Eleventh Annual Holiday Symposium, New Mexico State University, (1973), 1–126.

[62] (with M. Atiyah and L. Gårding) Lacunas for Hyperbolic Differential Operators with Constant Coefficients II, Acta Mathematica, **131**, (1973), 145–206.

[63] Some Aspects if Invariant Theory in Differential Geometry, in: *Differential Operators on Manifolds*, Edizioni Cremonese, (1975), 49–145.

[64] On the Shape of a Curve, Adv. in Math., **16**, (1975), 144–159.

[65] On the Gelfand-Fuks Cohomology, Proceedings of Symposia in Pure Math., **27**, (1975), 357–364.

[66] (with H. Shulman and J. Stasheff) On the de Rham Theory of Certain Classifying Spaces, Adv. in Math., **20**, (1976), 43–56.

[67] On Characteristic Classes in the Framework of Gelfand–Fuks Cohomology, Société Mathématique de France, (1976), 113–139.

[68] On Some Formulas for the Characteristic Classes of Group-Actions. in: *Differential Topology, Foliations and Gelfand–Fuks Cohomology*, Lec. Notes in Math., **652**, Springer–Verlag, (1977), 25–61.

[69] The Geometry and Representation Theory of Compact Lie Groups, London Math. Soc. Lec. Notes Ser., **34**, (1977), 65–90.

[70] On the Characteristic Classes of Groups of Diffeomorphisms, L'enseignment Mathématique, **XXIII**, (1977), 209–220.

[71] (with G. Segal) The Cohomology of the Vector Fields on a Manifold, Topology, **16**, (1977), 285–298.

[72] On Some Recent Development in Yang–Mills Theory, Harvard University Preprint, (1978), [previously unpublished].

[73] (with R. Seeley) Some Remarks on the Paper of Callias, Commun. Math. Phys., **62**, (1978), 235–245.

[74] Some Recollections from 30 Years Ago, in: *Constructive Approaches to Mathematical Models*, Academic Press, (1979), 33–39.

[75] Equivariant Morse Theory and the Yang–Mills Equation in Riemann Surfaces, in: *The Chern Symposium*, 1979, Springer–Verlag, (1980), 11-22.

[76] Morse Theoretic Aspects of Yang–Mills Theory, in: *Recent Developments in Gauge Theories*, Plenum Publishing, (1980), 7–28.

[77] An Equivariant Setting of the Morse Theory, L'Enseignement Mathématique, **XXVI**, (1980), 271–278.

[78] Marston Morse and his Mathematical Works, Bulletin (New Series) of the Amer. Math. Soc., **3**, (1980), 907–950.

[80] Lectures on Morse Theory, Old and New, Bulletin (New Series) of the Amer. Math. Soc., **7**, (1982), 331–358.

[81] (with M. Atiyah) The Yang–Mills Equations over Riemann Surfaces, Phil. Trans. R. Soc. Lond., **A 308**, (1982), 524–615.

[82] (with M. Atiyah) The Moment Map and Equivalent Cohomology, Topology, **23**, (1984), 1–28.

[83] On Topology and Other Things, Notices of the Amer. Math. Soc., **32**, (1985), 152-158.

[84] On Some Recent Interactions between Mathematics and Physics, Canad. Math. Bull., **28**, (1985), 129–164.

[85] On Characteristic Classes and Continuous Cohomology, in: *Supersymmetry and its applications: superstrings, anomalies and supergravity*, Cambridge University Press, (1986), 29–40.

[86] Jerusalem Lectures, Proceedings from the Jerusalem Winter Conference in Physics, Jerusalem, (1986), 1–38.

[87] On the Fixed Point Formula and the Rigidity Theorems of Witten, Lectures at Cargèse, (1987), 1–17.

[88] On Mathematics and Physics, Plenary Lectures at AAAS Meeting, Boston, (1988), 1–21.

[89] On Induced Representations, Proceedings of Symposia in Pure Mathematics, **48**, (1988), 1–13.

[90] On Topological Constraints on Analysis, in: *A Century of Mathematics in America, Part II*, AMS-MAA Invited Address, Providence, (1988), 527–542.

[91] (with C. Taubes) Ont he Rigidity Theorems of Witten, Jour. of the Amer. Math. Soc., **2**, (1989), 137–186.

[92] Morse Theory Indomitable, Proceedings from a Conference in honor of R. Thom, Paris, (1988), IHES, 99–114.

[93] Remarks upon Receiving the Steele Career Prize at the Columbus Meeting of the A.M.S., Columbus, (1990), 1–2.

Comments on Some of the Papers in Volume 3

Raoul Bott

Except for two purely expository reports, the papers here collected are all related to the algebraic topology aspects of foliations, the Gelfand–Fuks Cohomology of vector fields and the relationships between these two at first quite disparate subjects.

Contemplated now in retrospect, this collection of short papers reminds one of sketches for a large canvas. Yet the canvas is missing! Even now, twenty years later, the picture is too incomplete to be painted, and the subject will – like sleeping beauty – have to await some prince of a new idea to reinvigorate it.

Our Editor in his wisdom has provided us with two commentators who have admirably described the backdrop for these developments and the many crosscurrents in which they took place. I will therefore continue to keep my commentaries more or less subjective, as I have in the previous two volumes.

51. On a Topological Obstruction to Integrability.

I am pretty certain that this obstruction became evident to me while preparing a lecture for my topology and geometry course in 1968. I have the time-consuming habit of consulting notes or references only very rarely, and rather tend to launch into the general subject more or less improvisationally and I expect that this time I found myself doing characteristic classes right after the Frobenius Theorem and my customary short excursion into foliations. The simplest foliations are fibered manifolds where the "fibers" are the "leaves." It follows that the normal bundle of such a foliation is the pull-back of the tangent bundle of the base, and its characteristic classes therefore vanish in all dimensions greater than that of the base.

A general foliation of course has no global base, but the foliation does look like a fibering locally. This enables one to imitate the pullback construction and leads one to the notion of a "basic" connection of the normal bundle of any foliation. A straightforward curvature calculation within the Chern–Weil theory then produces the desired integrability criterion of the paper: *For a foliation $E \subset T$ the Pontryagin ring of T/E must vanish in dimension greater than 2 times the fiber dimension of T/E.*

This factor of 2 arises from the magic of the "curvature construction" of characteristic classes of Pontryagin, Chern and Weil, and is therefore applicable only to *twice differentiable foliation.*

I was of course delighted that this simple criterion answered a question of some year's standing and still exalt that this vanishing theorem depends so essentially on the marvelous features of the de Rham Theory. It cannot — to my knowledge — be obtained from purely topological considerations.

I presented this paper at the Chern Symposium in Berkeley and remember that the lecture was in the early afternoon. Also I remember that I had dozed off in my bedroom in the nearby faculty club and would have missed the lecture altogether, if Iz Singer, divining what had occurred, had not roused me in time to deliver it 15 minutes late.

52. On Topological Obstructions to Integrability.

This paper is the text of my report to the International Congress of 1970 at Nice. The Congress was the crowning event of an exciting summer for the whole family. We had spent the previous six weeks in a primitive mountain cabin at the Summer School in "Lesouche". This was the year Jaffe and Glimm presented their work on Constructive Field Theory and many of the young students there have since become leaders of this and related fields.

Cecile deWitt, the director of the summer school she had started some years earlier, had very kindly invited George Mackey and myself as mathematical observer-consultants. I believe Battelle footed the bill and both George and I are to this day grateful to both Cecile and Battelle for this exciting summer, literally in the shadow of Mt. Blanc.

After this adventure we motored to Nice where I was to present a one-hour lecture. I had great stage fright on my way to the podium. In fact, I remember deliberating along the way, whether a trip to the scaffold would not have been easier. But once launched, the lecture went well, and I found it exhilarating to address such a cosmopolitan gathering.

The congress itself was exciting, partly due to the spectacular location, partly due to the interesting new mathematical developments and partly due to the larger than life presence of its director, Jean Dieudonné. I remember an elegant garden party hosted by Mr. and Mrs. Dieudonné, but even more vivid is the remembrance of Dieudonné stomping about and slapping himself in a towering rage when he discovered that inappropriate measures had been taken to have Gromov's lecture read to us. I happened to be on the scene when the unfortunate messenger brought the bad news. It was amusing to watch all of us scampering for safety, but also inspiring to see him fly off the handle so. Here at last was a man in whose universe mathematics came first.

This report already contains references to André Haefliger's classifying spaces $B\Gamma_q$ for foliations – or rather their integrable homotopy classes. This profound new concept had by then found applications in many areas of foliation theory. In particular, one knew that $\pi_s(B\Gamma_q)$ *vanished* in dimension $\leq q$. In this context my vanishing theorem asserted that under the natural arrow $B\Gamma_q \rightarrow BGL_q$ – which now corresponds to passing from a foliation class to its normal bundle – the cohomology in dim $> 2q$ of BGL_q pulls back to zero. An immediate consequence of this interpretation was that new classes appeared in the homotopy theoretic fiber of this arrow, which hopefully could be used to prove that certain homotopy groups of $F\Gamma_q$ were *non-vanishing*. In fact, the principal new result of this report was that in the parallel holomorphic theory this technique actually enabled me to show that $\pi_{2q+1}(F\Gamma_q C)$ is highly non-trivial: this group surjects onto $\mathbb{C} \times \cdots \times \mathbb{C}$ ($2q$ copies) and thus is uncountable.

I had met André Haefliger casually at various conferences but I think it was 1969, when I came down to give a lecture on these matters at Princeton, that we really got to know each other and soon became very close friends. The one short joint paper [56], completely misrepresents the vast amount of joyful time we have spent together, often family with family, often in faraway places, thinking about these and related mathematical matters – as well as exploring life in general.

The joint note to the Bulletin [56] was actually to have been the preliminary announcement for that large canvas I alluded to earlier. That it never came to be is, as I said earlier, in part due to the incompleteness of the picture to be represented, but in part due also to the inertia that sets in after a research announcement is sent off. The temptation to think and work on other things – rather than to complete the tedious details of what had been announced – often proves too great.

53. (with J. Heitsch) A Remark on the Integral Cohomology of $B\Gamma_q$.

In this paper James Heitsch and I show that the vanishing theorem does not hold over $\mathbb{Z}$. I think this was an amalgamated paper in the sense that we had independently reached the same conclusions and joined forces for publication. James later became one of the masters in the difficult art of creating interesting foliations from the point of view of secondary characteristic classes.

54. On the Lefschetz Formula and Exotic Characteristic Classes.

André Haefliger pointed out in his commentary, that my early paper on vector fields in a sense set the stage for the vanishing theorem. After all, a vector field generates a 1-dimensional foliation! In this paper I remark that one can reverse this procedure and thereby evaluate certain secondary characteristic classes in the Haefliger spaces $F\Gamma_q$. Thus, in this setting the formulas of [41] can be used to prove some of the results in [52].

This paper was given in Rome and this was the occasion of my wife, Phyllis's, and my first trip to Italy. Not only that, it marked the first time we had none of our four children with us. We were enchanted by Rome, and later by Florence, although dealing with the Italian traffic aged me prematurely. We had come armed with a list of restaurants recommended by Louis Nirenberg which all proved exquisite. Louis had presented me with this list at a chance encounter in New York and I had been much amused by his confession that he never, I repeat, never, ventured out of the house without it.

55. (with P. Baum) Singularities of Holomorphic Foliations.

This paper extends our earlier paper on the singularities of meromorphic vector fields to holomorphic foliations and their singularities. The paper involves quite difficult technical questions and that these were surmounted so satisfactorily is entirely due to Paul's initiative. Paul also wrote up the paper as it stands. I believe I wrote up the final version of our earlier one.

56. (with A. Haefliger) On Characteristic Classes of Γ-foliations.

In this paper André and I describe what I like to call the "high road" to the "secondary" characteristic classes of a foliation. This road arises naturally from the very beginnings of the Gelfand–Fuks theory; that is, the continuous cohomology of the Lie algebra of C^∞ vector-fields on a manifold *with support at a point.*

In the 40's and 50's we had learned – mainly from Chern in my case – that on the principal bundle of frames P^1M of a manifold M, n tautologous 1-forms $\theta^1, \cdots, \theta^n$ appear quite naturally. What amongst other things we collectively learned from Gelfand and Fuks was that as we pass to the higher jet bundles P^kM, of k jets of maps of $\mathbb{R}^n$ to M, new tautologous forms appear and that at each stage the new tautologous forms enable one to describe the derivatives of those of the previous stage. Thus on P^2M, n^2 new 1-forms θ^i_j appear and $d\theta^i + \theta^i_j \wedge \theta^j = 0$, etc. Passing to the direct limit $n \to \infty$ one obtains a universal algebra of forms which in the $G-F$ scheme of things can be identified as the de Rham complex associated with the continuous cohomology of the Lie algebra of formal vector fields on $\mathbb{R}^n$ with *support concentrated at the origin.* In any case this algebra maps naturally into the de Rham complex of "$P^\infty M$". On the other hand, the homotopy type of $P^\infty M$ is the same as P^1M and if one divides by the maximal compact subgroup $O_n \subset GL(n)$, $P^\infty M/O_n$ is seen to be of the homotopy type of M itself! In this way one obtains an arrow from the relative version of the $G-F$ algebra into $\Omega^*\{P^\infty(M)/O_n\}$. Finally, the whole Yoga of the Haefliger universal space $B\Gamma_n$ enables one to lift this framework of ideas nearly word-for-word from M to $B\Gamma_n$. The resulting arrow in cohomology then produces the desired homomorphism of a certain universal and quite computable algebra – denoted by $H^*(WO_n)$ – into $H^*(B\Gamma_n)$. The image of this algebra constitutes the "high road" definition of secondary classes of foliation theory.

In our note we formulated these ideas in the more general context of pseudogroups Γ of diffeomorphisms, each of which now comes associated with a universal algebra $H^*(\mathfrak{a}(\Gamma), K)$ mapping to $H^*(B\Gamma_q)$.

The various concepts alluded to here are elucidated in many of the subsequent papers and lecture notes. E.g. [57], [59], [61], [63], [67], [68], [70]. In the final analysis what emerged here was that there should be a whole theory of continuous cohomology for the spaces $B\Gamma_n$, analogous to the beautiful theory of continuous cohomology as developed by Hochschild, Mostow, Van Est, and others, for finite-dimensional Lie groups, and presumably the computable object

$H^*(\mathfrak{a}(\Gamma); K)$ should be the corresponding continuous cohomology of $B\Gamma_n$.

Unfortunately, trying to make this analogy precise not only involves many technical problems with infinite dimensionality. Just proving that our arrow is injective into the continuous cohomology seemed then, and is even now, quite out of reach. For to prove injectivity one needs many examples. Thus in our work we sketch how to prove, in analogy to the famous prototype of these classes of Goodbillon–Vey–Roussary, that a certain very meager subset of our classes injects into $H^*(B\Gamma_n)$ in the real case, by studying homogeneous foliations.

A whole new order of examples would be needed to prove injectivity in general. In the complex analytic case much stronger injectivity results obtain, by virtue of my earlier examples [55], but again this world of examples is no match for the many classes in $H^*\{\mathfrak{a}(\Gamma_n C)\}$ to be distinguished.

57. (Notes by Lawrence Conlon) Lectures on characteristic classes and foliations.

These are beautifully written notes by Larry Conlon of lectures I delivered in Mexico in July 1971. In the last section I describe what I like to call the "low road" towards these secondary classes, and I do remember that this road became clear to me under the Eucalyptus trees of my outdoor study in La Jolla, where we used to summer for many years. (Now we tend to spend Christmas there – for as a serious consequence of these earlier summer visits, my eldest daughter now lives with her husband and four children in nearby Del Mar.) In any case, it was here that it became clear to me that the comparison of the basic connections with the orthogonal connection on the quotient bundle gave rise to a natural arrow from the algebra

$$WO_n = E(h_1, h_3, \cdots) \otimes \mathbb{Q}[c_1, \cdots]/I$$

where the h's have odd dimension $\leq n$, and the ideal I is spanned by polynomials of total weight $> n$, into the de Rham theory of the foliated manifold.

This procedure is then entirely in line with the work of Chern–Simons, and really also that of my earlier paper [38] with Chern where the "difference construction" also plays a vital role. At that time I do not think that I was aware of the relation of this WO_n to the Gelfand–Fuks Theory.

59. Some Remarks on Continuous Cohomology.

This report was given at an international conference in Kyoto, 1973, and marked my first visit to Japan. Of all the many impressions that overwhelm one when entering so different and remarkable a culture, two stand out, now, twenty years later. In Osaka I was taken to see a tea ceremony museum: A small, not very remarkable, building housed a variety of implements for the tea ceremony, and the director very kindly conducted the tour himself for this foreign guest. He stopped at one of the cases, and carefully lifted out a bamboo spoon. "This," he said, in a humble and quite matter-of-fact way, "is the most beautiful object in the world"! I was stunned and started to search in my mind for something in my own culture that I might conceivably hold out with such praise. Finally, the beginning of Mozart's Piano Concerto in D-minor came to mind – but I did not dispute the old gentleman.

The second vivid impression was a visit to a Geisha house. This actually occurred right at the beginning of our visit in Tokyo, when a small party of us was hosted by Kentaro Yano to this treat. Yano was an old friend from the Princeton days of 1949–51. When I objected to such an expensive pastime as a dinner in a fashionable Geisha house, he waved my concern aside. "I am a rich man," he said, "and in part that is your doing." It turned out that during our many lunches together in Princeton, where we often indulged in "light" mathematical problems and conundrums, he heard me use the term "elegant" in reference to a proposed solution. Upon returning to Japan, Yano started a column on mathematical puzzles in the newspapers which he called "Elegant Solutions." By 1973 his column was syndicated and the income from the column now far exceeded the meager salary of a Professor.

On a less frivolous note, the visit left us all with a very warm feeling for our Japanese hosts whose charming hospitality met us everywhere. I treasure the beautiful book on Japanese Art I received from Matsushima who had been a friend from earlier adventures in India, and also with great nostalgia recall the feast the Kodairas laid out for us in their own home.

As for the mathematics in this report – it concerned itself with André's and my dreams (see [56]), but by now laid out in a technically more final form. In all the categorical nonsense, the ideas of Graeme Segal (see [71]) are now also clearly discernible.

60. On the Chern–Weil Homomorphism and the Continuous Cohomology of Lie Groups.

This paper is rather a favorite of mine. During its composition I not only learned about continuous cohomology at first hand from a true master of the art – G. Hochschild – but it also finally became clear to me that there is a "semi-simplicial de Rham Theorem" of the most natural kind (see also [66]). Finally, it was nice to see that these ideas together with the abstract work of Dold–Puppe on nonadditive functors naturally replaces the invariant polynomials, with values in the Lie algebra, of the Chern–Weil theory with the continuous cohomology of G with values in the symmetric algebra of the Lie algebra. For compact groups these two spaces of course agree.

This work partly overlaps with independent work of Schulman and Kamber and Tondeur, and at the same time served as a model of what A. Haefliger and I hoped to achieve for $B\Gamma_n$.

Gerhardt Hochschild was an old friend from Princeton days, and at one point of our careers the University of Rochester made a joint offer to Mostow, Hochschild and me. That occurred when he was at Urbana and I was still at Ann Arbor.

Gerhardt always was a delightful companion, with many charming eccentricities. I think I have never seen him in anything but a grey shirt with open collar and darker grey Sears Roebuck pants. In the old days, a Camel cigarette would invariably dangle from his mouth, and I remember that he bought these cigarettes in cartons of cartons.

He also stockpiled reams of white paper on which he wrote in his meticulous hand. When it came to teaching, Gerhardt prepared his courses, again meticulously, on these sheets, sometimes years in advance. He likes a regulated life and one could count on him to turn up at the same coffee shop at the same time most days and that is where most of our discussions took place. Gerhardt is also as stubborn as they come. Though generous in the praise of others, he has very strict standards for himself. In the present instance, he refused be named a joint author. Later, I believe, he refused to accept a lifetime achievement award of the AMS.

61. (notes by M. Mostow and J. Perchik) Notes on Gelfand–Fuks Cohomology and Characteristic Classes.

and

71. (with G. Segal) The Cohomology of the Vector Fields on a Manifold.

These notes of a series of lectures in Las Cruces were compiled by two of my very able students, Mark Mostow and Jim Perchik. Regional conferences are essentially one-man shows and I remember being rather exhausted at the end of several consecutive days of 3-hour lectures. Still the friendly and attentive audience, as well as the distinctly "western" atmosphere, made this a very memorable occasion.

The notes are a much elaborated corrected and expanded version of my lectures and serve as a good introduction to this framework of ideas.

The Gelfand–Fuks cohomology of a manifold M is by definition the continuous cohomology of the Lie algebra LM of C^∞ vector-fields in M. Thus $H_c(LM)$ is computed from the Chain complex $A(M)$ with A^q the continuous alternating linear maps of q vector fields to $\mathbb{R}$, with the usual differential operator.

Gelfand and Fuks not only set the stage for computing $H_c(LM)$ but also carried out this program to evaluate their theory for a circle S^1. They found $H_c(LS^2)$, to be the tensor product of an exterior algebra with a generator in degree 3 times a polynomial ring on a generator in dim 2. That is, $H_c(LS^1) = E(w) \otimes \mathbb{R}(y)$.

Their highly original computation immediately presented the mathematical community with the problem of computing $H_c(LM)$ in general, and, maybe even more importantly, to fit this new cohomology into the hierarchy of traditional invariants. In particular, is this theory homotopy invariant?

My own keen interest in these questions must have started in the fall of 1972, and my contribution is pretty well described in the paper [65], although the point of view there described is also the focal point of these Las Cruces lectures delivered two years earlier in 1973.

In any case, I think it was in the fall of 1972 that I started to think about Gelfand–Fuks from scratch, and discovered a simple homotopy argument to show that the natural inclusion of $A_{pt}\mathbb{R}^n \to A\mathbb{R}^n$ of the continuous cochains with support at a p point in $\mathbb{R}^n$ into $A\mathbb{R}^n$ induces an isomorphism. (Recall that $H(A_{pt}(\mathbb{R}^n)) = H(WO_n)$ was the aspect of this theory which pertained to foliation.) But once $H_c(L\mathbb{R}^n)$ was known and seen to be a homotopy invariant, then the Meyer–Vietoris argument suggests a way to get at $H_c(LM)$ in general. The multiplicative behavior of $H_c(LM)$ under disjoint unions finally led me to what I called the main conjecture: That in fact $H_c(LM)$ is isomorphic to the usual singular cohomology of a certain function-space; the space of sections of a certain natural bundle Y_M over M.

The Meyer–Vietoris argument is, however, highly nontrivial in this context, in that difficult convergence questions arise, and these were resolved by Graeme Segal only a year or two later. See [71].

I had met Graeme in the early 1960's where he was a student of Michael Atiyah's at Oxford, and we have been friends, associates, collaborators ever since. In fact, Graeme's influence can be felt in nearly all my work and much of that of Michael Atiyah's as well.

Although very unlike my old friend, Arnold Shapiro in manner, Graeme is very much like Arnold in the sense that one can engage him profitably on a discussion of every aspect of mathematics. So on my many visits to Oxford – whatever the subject of burning interest might have been – Graeme was in the picture and we all would often appeal to him for clarification on this or that point. Graeme was always more functorial than Michael and I, and he could make the newest constructions of algebraic topology intelligible to us.

Just when Graeme had started to think about Gelfand–Fuks I don't know, but I do remember that I communicated my main conjecture to him right away and that its proof then became a goal for both of us. Maybe a better way of putting it would be to say that my main job in our collaboration was to keep after him until the convergence questions were cleared up.

For Graeme likes to stew over his mathematical ideas even more than André, Michael, and I do. Left to his own devices he will go on distilling them to the point where he finds them to be "obvious," whereupon he might easily forget them altogether.

The history of this "main conjecture" is explained in the introduction of [71]. Formulated independently by Fuks and me, it was eventually settled at around the same time and independently by André Haefliger using Sullivan's rational homotopy theory and by Trauber using the original Gelfand–Fuks filtration and quite a bit of homological algebra.

63. Some Sspects of Invariant Theory on Differential Geometry.

These are the notes compiled in retrospect from my lectures delivered in August of 1975 at Varenna under the auspices of (C.I.M.E.). This conference also was no picnic: a lecture every day by the three main speakers and seminars of various sorts in the afternoon.

Michael Atiyah and Claudio Procesi were the other main speakers, and I remember that Michael's room was referred to as the throne room, featuring a bathtub on a pedestal. We were housed in a beautiful Nunnery about which scurrilous – I am sure – rumors had it, that it had at times also served as a brothel.

These are essentially self-contained lectures treating all aspects of characteristic classes from the point of view suggested by the Gelfand–Fuks theory. Thus here one finds a natural notion of curvature of a manifold in the second order tangent bundle of M.

The scenery as well as the atmosphere were breathtaking. This was really only my second extended visit to Italy and I immediately fell under its spell. Especially memorable was a mushroom expedition followed by a luncheon feast organized by some of our younger mathematicians. The light, the food, the beauty and grace of our hosts left Michael and me completely enchanted.

Of course the beauty of the people and their elegance — especially on sunday afternoon — at times made me embarrassed about our collectively shoddy appearance. Somehow we were the antithesis of the Italian "Professore."

On my return trip through Milan, Sig Helgason and I desperately tried to catch the end of the Chess tournament in the outskirts, but got there too late. Maybe in recompense, I managed to visit Leonardo's *Last Supper* at a relatively quiet moment and I am under the spell of this vision to this day.

67. On Characteristic Classes in the Framework of Gelfand–Fuks Cohomology.

I considered it a great honor to be invited to give this address on the occasion of Cartan's 70th birthday, and so to be able to pay tribute in person to Cartan's great achievement of sustaining mathematics in every conceivable manner during his long and productive career.

It was also great fun to see Cartan appear at the "Meshui" organized by Duady and Hubbard in his honour, clad in a sweatshirt emblazoned with his own name.

I savoured his performance at that celebration all the more in view of the circumstances of our original meeting. That was in Mexico City during a magnificent reception at that famous conference organized by Lefschetz in 1956. Tequila was flowing freely and Spencer and J.H.C. Whitehead and I were in very good spirits indeed. As Serre told me later, after Cartan had in his usual quiet way observed our shenanigans for some time, he pointed to me and said, "Qui est ce fou?"

Henri Cartan was in excellent shape during all the festivities and one of the high points of the conference occurred at the very end when, with Cartan at the keyboard, we heard a chamber music concert devoted to the music of his tragically short-lived brother.

68. One Some Formulas for the Characteristic Classes of Group-Actions.

These are notes from my lectures at the Symposium in Rio de Janeiro in 1976, and contemplating them now brings back vivid memories of life in that great city, the heat of Brazil in January, the magnificent beaches, the youth of the inhabitants (some of us conjectured that all people over 50 were under house arrest!) and the many "feasts" and outings organized by the younger participants. This time I was surrounded by family. My wife, Phyllis, had come, André Haefliger and his wife, Minus, were also there, as was Graeme Segal as well as many of my younger students and friends. And of course the congress was organized by an old friend, Paul Schweitzer, from Princeton days.

These notes, as well as the paper [70] deal with an integrated version of the secondary characteristic classes of foliations. Indeed, these can be seen to map naturally to the continuous cohomology of the diffeomorphisms $Diff\ M$ of a manifold M, and my aim here was to use the semi-simplicial methods of the foregoing papers to describe explicit cocycles for them.

The correct context for these considerations turned out to be the "Compatible de Rham complex" of the S.S. description of $B\Gamma_n$, rather than the "straight de Rham complex" as used, say in [66]. Here then I was influenced by the ideas of Sullivan and DuPont. An appendix by Robert Brooks explains the continuous variation of the integrated Godlillon–Vey class in $H_c^2(Diff(S^1))$ discovered by Thurston, in this present framework. It had been my original hope that such a computation would be possible and I was pleased when Bob was actually able to carry it out.

At that stage of our understanding the main question to be resolved was precisely – which classes could be varied continuously and which could not. In his lecture, Heitch presented some beautiful new examples and clarified this subject in several directions.

Another highlight of the conference were André's superb set of lectures on de Rham theory, homotopy theory, and his solution of the "main conjecture" concerning Gelfand–Fuks cohomology.

But this bright picture was not without its shadows. Occasionally the prevailing poverty would arise to torment us, often quite amusingly, by vigorous gambits at thievery which few of us escaped.

The Bott and Haefliger families had just settled on the beach for a relaxing half hour of sunbathing and swimming when I noticed some horseplay in our vicinity. Having been forewarned, I immediately clutched my wallet and sounded the general alarm. But it was too late. Though all our wallets were still there, we soon discovered that André's brand new pants had disappeared. He had bought them fifteen minutes earlier!

And even the glorious ocean proved untrustworthy. On one of our early outings, a rogue wave smashed Herb Schulman into the sand breaking his collarbone. Luckily he recuperated soon enough, but the incident was a warning to all of us to treat the ocean with respect.

In all other ways the Conference was a great success and we were all sad to leave this sundrenched vivacious land for the somber winter awaiting us at home.

Working with Bott
(Rocking and Rolling with Raoul)

Paul Baum

While a graduate student at Princeton (somewhere in the early 1960s), I went to a lecture by Bott. He spoke on the Atiyah–Singer index theorem, which was quite new at the time. Here was a friendly and somewhat oversized man enthusiastically describing this marvelous index formula. He ended with a burst of great intensity, saying, "Well, I just wanted to tell you about this theory, which I find so beautiful."

This was my first in-person contact with Bott. Earlier (as a beginning graduate student under the patient guidance of Norman Steenrod), I had studied his papers applying Morse theory to the topology of Lie groups. I had greatly admired these papers, but I was even more delighted with the author himself. His talk on the Atiyah–Singer index theorem made a deep impression on me. I was fascinated by the mixture of topology and analysis, and I was also intrigued by the love of mathematics that came through so strongly in his talk.

About four years later I was in a very discouraged mood. I was now an assistant professor at Princeton and I had the feeling I was marching backwards. In work that everybody, including me, believed I had done, gaps and difficulties were appearing. Instead of moving ahead, I was wasting time and energy trying to repair arguments that I thought I had successfully finished one or two years before.

Into the midst of this gloom came Raoul. He gave a talk at IDA about his residue formula for holomorphic vector fields. Again, I was keenly interested in the mathematics he presented. I decided right then and there during his talk to simply abandon the problems I had been thinking about and start trying to further develop Bott's point of view on vector fields.

So Jeff Cheeger and I got to work to see what the residue formula was for a holomorphic vector field with nondegenerate submanifolds of zeroes. It took us some months to crack this – i.e., to see a clear, simple formula coming out of Bott's method. Once we had it, I wrote to Bott telling him our explicit answer for these residues. He wrote back informing me that he had already done this, but suggesting that Jeff and I write up the case of Killing vector fields. We did this, and the paper appeared in *Topology.* Eventually these ideas emerged in a somewhat changed form in work of Cheeger and Gromov.

Next, inspired by a comment of G. Washnitzer, I decided to see how Bott's theorem for holomorphic vector fields might extend to meromorphic vector fields. On a compact, complex-analytic manifold holomorphic vector fields are really quite rare. But meropmorphic vector fields are plentiful, so it seemed worthwhile to generalize Bott's theorem to the meromorphic case. By this time Cheeger had moved on to other things, so I found myself working alone on the meropmorphic vector field problem. In the late spring of 1968, I felt I had correctly generalized Bott's method to the meromorphic situation. I wrote to Bott about this and received a letter from him saying, "Dear Paul – We certainly are working in ||!" We were thinking about exactly the same points. Shortly after this, we met in Berkeley where we were both attending the meeting on global analysis. I outlined to Bott the formula and proof that I had in mind. To my astonishment, I found that I was actually slightly ahead of him. When he realized that my proof worked, he spread his arms wide and cheerfully shouted, "It works like a charm!" The technical trick on which the proof hinged he referred to as "a slightly mysterious *Kunstgriff*."

After I arrived at Brown in September 1968, there followed a period of about two years when I would sporadically visit Bott at Harvard, and we worked on a residue formula for singularities of holomorphic foliations. During this time I came to appreciate another of Bott's qualities – his humor. Bott can be very funny. We had many good laughs together.

Strangely, almost every time I visited Bott at Harvard it would rain. When I was leaving, Bott would say, "Good-bye Paul. See you on the next rainy day." We made some progress on the residue question. Our work here is closely related to the differential character theory of Cheeger–Chern–Simons. We proved a rigidity theorem and we formulated a rationality conjecture that appears to be a deep and difficult problem. At the Stanford differential geometry meeting, I learned that Simons had made the opposite conjecture. He was convinced that the residues in question are *not* always rational. The problem remains unsolved.

One rainy day I was with Raoul in his office in the Geographic Institute discussing these matters when a tall, thin, shy graduate student came in and said, "Professor Bott, where should I apply for a job?" I said, "Apply to Brown." The student was Bob MacPherson. His coming to Brown was one of the main events in the meteoric rise of the Brown Math Department.

Towards the end of my two-year collaboration with Bott, he was working on understanding the Haefliger classifying space for foliations and I had started working on Riemann–Roch with Bill Fulton and Bob MacPherson. I had a number of very interesting conversations with Raoul about the Haefliger classifying space. Approximately twenty years later, while working with Alain Connes, I recalled these conversations and they helped guide Alain and me towards the classifying space for proper actions that we need in our conjecture on the K-theory of group C^* algebras.

What did I learn from Bott? I learned – perhaps I should say relearned – from Bott things that I already knew: that there is great beauty and wonder and fun in mathematics; that the best mathematics is not necessarily the most difficult; that there is a mainstream to our subject that talented mathematicians like Bott have an instinctive feel for; that ideas should be presented clearly and without nonsense so that other mathematicians can understand and appreciate them.

I consider myself fortunate to have benefited from the clarity and beauty of Bott's mathematics, and from the warmth and joy of his personality.

Raoul Bott, Foliations, and Characteristic Classes An Appreciation

Lawrence Conlon

Let me begin this appreciation of Raoul's work on foliations with a very personal appreciation. My interest in foliated manifolds began when I attended his lectures in Mexico City (1971) on foliations and characteristic classes. I might well have been a purely passive participant and come away thinking that foliation theory was too hard for me, but it was my good fortune that Raoul designated me to write up the notes. Although daunting at first, this task introduced me to a field of research that has been and continues to be a profound delight.

The theory of foliated manifolds caught the interest of topologists in the early 1940s when George Reeb produced his famous foliation of S^3. Subsequent work by Reeb, Haefliger, Novikov, Sacksteder, Rosenberg, Roussarie, and many, many others laid the foundations of the *qualitative* theory of foliations, a challenging study with many surprises.

In the early 1970s, Bott initiated a new surge of interest in foliations, this time from the *quantitative* point of view, by his celebrated "vanishing theorem." As was true in Reeb's early work, Raoul's begins with a fresh but deceptively simple observation. The local structure of a foliation $\mathcal{F}$ provides a natural parallelism along the leaves in the normal bundle $\nu(\mathcal{F})$. This parallelism is locally absolute. It is easy to specify a connection ∇ (the Bott connection) on the normal bundle such that the natural parallelism coincides with the ∇-parallel transport along paths in the leaves. The observations that the parallelism along leaves is locally absolute implies that the curvature R of ∇ vanishes along the leaves. The Chern–Weil construction of Pontryagin classes, using this curvature tensor, then gives differential forms representing these classes that vanish in degrees $> 2q$, where q is the codimension of $\mathcal{F}$. In particular, given an $(n-q)$-plane distribution E on M^n with normal bundle $E^\perp$, the nonvanishing of a class of degree $> 2q$ in the Pontryagin ring $\text{Pont}^*(E^\perp) \subseteq H^*_{\text{DR}}(M)$ implies that the distribution E is not homotopic to an integrable distribution. Indeed, E is not even bundle-isomorphic to an integrable distribution.

A quite different proof of the Bott vanishing theorem was published by Bott, Shulman, and Stasheff in 1976. The idea was to define the Haefliger classifying space $B\Gamma_q$ as the geometric realization of a semisimplicial q-manifold. Thus, the universal Pontryagin classes for foliations of codimension q vanish in dimensions $> 2q$ for quite trivial reasons.

Bott's vanishing theorem is a very pretty result, but the real fun began when he and others realized that the vanishing of these classes at the *form* level makes possible the construction of secondary characteristic classes for foliations. These exotic classes were first described in the notes, mentioned above, for Raoul's lectures in Mexico City. In fact, in the actual lectures he only described what subsequently came to be known as the Godbillon–Vey class $\mathrm{gv}(\mathcal{F})$, asking whether there were any foliations for which it was nontrivial. He showed that the holomorphic analog of gv is generally nontrivial. Although included in the final write-up, the definitions of the other exotic classes was only formulated by Bott later in that same summer, after the note by Godbillon and Vey appeared in *Comptes Rendus.* This note independently defined $\mathrm{gv}(\mathcal{F})$, showed that it arises from the cohomology of the Lie algebra of formal vector fields on $\mathbf{R}$, and gave the now famous "Roussarie example," proving that gv is, indeed, a nontrivial class.

The exotic classes come from a universal cohomology algebra $H^*(\mathrm{WO}_q)$, where the foliation provides a canonical homomorphism

$$\lambda_{\mathcal{F}}\colon H^*(\mathrm{WO}_q) \to H^*_{\mathrm{DR}}(M),$$

depending only on a reasonable equivalence class containing $\mathcal{F}$. Indeed, this construction works for a general Haefliger structure $\mathcal{H}$, so there is a canonical homomorphism

$$\lambda_q\colon H^*(\mathrm{WO}_q) \to H^*(B\Gamma_q)$$

into the cohomology of the Haefliger classifying space. This immediately raises the question of the faithfulness of this representation. That is, are some of the exotic classes universally trivial and/or are there other universal classes? These questions, to the best of my knowledge, were never completely answered. Finally, another pair of researchers, Kamber and Tondeur, must be cited at this point. Their contribution to the theory of exotic classes of foliated bundles, also starting in the early 1970s and combining simplicial theory with curvature techniques, was foundational and truly massive.

These were ideas whose time had clearly come. As with the Godbillon–Vey class, it was seen that a seemingly unrelated theory, the Gel'fand–Fuks cohomology of the Lie algebra of formal vector fields, gave rise to the same universal algebra $H^*(\mathrm{WO}_q)$. Intrinsic reasons for this were described by Bott in his 1973 lectures at New Mexico State University and by Haefliger in his 1972 Séminaire Bourbaki. Another very important development in the Gel'fand–Fuks theory was made by Bott in the New Mexico Lectures and in his paper in the *Proceedings of the Symposium in Pure Mathematics*, held at Stanford University in the summer of 1973. This is the "Bott conjecture" that the continuous cohomology $H^*_{\mathrm{GF}}(M)$ of the Lie algebra of smooth vector fields on M is the ordinary cohomology $H^*\big(\Gamma(Y_nM)\big)$ of the space of sections of a certain bundle Y_nM associated to the frame bundle of M. Apparently, this conjecture had been formulated independently by Fuks. Affirmative solutions to the conjecture soon were obtained by several researchers, including Bott and Segal, Haefliger, Trauber, Stasheff, and Anderson.

Without question, Raoul's work lit a fire that quickly spread. Other fundamental research on the exotic classes came from many sources. For example, Thurston's demonstration of the continuous variation of $\mathrm{gv}(\mathcal{F})$ on S^3 led to considerable work by Heitsch as to which of the classes are rigid and which, like gv, admit continuous variation. Again, the Bott vanishing theorem was improved, for Riemannian foliations, by Pasternak, who showed that when the codimension q foliation $\mathcal{F}$ admits a bundlelike metric, the Pontryagin forms for $\nu(\mathcal{F})$, computed from a suitable Bott connection, vanish in degrees $> q$. This yielded new exotic classes for such foliations (Lazarov and Pasternak).

Of particular note is the work by many researchers on the conjecture by Moussu–Pelletier (1974) and Sullivan (1976) that $\mathrm{gv}(\mathcal{F}) = 0$ if all leaves grow nonexponentially. The list of contributors includes Herman, Mizutani, Morita, Tsuboi, Nishimori, Tsuchiya, Sergiescu, and Duminy. A key formula in most of this work is found in Raoul's paper, "On some formulas for the characteristic classes of group actions," published in the *Proceedings of the Rio Conference on Differential Topology, Foliations and Gel'fand–Fuks Cohomology* (Springer Lecture Notes, No. 652). This is an integral formula for the generalized Godbillon–Vey class that, as Raoul remarks, was known already to Thurston, but which only became accessible by Bott's proof. The complete solution by Duminy (never published) employed a measure-theoretic analysis proving that, in the absence of a "resilient leaf," $\mathrm{gv}(\mathcal{F}) = 0$. A resilient leaf is one that captures itself by a holonomy contraction. Such a leaf introduces a kind of "chaotic dynamics" into the foliation, being analogous to Smale's horseshoe in dynamical systems. As a result, the resilient leaf F and all leaves that approach L have exponential growth and the above conjecture is a corollary of Duminy's theorem. This initiated a whole new ergodic-theoretic attack on the exotic classes, spearheaded by Hurder, Katok, Heitsch, and others. In particular, by these methods Hurder obtained the vanishing of the generalized Godbillon–Vey class in arbitrary codimensions when all leaves grow subexponentially.

Finally, another development in foliation theory, Conne's theory of operators elliptic along the leaves of a foliated space and the C^* algebras of foliated spaces, has led to generalization of yet other work of Bott in the form of a Lefschetz fixed-point theory for foliations (Heitsch and Lazarov) and to new insight into characteristic classes and ergodic theory.

These last remarks have been intended as a sampling of the furious activity inspired by Raoul Bott's work on foliations and characteristic classes. Of necessity, this sampling is far from complete and my apologies are offered to all whose work has not been mentioned. While the activity is no longer quite so furious, Raoul's legacy remains and foliations continue to inspire research in algebraic topology, geometric topology, ergodic theory, and operator theory.

Raoul Bott and Foliation Theory in the 1970s

André Haefliger

As Lawrence Conlon has already given an excellent survey, published in this volume, of the work of Raoul Bott and of its more recent developments, I shall focus on his paper, "An obstruction to complete integrability," to put it into its historical perspective and emphasize its impact in the development of foliation theory in the beginning of the 1970s.

The basic concepts of foliation theory were established in the 1940s by Ehresmann and Reeb. The very first striking fact was the discovery by Reeb in 1944 of a C^∞-foliation of the 3-sphere by surfaces; the existence of such a foliation was a surprise to many people. In the introduction of his thesis, Reeb asks the following fundamental question:

"Assume that a differentiable n-manifold M admits a continuous field of p-planes, $1 < p < n$ (i.e., a subbundle of rank p of its tangent bundle), does it admit a completely integrable field of p-planes (i.e., a field of p-planes that is tangent to a smooth foliation with leaves of dimension p)?"

Reeb also notes that there are compact n-manifolds M with a simply connected boundary admitting continuous fields of $(n-1)$-planes that are tangent to the boundary, but such a field cannot be completely integrable (this is a direct consequence of Reeb's stability theorem).

In the following years, many examples of codimension-one foliations were constructed, in particular on all 3-manifolds (Lickorish, *Ann. of Math.* **82** [1965], 414–420), and eventually, in 1968, John Wood proved (*Ann. of Math.* **89** [1969], 336–358) that every field of 2-planes on a 3-manifold is homotopic to an integrable one. Later on, Blaine Lawson constructed the first examples of smooth foliations of codimension one on many higher-dimensional spheres (*Ann. of Math.* **94** [1971], 494–503), and this was extended to all odd-dimensional spheres by Itiro Tamura (*Comm. Math. Helv.* **47** [1972], 73–78).

The discovery by Raoul Bott in 1968 of obstructions to deforming a field of p-planes into a smooth, completely integrable one was a very important event; this result had a great influence on the development of foliation theory in the 1970s and its proof initiated a revival of the use of differential forms and connection theory in many parts of mathematics.

This work came in fact after a previous seminal paper of Bott written in the same vein in 1966: "Vector fields and characteristic numbers" (*Michigan Math. Journal* **14** [1967], 231–244), which can be considered the first paper of Bott on foliation theory. I quote from his introduction: "A well-known theorem of Heinz Hopf asserts that on a compact manifold the properly counted number of zeros of a vector field equals the Euler number of the manifold. The purpose of this paper is to show that when a vector field satisfies certain differential equations, then there are other relations between the characteristic numbers of the manifold and local invariants of the vector field near its zero."

Indeed, a striking corollary was that all Chern numbers of a compact, complex manifold M vanish if there is on M a holomorphic vector field X (hence a holomorphic foliation of complex dimension one – the holomorphic character forces, in particular, the complete integrability of the field of real 2-planes tangent to the leaves). More generally, if X has nondegenerated singular points, then the Chern numbers are determined by the singular points. In fact, as mentioned by Bott in the introduction of his paper, a formula of this type was conjectured by V. Guillemin, who derived some special cases from the Atiyah–Bott fixed-point formula; then Atiyah and Bott realized that the general theorem could be deduced from their fixed-point formula. I quote again from the introduction: "My aim here is to present a more direct proof in the framework of the theory of characteristic classes as developed by Chern, Pontryagin and Weil." This beautiful paper already contains in germ the vanishing theorem of characteristic classes of the normal bundle of a foliation as well as the various residue formulas that are a consequence of it and were developed later on by Bott in collaboration with P. Baum. Also, the Chern–Weil theory is applied here for the first time in this context.

Bott realized two years later that his method could be applied more generally to a differentiable (resp. holomorphic) foliation, proving that any polynomial in the Pontryagin classes (resp. Chern classes) of the normal bundle of a foliation of real (resp. complex) codimension q should vanish in degree $> 2q$. More precisely, for a connection on the normal bundle to the leaves that is flat in the direction of the leaves (called nowadays a Bott connection), the corresponding differential forms given by the Chern–Pontryagin–Weil theory vanish identically. The proof of this theorem, which occupies not more than a page, is a model of simplicity and efficiency. It could have been discovered twenty years earlier, as the theory of characteristic classes using connections and differential forms was perfectly well understood by André Weil in 1949 (*Collected Papers*, Springer-Verlag, Vol. 1, 422–436) and beautifully explained by H. Cartan in the "Colloque de Topologie," Bruxelles 1950, pp. 15–27 and 57–71. Moreover, in the same volume and in between the two papers of H. Cartan, Ehresmann notes in his article "Les conexions infinitésimales dans un espace fibré différentiable" on p. 38 that the normal bundle to a leaf of a foliation is naturally flat. The simplicity of this necessary condition for complete integrability is also quite surprising, and as far as I know all known obstructions to complete integrability are formal consequences of this Bott vanishing theorem (stated in its strong form, namely the actual vanishing of the differential form of degree $> 2q$ representing the polynomial in the Pontryagin classes; see Shulman's thesis, Berkeley, 1972).

My interest in foliation theory arose again when I learned from A. Phillips about the preprint of this paper of Bott: "On a topological obstruction to integrability," published two years later in the *Proceedings of the Symposia in Pure Mathematics*, AMS, Vol. XVI. The result of Bott was suggesting that there might be a nontrivial obstruction theory for deforming a field of planes of codimension q into a completely integrable one. I was already familiar with the notion of homotopy of Γ_q-structures (or equivalently of foliated microbundles of rank q) while I was writing my thesis in 1958, and I realized soon in 1969 that one had a classifying space $B\Gamma_q$ for those foliated microbundles, and that the vanishing theorem of Bott could be phrased in this framework. In that time, using the Gromov–Phillips transversality theorem, I could deduce from that an obstruction theory for the existence of foliations only on open manifolds, but W. Thurston proved three years later the much deeper theorem that this was also valid on closed manifolds.

In one of the plenary sessions of the International Congress at Nice in 1970, Bott reported on his theorem and the classifying space for foliations. At this same congress, I.M. Gel'fand was also invited to give one of the main talks. His address, read by someone else, reported among other topics on his joint work with D.B. Fuks on the cohomology of the Lie algebra of smooth and of formal vector fields on R^q. For instance, he stated that the latter was isomorphic to the cohomology of the restriction of the universal bundle $EU_q \to BU_q$ above the $2q$-skeleton of BU_q. Apparently, no one noticed at that time that there was a deep connection between those two talks.

The decisive step in that direction came the following spring of 1971 when Claude Godbillon, a former student of Reeb, and Jacques Vey, who had just studied the papers of Gel'fand and Fuks on the cohomology of the Lie algebra of formal vector fields, met in Strasbourg and discovered their famous invariant, establishing the first link between Gel'fand–Fuks cohomology and foliation theory. Roussarie showed that this invariant was not trivial, and soon after, Thurston showed that it could take all real values (*Bull. AMS* **78** [1972], 511–514). In the months following the publication of the note of Godbillon–Vey (*C. R. Acad. Sci. Paris* **273** [1971], 92–95), it became clear independently to many people (Raoul and myself, Malgrange, Bernstein-Rozenfeld) that there was a natural homomorphism of the Gel'fand–Fuks cohomology of different kinds into the real cohomology of the classifying spaces of foliations of the corresponding kinds. In particular, the Bott vanishing theorem was valid already on the level of Gel'fand–Fuks cohomology.

Independently of that, Bott had realized before that his vanishing theorem was forcing the existence of a whole series of so-called secondary characteristic classes (in particular, he had discovered an analog of the Godbillon–Vey invariant in the complex case and had proved that it was not trivial). One should also recall that around the same time, Chern and Simons (*Proc. Nat. Acad. Sci. USA* **68** [1971], 791–794) had constructed their famous invariant whose existence is also a consequence of the vanishing of a Pontryagin form.

One of the main results of Gel'fand–Fuks mentioned above is that the cohomology of the Lie algebra of formal vector fields on R^q is isomorphic to the cohomology of the differential graded commutative algebra $\widehat{W}_n$ quotient of the free algebra generated by $2q$ generators u_i and c_i of degrees $2i-1$ and $2i$, respectively, $i = 1, \ldots, q$, with $du_i = c_i$, by the ideal generated by the polynomials in the c_i's of degree $> 2q$. This is the result of a nontrivial computation (for an excellent account, see the talk of C. Godbillon at the Bourbaki Seminar, No. 421, 1972), and I still find it rather mysterious that the final result depends only on the analog of the vanishing theorem of Bott. It means that, whatever functorial construction you imagine to build up invariants of foliations (with trivialized normal bundle) using differential forms, you will not get anything else than what is in the algebra of secondary characteristic classes. This is not true, for example, for foliations with a symplectic transversal structure.

The more direct approach of Bott had a tremendous influence and was the source of numerous generalizations in many different directions. It generated a considerable literature. Let us mention, among many others, the names of Bott, of course, Bernstein-Rozenfeld, Fuks, Heitsch, Kamber and Tondeur, Lazarof, Moussu, Pasternak, Schweitzer, Shulman, Stasheff, Vey, etc., and later on S. Hurder, who pushed the theory of secondary characteristic classes very far and extended as well the deep work of Duminy on the Godbillon-Vey invariant to other classes.

Coming back to the beginning of the 1970s, the philosophy developed by Raoul and me was that, in the various contexts, the Gel'fand–Fuks cohomology could be considered as the continuous (or differentiable) cohomology of the classifying space for foliations (in a sense analogous to that of Van Est), and that features of the former should reflect those of the latter, namely, the real world.

In parallel to those developments, between 1970 and 1973 Mather and Thurston proved very deep theorems on the homological properties of $B\overline{\Gamma}_q$, the classifying space of framed foliations of class C^r of codimension q (for references, see the reports of W. Thurston and J. Mather in the *Proceedings of the ICM Paris* [1974], Vol. 1, pp. 92–95, and Vol. 2, pp. 35–37). In particular, they showed that the natural map of $B\overline{\mathrm{Diff}(R^q)}$ in the q-fold loop space of $B\overline{\Gamma}_q$ induces an isomorphism in integral homology, where $B\overline{\mathrm{Diff}(R^q)}$ is the classifying space for a C^r-foliated trivialized bundle with fiber R^q and compact support. In view of that, the study of $B\overline{\Gamma}_q$ is essentially reduced to the study of the homology properties of the group of diffeomorphisms of R^q with compact support. For instance, the $(q+1)$-connectivity of $B\overline{\Gamma}_q$ is deduced from the theorem of Mather about the simplicity of the group of diffeomorphisms with compact support of R^q.

According to the philosophy mentioned above, the result of Thurston–Mather suggested that the cohomology of the Lie algebra of smooth vector fields on R^q with compact supports is isomorphic to the cohomology of the q-fold loop space on a wedge of spheres (Vey had shown that the minimal model, in the sense of Sullivan, of $\widehat{W}_n$ did correspond to a wedge of spheres). This conclusion was in contradiction with a finiteness result of Gel'fand–Fuks; this led me to have a closer look at their computations and I discovered a mistake in their argument.

At the end of 1972, Thurston proved more generally that the classifying space for foliated bundles with fiber a smooth manifold M has the same integral homology as the space of sections of a bundle with fiber $B\Gamma_q$. The translation of this theorem in continuous cohomology is the conjecture that Bott made in 1973; he came to it from a more direct approach that was the basis of his proof with G. Segal of this conjecture.

As I said above, for the problem of deforming a field of tangent planes of codimension q into a smooth, completely integrable one, no other obstructions than the one given by the vanishing theorem of Bott (in its strong form) is known. Thurston has proved, for instance, that any continuous field of p-planes whose complementary subbundle is trivial is homotopic to a field tangent to a C^∞-foliation. He has also asserted (*loc. cit.* above) that any continuous field of p-planes on a smooth manifold is homotopic to a field tangent to a Lipschitz foliation whose leaves are smooth manifolds; recently, Tsuboi has been able to replace Lipschitz by C^1 (*Ann. of Math.* **130** [1989], 227–271). Hence there is no obstruction to integrability in the topological and even in the C^1 case. This shows that more differentiability than C^1 is essential in the vanishing theorem of Bott.

Secondary characteristic classes give obstructions for the related problem of deforming a field of planes into an integrable one without modifying it around a closed set where it is already completely integrable (there is only one exception of the kind given in the example of Reeb mentioned in the beginning; see Thurston, *Ann. of Math.* **104** [1976], 249–268). In particular they give obstructions for constructing a cobordism between two foliations. A typical open problem is to know whether a codimension-one foliation on a 3-manifold is cobordant to zero when its Godbillon–Vey invariant vanishes. T. Tsuboi has obtained interesting partial results in that direction. Up to now, it seems that all knowledge about the nontriviality of the cohomology of the classifying space for framed smooth foliations follows from evaluating secondary characteristic classes on specific examples combined with formal constructions. This confirms that, up to now, the vanishing theorem of Bott plays the leading role.

The Papers of Raoul Bott

Reprinted from *Proceedings of Symposia in Pure Mathematics*, Vol. XVI, pp. 127-131,

ON A TOPOLOGICAL OBSTRUCTION TO INTEGRABILITY

RAOUL BOTT

1. **Introduction.** A subbundle of the tangent bundle to a C^∞-manifold is called integrable if its space of smooth sections is closed under the bracket operation. The primary purpose of this note is to show that in general subbundles of the tangent bundle need not be isomorphic to integrable ones. Thus in particular we answer a question raised by several people, see [3], for instance, as to whether every subbundle of the tangent bundle is deformable into an integrable one.

Our counterexample is based on the following

Integrability criterion. A subbundle E of a tangent bundle T is integrable only if the ring Pont T/E *generated by the real Pontrayagin classes of* T/E *vanishes in dimensions greater than* $2 \times \dim(T/E)$:

$$\text{Pont}^k(T/E) = 0 \quad \text{if } k > 2\dim(T/E). \tag{1.1}$$

Note that Pont (T/E) may also be interpreted as Pont $(T - E)$ and so depends only on the isomorphism classes of T and E. Hence if Pont (T/E) does not satisfy (1.1) then no subbundle of E' of T, isomorphic to E, is integrable.

The vanishing theorem (1.1) is made plausible by the following consideration: Suppose that

$$\begin{array}{c} Y \\ \downarrow \pi \\ X \end{array}$$

is a C^∞ fibering and let E be the bundle along the fibers of π. Then E is trivially integrable and fits into the exact sequence:

$$0 \to E \to T \to \pi^* TX \to 0, \qquad T = TY;$$

over Y. By naturality and the obvious vanishing of $H^*(X)$ in dim > dim X, we conclude that

$$\text{Pont}^k(T/E) = 0 \qquad \text{for } k > \dim(T/E). \tag{1.2}$$

In general an integrable E of course does not arise in this manner. However, it turns out that if one utilizes the differo-geometric construction of Pont(T/E) then the vanishing phenomenon (1.2) is easily seen to persist—but only in the smaller range of dimensions given by (1.1). This approach naturally yields only information about the real cohomology, and I don't know whether (1.1) is true over the integers.

The Euler class $e(T/E)$ is also not covered by this method, so that for instance I do not know whether $e(T/E)^2$ or $e(T/E)^3$ have to vanish for an integrable E. The fourth power $e(T/E)^4$ of course has to because $e(T/E)^2 \in \text{Pont}\,(T/E)$.

In the complex analytic situation one obtains perfect agreement with the case of a fibering. Indeed (1.1) is then strengthened to yield the following:

Complex integrability criterion. A holomorphic subbundle E of a holomorphic tangent bundle T is integrable only if the ring Chern (T/E), *generated by the real* Chern-*classes of* (T/E) *vanishes in dimensions greater than* $2 \times \dim_{\mathbf{C}}(T/E)$:

$$\text{Chern}^k(T/E) = 0, \qquad k > \dim_{\mathbf{R}}(T/E). \tag{1.3}$$

The proof of these propositions will be given in §2, and in §3 I will discuss the case of integrable subbundles with singularities. Here I would like to conclude with some simple corollaries to these integrability conditions which also yield the counter example alluded to earlier.

COROLLARY 1.4. *Let X be a compact analytic manifold which admits a nonvanishing holomorphic vector field. Then all the* Chern *numbers of X vanish.*

PROOF. Let E be the trivial subbundle generated by the vector-field. It is obviously integrable. Hence by (1.3) $\text{Chern}^{2n}(T/E) = 0$, $2n = \dim_{\mathbf{R}} X$. Because E is trivial Chern (T/E) = Chern (T) so that the corollary follows. I have noted this result earlier (see [1]) in a different connection.

COROLLARY 1.5. *Let T be the tangent bundle of P_n, the projective space of n complex dimensions, with n odd and > 1. Then T contains a holomorphic subbundle E of codimension* 1 *but contains no integrable holomorphic subbundle of codimension* 1.

PROOF. To construct E interpret P_n as the projective space $P(V)$ of one-dimensional subspaces of $V = \mathbf{C}^{n+1}$ and let

$$v \to S \to \tilde{V} \to Q \to 0$$

be the usual exact sequence of bundles over $P(V)$. (Thus $\tilde{V} = P(V) \times V$, and S is the subbundle of $\tilde{V}$ given by $S = \{(l, v) | l \in P(V), v \in V, v \in l\}$. The tangent bundle T of $P(V)$ is then well known to be given by $T = \text{Hom}\,(S, Q)$, and the dual to S in the hyperplane bundle H of $P(V)$. Now choose a basis $\{v_i\}$ of V^* and construct the *nondegenerate* 2-form;

$$\omega = v_1 \wedge v_2 + v_3 \wedge v_4 + \dots + v_n \wedge n_{n+1} \qquad (n \text{ is odd!}).$$

This form is then seen to induce a holomorphic surjection

$$T \xrightarrow{\omega_*} H^2 \to 0,$$

by the formula

$$\omega_*(t)[v] = \omega(v \wedge tv),$$

where $l \in P(V)$, $t \in T_l = \text{Hom}(l, V/l)$ and $v \in l$.

We take for E the kernel of ω_* so that

$$T/E \simeq H^2. \tag{1.6}$$

So much for the first assertion. To prove the second one assume $E \subset T$ holomorphic, integrable, and of codimension 1. Then by our criterion $c_1(T/E)^2 = 0$. Because $n > 1$, this implies $c_1(T/E) = 0$. But this contradicts the fact that $c_n(T) = c_{n-1}(E) \cdot c_1(T/E)$ is nonzero. Q.E.D.

A similar argument, using (1.1) in the form $e(T/E)^4 = 0$, yields the following:

COROLLARY 1.7. *Let T be the tangent space to P_n with n odd and > 3. Then T contains a real subbundle of codimension 2, but contains no integrable real subbundle of codimension 2.*

In the complex case one can actually strengthen (1.5) as follows:

COROLLARY 1.8. *Let X be any compact analytic manifold whose tangent bundle T is positive. Then T has no integrable analytic subbundles.*

This follows from the fact that every quotient bundle of a positive bundle is positive—see [2] for instance. Hence if E is a holomorphic subbundle of T, then T/E is positive whence $c_1^n(T/E) > 0$, $n = \dim_C X$, contradicting the integrability of E. Because complex projective space has a positive tangent bundle, this assertion strengthens 1.5.

A more direct proof of (1.8) for the complex projective spaces was pointed out to me by Joel Pasternack. He shows by a combinatorial argument that in $H^*(P_n)$, there is no nontrivial factorization of $c(T)$ into a product $c(T) = c(E) \cdot c(T/E)$ if $\mathrm{Chern}^k(T/E) = 0$ for $k > \dim T/E$.

2. The proof of the criteria. Let us consider the real case alone; The complex analytic one is quite analogous. Also I will assume familiarity with the theory of real characteristic classes. Suppose then that $E \subset T$ is integrable and let $Q = T/E$. The dual Q^* is then contained in T^* and the Frobenius theorem implies that X has a coordinate covering $\{U, x_i^U\}$, $i = 1, \ldots, n$; such that on $U \in \{U\}$, Q^* is spanned by forms

$$dx_1, \ldots, dx_k, \quad k = \text{fiber} - \dim Q^*$$

where the x_i^U are the coordinates for U.

One now defines a connection D on Q^* in the following manner. Let D_U be the connection for $Q^*|U$, for which

$$D_U \cdot dx_1^U \equiv 0, \quad i = 1, \ldots, k. \tag{2.1}$$

Next choose a partition of unity λ_U for $\{U\}$ and set

$$D = \sum \lambda_U D_U.$$

It follows from (2.1) that the connection form θ_U, of D relative to the frame $dx_1^U, \ldots, dx_k^U$ on U, takes the form:

$$\theta_U = \sum_V \lambda_V dg_{UV} \cdot g_{VU}, \tag{2.2}$$

where g_{UV} are the transections between the frames dx_1^U, θdx_i^V, $i = 1, \ldots k$ on $U \wedge V$: symbolizing

$$dx^U = g_{UV} dx^V. \tag{2.3}$$

Differentiating this relation yields

$$0 = dg_{UV} \cdot dx^V \tag{2.4}$$

from which it follows easily that the forms of θ_U are all sections of Q^*. From this it follows in turn that the *curvature forms of D all lie in the ideal, I, generated by the sections of Q^*.*

Now the ideal I clearly has the property that if $a_1, \ldots, a_{k+1}$ are any $k + 1$ elements in I then $a_1 \wedge a_2 \ldots \wedge a_{k+1} = 0$. Finally. recalling that Pont Q^* is generated by all possible symmetric polynomials applied to the curvature of D and that Pont Q^* = Pont Q, the desired criterion (1.1) follows.

Note that the Euler class of Q^* can in general not be realized in this manner. It appears as a characteristic class of the curvature only relative to a Riemannian connection.

3. **Integrable subbundles with singularities.** Suppose now that v is a holomorphic vector field on the compact analytic manifold X which vanishes only to the first order at a (necessarily) finite number of points $\{P\}$. The differential of V then induces complex linear maps $L_P \cdot T_P \to T_P$ at each of these points, and in [**1**] I gave a formula for any Chern number $c^\alpha(X)$ of X in terms of the eigenvalues of the L_P.

Precisely, if $c_i(L)$ is defined by

$$\det(1 + tL) = \sum t^i c_i(L)$$

and $c^\alpha(L) = c_1^{\alpha_1}(L) \ldots c_n^{\alpha_n}(L)$, it was shown in [**1**] that

$$\sum_P \frac{c^\alpha(L)}{\det L} = c^\alpha(X), \quad L = L_P. \tag{3.1}$$

This formula is clearly a refinement of our Corollary 1 and leads one to conjecture that if $E \to T$ is a homomorphism of bundles whose image is an integrable subbundle of $T = TX$ outside of a singular set $\Sigma \subset X$, then the classes of $\mathrm{Pont}^k(T - E)$; $k > 2\{\dim T - \dim E\}$; should be computable in terms of local invariants of E near the singular set Σ. A first step confirming this conjecture is given by the following very recent result of P. Baum and myself which generalizes (3.1) to the case of line bundles.

Let X be a compact complex analytic manifold with tangent-bundle $T = TX$, and let

$$E \xrightarrow{\lambda} T \tag{3.2}$$

be a holomorphic map of the line bundle E into T which vanishes only to the first order at the points $\{P\}$. The differential of λ then induces homomorphisms

$L_P: E_P \to \mathrm{End}(T_P)$ so that for every nonzero $e \in E_P$, $c^\alpha(L_P(e))$ is well defined, and if $|\alpha| = \alpha_1 + 2\alpha_l + \dots + n\alpha_n = \dim_C X$, $c^\alpha[L_P(e)]/\det L_P(e)$ is independent of $e \neq 0 \in E_P$. Denoting this ratio by $c^\alpha(L_P)/\det(L_P)$, we show

THEOREM I (BAUM-BOTT). *In the situation just described the* Chern *numbers of* $T - E$ *are related to the zeroes of* λ *by the formula*

$$\sum_{\{P\}} \frac{c^\alpha(L)}{\det L} = c^\alpha(T - E), \quad L = L_P. \tag{3.2}$$

The proof of this relation is considerably harder than the proof of (3.1) and will be reported on elsewhere. One may, of course, interpret (3.2) as generalizing (3.1) to meromorphic vector fields. Indeed the transpose of λ is a section of $T \otimes E^*$ which vanishes precisely at the zeroes of λ, so that if s is a holomorphic section of E^* which does not vanish at the zeroes of λ, then $v = \lambda^t/s$ is a meromorphic vector-field whose infinitesimal numbers $c^\alpha(L_P)/\det L_P$ are well defined and agree with the infinitesimal numbers defined for λ. Thus (3.2) expresses the relations that must exist between the local numbers at the zeroes of a meromorphic vector field of this type on X.

The complete generalization of (3.2) to higher dimensional E seems quite difficult. However, the following is not hard to prove:

THEOREM II. *Let* $\lambda: E \to T$ *be an injection outside of the submanifold* $\Sigma \subset X$, *and let its image be integrable there. Then every* $\phi \in \mathrm{Chern}^k(T - E)$, *with* $k > \dim_C(T - E)$ *is in the image of the umkehrungs homomorphism* $\iota_*: H^*(\Sigma) \to H^*(X)$.

Thus every such $\phi = \iota_* \psi$ for some $\psi \in H^{k-\sigma}(\Sigma)$, $\sigma \in \dim_R \Sigma$, and what remains to be done, is to understand the recipe for ψ in terms of ϕ and the zero behavior of λ on Σ.

REFERENCES

1. Raoul Bott, *Vector fields and characteristic numbers*, Michigan Math. J. **14** (1967), 231–244.
2. Raoul Bott and S. S. Chern, *Hermitian vector bundles and the equidistribution of the zeroes of their holomorphic sections*, Acta Math. **114** (1965), 71–112.
3. S. S. Chern, *The geometry of G-structures*, Bull. Amer. Math. Soc. **72** (1966), 167–219.

HARVARD UNIVERSITY

Actes, Congrès intern. math., 1970. Tome 1, p. 27 à 36.

ON TOPOLOGICAL OBSTRUCTIONS TO INTEGRABILITY

by RAOUL BOTT [1]

§ 1. Introduction.

In this lecture I would like to describe the state of the art in the problem of " foliating " a manifold or, as I prefer to view it, the problem of constructing integrable fields on a manifold. This subject has seen some interesting developments in the past two years and is also contemporary in the sense that, as you will see, it leads to " huge spaces ". By a huge space I mean here simply one whose homotopy groups are not finitely generated in every dimension. In the past we—and I think quite rightly—have shied away from such objects, but recently they have cropped up in various contexts: notably in the index theory associated to Von Neumann algebras of type II, and also in the localization of spaces at a given prime, and I am confident that in the future these " huge " spaces will enter into many of the analysis inspired problems in topology.

§ 2. Integrability.

Let me start by recalling the basic facts concerning the local theory of integrability. Consider a C^∞-manifold M and let TM denote its field of tangent planes. By a section of TM one means a smooth function $p \to X_p$ which attaches to each $p \in M$ a tangent vector at p. These are therefore the " vector-fields " or " infinitesimal motions " of M. If x, y are any two such sections their Lie bracket $[x, y]$ is again a well determined vector-field on M and the bracket operation satisfies the Jacobi-identity:

$$[x, [y, z]] = [[x, y], z] + [y, [x, z]]. \tag{2.1}$$

By a field of tangent k-planes on M one means a smooth family $E = \{ E_p ; p \in M \}$ of k-subspaces of T_pM. In short a k-dimensional " sub-bundle " of TM, and such a field is called integrable if its space of sections is closed under the bracket:

$$x, y \in \Gamma(E) \Rightarrow [x, y] \in \Gamma(E). \text{ } (^2) \tag{2.2}$$

The term integrable is here justified by the well-known theorem of Frobenius [7], Clebsch-Deahna to the effect that if E is integrable, then locally E is generated by

(1) This work was partially supported by National Science Foundation grant GP 9566.
(2) $\Gamma(E)$ denotes the set of smooth sections of E.

parallell translation—relative to some coordinate system—from a fixed k-plane E_0. Quite equivalently this may also be put in the following way:

There exists a covering $\{U_\alpha\}$ *of* M *by coordinate patches* U_α, *with coordinates* $\{x_1^\alpha, \ldots, x_n^\alpha\}$ *such that on* U_α, E *consists of the planes tangent to the slices*

$$x_{k+1}^\alpha = c_1 \ldots x_n^\alpha = c_q, \qquad q = n - k.$$

These slices are therefore local integral manifolds of maximal dimension, which fiber U_α into submanifolds of codimension q.

It follows that if one defines

$$f_\alpha : U_\alpha \to \mathbb{R}^q$$

by the formula

$$f_\alpha(p) = \{x_{k+1}^\alpha(p), \ldots, x_n^\alpha(p)\}$$

then f_α defines a " submersion " of U_α in $\mathbb{R}^q$, in the sense that the differential of f_α,

$$df_\alpha : T_p U_\alpha \to T_{f(p)}\mathbb{R}^q$$

is *onto at each point of* U_α, and our previous slices now are simply the *fibers*, $f_\alpha^{-1}(p)$, of f_α.

The $\{f_\alpha\}$ may therefore be thought of as a system of maximal local integrals of E, which completely describe E.

Now using the implicit function theorem, it is easy to see that because f_α and f_β are both submersions, one can, for each $x \in U_\alpha \cap U_\beta$, find diffeomorphisms:

$$g_{\alpha\beta}^x : W_\beta^x \to W_\alpha^x,$$

of a neighborhood of $f_\beta(x) \in \mathbb{R}^q$ into a neighborhood of $f_\alpha(x) \in \mathbb{R}^q$, such that near x

$$g_{\alpha\beta}^x \circ f_\beta = f_\alpha. \tag{2.3}$$

Finally, it follows from (2.3) and again the submersion property of f_α that for points near $x \in U_\alpha \cap U_\beta \cap U_j$:

$$g_{\alpha\beta}^x \circ g_{\beta j}^x = g_{\alpha j}^x. \tag{2.4}$$

I have written these equations mainly for future reference. At this point, I want you essentially only to understand that integrable subbundles E of TM can either be described by the integrability condition (2.2), or by a system of local integrals $\{f_\alpha\}$ of E which are local submersions of M in $\mathbb{R}^q$. Then, in particular any global submersion $f: M \to N$ of one manifold on the other defines an integrable field or " foliation " on M. Thus, for instance, if f is a fibration, then the field of tangents along the fiber is always integrable. Integrable fields generated in this way may be thought of as the most trivial examples.

To show you what may happen in more interesting cases let me remind you of two classical examples.

The first is the foliation on the torus $\mathbb{R}^2/\mathbb{Z}$ induced by the " foliation " of $\mathbb{R}^2$ by lines of a given slope m:

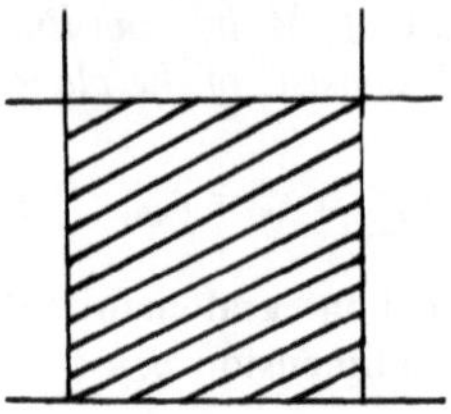

FIG. 1.

Thus, here I am drawing the " leaves ", i. e., *the maximal integral submanifolds* of the line field. If m is rational these leaves are all circles. If m is irrational they are all dense in T.

Next let me show you the beautiful Reeb foliation of the three sphere: First foliate the strip $|y| \leq 1$ in $\mathbb{R}^2$ as indicated in Figure 2:

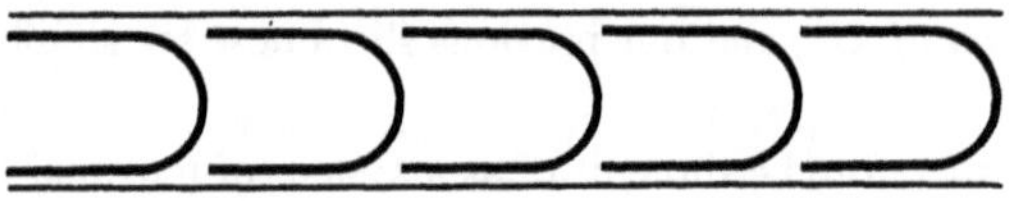

FIG. 2.

Next rotate this figure about the x-axis to obtain a foliation of a cylinder. There after identity points which differ by a integer x coordinate.

The result is a foliation of the anchor-ring,

FIG. 3.

whose leaves are either planes coiling up to the bounding torus, or the bounding torus itself. Now if we take, for S^3 the set in complex 2-space $\mathbb{C}_2$, given by

$$|z_1|^2 + |z_2|^2 = 1,$$

it is easy to see that S^3 is the union of two anchor rings

$$S^3 = A_1 \cup A_2,$$

given by the equations $|z_1| \leq |z_2|$ and $|z_2| \leq |z_1|$ which intersect in the torus $T = \{ |z_1| = |z_1| = 1/2 \}$. The foliations just described on A_1 and A_2 therefore fit together to form a foliation of S_3, which has one compact leaf, namely the torus T. All the other leaves are non-compact and curl up around this torus in opposite directions as we approach T from outside and inside. One may use this fact to show that this foliation though C^∞, is not *analytic.*

Concerning the higher spheres we know very little, in fact, we do not know whether any odd sphere S^n, of dim ≥ 3 admits an integrable $(n-1)$ field (*). One only has A. Haefliger's beautiful result that: *analytic integrable $(n-1)$-fields exist on a compact n-manifold only if its fundamental group is infinite.*

Another question which arises immediately in connection with this example is the existence of a compact leaf, and in this regard we have another beautiful result, due to Novikov, which asserts that every integrable 2-field on S^3 has a compact leaf. For 1-fields on S^3 it is not known whether a compact leaf has to exist. In fact, this is the famous Seifert problem. But these interesting and deep questions are really not pertinent to the problem (2.5) and I will have to leave them without further comment.

§ 3. On the nature of the global problem.

It is clear from the preceding that locally one can always construct integrable q-fields on a manifold M. The question therefore arises as to what difficulties one encounters in trying to construct a global field.

Now first of all, observe that difficulties will arise, because in general M does not admit a q-field, integrable or not. For instance, as is very well known, the 2-sphere S^2 admits no smooth line-field. On the other hand, the nature of this first question " does M admit a q-field ? " has been understood and much studied for many years. In particular, it has been converted into a purely homotopy-theoretic question.

Let me describe this translation to you, as it also points the way for our more refined question.

Please keep in mind during this development, that the homotopy theorist is a most singleminded person who treats only questions which can be phrased in terms of homotopy classes of continuous maps. Hence to please him we must convert all our geometric information into spaces and maps. In the present context this is not hard to do.

First of all one forms the Grassmanian variety

$$G_m(\mathbb{R}^N) = \{ A \subset \mathbb{R}^N \} \tag{3.1}$$

consisting of the set of m-subspaces of $\mathbb{R}^N$, topologized by the requirement that two such subspaces A and B are close, if and only if the unit spheres of A and B are close in $\mathbb{R}^N$. Next one includes $\mathbb{R}^N \subset \mathbb{R}^{N+1}$ in a standard manner and takes the limit of the compact spaces $G_m(\mathbb{R}^N)$ under the induced inclusions, to obtain the space

$$G_m = \lim_{N\to\infty} G_m(\mathbb{R}^N). \tag{3.2}$$

(*) Added November 10, 1970. Quite recently B. Lawson has constructed such foliations on all spheres of dimension $2^k + 3$.

This " infinite Grassmanian " is of fundamental importance in topology, because it *classifies* the vector-bundle functor; that is, there is a natural m-vector bundle E_m over G_m, with the property that for any reasonable space X the set of isomorphism classes of m-vector bundles over X, say $\mathrm{Vect}_m(X)$, is naturally in one to one correspondance with the homotopy classes $[X, G_m]$ of maps of X into G_m.

(3.3) $$\mathrm{Vect}_m(X) \simeq [X, G_m].$$

This correspondence assigns to a map $f: X \to G_m$ the pullback $f^{-1}E_m$ of E_m to X.

In case some of you are lost at this point, let me describe for you a particular consequence of (3.3) in quite elementary terms.

First of all note that an imbedding of M in a Euclidean space, $M \subset \mathbb{R}^N$ induces a map

(3.4) $$\gamma: M \to G_m.$$

Indeed, simply let $\gamma(p)$ equal the subspace of $\mathbb{R}^N$ *parallel to the tangent plane to M at p.*

Now it turns out that these maps all belong to the same homotopy class $\gamma_M \in [M, G_m]$, and that this homotopy class which we refer to as the Gauss map of M, corresponds to the tangent bundle under the isomorphism (3.3).

The class γ_M is the first and fundamental homotopy theoretic invariant of the differentiable structure on M. Incidentally γ_M, also serves to define the *Pontrjagin ring* of M. This is the image of the cohomology $H^*(G_m; \mathbb{Q})$ under γ_M^* in $H^*(M; \mathbb{Q})$. In fact, quite generally, if E is any vector bundle over X, one defines its rational Pontrjagin ring by the formula

(3.5) $$\mathrm{Pont}\,(E) = f_E^* H^*(G_m; \mathbb{Q})$$

where $f_E: X \to G_m$ is the map corresponding to E under the isomorphism (3.3).

But to return to our problem of finding a k-plane field on M. The class γ_M is very pertinent to this question because, as is actually not hard to see, *constructing a k-field on M amounts to giving a " lifting " $\tilde{\gamma}$ of the Gauss map in the following diagram:*

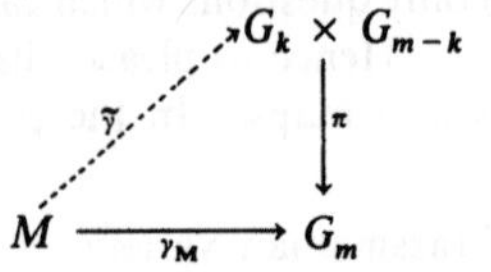

Here π is induced by the direct sum maps

$$G_k(\mathbb{R}^N) \times G_{m-k}(\mathbb{R}^{N'}) \to G_m(\mathbb{R}^{N+N'})$$

sending (A, B) to $A + B$.

Problems of the type

(3.7)

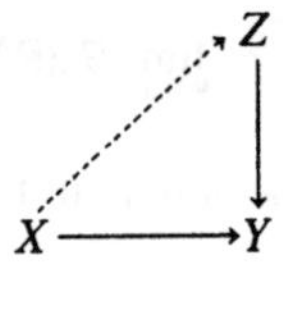

where the solid arrows are given homotopy classes of maps and a map from X to Z is sought which makes the diagram homotopy commutative, are called lifting problems and one has by now quite standard methods of treating them. Because, as we have just noted, the problem of constructing a k-field on M can be translated into such a lifting problem for the Gauss map, it is natural to ask whether our more refined question concerning the existence of integrable fields has a similar translation into a further lifting of γ_M. Now in the last two years Haefliger [5] and Milnor [7], using different approaches, but both based on deep results of Phillips [8, 9 and 10] and more generally Gromov [3], have essentially clarified the status of this question. Let me very briefly summarize Haefliger's point of view here.

Recall that an integrable E gave rise to *local* submersions

$$f_\alpha : U_\alpha \to \mathbb{R}^q,$$

and transition functions $g_{\alpha\beta}$ satisfying the equations (2.3) and (2.4). Haefliger now drops the condition that f_α be *submersions*, and considers more general systems $(f_\alpha, g_{\alpha\beta})$ satisfying only (2.3), and (2.4). Under a suitable equivalence relation, these systems give rise to a set-valued functor $\mathscr{H}_q(M)$, which one should think of as homotopy classes of foliations with singularities. The virtue of this construction is first of all that $\mathscr{H}_q$ makes sense on *all-spaces* (not just on manifolds!) is homotopy invariant and satisfies the " Meyer-Vietoris " condition of E. Brown [2]. Hence by Brown's general existence theorem there exists a space $B\Gamma_q$ which " classifies " $\mathscr{H}_q$. That is, there is a natural correspondence:

$$\mathscr{H}_q(X) = [X, B\Gamma_q]. \tag{3.8}$$

The space $B\Gamma_q$ thus plays the same role *relative* to $\mathscr{H}_q$ as the space G_q plays relative to the isomorphism classes of vector-bundles $\mathrm{Vect}_q(X)$. Furthermore, passing from Haefliger's " cocycle " $\{ f_\alpha, g_{\alpha\beta} \}$ to the differential $dg_{\alpha\beta}$ gives rise to a map

$$\nu : B\Gamma_q \to G_q \tag{3.9}$$

which expresses the fact that each element of $\mathscr{H}_q(M)$ has an associated " quotient-bundle ".

The construction of $\mathscr{H}_q$ and hence $B\Gamma_q$ now naturally leads to the questions:

A. How does the functor $\mathscr{H}_q(M)$ differ from the classes of integrable fields on M under a suitable equivalence relation?
B. To what extent does the homotopy of $B\Gamma_q$ differ from that of G_q?

For both these problems the Phillips-Gromov generalization of the Smale-Hirsch immersion theory it of fundamental importance. Essentially is enables one to push all the singularities of a " Haefliger structure " on open manifolds off to infinity. As a consequence on open manifolds any Haefliger structure compatible with the Gauss map is homotopic to an honest foliation! The precise result is as follows:

THEOREM I (Haefliger, Milnor). — *Let $\mathscr{E}_q(M)$ denote the classes of integrable plane fields on M of codimension q; under the following equivalence relation: two such fields E and E' are equivalent if and only if there exists a field $\mathscr{E}$ of codimension q on $M \times I$,* *which is transversal to all the slices* $M \times$ const, *and induces* E (resp. E') *on the slice* $M \times 0$ (resp. $M \times 1$).

Then on open manifolds

(3.10) $$\mathscr{E}_q(M) = \textit{homotopy classes of liftings of } \gamma_M$$

in the diagram

(3.11) $$\begin{array}{ccc} & & G_{m-q} \times B\Gamma_q \\ & \nearrow & \downarrow {\scriptstyle 1 \times \nu} \\ & & G_{m-q} \times G_q \\ & & \downarrow {\scriptstyle \pi} \\ M & \xrightarrow{\gamma_M} & G_m \end{array}$$

Concerning the second problem these same methods lead to the result.

Theorem II (Haefliger, Milnor). — *The map* $\nu : B\Gamma_q \to G_q$ *induces isomorphisms in homotopy in dimension* $\leq q$ *and is onto in dimension* $\leq q + 1$.

Thus, in particular, combining these two theorems we see that if M is open and of the homotopy type of a complex of dimension $\leq q + 1$ then every plane field of codimension q on M is homotopic to an integrable one.

To summarize the situation, these developments show that first of all on open manifolds our problem reduces to a lifting problem, and secondly that in low dimensions integrability induces no new difficulty. In short, these theorems are both of the existence type.

I would finally like to report on the meager crop of *nonexistence* theorems which are at present known.

§ 4. Some global obstructions to integrability.

Classical obstruction-theory teaches one that a complete understanding of the obstructions to lifting a map from X to Y,

$$\begin{array}{ccc} & & Y \\ & \nearrow & \downarrow {\scriptstyle \pi} \\ M & \longrightarrow & X \end{array}$$

involves, first of all, the homotopy groups of the " homotopy-theoretic fiber " of π. This is the space F which occurs as the inverse image of a point p in X under π, when π is replaced by a fibering in its homotopy class.

For instance if $F\Gamma_q$ denotes this fiber for the map $\nu : B\Gamma_q \to G_q$, so that we have the exact " sequence ":

(4.1) $$F\Gamma_q \to B\Gamma_q \to G_q,$$

then Theorem II is quite equivalently expressed by the statement

(4.2) $$\pi_r(F\Gamma_q) = 0, \quad \text{for} \quad r \leq q.$$

The homotopy groups of the fiber are important because there is no impediment to lifting over successive skeletons as long as these homotopy groups are zero, while, in general the obstruction to lifting from t to the $(t+1)$st skeleton is in $H^{t+1}(M, \pi_t(F))$. Of particular interest therefore is the first nonvanishing homotopy group of F.

Now in many of the classical lifting problems one could get at this information because the universal spaces X and Y were given explicitly by some relatively easy constructions. For instance, in the classical problem $G_k \times G_{m-k} \xrightarrow{\pi} G_m$ all the spaces can be treated directly.

In the present instance, and this is really typical of all the more subtle modern universal spaces such as B_{Top}, B_{PL}, etc., the space $B\Gamma_q$ is not really known to us in any manageable manner, and one can therefore get at this type of information only by very roundabout methods.

At present only the following results are known about the higher homotopy of $F\Gamma_q$. First of all, J. Mather [6] has very recently constructed a surjection (*):

(4.3) $$\mathrm{Diff}_0(\mathbb{R}^1) \to \pi_2(F\Gamma_1) \to 0.$$

On the other hand, one can use the integrability condition which I noticed two years ago to prove that:

(4.4) *For $q \geq 2$, some $\pi_k(F\Gamma_q)$ is nonzero, and in fact not finitely generated.*

Let me remark briefly how this first nonexistence—or obstruction—result comes about.

First I recall the integrability criterion alluded to earlier [1].

INTEGRABILITY CRITERION: *A sub-bundle E of the tangent bundle TM is integrable only if the ring* Pont (T/E) *generated by the rational Pontrjagin classes of T/E vanishes in dimension greater than* $2 \times \dim (T/E)$

(4.5) $$\mathrm{Pont}^k(T/E) = 0 \quad \text{if} \quad k > 2 \dim T/E.$$

The proof of this proposition is very direct, provided only that one uses the geometric definition due to Pontrjagin, Chern, Weil of the Pontrjagin classes as real cohomology classes represented by differential forms. Indeed, to give a clue to the initiated in this geometric framework, the infinitesimal integrability condition can be exploited to define a connection on T/E which is flat *along the leaves*, and then the result follows immediately. Essentially the same construction can be used to strengthen this criterion as follows:

THEOREM III. — *The homomorphism*

(4.6) $$\nu^* : H^*\{G_q; \mathbb{Q}\} \to H^*(B\Gamma_q; \mathbb{Q})$$

is zero in dimensions greater than $2q$.

Now the rational cohomology of G_q is well known to be a polynomial algebra $\mathbb{Q}[P_1, \ldots, P_{[q/2]}]$ in the universal Pontrjagin classes $P_i \in H^{4i}(G_q, \mathbb{Q})$, and is therefore,

(*) $\mathrm{Diff}_0(\mathbb{R}^1)$ denotes the group of diffeomorphisms of $\mathbb{R}^1$ with compact support.

in particular, non-trivial in positive dimensions provided $q \geq 2$. By a standard spectral sequence argument it follows therefore that $\pi_k(F\Gamma_q)$ must be nontrivial for some k. To obtain the nonfinite generation, one still has to show that if one uses $\mathbb{Z}_p$ coefficients then:

$$\nu^* : H^*(G_q, \mathbb{Z}_p) \to H^*(B\Gamma_q; \mathbb{Z}_p) \tag{4.7}$$

is injective.

To prove this one merely has to construct many examples of integrable fields E whose quotient bundles T/E have large mod p Pontrjagin rings, and such examples are easy to construct by taking E to be the horizontal space of flat vector bundles.

A question which seems to me of great interest is whether some of the groups $\pi_k(F\Gamma_q)$ are uncountable or not. In particular, one can relativize the integrability criterion to obtain homomorphisms of certain homotopy groups of $F\Gamma_q$ into the Reals and I would dearly like to know whether they are onto. The first case of interest occurs when $q = 3$ and in this situation the relative invariant gives rise to a homomorphism

$$\theta : \pi_7(F\Gamma_3) \to \mathbb{R}. \tag{4.8}$$

Let me now conclude with a very brief remark about the complex analytic case, where some of these questions can be settled.

As is pointed out in Haefliger's paper [5], the space $B\Gamma_q$ should be thought of as the classifying space associated to the groupoid of *germs of diffeomorphisms of* $\mathbb{R}^q$. (Recall that the $g^x_{\alpha\beta}$ were local diffeomorphisms of $\mathbb{R}^q$). A corresponding construction for germs of *complex-analytic automorphism of* $\mathbb{C}^q$ is possible, and leads to a space $B\Gamma_q\mathbb{C}$. One also has a corresponding fibering

$$F\Gamma_q\mathbb{C} \to B\Gamma_q\mathbb{C} \xrightarrow{\nu_{\mathbb{C}}} G\mathbb{C}_q \tag{4.9}$$

where now $G\mathbb{C}_q$ denotes the Grassmanian of complex subspaces of $\mathbb{C}^\infty$.

In this situation one can compute the relative invariants alluded to earlier and is then led to the

THEOREM IV. — *The homomorphism*

$$\nu^*_{\mathbb{C}} : H^*\{ G\mathbb{C}_q ; \mathbb{R}\} \to H^*(B\Gamma_q\mathbb{C}; \mathbb{R}) \tag{4.10}$$

is zero in $\dim \geq 2q + 1$.

Furthermore there exists a relative invariant θ_q *which maps* $\pi_{2q+1}(F\Gamma_q\mathbb{C})$ *onto* $\underbrace{\mathbb{C} \times \ldots \times \mathbb{C}}_{d(q)}$

$$\pi_2(F\Gamma_q\mathbb{C}) \xrightarrow{\theta_q} \underbrace{\mathbb{C} \times \ldots \times \mathbb{C}}_{d(q)} \to 0 \tag{4.11}$$

where $d(q) = \dim_{\mathbb{R}} H^{2(q+1)}(G\mathbb{C}_q, \mathbb{R})$.

In this case at least, I have therefore fulfilled my promise to introduce you to some genuinely huge spaces.

BIBLIOGRAPHY

[1] Raoul Bott. — On a topological obstruction to integrability, *Proceedings of Symposia in Pure Mathematics*, vol. XVI, Global Analysis, AMS, pp. 127-131.
[2] E. H. Brown. — Abstract homotopy theory, *Trans. Amer. Math. Soc.*, 119 (1965), pp. 79-85.
[3] M. L. Gromov. — *Izv. Akad. Nauk SSSR*, 33 (1969), pp. 707-734.
[4] A. Haefliger. — Structures feuilletées et cohomologie à valeur dans un faisceau de groupoïdes, *Comment. math. helvet.*, 32 (1958), pp. 248-329.
[5] —. — Feuilletages sur les variétés ouvertes, *Topology*, vol. 9, pp. 183-194.
[6] J. Mather. — *On Haefliger's classifying space* (in preparation).
[7] J. Milnor. — *Foilations and foliated vector bundles*, M. I. T., mimeographed notes, 1970.
[8] A. Phillips. — Submersions of open manifolds, *Topology*, 6 (1967), pp. 171-206.
[9] —. — Foliations on open manifolds. II, *Comment. math. helvet*, 44 (1969), pp. 367-370.
[10] —. — Smooth maps transverse to a foliation, to appear.

Department of Mathematics,
Harvard University
2 Divinity Avenue,
Cambridge, Mass. 02138
(U. S. A.)

Topology Vol. 11, pp. 141–146. Pergamon Press, 1972.

A REMARK ON THE INTEGRAL COHOMOLOGY OF $B\Gamma_q$

RAOUL BOTT† and JAMES HEITSCH

(*Received* 19 *November* 1971)

§1. INTRODUCTION

IN A RECENT paper [2], A. Haefliger constructed classifying spaces for foliations of co-dimension q on open manifolds. In the case of real foliations, he calls the space $B\Gamma_q$ and he obtains a map $\nu: B\Gamma_q \to BO_q$. In [1], the first author showed that the rational pontrjagin ring of the quotient bundle of a foliation vanishes in high dimensions. This result may be interpreted as follows.

THEOREM 1. $\nu^*: H^k(BO_q; \mathbf{Q}) \to H^k(B\Gamma_q; \mathbf{Q})$ *is the zero map if* $k > 2q$.

In this paper we provide a negative answer to the question raised in [1] as to whether these results hold over the integers. We do this by constructing enough examples of foliations whose integral pontrjagin rings are non-zero mod p in high dimensions for all odd primes p. These foliations are the "horizontal" foliations of bundles with discrete structural groups. Translating this result to a statement about the map ν we show that any infinite cyclic subgroup $G \subset H^k(BO_q; \mathbf{Z})$ injects under ν^*. But if $k > 2q$ Theorem 1 implies that the natural map $\rho: H^k(B\Gamma_q; \mathbf{Z}) \to H^k(B\Gamma_q; \mathbf{Q})$ sends $\nu^*(G)$ to zero. Finally we show that whenever the above situation holds the homology of the second space must be infinitely generated in dimension $k - 1$. This gives our main theorem.

THEOREM 3. $H_{4k-1}(B\Gamma_q; \mathbf{Z})$ *is not finitely generated if* $2k > q$ *and* $q \geq 2$.

We also show:

THEOREM 2. $\nu^*: H^*(BO_q; \mathbf{Z}_p) \to H^*(B\Gamma_q; \mathbf{Z}_p)$ *is injective for* $q \geq 2$ *and* p *an odd prime.*

Note that the above theorems imply that $\pi_k(F\Gamma_q)$ must be non-finitely generated for some k where $F\Gamma_q$ is the fiber of ν.

Actually much more can be shown than is done here. There are spaces $BS\Gamma_q$ and $BU\Gamma_q$ associated with BSO_q and BU_q respectively with properties similar to those of $B\Gamma_q$. See [3]. Since the bundles used here all carry a natural complex structure (and so are orientable) one obtains theorems corresponding to Theorem 2 and Theorem 3 by replacing BO_q and $B\Gamma_q$ by BU_q and $BU\Gamma_q$ (or BSO_q and $BS\Gamma_q$) throughout the paper. By arguments similar to those presented here one can show that ν^* is injective in Z cohomology for all these spaces. The fact that ν^* is injective on non-torsion elements is a corollary of Theorem 2. Since BU_q is torsion free, nothing more is required for this case. To show the result for

† The work of the first author was supported in part by the N.S.F. under grant GP-65650.

BO_q and BSO_q one need only show that ν^* is injective in Z_2 cohomology, since BO_q and BSO_q have only order two torsion. This can be done by constructing foliations on bundles over Lens spaces and using these to show that there exist maps f_1 and f_2 making the diagrams

$$\begin{array}{ccc} & & B\Gamma_1 \\ & \overset{f_1}{\nearrow} & \downarrow \nu \\ BZ_2 & \xrightarrow[i_1]{} & BO_1 \end{array} \qquad \begin{array}{ccc} & & BS\Gamma_2 \\ & \overset{f_2}{\nearrow} & \downarrow \nu \\ BZ_m & \xrightarrow[i_2]{} & BSO_2 \end{array}$$

commute, where i_1 and i_2 are the natural maps. The result now follows from this information and commutative diagrams such as the following

$$\begin{array}{ccc} B\Gamma_1 \times \cdots \times B\Gamma_1 & \longrightarrow & B\Gamma_q \\ \downarrow \nu \times \cdots \times \nu & & \downarrow \nu \\ BO_1 \times \cdots \times BO_1 & \xrightarrow{\oplus} & BO_q \end{array}$$

where $\oplus$ is the Whitney sum map. One obtains corresponding statements about the non-finite generation of the homologies of $BS\Gamma_q$ and $BU\Gamma_q$ and the homotopy groups of $FS\Gamma_q$ and $FU\Gamma_q$. This is done in [4]. Some of these extensions have also been obtained by John Wood [5].

§2. HORIZONTAL FOLIATIONS AND THE COUNTEREXAMPLES

We begin with some general remarks about horizontal foliations. Let X be a manifold and $\pi: E \to X$ a vector bundle over X with discrete structural group G. Locally E is a product and each product neighborhood has a natural horizontal foliation. Since G is discrete, the change of coordinates maps are locally constant and so preserve these local foliations. This gives a foliation of E called the horizontal foliation. The quotient bundle η to the horizontal foliation on E is isomorphic to $\pi^*(E)$. Also, the zero section $j: X \to E$ is a leaf of the foliation and $j^*(\eta)$ is isomorphic to E. Let $f: X \to BO_q$ classify E. Then $f \circ \pi$ classifies $\pi^*(E)$ and by a result of Haefliger, [2] Corollary 2.3, there is a bundle map $\hat{f}: E \to B\Gamma_q$ such that $\nu \circ \hat{f} = f \circ \pi$. Letting $\bar{f} = \hat{f} \circ j$ we obtain a commutative diagram of bundle maps

(*) $$\begin{array}{ccc} & & B\Gamma_q \\ & \overset{\bar{f}}{\nearrow} & \downarrow \nu \\ X & \xrightarrow[f]{} & BO_q \end{array}$$

We will use this diagram in the next section.

We have the following situation: a manifold X and a vector bundle $\pi: E \to X$ such that $\pi^*(E)$ is the quotient bundle of a foliation on E. Since π^* is an isomorphism in cohomology, the pontrjagin ring of $\pi^*(E)$ is the same as that of E. Thus to construct foliations whose quotient bundles have integral pontrjagin rings which do not vanish in high dimensions, we need only construct bundles with discrete structural groups whose mod p pontrjagin rings do not vanish in high dimensions for some prime p.

Let $X = S^{2n-1}/\mathbf{Z}_p$ where p is an odd prime, $S^{2n-1} \subset \mathbf{C}^n$ and $\mathbf{Z}_p$ acts through $U_1 \times \cdots \times U_1$. Let L be the basic complex line bundle over X. L has structural group Z_p. The first Chern class of L, $c_1(L)$, generates $H^2(X; \mathbf{Z}_p)$ and an easy argument using the Gysin sequence shows $(c_1(L))^{n-1} \neq 0$, and $H^*(X; \mathbf{Z}_p) \supset \mathbf{Z}_p[c_1]/((c_1)^n)$. The pontrjagin class of L is $1 - (c_1)^2$ and the ring generated by it has non-zero elements in dimension $2[n/2]$. Choosing n large we are done.

§3. TRANSLATION TO HAEFLIGER'S SPACE

We use the results of the previous section to prove Theorem 2. Before doing so we need a lemma which is a special case of Theorem 2.

LEMMA. *v^*: $H^*(BO_2; \mathbf{Z}_p) \to H^*(B\Gamma_2; \mathbf{Z}_p)$ is injective for p an odd prime.*

Proof. Recall the commutative diagram $(*)$ of Section 2. Let $X = S^{2n-1}/\mathbf{Z}_p$ and $E = L$. Then $H^*(X; \mathbf{Z}_p) \supset \mathbf{Z}_p[c_1(L)]/(c_1(L))^n$. Since $H^*(BO_2; \mathbf{Z}_p) = \mathbf{Z}_p[p_1]$ and $f^*(p_1) = -(c_1(L)^2)$, f^* is injective up to dimension $2[n/2]$. The diagram $(*)$ commutes, so by taking n arbitrarily large the result follows.

Proof of Theorem 2. Let $\oplus$: $BO_2 \times \cdots \times BO_2 \to BO_{2q}$ be the Whitney sum map and i: $BO_{2q} \to BO_{2q+1}$ the inclusion map. Also, let g: $B\Gamma_2 \times \cdots \times B\Gamma_2 \to B\Gamma_{2q}$ be the map which classifies the product of q copies of the universal Γ_2 structure as a Γ_{2q} structure and let h: $B\Gamma_{2q} \to B\Gamma_{2q+1}$ be the map which classifies the universal Γ_{2q} structure as a Γ_{2q+1} structure. The diagram

$$\begin{array}{ccccc} B\Gamma_2 \times \cdots \times B\Gamma_2 & \xrightarrow[g]{} & B\Gamma_{2q} & \xrightarrow[h]{} & B\Gamma_{2q+1} \\ \downarrow{\scriptstyle v \times \cdots \times v} & & \downarrow{\scriptstyle v} & & \downarrow{\scriptstyle v} \\ BO_2 \times \cdots \times BO_2 & \xrightarrow[\oplus]{} & BO_{2q} & \xrightarrow[i]{} & BO_{2q+1} \end{array}$$

commutes. The right square commutes since $i \circ v$ and $v \circ h$ classify the same bundle. The left square commutes for the same reason. It is well known that $(\oplus)^*$ and i^* are injective in Z_p cohomology, and it follows from the Künneth formula and the Lemma that $(v \times \cdots \times v)^*$ is injective in Z_p cohomology. The result now follows.

COROLLARY. *If G is an infinite cyclic subgroup of $H^*(BO_q; \mathbf{Z})$ then v^* is an injection on G.*

Proof. This is a consequence of the previous theorem, the commutative diagram

$$\begin{array}{ccc} H^*(BO_q; \mathbf{Z}) & \xrightarrow{v^*} & H^*(B\Gamma_q; \mathbf{Z}) \\ \downarrow{\scriptstyle \alpha} & & \downarrow{\scriptstyle \bar{\alpha}} \\ H^*(BO_q; \mathbf{Z}_p) & \xrightarrow{v^*} & H^*(B\Gamma_q; \mathbf{Z}_p) \end{array}$$

where the vertical homomorphisms are the natural maps, and the fact that α is just tensoring with $\mathbf{Z}_p$.

§4. PROOF OF THEOREM 3

Theorem 3 is a corollary of the following more general theorem.

THEOREM 4. *Let $\nu: X \to Y$ be a map between* CW *complexes and let $\rho: H^*(X; \mathbf{Z}) \to H^*(X; \mathbf{Q})$ be the natural map. Suppose that there is an infinite cyclic subgroup $G \subset H^k(Y; \mathbf{Z})$ such that ν^* is injective on G and $\rho \circ \nu^*(G) = 0$. Then $H_{k-1}(X; \mathbf{Z})$ is not finitely generated.*

Proof. For any CW complex and Z module M there is a functorial split exact sequence

$$0 \to \mathrm{Ext}(H_{k-1}(X; \mathbf{Z}), M) \to H^k(X; M) \to \mathrm{Hom}(H_k(X; \mathbf{Z}), M) \to 0.$$

If $H_{k-1}(X; \mathbf{Z})$ is finitely generated then $\mathrm{Ext}(H_{k-1}(X; \mathbf{Z}), \mathbf{Z})$ consists entirely of torsion elements. Consider the diagram

$$\begin{array}{ccccc}
H^k(Y; \mathbf{Z}) & \xrightarrow{\nu^*} & H^k(X; \mathbf{Z}) & \xrightarrow{\rho} & H^k(X; \mathbf{Q}) \\
& & \downarrow r & & \downarrow s \\
& & \mathrm{Hom}(H_k(X; \mathbf{Z}), \mathbf{Z}) \oplus \mathrm{Ext}(H_{k-1}(X; \mathbf{Z}), \mathbf{Z}) & & \mathrm{Hom}(H_k(X; \mathbf{Z}), \mathbf{Q}) \\
& & & \searrow w \qquad \nearrow t & \\
& & & \mathrm{Hom}(H_k(X; \mathbf{Z}), \mathbf{Z}) \otimes \mathbf{Q} &
\end{array}$$

where r and s are the natural isomorphisms, $w(a) = a \otimes 1$ and t is the natural map. The diagram commutes on the chain level so it commutes. Chasing one way we have $\rho \circ \nu^*(G) = 0$. Since the kernel of $w \circ r$ is the torsion part of $H^k(X; \mathbf{Z})$ and t is an injection, $s^{-1} \circ t \circ w \circ r \circ \nu^*(G) \neq 0$ a contradiction. Thus $H_{k-1}(X; \mathbf{Z})$ is not finitely generated.

APPENDIX

For those familiar with Haefliger's interpretation of $B\Gamma_q$ as the Milnor classifying-space of the topological semigroup Γ_q of germs of diffeomorphisms of $\mathbf{R}^q$, (see [3]) one can also explain Theorem 2 in the following more functorial manner.

First recall that in this model for $B\Gamma_q$, the map

$$\nu: B\Gamma_q \to BGL_q$$

is induced by the continuous map of topological semigroups:

$$\tilde{\nu}: \Gamma_q \to GL_q,$$

which sends a germ $\gamma \in \Gamma_q$, to $d\gamma$ evaluated at the source of γ.

On the other hand, there is also an obvious inclusion

$$\tilde{\imath}: GL_q \to \Gamma_q$$

which assigns to $g \in GL_q$ the germ at 0 of the diffeomorphism of R^q induced by g via the standard action of GL_q. Note, however, that this inclusion is continuous only if GL_q is given its discrete topology! Let us write GL_q^δ for this object. Then we have the following diagram

$$GL_q^\delta \xrightarrow{\imath} \Gamma_q \xrightarrow{\nu} GL_q$$

of maps of topological semigroups, which via the Milnor-construction induces maps

$$BGL_q^\delta \xrightarrow{\imath} B\Gamma_q \xrightarrow{\nu} BGL_q.$$

To prove Theorem 2 it is therefore sufficient to prove a corresponding result for the composition:

$$\nu \circ \imath \colon BGL_q^\delta \to BGL_q.$$

For this purpose let $T = \mathbf{R}/\mathbf{Z} \times \cdots \times \mathbf{R}/\mathbf{Z}$ (l-factors) $l = [q/2]$; be a maximal torus of GL_q and let

$$T_{\mathbf{Q}} = \mathbf{Q}/\mathbf{Z} \times \cdots \times \mathbf{Q}/\mathbf{Z}$$

be the corresponding product of the rationals mod $\mathbf{Z}$ in the discrete topology.

Clearly one then has the diagram:

$$\begin{array}{ccc} H^*(BGL_q^\delta) & \xleftarrow{(\nu\circ\imath)^*} & H^*(BGL_q) \\ \downarrow & & \downarrow j^* \\ H^*(BT_{\mathbf{Q}}) & \xleftarrow{\alpha^*} & H^*(BT). \end{array}$$

Furthermore it is well-known that except at the prime 2, j^* is an isomorphism onto the ring of invariants in $H^*(BT)$ under the Weyl-group of GL_q. Hence to prove non-vanishing theorems for ν^* it is sufficient to prove them for α^*.

On the other hand, the injectivity of α^* is quite easy to establish. Consider first the one dimensional case: $T = \mathbf{R}/\mathbf{Z}$, $T_{\mathbf{Q}} = \mathbf{Q}/\mathbf{Z}$. One has $T_{\mathbf{Q}} = \lim_{\to} \mathbf{Z}/n\mathbf{Z}$. Because the indexing set of this limit is $\mathbf{Z}^+$, the infinite telescope $\lim_{n\to} B\mathbf{Z}/n\mathbf{Z}$ represents $BT_{\mathbf{Q}}$. Finally because $\mathbf{Z}/n\mathbf{Z}$ is finite, the cohomology of $B\mathbf{Z}/n\mathbf{Z}$ is finite in each dimension >0 so that we can use the inverse limit to compute the reduced cohomology of $H^+(BT_{\mathbf{Q}})$:

$$H^+(BT_{\mathbf{Q}}) = \lim_{\leftarrow} H^+(B\mathbf{Z}/n\mathbf{Z}).$$

Now the reduced cohomology of $B\mathbf{Z}/n\mathbf{Z}$ (over $\mathbf{Z}$ throughout) is the ideal $\mathbf{Z}/n\mathbf{Z}[x]^+$ generated by x in $\mathbf{Z}/n\mathbf{Z}[x]$, with x the pullback under α of the generator of $H^2(BT)$. It follows that

$$H^+(BT_{\mathbf{Q}}) \simeq \hat{\mathbf{Z}}[x]^+,$$

where

$$\hat{\mathbf{Z}} = \lim_{\leftarrow} \mathbf{Z}/n\mathbf{Z},$$

is the profinite completion of $\mathbf{Z}$.

In short, then, when $\dim T = 1$, α^* is not only injective but induces an isomorphism of the profinite completion of $H^+(BT)$ with $H^+(BT_{\mathbf{Q}})$:

$$\hat{H}^+(BT) \simeq H^+(BT_{\mathbf{Q}}),$$

and under this isomorphism α^* corresponds to the canonical injection of $H^+(BT)$ in its profinite completion.

Using this and the Künneth theorem, the injectivity of α^* is then easily seen also in the general case. Actually, the following precise generalization of the 1-dimensional case is valid:

PROPOSITION. *The inclusion* $T_{\mathbf{Q}} \to T$ *induces an isomorphism of*

$$\hat{H}^+(BT) = \hat{\mathbf{Z}}[x_1, \ldots, x_l]^+$$

with $H^+(BT_{\mathbf{Q}})$.

We sketch a proof of this fact along lines suggested by D. Quillen. Consider the diagram:

$$\begin{array}{ccccccccc} 0 & \longrightarrow & \mathbf{Z} & \longrightarrow & \mathbf{Q} & \longrightarrow & \mathbf{Q}/\mathbf{Z} & \longrightarrow & 0 \\ & & \downarrow & & \downarrow & & \downarrow & & \\ 0 & \longrightarrow & \hat{\mathbf{Z}} & \longrightarrow & \hat{\mathbf{Z}} \otimes_{\mathbf{Z}} \mathbf{Q} & \longrightarrow & \hat{\mathbf{Z}} \otimes_{\mathbf{Z}} \mathbf{Q}/\mathbf{Z} & \longrightarrow & 0 \end{array}$$

where the lower sequence is obtained by tensoring with $\hat{\mathbf{Z}}$. Because $\hat{\mathbf{Z}}$ is torsion free, this lower sequence is also exact. Furthermore, the last arrow on the right is an isomorphism because $\mathbf{Q}/\mathbf{Z} = \lim \mathbf{Z}/n\mathbf{Z}$ and $\hat{\mathbf{Z}} \otimes \mathbf{Q}/\mathbf{Z} = \lim(\hat{\mathbf{Z}}/n\hat{\mathbf{Z}} \simeq \mathbf{Z}/n\mathbf{Z})$. Now $H^+(BT_{\mathbf{Q}}; \mathbf{Q}) = H^+(BT_{\mathbf{Q}}; \hat{\mathbf{Z}} \otimes_{\mathbf{Z}} \mathbf{Q}) = 0$. It follows, therefore, that

$$H^+(BT_{\mathbf{Q}}; \mathbf{Z}) \simeq H^+(BT_{\mathbf{Q}}; \hat{\mathbf{Z}}),$$

and the proposition will now follow once we prove that

$$H^*(BT_{\mathbf{Q}}; \hat{\mathbf{Z}}) = \hat{\mathbf{Z}}[x_1, \ldots, x_l].$$

Finally, this formula is established by induction on dim T, via the spectral sequence of the product

$$B(\mathbf{Q}/\mathbf{Z})^k \times B\mathbf{Q}/\mathbf{Z} = B(\mathbf{Q}/\mathbf{Z})^{k+1}.$$

In fact, E_2 of this spectral sequence is $H^*(B\mathbf{Q}/\mathbf{Z}; \hat{\mathbf{Z}}[x_1, \ldots, x_k])$ by the induction hypothesis and hence is already equal to $\hat{\mathbf{Z}}[x_1, \ldots, x_{k+1}]$. On the other hand, the differentials vanish, for instance, because there are only even dimensional terms. Q.E.D.

A final remark to explain this somewhat labored approach. The point is that if one computes via the inverse limit, all kinds of "error terms" due to the Künneth formula appear in each finite approximation and it is then cumbersome to show that they disappear in the inverse limit.

REFERENCES

1. R. BOTT: On a Topological Obstruction to Integrability, *Proceedings of Symposia in Pure Mathematics*, Vol. 10 (1970), pp. 127–131.
2. A. HAEFLIGER: Fuelletages sur les varietes ouvertes, *Topology* **9** (1970), 183–194.
3. A. HAEFLIGER: Homotopy and Integrability, *Proceedings of the Amsterdam Summer Institut on Manifolds*, Springer-Verlag Lecture Notes, No. 197.
4. J. HEITSCH: The Cohomologies of Classifying Spaces for Foliations (thesis), University of Chicago (1971).
5. J. WOOD, private communication.

Harvard University
University of Chicago
DePaul University
University of California, Berkeley

Istituto Nazionale di Alta Matematica
Symposia Mathematica
Volume X (1972)

ON THE LEFSCHETZ FORMULA AND EXOTIC CHARACTERISTIC CLASSES (*) (**)

RAOUL BOTT

The purpose of this note is to explain the relation of recently found exotic classes in the cohomology of the Haefliger classifying space $B\Gamma_q\mathbf{C}$ and the fixed point indexes which occur in the generalized infinitesimal Lefschetz formula for complex manifolds.

To motivate our point of view recall first of all the setting of the classical Lefschetz theorem. Consider then a smooth self map

$$f\colon M \to M$$

of a compact differentiable manifold M. The Lefschetz number $L(f)$ is then defined as the alternating sum:

$$L(f) = \sum (-1)^i \operatorname{trace} H^i(f) ,$$

where $H^i(f)$ is the homomorphism induced by f in the de Rham cohomology $H^i(M)$, and the Lefschetz theorem in its crudest form asserts that if $L(f) \neq 0$ then f must have a fixed point. The more delicate forms in one way or another express the number $L(f)$ as a sum

$$L(f) = \sum_F \mu_f(F) ,$$

where F runs over the components of the fixed points of f, and $\mu_f(F)$ are the «fixed point indexes» of F relative to f. When these fixed point sets are well behaved their indexes furthermore have a geometric

(*) I risultati conseguiti in questo lavoro sono stati esposti nella conferenza tenuta il 26 maggio 1971.

(**) This work was supported in part by National Science Foundation grant GP-7952X3.

interpretation. For instance, if F is an isolated point P, then the map

$$\sigma_f^P\, x \to x - f(x) \qquad (x \text{ near } P \text{ but } x \neq P)$$

defines a homotopy element of $\pi^{n-1}(S^{n-1})$, $n = \dim M$, and with correct orientation conventions one has the relation:

$$\mu_f(P) = \varphi(\sigma_f^P)$$

where φ is a generator of $H^{n-1}(S^{n-1})$.

In short then, the « variance » of $\mu_f(P)$ is given by a universal cohomology class φ in a certain universal space—S^{n-1} in this case—which describes the local geometry.

Let us turn now to the generalized Lefschetz theorem for complex manifolds as given in [2]. Here one assumes that everything in sight as complex analytic interprets $H^i(M)$ as the ith *cohomology* of the *sheaf of germs* of *holomorphic functions*, and $H^i(f)$ correspondingly as the homomorphism induced by f on this functor.

Our generalized Lefschetz formula then still takes the form:

$$L(f) = \sum_F \mu(F)\,, \tag{1}$$

but of course the local indexes have a quite different flavor and are in fact complex numbers. For instance if P is a nondegenerate fixed point then this holomorphic $\mu_f(P)$ has the value:

$$\mu_f(P) = 1/\det_P(1 - df_P)\,.$$

A question which has always puzzled me is whether this holomorphic multiplicity $\mu_f(P)$ is again expressible in the form

$$\mu_f(P) = \varphi\{\sigma_f(P)\}$$

where $\sigma_f(P)$ is a homotopy element of some universal space B describing the maps f near p, and $\varphi \in H^{n-1}(B)$. At the moment I am not clear as to whether this is in fact possible—or indeed useful. But in the infinitesimal analogue of this situation, such an interpretation arises naturally, and there the classifying space, $B\Gamma_{n-1}\mathbf{C}$, of complex foliations of codimension $(n-1)$ roughly plays the role of the universal mapping space B.

Let me turn to this infinitesimal case now. Here one starts with a holomorphic vector field X on M. To derive an infinitesimal analogue to the formula

$$L(f) = \sum \mu_f(F)$$

one considers the one-parameter family

$$f_t = \exp(tX)$$

obtained by integrating X (on the compact manifold M) and then considering the fixed point formula as a function of t.

For instance, if X is nondegenerate in the sense that at any zero P of X the linear part of X has eigenvalues $\lambda^P = \{\lambda_1^P, \ldots, \lambda_n^P\}$ none of which vanish, then f_t will have only nondegenerate fixed points for small t and our formula takes the form

$$L(f_t) = \sum_P \prod_1^n \frac{1}{(1 - \exp[t\lambda_i])} \qquad \lambda_i = \lambda_i^P.$$

Expanding the term $\prod_1^n 1/(1-\exp[t\lambda_i])$ in powers of t, yields a Laurent Series

$$(2) \quad \prod_1^n (1/1 - \exp[t\lambda_i]) = \left(\frac{1}{\lambda_1 \ldots \lambda_n}\right) \cdot \left\{\frac{a_{-n}}{t^n} + \frac{a_{-n+1}}{t^{n-1}} + \ldots + a_0 + ta_1 + \ldots\right\}$$

where the a_i are certain symmetric polynomials in the $\{\lambda_i^P\}$, which in fact are easily seen to be the so-called *Todd polynomials*. On the other hand the left hand side $L(f_t)$ of our relation is again easily seen to be smooth as a function of t, and in particular $L(f_0) = L$ (identity) is—by definition—the arithmetic genus, $\sum (-1)^i \dim H^i(M)$, of M. Now by the Riemann-Roch theorem this arithmetic genus is given by precisely the Todd number of M. Note here that this number is defined as the integral

$$\text{Todd}\,(M) = \int_M a_0(x_1, \ldots, x_n)\,; \qquad \prod_1^n (1 + tx_i) = c(M)$$

over M, of the class obtained from the polynomial $a_0(x_1, \ldots, x_n)$ when the elementary symmetric functions in the x_i's are interpreted as the Chern classes of M. If we put these two facts together one clearly arrives at the following

PROPOSITION: *Let X be a nondegenerate vector-field on M with eigenvalues $\{\lambda^P\}$ at the zeroes $\{P\}$. Also let a_r be the Todd polynomials as defined by* (2). *Then*

$$\sum \frac{a_i(\lambda^P)}{\lambda_1^P \ldots \lambda_n^P} = \begin{cases} 0 & \text{if } i < 0 \\ \text{Todd}\,(M) & \text{if } i = 0\,. \end{cases}$$

Actually a more general result is true, one in which the polynomials a_i can be replaced by arbitrary symmetric polynomials φ. To describe this situation more precisely let me first introduce the following notation: c^α shall denote a monomial of the type

$$c^\alpha = c_1^{\alpha_1} \cdot c_2^{\alpha_2} \ldots c_q^{\alpha_q}$$

in variables $c_1, \ldots, c_q$, and the weight of c^α, denoted by $|\alpha|$ is defined by

$$|\alpha| = \alpha_1 + 2\alpha_2 + \ldots + q\alpha_q .$$

If M is a compact complex manifold of $\dim_{\mathbf{C}} = n$, and c^α is monomial of weight n in $c_1 \ldots c_n$, then $c^\alpha(M)$ denotes the Chern number $\int_M c^\alpha$ of M, and if $\lambda = \{\lambda_i\}$ is a system of r numbers $r \leqslant n$, then we define $c_i(\lambda)$ as the ith elementary symmetric function in the λ's and $c^\alpha(\lambda)$ correspondingly. With this understood we have the following extension of the formula (2). (See [3] or [4]).

PROPOSITION: *Under the conditions on X and M described above, one has the formulae*

$$\sum \frac{c^\alpha(\lambda^P)}{c_n(\lambda^P)} = \begin{cases} 0 & \text{if } |\alpha| < n \\ c^\alpha(M) & \text{if } |\alpha| = n \end{cases} \tag{3}$$

for all monomials c^α in the letters $c_1, \ldots, c_n$.

I think of this relations as the proper infinitesimal analogue of (1). It can be proved in two quite different ways. On the one hand we can apply the generalized Lefschetz theorem and the Riemann-Roch theorem to f_t and the whole family of vector-bundles naturally associates to M. Then using the obvious extension of our previous argument the formula (3) is proved for a linearly independent set of monomials c^α. (See [1] for this approach).

On the other hand the relation (3) can also be derived in a much more elementary way which also naturally leads to the theory of characteristic classes of foliation theory ([1]).

The clue to this relationship is simply that 1-dimensional sub-bundles of the tangent bundle TM are always integrable and hence define a foliation on M. As we are dealing with the less familiar complex case here, let me first define Haefliger's notion of a complex foliation $\mathcal{F}$

([1]) One can then turn the argument about and use (3) to derive Riemann-Roch—but unfortunately only under the very restrictive condition that M admits a vector field X of the type under discussion.

of codimension q on M quite precisely (see [12] for instance), and briefly explain his point of view.

Given a smooth manifold M—which need not be holomorphic—such a foliation $\mathcal{F}$ is defined by a covering $U=\{U_\alpha\}$ and C^∞ submersions—the local charts of $\mathcal{F}$—

$$f_\alpha\colon U_\alpha \to \mathbf{C}^q$$

such that for any $x \in U_\alpha \cap U_\beta$ the germs of f_α and f_β at x, are related by the formula:

$$f_\alpha^x = \gamma_{\alpha\beta}^x \cdot f_\beta^x\,,$$

with $\gamma_{\alpha\beta}^x$ a germ of a *complex analytic map* of some neighborhood of $f_\beta(x)$ to some neighborhood of $f_\alpha(x)$. Two such $\mathcal{F}$'s are now called equivalent if their local charts combine to form local charts for the union of the two coverings. Finally $\mathcal{F}$ and $\mathcal{F}'$ are called homotopic if there is a foliation $\mathcal{F}'$ on $M\times I$ which induces $\mathcal{F}$ and $\mathcal{F}'$ on $M\times 0$ and $M\times 1$ respectively. Now in Haefliger's theory it is shown that on open manifolds these homotopy classes agree with a contravariant functor from spaces to sets which satisfies the representability axioms of Brown and so is represented by a universal space $B\Gamma_q\mathbf{C}$. In short, every $\mathcal{F}$ on M induces a unique homotopy class of maps

$$\varrho_{\mathcal{F}}\colon M \to B\Gamma_q\mathbf{C}\,.$$

Now every foliation $\mathcal{F}$ has a well-defined normal bundle $\nu(\mathcal{F})$ which is obtained by patching the inverses of the tangent bundles to $\mathbf{C}^q$ under f_α, that is $f_\alpha^{-1}\circ T\mathbf{C}^q$ on U_α, via the differential of the $\gamma_{\alpha\beta}$. This correspondance induces a map

$$\nu\colon B\Gamma_q\mathbf{C} \to BGL_q(\mathbf{C})$$

and one denotes the homotopy theoretic fiber of this map by $F\Gamma_q\mathbf{C}$. Thus one has the usual long exact sequence:

$$(4)\qquad \to \pi_*(F\Gamma_q\mathbf{C}) \to \pi_*(B\Gamma_q\mathbf{C}) \to \pi_*(BGL_q\mathbf{C}) \xrightarrow{\delta} \pi_{*-1}(F\Gamma_q\mathbf{C})\,.$$

In Hefliger's theory $\pi_*(F\Gamma_q\mathbf{C})$ thus describes the basic obstruction on open manifolds to lifting a given complex q-bundle E to a foliation $\mathcal{F}$ with normal bundle E, and $H^*(B\Gamma_q\mathbf{C})$ plays the role of real characteristic classes of complex foliations.

Let us now return to our vector-field X on the complex manifold

M. By the Frobenius theorem X defines a foliation $\mathcal{F}$ of codimension $(n-1)$ on the complement of its zero-set $M' = M - \text{zero}(X)$. Thus X defines a map

$$\varrho\colon M' \to B\Gamma_{n-1}\mathbf{C}, \qquad n = \dim_{\mathbf{C}} M .$$

Now for each zero P let S_P be a small punctured disc about P in M'. Then S_P is a $(2n-1)$ homotopy sphere and μ restricted to S_P gives rise to an element $\varrho(P)$ of $\pi_{2n-1}(B\Gamma_{n-1}\mathbf{C})$. This element is not quite the σ we need. Rather observe that, because $\pi_{2n-1}(BGL_{n-1})$ is finite of order $k_n = (n-1)!$, the element $k_n \cdot \varrho(P)$ admits a lifting to $\pi_{2n-1} \cdot (F\Gamma_{n-1}\mathbf{C})$ and we define

$$\sigma(P) \in \pi_{2n-1}(F\Gamma_{n-1}\mathbf{C}) \tag{5}$$

to be such a lifting. Note that $\sigma(P)$ is uniquely determined by $\varrho(P)$ up to an element of finite order because $\pi_{2n}(BGL_{n-1})$ is also finite.

With this understood we have the following theorem:

THEOREM 1.: *For each* $c^\alpha = c_1^{\alpha_1} \dots c_{n-1}^{\alpha_{n-1}}$ *the space* $F\Gamma_{n-1}\mathbf{C}$ *has an exotic characteristic class* $\tau c^\alpha \in H^{2n-1}(F\Gamma_{n-1}\mathbf{C}; \mathbf{C})$ *with the following properties:*

a) τc^α *is transgressive in the fibering*

$$F\Gamma_q\mathbf{C}^2 \to B\Gamma_q\mathbf{C} \xrightarrow{\nu} BGL_q \qquad q = n-1$$

and transgresses to $c^\alpha \in H^{2n}(BGL_q)$,

b) τc^α *is the universal class relative to* c^α *in the infinitesimal Lefschetz formula* (3). *That is, whenever* P *is a zero of* X *on* M*—subject to our conditions—then*

$$\tau c^\alpha[\sigma(P)] = k_n \cdot c^\alpha(\lambda^P)/c^n(\lambda^P) . \tag{6}$$

REMARKS: The c^α here of course play the role of the universal Chern-classes in BGL_{n-1}. The transgressiveness of the c^α with $|\alpha| = 2n$ furthermore implies that these classes disappear under $\nu^*\colon H^*(BGL_{n-1}) \to H^*(B\Gamma_{n-1}\mathbf{C})$. Thus the present theorem may be thought of as a refinement of my vanishing theorem of several years ago ([5]) which states that

$$\nu^*\colon H^*(BGL_q; \mathbf{R}) \to H^*(B\Gamma_q\mathbf{C}; \mathbf{R})$$

is zero in $\dim > 2q$.

The formula (6) solves our original problem of interpreting the fixed point indexes by universal classes. Read the other way the theorem makes precise the result announced in [6] that $\pi_{2n-1}[F\Gamma_{n-1}\mathbf{C}]$ admits a surjection onto $\mathbf{C}^r$ $r = \dim H^{2n}(BGL_{n-1})$. Indeed this follows easily by constructing an X on $\mathbf{C}P_n$ say, with prescribed eigenvalues at one of its zeroes. Put differently again, it shows that $F\Gamma_{n-1}\mathbf{C}$ has a large number of exotic classes which can be moved continuously.

A natural question which now arises, is the extent to which the homology class of the element $\varrho(P)$ in $H_{2n-1}(B\Gamma_{n-1}\mathbf{C};\mathbf{Z})$ already determines the fixed point multiplicity. Note that no class $\varphi \in H^{2n-1}\cdot(B\Gamma_{n-1}\mathbf{C};\mathbf{C})$ could play the role of «representing» c^α because the foliation by X being defined on all of M' we must have the relation

$$\sum_P \varphi[\varrho(P)] = 0$$

for any class in $B\Gamma_{n-1}\mathbf{C}$. Thus the best that could be expected is to produce for each c^α a class $\partial c^\alpha \in H^{2n-1}(B\Gamma_{n-1}\mathbf{C};\ \mathbf{C}/\mathbf{Z})$ such that one has the universally valid formula

$$\partial c^\alpha(\varrho(P)) \equiv c^\alpha(\lambda^P) \bmod \mathbf{Z}\,. \tag{7}$$

Such classes do exist. Indeed one has the following theorem:

THEOREM 2: *Each c^α as in Theorem 1, determines exotic classes*

$$\partial\mathfrak{J}c^\alpha \in H^{2n-1}(B\Gamma_{n-1}\mathbf{C};\mathbf{R})\,,$$

and

$$\partial\mathfrak{R}c^\alpha \in H^{2n-1}(B\Gamma_{n-1}\mathbf{C};\ \mathbf{R}/\mathbf{Z})$$

such that if we set

$$\partial c^\alpha = \partial\mathfrak{R}c^\alpha + i\,\partial\mathfrak{J}c^\alpha\,.$$

Then (7) *holds universally.*

These are the main results that I wanted to report on here and I will conclude with some very brief comments on their proof.

The study of the cohomology of spaces like $B\Gamma_q$ has made great strides during the six months since this talk was delivered, and is still very much in flux. My own approach was first of all motivated by the work of Chern and Simons [9], but after the work of Godbillon-Vey [10] pointed the direction toward the formal vector-field computations of Gelfand-Fuks [11], A. Haefliger and I independently and at the same time developed different but as it turned out, essentially equivalent procedures for generalizing the Godbillon-Vey construction. As Haefliger and I plan to report on this soon [1], in the framework of the formal vector-fields, I will here—very briefly— outline my procedure, which is inspired by the Chern-Simons principle of constructing secondary characteristic classes by comparing two types of connections. Actually the $\mathbf{R}/\mathbf{Z}$ classes of Theorem 2 are predicted as a direct consequence of the vanishing theorem in the general theory of Chern-Simons, and I do not yet understand whether one can predict more of these classes than they do.

Now to my recipe.

For this purpose think of

$$F\Gamma_q\mathbf{C} \xrightarrow{i} B\Gamma_q\mathbf{C} \xrightarrow{\nu} BGL_q$$

as a fibering of smooth manifolds. Also imagine that under ν^{-1} the universal bundle E over BGL_q becomes the normal bundle $\nu(\mathcal{F})$ of the universal foliation over $B\Gamma_q\mathbf{C}$:

$$\nu^{-1}E = \nu(\mathcal{F}) .$$

Now then, the normal bundles of foliations have associated to them a natural class of basic or *torsion free* connections α which have the property that the homomorphism which they induce

$$\alpha_{\#}: \mathbf{Z}[c_1, \ldots, c_q] \to A(B\Gamma_q\mathbf{C}) \quad (^2)$$

vanishes on the ideal generated by monomials of weight $> q$. (This is the more precise statement of the vanishing theorem. Of course in all of this the spaces $B\Gamma_q\mathbf{C}$, etc. are not meant to be taken literally—for actual proofs one works on finite manifolds only—i.e., with the functors they define. See [7] for such a treatment in the real case. With this understood $\alpha_{\#}$ of course sends c_i to the Chern form of $\nu(\mathcal{F})$ in the basic connection α).

Now then, we can also endow $\nu(\mathcal{F})$ with a Hermitian connection β, which in fact may be assumed to *have come from such a connection on* BGL_q. One therefore gets another homomorphism:

$$\beta_{\#}: \mathbf{Z}[c_1, \ldots, c_q] \to A(B\Gamma_q\mathbf{C}) .$$

On the other hand by the homotopy formula in the theory of charac-

(2) A denotes the de Rham complex with complex coefficients.

teristic classes there exist forms $h_i \subset A(B\Gamma_q\mathbf{C})$ depending functionally on α, β such that

$$\alpha_{\#} c_i - \beta_{\#} c_i = dh_i \,. \tag{8}$$

Finally note that the Chern forms for a Hermitian structure are actually *real*, that is

$$\beta_{\#} c_i = \overline{\beta_{\#} c_i} \,. \tag{9}$$

Now putting these three pieces of structure together we define a complex WU_q and a map

$$\hat{\alpha}_{\#} \colon WU_q \to A(B\Gamma_q\mathbf{C})$$

in the following manner: Let $\mathbf{Z}_q[c_1, \ldots, c_q]$ be the ring $\mathbf{Z}[c_1, \ldots, c_q]/$(elements of weight $> q$). Then set,

$$WU_q = \mathbf{Z}_q[c_1, \ldots, c_q] \otimes \mathbf{Z}_q[\bar{c}_1, \ldots, \bar{c}_q] \otimes E(h_1, \ldots, h_q) \,,$$

where the h_i are exterior generators, and define d in this complex by declaring all the c_qs and $\bar{c}_q$s to be cycles, while

$$d(1 \otimes 1 \otimes h_i) = c_i \otimes 1 \otimes 1 - 1 \otimes \bar{c}_i \otimes 1 \;;$$

or more informally

$$dh_i = c_i - \bar{c}_i \,.$$

Finally map WU_q into $B\Gamma_q\mathbf{C}$ by sending c_i to $a_{\#} c_i$; $\bar{c}_i$ to $\overline{\alpha_{\#} c_i}$ and h_i to the forms h_i of (8). Concerning this construction one then has the following result:

THEOREM 3: *The induced map*

$$\alpha^* \colon H^*(WU_q) \to H^*(B\Gamma_q; \mathbf{C})$$

is independent of the choices involved and so defines canonical exotic classes in $H^*(B\Gamma_q; \mathbf{C})$.

REMARK: From this point of view the earlier classes $\partial \mathfrak{J} c^\alpha$ correspond to the $\hat{\alpha}_{\#}$ images of the following cycles. If $c^\alpha = c_{i_1} \cdot c_{i_2} \ldots c_{i_k}$ then $\partial \mathfrak{J} c_\alpha$ corresponds to

$$\frac{1}{k} \hat{\alpha}_{\#} \{h_i \, c_{i_2} \ldots c_{i_k} + \bar{c}_{i_1} h_{i_2} c_{i_3} \ldots c_{i_k} + \ldots + \bar{c}_{i_1} \ldots \bar{c}_{i_{k-1}} h_{i_k}\} \,.$$

To construct exotic classes on $F\Gamma_q$ one can proceed even more directly. Note that $\beta_\# c_i$ can be factored through ν^* because the Hermitian structure was chosen on E. Hence restricting (8) to a fiber $F\Gamma_q\mathbf{C}$, one obtains the relation

$$i^*\alpha_\# c_i = di^* h_i . \tag{10}$$

Now let W_q be the complex

$$W_q = \mathbf{Z}_q[c_i, \ldots, c_q] \otimes E(h_1, \ldots, h_q)$$

$$dh_i = c_i$$

and use $i^*\alpha_\#$ as well as (10) to construct a chain map

$$i^*\hat{\alpha}_\# W_q \to A(F\Gamma_q) .$$

Then just as before one has the

THEOREM 4: *The induced map*

$$\widehat{i^*\alpha_*} : H^*(W_q) \to H^*(F\Gamma_q\mathbf{C}; \mathbf{C})$$

is independent of the choices involved and so defines canonical exotic classes in $F\Gamma_q\mathbf{C}$. Furthermore in this description

$$\mathfrak{J}c^\alpha = \widehat{i^*\alpha_*}(h_{i_1} c_{i_2} \ldots c_{i_k}) .$$

To what extent the « other » classes in the $H^*(WU_q)$ or $H^*(W_q)$ map into nontrivial elements in $B\Gamma_q\mathbf{C}$ and $F\Gamma_q\mathbf{C}$ is not quite understood as yet nor do we know whether any of these « other » classes take continuous values.

In conclusion let me remark that $H^*(W_q)$ is empirically seen to be precisely the cohomology of the formal vector-fields over $\mathbf{C}^q$ in the theory of Gelfand-Fuks, and it from this point of view that Haefliger obtains his exotic classes. This is also the point of view which we hope to describe in our forthcoming note.

Testo pervenuto il 27 marzo 1972.
Bozze licenziate il 4 agosto 1972.

BIBLIOGRAPHY

[1] M. F. Atiyah and R. Bott, *A Lefschetz fixed point formula for elliptic complexes*, I, Annals of Math., 86 (1967), 374-407.

[2] M. F. Atiyah and I. M. Singer, *The index of elliptic operators*, III, Annals of Math., 87 (1968), 546-604.

[3] P. Baum and R. Bott, *On the zeroes of meromorphic vector fields*, Essays on Topology and Related Topics, Mémoires dédiées à Georges de Rham, Springer-Verlag, Berlin, 1970, 29-47.

[4] R. Bott, *Vector fields and characteristic numbers*, Mich. Math. J., 14 (1967) 231-244.

[5] R. Bott, *On a topological obstruction to integrability*, Proceedings of Symposia in Pure Math., vol. 16, Global Analysis, Amer. Math. Soc., 127-131.

[6] R. Bott, *On topological obstructions to integrability*, Proceedings of the International Congress of Mathematicians, Nice, September 1970, 27-36.

[7] R. Bott, *Lectures on Foliations*, Mexico, 1971. To be published by Springer.

[8] R. Bott and A. Haefliger, *On exotic classes in certain infinite Lie groups*. To be published.

[9] S. Chern and J. Simons, *Some cohomology classes in principal fiberbundles and their applications to Riemannian geometry*, Proc. Nat. Acad. Sci. USA, 68 (1971), 791-794.

[10] Godbillon-Vey, *Un invariant des feuilletages de codimension* 1, Comptes Rendus Acad. Sci. Paris, June 1971.

[11] I. M. Gelfand and D. B. Fuks, *The cohomology of the Lie algebra of tangent vector fields on a smooth manifold, I and II*, I. Functional Analysis, 3 (1969), 32-52 and 4 (1970), 23-32.

[12] A. Haefliger, *Feuilletages sur les variétés ouvertes*, Topology, 9 (1970), 183-194.

J. DIFFERENTIAL GEOMETRY
7 (1972) 279-342

SINGULARITIES OF HOLOMORPHIC FOLIATIONS

PAUL BAUM & RAOUL BOTT

To S. S. Chern & D. C. Spencer on their 60th birthdays

0. Introduction

The purpose of this note is twofold. First, we give a simpler and more natural proof of our meromorphic vector-field theorem of [5]; and second, we give a theorem on singularities of holomorphic foliations which includes the meromorphic vector-field theorem as a special case. We have tried to make the exposition as elementary and self-contained as possible.

To recall the result of [5], let M be a complex analytic manifold. Set $n = \dim_C M$. Assume $n \geq 2$. Let T be the holomorphic tangent bundle of M, L be a holomorphic line bundle on M, and $\eta: L \to T$ be a holomorphic vector-bundle map. Let $X_1, \cdots, X_n$ be indeterminates, and φ be a polynomial in $X_1, \cdots, X_n$ with complex coefficients:

$$\varphi \in C[X_1, \cdots, X_n] . \tag{0.1}$$

Assume that φ is symmetric and homogeneous of degree n. Given an isolated zero p of η, define a number $\varphi(\eta, p)$ as follows. About p choose a nonvanishing holomorphic section s_p of L. Also about p, choose a complex-analytic coordinate system $z_1, \cdots, z_n$ with origin at p. The vector-field $\eta(s_p)$ is then well-defined near p, and there has the expansion

$$\eta(s_p) = \sum_{i=1}^{n} a_i \partial / \partial z_i , \tag{0.2}$$

where the a_i are holomorphic functions near p.

Form the matrix A of partial derivatives: $A = \| \partial a_i / \partial z_j \|$. Let $\sigma_1, \sigma_2, \cdots, \sigma_n$ be the elementary symmetric functions in $X_1, \cdots, X_n$. Define $\sigma_i(A)$ by

$$\det (I + tA) = 1 + t\sigma_1(A) + \cdots + t^n \sigma_n(A) . \tag{0.3}$$

Thus each $\sigma_i(A)$ is a function near p. Since φ is symmetric, there is a unique polynomial $\tilde{\varphi}$ such that

---

Communicated by R. Bott. The work of the first author was partially supported by the National Science Foundation Grant GP 22580, and that of the second author by the National Science Foundation Grant GP 31359X.

(0.4) $$\varphi = \tilde{\varphi}(\sigma_1, \cdots, \sigma_n) .$$

Define $\varphi(A)$ by

(0.5) $$\varphi(A) = \tilde{\varphi}(\sigma_1(A), \cdots, \sigma_n(A)) .$$

Then $\varphi(\eta, p)$ is defined to be the value at p of the Grothendieck residue symbol.

(0.6) $$\varphi(\eta, p) = \mathrm{Res}_p \begin{bmatrix} \varphi(A)dz_1 \cdots dz_n \\ a_1, \cdots, a_n \end{bmatrix} .$$

If p is a nondegenerate zero of η, i.e., if $\det \|(\partial a_i/\partial z_j)(p)\| \neq 0$, let $\lambda_1, \cdots, \lambda_n$ be the eigenvalues of $\|(\partial a_i/\partial z_j)(p)\|$. From the general properties of the Grothendieck residue given in [18] it then follows that in this case,

(0.7) $$\varphi(\eta, p) = \varphi(\lambda_1, \cdots, \lambda_n)/(\lambda_1 \cdots \lambda_n) .$$

More generally, here is an explicit algorithm for computing the right-hand side of (0.6).

Since the origin is an isolated zero of the a_i, there exist positive integers $\alpha_1, \cdots, \alpha_n$ with $z_i^{\alpha_i}$ in the ideal generated by $a_1, \cdots, a_n$. Hence there exist holomorphic functions b_{ij} near p with

(0.8) $$z_i^{\alpha_i} = \sum_{j=1}^{n} b_{ij} a_j .$$

One then has

(0.9) $$\mathrm{Res}_p \begin{bmatrix} \varphi(A)dz_1 \cdots dz_n \\ a_1, \cdots, a_n \end{bmatrix} = \mathrm{Res}_p \begin{bmatrix} \varphi(A) \det \|b_{ij}\| dz_1 \cdots dz_n \\ z_1^{\alpha_1}, \cdots, z_n^{\alpha_n} \end{bmatrix} .$$

The right-hand side of (0.9) is now evaluated by expanding $\varphi(A) \det \|b_{ij}\|$ in a power series in the z_i. The coefficient of $dz_1 \cdots dz_n/(z_1 \cdots z_n)$ in the resulting Laurent series for $\varphi(A) \det \|b_{ij}\| dz_1 \cdots dz_n/(z_1^{\alpha_1} \cdots z_n^{\alpha_n})$ is the desired answer.

This algorithm was derived for us by R. Hartshorne. It is an immediate consequence of the general properties of the Grothendieck residue given in [18].

It can be easily checked that $\varphi(\eta, p)$ does not depend on the choices made in defining it. Hence $\varphi(\eta, p)$ is a well-defined local number depending only on φ and the local behavior of η near p.

The result of [5] is:

Theorem 1. *Let M be a compact complex-analytic manifold, $\eta: L \to T$ be a holomorphic vector-bundle map with isolated zeroes, and φ be symmetric and homogeneous of degree n. Consider the virtual bundle $T - L$. Then*

(0.10) $$\varphi(T - L)[M] = \sum_{p \in \mathrm{Zero}(\eta)} \varphi(\eta, p) .$$

Remarks. (a) Let $c_1(T - L), \cdots, c_n(T - L)$ be the Chern classes of

$T - L$, taken in $H^*(M; C)$. Then, as is customary, $\varphi(T - L)$ is defined by

$$(0.11) \qquad \varphi(T - L) = \tilde{\varphi}(c_1(T - L), \cdots, c_n(T - L)) ,$$

where $\tilde{\varphi}$ is as in (0.4). Since φ is homogeneous of degree n, $\varphi(T-L) \in H^{2n}(M; C)$. $\varphi(T - L)[M]$ denotes $\varphi(T - L)$ evaluated on the canonical generator of $H_{2n}(M; C)$.

(b) If M is a submanifold of complex projective space, then by tensoring T with a sufficiently high power of the hyperplane bundle H, $\dim_C \Gamma(T \otimes H^r)$ can be made arbitrarily large. Here $\Gamma(T \otimes H^r)$ denotes the vector-space of all holomorphic sections of $T \otimes H^r$. Furthermore, almost all sections of $T \otimes H^r$ will have only isolated zeroes when r is large enough. A section of $T \otimes H^r$ gives a vector-bundle map $(H^r)^* \to T$. Thus there are many examples to which Theorem 1 applies.

We now take the point of view that Theorem 1 is really a theorem about holomorphic foliations with singularities. To see this, let us use the notation convention that whenever E is a holomorphic vector-bundle, $\underline{E}$ shall denote the sheaf of germs of holomorphic sections of E. Then at the sheaf level η is injective.

$$(0,12) \qquad \eta: \underline{L} \to \underline{T} \quad \text{is injective} .$$

Set $\xi = \eta(\underline{L})$ and $Q = \underline{T}/\xi$. Observe that ξ is an *integrable* subsheaf of $\underline{T}$ in the sense that for each $x \in M$, the stalk ξ_x is closed under the bracket operation for vector-fields. On M-Zero (η) we have a one-dimensional foliation, in the usual sense, of M-Zero (η). On M, however, we have a foliation with singularities. ξ can be thought of as the tangent sheaf of the foliation with singularities. If $\mathcal{O}$ is the structure sheaf of M, then the singularities occur precisely at those points $p \in M$ such that Q_p is not a free $\mathcal{O}_p$-module.

The exactness of

$$(0.13) \qquad 0 \to \underline{L} \to \underline{T} \to Q \to 0$$

implies that $c_i(Q) = c_i(T - L)$, $i = 1, \cdots, n$. Hence (0.10) can be rewritten

$$(0.14) \qquad \varphi(Q)[M] = \sum_{p \in \mathrm{Zero}(\eta)} \varphi(\eta, p) .$$

So we conclude that Theorem 1 computes the Chern numbers of Q in terms of local information at the singularities of the foliation.

Pass now to higher dimensional foliations. Define a subsheaf $\xi \subset \underline{T}$ to be *integrable* if

(i) ξ is coherent,

(ii) for each $x \in M$, ξ_x is closed under the bracket operation for vector-fields.

Set $Q = \underline{T}/\xi$ and $S = \{x \in M \mid Q_x$ is not a free $\mathcal{O}_x$-module$\}$. S is a closed holomorphic subvariety of M. S will be referred to as the *singular set*. On $M - S$ there is a unique holomorphic sub-vector-bundle F of $T \mid M - S$ such that

$$(0.15) \qquad \underline{F} = \xi \mid M - S .$$

We assume that $\dim_C F_x$ is constant throughout $M - S$. This is automatically the case if M is connected. $\dim_C F_x$ will be denoted by k and will be referred to as the leaf dimension of ξ. We shall always assume

$$(0.16) \qquad 1 \leq k < n .$$

Given $p \in M - S$, the well-known theorem of Frobenius asserts that there exists a complex-analytic coordinate system $z_1, \cdots, z_n$ defined on an open neighborhood U_p of p such that

$$(0.17) \qquad \partial/\partial z_1, \cdots, \partial/\partial z_k \qquad \text{is a frame of } F \mid U_p .$$

A leaf of this foliation of $M - S$ will be called a leaf of ξ.

It is natural to assume that ξ satisfies the following condition:

(0.18) Let U be an open subset of M, and γ a holomorphic section of $T \mid U$. Suppose that $\gamma(x) \in F_x$ for each $x \in U \cap (M - S)$. Then at each $p \in U \cap S$ the germ of the holomorphic vector-field γ is in ξ_p.

A ξ which satisfies this condition will be said to be *full*. In the situation of Theorem 1, $\eta(\underline{L})$ is full.

If ξ is integrable and $\dim_C S \leq n - 2$, then there is a unique sheaf $\hat{\xi}$ such that $\hat{\xi}$ is both full and integrable, and

$$(0.19) \qquad \hat{\xi} \mid M - S = \xi \mid M - S .$$

To define $\hat{\xi}$, let F be as in (0.15). Define $\hat{\xi}$ by

$$(0.20) \quad \Gamma(\hat{\xi}, U) = \{\gamma \in \Gamma(T \mid U) \mid \gamma(x) \in F_x \text{ whenever } x \in U \cap (M - S)\} .$$

In (0.20), U is any open set of M, $\Gamma(\hat{\xi}, U)$ denotes the continuous sections of $\hat{\xi} \mid U$, and $\Gamma(T \mid U)$ denotes the holomorphic sections of $T \mid U$. Thus by restricting attention to full integrable sheaves we rule out artificial singularities and deal only with genuine foliation singularities.

Given a full integrable subsheaf ξ of $\underline{T}$ we would like to compute Chern polynomials $\varphi(Q)$ in terms of local information at the singular set S. Let Z be a connected component of S. Recall that if M is compact, there is the homomorphism μ_*:

$$(0.21) \qquad \mu_*: H_j(Z; C) \to H^{2n-j}(M; C) \qquad j = 0, 1, \cdots, 2n .$$

$\mu_* = \alpha i_*$ where $i_*: H_j(Z; C) \to H_j(M; C)$ is induced by the inclusion of Z in M, and $\alpha: H_j(M; C) \to H^{2n-j}(M; C)$ is the Poincaré duality isomorphism.

We then have:

Theorem 2 (*Residue existence theorem*). *Let M be a complex-analytic manifold, ξ be a full integrable subsheaf of $\underline{T}$, and k be the leaf dimension of ξ. Set $Q = \underline{T}/\xi$, and let $\varphi \in C[X_1, \cdots, X_n]$ be a symmetric polynomial which is homogeneous of degree l, where $n = \dim_C M$ and $n - k < l \leq n$. Let Z be a connected component of the singular set S, and assume that Z is compact. Then there exists a homology class* $\mathrm{Res}_\varphi(\xi, Z) \in H_{2n-2l}(Z; C)$ *such that*

(0.22) $\mathrm{Res}_\varphi(\xi, Z)$ *depends only on φ and on the local behavior of the leaves of ξ near Z,*

(0.23) *if M is compact, then* $\sum_Z \mu_* \mathrm{Res}_\varphi(\xi, Z) = \varphi(Q)$.

Remarks. (a) If M is compact then clearly every connected component of S must be compact. In (0.23) the sum is taken over all the connected components of S.

(b) Let $\sigma_1, \cdots, \sigma_n$ be the elementary symmetric functions of $X_1, \cdots, X_n$. Since φ is symmetric and homogeneous of degree l, there is a unique polynomial ϕ in $\sigma_1, \cdots, \sigma_l$ with $\phi(\sigma_1, \cdots, \sigma_l) = \varphi$. Let $c_1(Q), \cdots, c_n(Q)$ be the Chern classes of Q. Then $\varphi(Q)$ is defined by setting $\varphi(Q) = \phi(c_1(Q), \cdots, c_l(Q))$.

(c) Let U be an open subset of M with $U \supset Z$. $\mathrm{Res}_\varphi(\xi, Z)$ is a local matter so $\mathrm{Res}_\varphi(\xi, Z)$ depends only on φ and on $\xi | U$.

(d) This is just an existence theorem. It asserts that $\mathrm{Res}_\varphi(\xi, Z)$ exists and has the desirable properties (0.22) and (0.23). But it does not give an explicit formula for $\mathrm{Res}_\varphi(\xi, Z)$ in terms of local information near Z.

To think about the problem of explicitly computing $\mathrm{Res}_\varphi(\xi, Z)$, one must confront the question: "What is the 'generic' singularity of a foliation?" Put otherwise: "What sort of a singularity is it reasonable to expect?" This appears to be a delicate question whose complete answer has eluded us. We have therefore only considered the case when the singular set satisfies certain natural dimension conditions. When $k = 1$, these conditions reduce to asserting that the singular set consists of isolated points.

In general, observe that a connected component Z of S comes endowed with a filtration. For given $p \in Z$ choose holomorphic vector-fields $\gamma_1, \cdots, \gamma_r$ defined on an open neighborhood U_p of p in M such that

(0.24) For all $x \in U_p$, the germs at x of the holomorphic vector-fields $\gamma_1, \cdots, \gamma_r$ are in ξ_x and span ξ_x as an $\mathcal{O}_x$-module.

Define a subspace $V_p(\xi) \subset T_p$ by letting $V_p(\xi)$ be the sub-vector-space of T_p spanned by $\gamma_1(p), \cdots, \gamma_r(p)$. $V_p(\xi)$ depends only on p and ξ, and is independent of the choice of $\gamma_1, \cdots, \gamma_r$. Set

(0.25) $$Z^{(i)} = \{p \in Z \mid \dim_C V_p(\xi) \le k - i\}\,, \qquad i = 1, \cdots, k\,.$$

Then

(0.26) $$Z = Z^{(1)} \supset \cdots \supset Z^{(k)}$$

is a filtration of Z. Each $Z^{(i)}$ is a closed holomorpic subvariety of M.

Our dimension conditions on Z are:

(0.27) $$\dim_C Z = k - 1\,,$$

(0.28) $$\dim_C Z^{(2)} < k - 1\,.$$

If (0.27) is valid for Z, a point $p \in Z$ will be said to be *regular* if there exist an open neighborhood U_p of p in M and complex-analytic coordinates $z_1 \cdots, z_n$ defined on U_p such that

(0.29) $$U_p \cap Z = \{x \in U_p \mid z_k(x) = \cdots = z_n(x) = 0\}\,.$$

Let N be the set of all points p in Z, which are not regular. N is a closed holomorphic subvariety of M with $\dim_C N < k - 1$.

Elsewhere [4] a proof will be given of the following theorem which to some extent describes the structure of a singularity for which (0.27) and (0.28) are valid: Given such a Z, let $p \in Z - (Z^{(2)} \cup N)$. The theorem asserts that in the vicinity of p the foliation singularity is the "pull-back" via a submersion of an isolated zero of a holomorphic vector field. The submersion maps a neighborhood of p in M onto a neighborhood of the origin in C^{n-k+1}.

(0.30) Theorem. *Let M be a complex-analytic manifold, ξ be a full integrable subsheaf of $\underline{T}$, and Z be a connected component of the singular set S. Assume that $\dim_C Z = k - 1$ and $\dim_C Z^{(2)} < k - 1$. Let $p \in Z - (Z^{(2)} \cup N)$. Then there exist an open neighborhood U_p of p in M, complex-analytic coordinates $z_1, \cdots, z_n$ defined on U_p, holomorphic functions $a_k, \cdots, a_n$ on U_p, and a positive real uumber ε such that:*

(0.31) $$Z \cap U_p = \{x \in U_p \mid z_k(x) = \cdots = z_n(x) = 0\}\,.$$

(0.32) $$x \to (z_1(x), \cdots, z_n(x)) \quad \textit{maps } U_p \textit{ onto}$$
$$\{(\zeta_1, \cdots, \zeta_n) \in C^n \mid |\zeta_i| < \varepsilon,\ i = 1, \cdots, n\}\,.$$

(0.33) $$Z \cap U_p = \{x \in U_p \mid a_k(x) = \cdots = a_n(x) = 0\}\,.$$

(0.34) *If $1 \le j \le k - 1$ and $k \le i \le n$, then $\partial a_i / \partial z_j$ vanishes throughout U_p,*

(0.35) *At each $x \in U_p$ the germs of the holomorphic vector-fields $\partial/\partial z_1, \cdots, \partial/\partial z_{k-1}, \sum_{i=k}^{n} a_i \partial/\partial z_i$ are in ξ_x and span ξ_x as on $\mathcal{O}_x$-module.*

Remarks. (a) (0.34) implies that for $x \in U_p$, $a_i(x)$ depends only on $z_k(x), \cdots, z_n(x)$. Thus the submersion referred to above is

$$x \to (z_k(x), \cdots, z_n(X)) .$$

(b) Several examples of foliation singularities for which (0.27) and (0.28) are valid will be described in § 11 below.

Let $\deg \varphi = n - k + 1$. Assume that Z is compact and satisfies (0.27) and (0.28). Let $Z_1, \cdots, Z_s$ be the irreducible complex-analytic components of Z of dimension $k - 1$. Denote by $[Z_i]$ the element of $H_{2k-2}(Z; C)$ given by the fundamental cycle of Z_i. Then $[Z_1], \cdots, [Z_s]$ is a vector-space basis for $H_{2k-2}(Z; C)$. To each Z_i associate a complex number $\#(\varphi, \xi, Z_i)$ as follows. Choose $p \in Z_i - (Z^{(2)} \cup N)$. Choose a neighborhood U_p of p and $z_1, \cdots, z_n$, $a_k, \cdots, a_n$, ε as in (0.31)–(0.35). Form the $(n - k + 1) \times (n - k + 1)$ matrix A of partial derivatives:

$$A = \|\partial a_i/\partial z_j\| , \quad k \le i, j \le n . \tag{0.36}$$

If $\det \|(\partial a_i/\partial z_j)(p)\| \neq 0$, then let $\lambda_1, \cdots, \lambda_{n-k+1}$ be the eigenvalues of $\|(\partial a_i/\partial z_j)(p)\|$. In this case,

$$\#(\varphi, \xi, Z_i) = \varphi(\lambda_1, \cdots, \lambda_{n-k+1}, 0, \cdots, 0)/(\lambda_1 \cdots \lambda_{n-k+1}) . \tag{0.37}$$

More generally, let $\sigma_1, \cdots, \sigma_n$ be the elementary symmetric functions in the indeterminates $X_1, \cdots, X_n$. For $i = 1, \cdots, n - k + 1$ define $\sigma_i(A)$ by

$$\det (I + tA) = 1 + t\sigma_1(A) + \cdots + t^{n-k+1}\sigma_{n-k+1}(A) . \tag{0.38}$$

Since $\deg \varphi = n - k + 1$, there is a polynomial $\tilde{\varphi}$ in $\sigma_1, \cdots, \sigma_{n-k+1}$ with

$$\varphi = \tilde{\varphi}(\sigma_1, \cdots, \sigma_{n-k+1}) , \quad \deg \varphi = n - k + 1 . \tag{0.39}$$

Define $\varphi(A)$ by

$$\varphi(A) = \tilde{\varphi}(\sigma_1(A), \cdots, \sigma_{n-k+1}(A)) . \tag{0.40}$$

Thus $\varphi(A)$ is a holomorphic function on U_p.

Let $D_p = \{x \in U_p \mid z_1(x) = z_1(p), \cdots, z_{k-1}(x) = z_{k-1}(p)\}$. D_p is a holomorphic normal disc to Z_i at p. Restrict $\varphi(A)$ to D_p and define $\#(\varphi, \xi, Z_i)$ to be the value at p of the Grothendieck residue symbol

$$\#(\varphi, \xi, Z_i) = \mathrm{Res}_p \begin{bmatrix} \varphi(A) dz_k \cdots dz_n \\ a_k, \cdots, a_n \end{bmatrix} . \tag{0.41}$$

Then $\sum_{i=1}^{s} \#(\varphi, \xi, Z_i)[Z_i]$ is a well-defined homology class depending only on φ and the local behavior of the leaves of ξ near Z.

Theorem 3. *Let M be a complex manifold, ξ be a full integrable sub sheaf of $\underline{T}$, S be the singular set of ξ, and Z be a connected component of S. Assume that Z is compact,* $\dim_C Z = k - 1$, $\dim_C Z^{(2)} < k - 1$. *Let $Z_1, \cdots, Z_s$ be the* *irreducible complex-analytic components of Z of dimension $k - 1$. Let* $\deg \varphi = n - k + 1$. *Then*

$$\text{(0.42)} \qquad \operatorname{Res}_\varphi(\xi, Z) = \sum_{i=1}^{s} \#(\varphi, \xi, Z_i)[Z_i] .$$

Remark. Suppose $k = 1$. Then from (0.14) it is clear that Theorem 2 and Theorem 3 combine to imply Theorem 1. Hence Theorem 2 and Theorem 3 together constitute a result on holomorphic foliations, which includes Theorem 1 as a special case.

We turn now to the question of computing $\operatorname{Res}_\varphi(\xi, Z)$ when $n - k + 1 < \deg \varphi \leq n$. Here we have been unable to find an explicit formula for $\operatorname{Res}_\varphi(\xi, Z)$. However, we have discovered that $\operatorname{Res}_\varphi(\xi, Z)$ has a rigidity property. This rigidity[1] appears to be the most relevant fact about these $\operatorname{Res}_\varphi(\xi, Z)$.

Theorem 4 (*Rigidity thorem*). *Let M be a complex manifold. Assume that* $n - k + 1 < \deg \varphi \leq n$. *Let U be an open subset of M, and* $[a, b]$ *be a closed interval of real numbers. For* $t \in [a, b]$ *let* $\{\xi_t\}$ *be a* C^∞ *1-parameter family of full integrable subsheaves of* $\underline{T} | U$. *Let* $Z_t = \{x \in U | (\underline{T}/\xi_t)_x$ *is not a free* $\mathcal{O}_x$-*module*$\}$. *Assume that each* Z_t *is compact and connected, and also that there is a fixed compact subset B of U with*

$$\text{(0.43)} \qquad Z_t \subset B \qquad \textit{for all } t \in [a, b] .$$

Let $i_*: H_*(Z_t; C) \to H_*(U; C)$ *be induced by the inclusion of* Z_t *in U. Then*

$$\text{(0.44)} \qquad i_* \operatorname{Res}_\varphi(\xi_a, Z_a) = i_* \operatorname{Res}_\varphi(\xi_b, Z_b) .$$

An immediate corollary of Theorem 4 is

(0.45) Corollary. *Let* $M, U, [a, b], \{\xi_t\}, \varphi$ *be as above. Assume, in addition, that there is a fixed compact connected subvariety Z of U with*

$$\text{(0.46)} \qquad Z_t = Z \qquad \textit{for all } t \in [a, b] .$$

Then

$$\text{(0.47)} \qquad \operatorname{Res}_\varphi(\xi_a, Z) = \operatorname{Res}_\varphi(\xi_b, Z) .$$

Remarks. (a) In Theorem 4 and Corollary (0.45) no special assumption is made on Z_t other than that Z_t be compact and connected. In particular, it is *not* required that (0.27) and (0.28) be valid for Z_t.

(b) Theorem 4 and Corollary (0.45) show that the two cases $\deg \varphi = n - k + 1$ and $\deg \varphi > n - k + 1$ are quite different. If $\deg \varphi = n - k + 1$, then there are many examples where $\operatorname{Res}_\varphi(\xi_t, Z)$ is not constant in t.

[1] Independently and in a slightly different context, a similar rigidity theorem has been recently noted by James L. Heitsch, *Deformations of secondary characteristic classes*, to appear in Topology.

Thorem 4 suggests a conjecture. Let Q denote the rational numbers. The inclusion $Q \subset C$ gives inclusions

$$Q[X_1, \cdots, X_n] \subset C[X_1, \cdots, X_n] , \tag{0.48}$$

$$H_*(Z; Q) \subset H_*(Z; C) . \tag{0.49}$$

Rationality conjecture. *Let M be a complex manifold, ξ be a full integrable subsheaf of $\underline{T}$, and Z be a compact connected component of the singular set S. If $n - k + 1 < \deg \varphi \leq n$ and $\varphi \in Q[X_1, \cdots, X_n]$, then*

$$\mathrm{Res}_\varphi (\xi, Z) \in H_*(Z; Q) . \tag{0.50}$$

Remark. This conjecture, if true, would again point up a very sharp difference between the two cases where $\deg \varphi = n - k + 1$ and $\deg \varphi > n - k + 1$.

Two special situations deserve special comment. If the singular set S is empty, then Theorem 2 becomes the vanishing theorem of [5] and [9].

Vanishing theorem. *Let M be a complex manifold. On M, let F be an integrable holomorphic sub-vector-bundle of T. Then*

$$\varphi(T/F) = 0 \tag{0.51}$$

for all φ with $n - k < \deg \varphi \leq n$.

Remarks. (a) In this vanishing theorem, M is not required to be compact.

(b) If the foliation of M is a fibration, then (0.51) is obvious. For in this case let X be the base of the fibration and let $\pi: M \to X$ be the projection of M onto X. Then

$$T/F = \pi^!(TX) , \tag{0.52}$$

where TX is the holomorphic tangent bundle of X and $\pi^!(TX)$ is the pull-back by π of TX. Let $\pi^*: H^*(X; C) \to H^*(M; C)$ be the cohomology map induced by π. Then (0.52) implies

$$\varphi(T/F) = \pi^*\varphi(TX) . \tag{0.53}$$

Since $\dim_C X = n - k$, $\varphi(TX)$ vanishes whenever $\deg \varphi > n - k$. Hence (0.51) is evident in this case.

(c) Compact complex manifolds very rarely foliate without singularities. For example, (0.51) can be used to prove that there is no holomorphic foliation (without singularities) of CP^n. Foliations with singularities, however, exist in great abundance.

A second special case of interest is the case when (0.27) and (0.28) are valid for Z and in addition to this $Z^{(2)}$ and N are empty. Here we can give an explicit formula for $\mathrm{Res}_\varphi (\xi, Z)$ for all φ with $n - k < \deg \varphi \leq n$. See § 11 below.

Finally, let us remark that the local classes $\mathrm{Res}_\varphi (\xi, Z)$ are functorial in an appropriate sense. Once this is made precise, it becomes clear that the work of this note is very closely related to the problem of computing the homotopy and homology of the foliation classifying spaces $B\Gamma_q^C$ introduced by A. Haefliger [15]. This will be commented on in § 12 below.

The paper is divided into 12 sections with the following titles:

1. Connections and curvature
2. Partial connections
3. Proof of the vanishing theorem
4. Exact sequences
5. Z-sequences
6. Coherent-real analytic sheaves
7. Proof of the residue existence theorem
8. Proof of Theorem 1
9. Proof of Theorem 3
10. Proof of the rigidity theorem
11. Examples
12. On the space $B\Gamma_q^C$

We thank P. Griffiths, R. Hartshorne, and R. MacPherson for many helpful comments and suggestions. L. Illusie [19] has, independently, done some work quite analogous to ours in the algebraic category.

1. Connections and curvature

Some standard facts on connections and curvature are very briefly reviewed here. For a careful detailed treatment see [10]. The matters considered here are purely C^∞, so in this section let M be a C^∞ manifold. Set $m = \dim_R M$, and let n be the largest integer with $n \leq m/2$. Let T_RM be the usual C^∞ tangent bundle of M, which is a real vector bundle. We wish to consider only complex vector-bundles, so let τ be the complexification of T_RM, i.e.,

$$\tau = C \underset{R}{\otimes} T_RM \ . \tag{1.1}$$

If E is a C^∞ complex vector-bundle on M, then $C^\infty(E)$ denotes the space of all C^∞ sections of E. E^* denotes the bundle dual to E. Λ^iE denotes the i-th exterior power of E.

On M we have the de Rham complex of all C^∞ complex-valued differential forms on M:

$$0 \longrightarrow A^0 \xrightarrow{d} A^1 \xrightarrow{d} \cdots \xrightarrow{d} A^m \longrightarrow 0 \ . \tag{1.2}$$

A^0 is the set of all smooth functions from M to C. For $i \geq 1$, $A^i = C^\infty(\Lambda^i\tau^*)$. d is the usual de Rham differentiation operator.

(1.3) $H^i(M; C) = \text{Kernel}\{d: A^i \to A^{i+1}\}/\text{Image}\{d: A^{i-1} \to A^i\}$.

If $\omega \in A^i$ has $d\omega = 0$, then we denote by $[\omega]$ the element of $H^i(M; C)$ determined by ω:

(1.4) $$[\omega] \in H^i(M; C) .$$

(1.5) Definition. Let E be a C^∞ complex vector-bundle on M. *A connection* for E is a C-linear map D from $C^\infty(E)$ to $C^\infty(\tau^* \otimes E)$ such that

(1.6) $$D(fs) = df \otimes s + fDs ,$$

whenever $f \in A^0$ and $s \in C^\infty(E)$.

Remark. E always has many connections.

If D is a connection for E, then for each $i > 0$, D induces a unique C-linear map, also denoted by D:

(1.7) $$D: C^\infty(\Lambda^i\tau^* \otimes E) \to C^\infty(\Lambda^{i+1}\tau^* \otimes E)$$

such that

(1.8) $$D(\omega \otimes s) = d\omega \otimes s + (-1)^i \omega Ds ,$$

whenever $\omega \in A^i$ and $s \in C^\infty(E)$.

There is a unique C^∞ vector-bundle map $K(D)$:

(1.9) $$K(D): E \to \Lambda^2\tau^* \otimes E$$

such that for all $s \in C^\infty(E)$,

(1.10) $$DDs = K(D)s .$$

$K(D)$ is the curvature of D.

Let U be an open subset of M. If $s \in C^\infty(E)$ vanishes on U, then Ds also vanishes on U. From this remark it follows immediately that D restricts to give a connection for $E \mid U$:

(1.11) $$D: C^\infty(E \mid U) \to C^\infty(\tau^* \otimes E \mid U) .$$

On U, let $e_1, \cdots, e_r$ be a C^∞ frame of E. A matrix $\theta = \|\theta_{ij}\|$ of 1-forms is determined by

(1.12) $$De_i = \sum_{j=1}^{r} \theta_{ij} \otimes e_j .$$

θ is the connection matrix of D with respect to the frame $e_1, \cdots, e_r$. Set $\kappa = d\theta - \theta \wedge \theta$. Then

(1.13) $$\kappa_{ij} = d\theta_{ij} - \sum_{\alpha=1}^{r} \theta_{i\alpha} \wedge \theta_{\alpha j} .$$

$\kappa = \|\kappa_{ij}\|$ is the curvature matrix of D with rsepect to $e_1, \cdots, e_r$, so that

(1.14) $$K(D)e_i = DDe_i = \sum_{j=1}^{r} \kappa_{ij} \otimes e_j .$$

If $e'_1, \cdots, e'_r$ is another C^∞ frame of E on U, let $A = \|a_{ij}\|$ be determined by

(1.15) $$e'_i = \sum_{j=1}^{r} a_{ij} e_j .$$

Let κ' be the curvature matrix of D with respect to $e'_1, \cdots, e'_r$. Then

(1.16) $$\kappa' = A\kappa A^{-1} .$$

Let $\sigma_1, \cdots, \sigma_n$ be the elementary symmetric functions of $X_1, \cdots, X_n$. Define $\sigma_1(\kappa), \cdots, \sigma_n(\kappa)$ by

(1.17) $$\det (I + t\kappa) = 1 + t\sigma_1(\kappa) + \cdots + t^n\sigma_n(\kappa) .$$

$\sigma_j(\kappa)$ is then a $2j$-form on U. Note that if $r < n$, then $\sigma_j(\kappa) = 0$ whenever $r < j \leq n$. (1.16) implies

(1.18) $$\sigma_j(\kappa) = \sigma_j(\kappa') , \qquad j = 1, \cdots, n .$$

Hence by choosing local frames for E throughout M a well-defined differential form $\sigma_j(K(D))$ is obtained on M. $\sigma_j(K(D))$ is closed, i.e.,

(1.19) $$d\sigma_j(K(D)) = 0 .$$

Let $c_1(E), \cdots, c_n(E)$ be the Chern classes of E taken in $H^*(M; C)$. Note that if $r < n$, then $c_j(E) = 0$ whenever $r < j \leq n$. The Chern-Weil theory of characteristic classes [10] asserts that the element of $H^{2j}(M; C)$ determined by $\sigma_j(K(D))$ is $(2\pi/\sqrt{-1})^j c_j(E)$, i.e.,

(1.20) $$[\sigma_j(K(D))] = (2\pi/\sqrt{-1})^j c_j(E) , \qquad j = 1, \cdots, n .$$

In particular, if $\tilde{D}$ is another connection for E, then $[\sigma_j(K(D))] = [\sigma_j(K(\tilde{D}))]$.

Assume $l \leq n$. If $\varphi \in C[X_1, \cdots, X_n]$ is symmetric and homogeneous of degree l, set $\varphi = \tilde{\varphi}(\sigma_1, \cdots, \sigma_l)$. Define $\varphi(E) \in H^{2l}(M; C)$ by

(1.21) $$\varphi(E) = \tilde{\varphi}(c_1(E), \cdots, c_l(E)) .$$

Let D be a connection for E, and set $K = K(D)$. On M define a $2l$-form $\varphi(K)$ by

(1.22) $$\varphi(K) = \tilde{\varphi}(\sigma_1(K), \cdots, \sigma_l(K)) .$$

(1.9) implies

(1.23) $$d\varphi(K) = 0 .$$

(1.20) implies

(1.24) $$[\varphi(K)] = (2\pi/\sqrt{-1})^l\varphi(E) .$$

2. Partial connections

As in § 1, let M be a C^∞ manifold, and E a C^∞ complex vector-bundle on M. If H is a C^∞ sub-vector-bundle of τ, then H^* is a quotient bundle of τ^*. Denote by $\rho\colon \tau^* \to H^*$ the projection of τ^* onto H^*.

(2.1) Definition. A *partial connection* for E is a pair (H, δ) where H is a C^∞ sub-vector-bundle of τ and δ is a C-linear map from $C^\infty(E)$ to $C^\infty(H^* \otimes E)$ such that

(2.2) $$\delta(fs) = \rho(df) \otimes s + f\delta s ,$$

whenever $f \in A^0$ and $s \in C^\infty(E)$.

Remark. Let (H, δ) be a partial connection for E, and U an open subset of M. If $s \in C^\infty(E)$ vanishes on U, then δs also vanishes on U. From this it follows that (H, δ) restricts to give a partial connection for $E \mid U$:

(2.3) $$\delta\colon C^\infty(E \mid U) \to C^\infty(\tau^* \otimes E \mid U) .$$

(2.4) Definition. Let (H, δ) be a partial connection for E, and D a connection for E. *D extends δ* if the diagram

$$\begin{array}{ccc} C^\infty(E) & \xrightarrow{\ D\ } & C^\infty(\tau^* \otimes E) \\ & \searrow^{\delta} & \downarrow{\scriptstyle \rho\otimes 1} \\ & & C^\infty(H^* \otimes E) \end{array}$$

is commutative.

(2.5) Lemma. *Let (H, δ) be a partial connection for E. Then there exists a connection D for E such that D extends δ.*

Proof. Cover M by open sets $\{U_\alpha\}$ such that on each U_α there is a C^∞ frame $e_1^\alpha, \cdots, e_r^\alpha$ of E. Define $\gamma_{ij}^\alpha \in C^\infty(H^* \mid U_\alpha)$ by

(2.6) $$\delta e_i^\alpha = \sum_{j=1}^r \gamma_{ij}^\alpha \otimes e_j^\alpha .$$

Choose $\theta_{ij}^\alpha \in C^\infty(\tau^* \mid U_\alpha)$ such that

(2.7) $$\rho(\theta^\alpha_{ij}) = \gamma^\alpha_{ij} .$$

Define a connection D_α for $E \mid U_\alpha$ by

(2.8) $$D_\alpha e^\alpha_i = \sum_{j=1}^{r} \theta^\alpha_{ij} \otimes e^\alpha_j .$$

Then on U_α the diagram

(2.9) $$\begin{array}{ccc} C^\infty(E \mid U_\alpha) & \xrightarrow{D_\alpha} & C^\infty(\tau^* \otimes E \mid U_\alpha) \\ & \searrow^{\delta} & \downarrow{\scriptstyle \rho \otimes 1} \\ & & C^\infty(H^* \otimes E \mid U_\alpha) \end{array}$$

is commutative.

Let $\{\psi_\alpha\}$ be a partition of unity subordinate to the cover $\{U_\alpha\}$. Define a connection D for E by

(2.10) $$D = \sum_\alpha \psi_\alpha D_\alpha .$$

D extends δ.

(2.11) Lemma. *Let (H, δ) be a partial connection for E, and $s \in C^\infty(E)$ be such that:*

(2.12) $$s(x) \neq 0 \qquad \textit{for all } x \in M ,$$

(2.13) $$\delta s = 0 .$$

Then there exists a connection D for E with

(2.14) $$D \textit{ extends } \delta ,$$

(2.15) $$Ds = 0 .$$

Proof. Proceed as in the proof of Lemma (2.5) except that e^α_1 is required to be $s \mid U_\alpha$, and θ^α_{ij} is required to be zero.

Remarks. We have the evident pairing $C^\infty(H) \times C^\infty(H^*) \to A^0$. Hence $u \in C^\infty(H)$ determines a map $i(u)$ from $C^\infty(H^*)$ to A^0:

(2.16) $$i(u) \colon C^\infty(H^*) \to A^0 .$$

Similarly, u determines a map, also denoted by $i(u)$, from $C^\infty(H^* \otimes E)$ to $C^\infty(E)$:

(2.17) $$i(u) \colon C^\infty(H^* \otimes E) \to C^\infty(E) .$$

Note also that if $f \in A^0$, then by applying u to f we obtain $u[f] \in A^0$:

(2.18) $$u[f] = i(u)\rho(df)\ .$$

Let (H, δ) be a partial connection for E. Then

(2.19) $$i(u_1 + u_2)\delta s = i(u_1)\delta s + i(u_2)\delta s\ ,$$

(2.20) $$i(fu)\delta s = fi(u)\delta s\ ,$$

(2.21) $$i(u)\delta(s_1 + s_2) = i(u)\delta s_1 + i(u)\delta s_2\ ,$$

(2.22) $$i(u)\delta(fs) = u[f]s + fi(u)\delta s\ ,$$

whenever $u, u_1, u_2 \in C^\infty(H)$, $s, s_1, s_2 \in C^\infty(E)$, and $f \in A^0$.

3. Proof of the vanishing theorem

Let M be a complex-analytic manifold. As in (1.1) set $\tau = C \otimes_R T_R M$. Then there are the standard splittings:

(3.1) $$\tau = T \oplus \bar{T}\ ,$$

(3.2) $$\tau^* = T^* \oplus \bar{T}^*\ .$$

T is the holomorphic tangent bundle of M. $\bar{T}$ is the anti-holomorphic tangent bundle of M. A C^∞ section of T^* is a 1-form of type (1,0). A C^∞ section of $\bar{T}^*$ is a 1-form of type (0,1).

Let U be an open subset of M. On U let $z_1, \cdots, z_n$ be a complex-analytic coordinate system. Then on U:

(3.3) $\partial/\partial z_1, \cdots, \partial/\partial z_n$ is a holomorphic frame of T,

(3.4) $dz_1, \cdots, dz_n$ is a holomorphic frame of T^*.

Let E be a holomorphic vector-bundle on M. If U is an open subset of M, then $\Gamma(E \mid U)$ will denote the space of all holomorphic sections of $E \mid U$. Since E is holomorphic there is the $\bar{\partial}$ operator:

(3.5) $$\bar{\partial}\colon C^\infty(E) \to C^\infty(\bar{T}^* \otimes E)\ .$$

Setting $H = \bar{T}$ and $\delta = \bar{\partial}$, we then have a partial connection $(\bar{T}, \bar{\partial})$ for E. Note that:

(3.6) $$\Gamma(E \mid U) = \text{Kernel}\,\{\bar{\partial}\colon C^\infty(E \mid U) \to C^\infty(\bar{T}^* \otimes E \mid U)\}\ .$$

A connection for E which extends $(\bar{T}, \bar{\partial})$ is said to be a connection of type (1,0). A straightforward argument shows that a connection D for E is of type (1,0) if and only if D has the following property:

(3.7) Whenever $e_1, \cdots, e_r$ is a holomorphic frame for E, the connection matrix $\|\theta_{ij}\|$ of D with respect to this frame has each θ_{ij} of type (1,0).

The bracket operation for C^∞ sections of T satisfies:

$$[u_1 + u_2, u_3] = [u_1, u_3] + [u_2, u_3] , \tag{3.8}$$

$$[fu_1, u_2] = -u_2[f]u_1 + f[u_1, u_2] , \tag{3.9}$$

$$[u_1, u_2 + u_3] = [u_1, u_2] + [u_1, u_3] , \tag{3.10}$$

$$[u_1, fu_2] = u_1[f]u_2 + f[u_1, u_2] , \tag{3.11}$$

whenever $u_1, u_2, u_3 \in C^\infty(T)$ and $f \in A^0$.

Recall also that if U is an open subset of M, and $z_1, \cdots, z_n$ is a complex-analytic coordinate system on U, then

$$[\partial/\partial z_i, \partial/\partial z_j] = 0 \qquad 1 \leq i, j \leq n . \tag{3.12}$$

(3.13) Definition. A holomorphic sub-vector-bundle F of T is *integrable* if $C^\infty(F)$ is closed under the bracket operation.

Remark. A holomorphic sub-vector-bundle F of T is integrable if and only if:

(3.14) whenever U is an open subset of M, and $\gamma_1, \gamma_2 \in \Gamma(F \mid U)$, then $[\gamma_1, \gamma_2] \in \Gamma(F \mid U)$.

Assume now that F is an integrable holomorphic sub-vector-bundle of T. Form the quotient bundle T/F and denote by $\eta\colon T \to T/F$ the projection of T onto T/F. Let $u \in C^\infty(F)$ and $s \in C^\infty(T/F)$. Choose $\tilde{s} \in C^\infty(T)$ such that

$$\eta(\tilde{s}) = s . \tag{3.15}$$

Then, since $C^\infty(F)$ is closed under bracket,

$$\eta[u, \tilde{s}] \qquad \text{depends only on } u \text{ and } s. \tag{3.16}$$

Denote $\eta[u, \tilde{s}]$ by $\langle u, s\rangle$. Then from (3.8)–(3.11) it is clear that:

$$\langle u_1 + u_2, s\rangle = \langle u_1, s\rangle + \langle u_2, s\rangle , \tag{3.17}$$

$$\langle fu, s\rangle = f\langle u, s\rangle , \tag{3.18}$$

$$\langle u, s_1 + s_2\rangle = \langle u, s_1\rangle + \langle u, s_2\rangle , \tag{3.19}$$

$$\langle u, fs\rangle = u[f]s + f\langle u, s\rangle , \tag{3.20}$$

whenever $u, u_1, u_2 \in C^\infty(F)$, $s, s_1, s_2 \in C^\infty(T/F)$, and $f \in A^0$.

Comparing (3.17)–(3.20) to (2.19)–(2.22) and noting that T/F is a holomorphic vector-bundle on M, we then have

(3.21) Proposition. *Let F be an integrable holomorphic sub-vector-bundle of T. Then there exists a unique partial connection $(F \oplus \bar{T}, \delta)$ for T/F such that*

$$i(u)\delta s = \langle u, s\rangle \,, \tag{3.22}$$

$$i(v)\delta s = i(v)\bar{\partial} s \,, \tag{3.23}$$

whenever $u \in C^\infty(F)$, $v \in C^\infty(\bar{T})$, *and* $s \in C^\infty(T/F)$.

(3.24) Definition. A *basic connection* for T/F is a connection for T/F which extends $(F \oplus \bar{T}, \delta)$.

Remarks. (a) A connection D for T/F is basic if and only if

$$i(u)D(\eta\gamma) = \eta[u, \gamma] \quad \text{whenever } u \in C^\infty(F) \text{ and } \gamma \in C^\infty(T), \tag{3.25}$$

$$D \text{ is of type } (1, 0) \,. \tag{3.26}$$

(b) By Lemma (2.5) a basic connection D exists for T/F.

(3.27) Proposition. *Let M be a complex manifold, and F an integrable holomorphic sub-vector-bundle of T. Set $n = \dim_C M$, $k = \dim_C F_x$. Let $\varphi \in C[X_1, \cdots, X_n]$ be symmetric and homogeneous of degree l, where $n - k < l \leq n$. Let D be a basic connection for T/F, and set $K = K(D)$. Then*

$$\varphi(K) = 0 \,. \tag{3.28}$$

Proof. Given $p \in M$, let U be an open neighborhood of p in M such that on U there is a complex-analytic coordinate system $z_1, \cdots, z_n$ with

$$\partial/\partial z_1, \partial/\partial z_2, \cdots, \partial/\partial z_k \in \Gamma(F \mid U) \,. \tag{3.29}$$

Let $A(U)$ be the set of all C^∞ complex-valued differential forms on U. $A(U)$ is a ring under the usual addition and wedge product of differential forms. In $A(U)$, let $I(F, U)$ be the ideal generated by $dz_{k+1}, \cdots, dz_n$. This ideal has the properties:

(3.30) If $\omega \in I(F, U)$, then $d\omega \in I(F, U)$.

(3.31) If $\omega_1, \cdots, \omega_{n-k+1}$ are any $n - k + 1$ elements of $I(F, U)$, then $\omega_1 \wedge \cdots \wedge \omega_{n-k+1} = 0$.

Let $\eta: T \to T/F$ be the projection, and $\theta = \|\theta_{ij}\|$ the connection matrix of D with respect to the frame $\eta\partial/\partial z_{k+1}, \cdots, \eta\partial/\partial z_n$. D is basic, so (3.26) implies that each θ_{ij} is of type (1,0). (3.25) and (3.12) imply that for each θ_{ij}

$$0 = i(\partial/\partial z_1)\theta_{ij} = \cdots = i(\partial/\partial z_k)\theta_{ij} \,. \tag{3.32}$$

Hence each θ_{ij} is in $I(F, U)$. Let $\kappa = \|\kappa_{ij}\|$ be the curvature matrix of D with respect to $\eta\partial/\partial z_{k+1}, \cdots, \eta\partial/\partial z_n$. From (3.30) and (1.13) it is clear that each κ_{ij} is in $I(F, U)$:

$$\kappa_{ij} \in I(F, U) \,. \tag{3.33}$$

As in (1.17) define $\sigma_1(\kappa), \cdots, \sigma_n(\kappa)$ by

$$(3.34) \qquad \det(I + t\kappa) = 1 + t\sigma_1(\kappa) + \cdots + t^n\sigma_n(\kappa) .$$

Set $\varphi = \tilde{\varphi}(\sigma_1, \cdots, \sigma_l)$, and $l = \deg \varphi$. Then on U,

$$(3.35) \qquad \varphi(K) \,|\, U = \tilde{\varphi}(\sigma_1(\kappa), \cdots, \sigma_l(\kappa)) .$$

Since $l \geq n - k + 1$, (3.31) and (3.33) now imply that $\varphi(K)$ vanishes on U. This proves (3.28).

Due to (1.24), (0.51) is now evident.

4. Exact sequences

Some well-known facts about connections and exact sequences of vector bundles are collected here. As in § 1, the matters considered here are purely C^∞. So in this section let M be a C^∞ manifold. Let $m = \dim_R M$, and let n be the largest integer with $n \leq m/2$.

If E is a C^∞ complex vector-bundle on M, let $c(E)$ denote the total Chern class of E in $H^*(M; C)$, so that

$$(4.1) \qquad c(E) = 1 + c_1(E) + \cdots + c_n(E) .$$

Note that in the ring $H^*(M; C) = H^0(M; C) \oplus H^1(M; C) \oplus \cdots \oplus H^m(M; C)$, $c(E)$ is invertible.

If E_1, E_0 are two C^∞ vector-bundles on M, then the total Chern class of the virtual bundle $E_0 - E_1$ is defined by

$$(4.2) \qquad c(E_0 - E_1) = c(E_0)/c(E_1) .$$

Thus the Chern classes $c_1(E_0 - E_1), \cdots, c_n(E_0 - E_1)$ are determined by

$$(4.3) \qquad c_j(E_0 - E_1) \in H^{2j}(M; C) ,$$

$$(4.4) \qquad c(E_0)/c(E_1) = 1 + c_1(E_0 - E_1) + \cdots + c_n(E_0 - E_1) .$$

More generally, let $E_q, E_{q-1}, \cdots, E_0$ be C^∞ complex vector-bundles on M. Set $\varepsilon(i) = (-1)^i$. Then the total Chern class of the virtual bundle $\sum_{i=0}^q (-1)^i E_i$ is defined by

$$(4.5) \qquad c\left(\sum_{i=0}^q (-1)^i E_i\right) = \prod_{i=0}^q (c(E_i))^{\varepsilon(i)} , \qquad \varepsilon(i) = (-1)^i .$$

Set $\zeta = \sum_{i=0}^q (-1)^i E_i$. Thus the Chern classes $c_1(\zeta), \cdots, c_n(\zeta)$ are determined by

$$(4.6) \qquad c_j(\zeta) \in H^{2j}(M; C) ,$$

$$(4.7) \qquad \prod_{i=0}^q (c(E_i))^{\varepsilon(i)} = 1 + c_1(\zeta) + \cdots + c_n(\zeta) .$$

Let $\varphi \in C[X_1, \cdots, X_n]$ be symmetric and homogeneous of degree l. Assume $l \leq n$. Set $\varphi = \tilde{\varphi}(\sigma_1, \cdots, \sigma_l)$. Then $\varphi(\zeta)$ is defined by

$$\varphi(\zeta) = \tilde{\varphi}(c_1(\zeta), \cdots, c_l(\zeta)) . \tag{4.8}$$

Hence $\varphi(\zeta) \in H^{2l}(M; C)$.

Suppose now that $D_q, D_{q-1}, \cdots, D_0$ are connections for $E_q, E_{q-1}, \cdots, E_0$ respectively. Set $K_i = K(D_i)$, and define differential forms $\sigma_j(K_q|K_{q-1}|\cdots|K_0)$ by

$$\sigma_j(K_q|K_{q-1}|\cdots|K_0) \in A^{2j} , \qquad j = 1, \cdots, n , \tag{4.9}$$

$$\prod_{i=0}^{q} (\det (I + K_i))^{\epsilon(i)} = 1 + \sigma_1(K_q|K_{q-1}|\cdots|K_0) + \cdots + \sigma_n(K_q|K_{q-1}|\cdots|K_0) . \tag{4.10}$$

(1.19) implies

$$d\sigma_j(K_q|K_{q-1}|\cdots|K_0)) = 0 . \tag{4.11}$$

(1.20) and (4.7) imply

$$[\sigma_j(K_q|K_{q-1}|\cdots|K_0)] = (2\pi/\sqrt{-1})^j c_j(\zeta) . \tag{4.12}$$

As above let φ be symmetric and homogeneous of degree $l \leq n$. Set $\varphi = \tilde{\varphi}(\sigma_1, \cdots, \sigma_l)$ and $\omega_j = \sigma_j(K_q|K_{q-1}|\cdots|K_0)$. Define a $2l$-form $\varphi(K_q|K_{q-1}|\cdots|K_0)$ on M by

$$\varphi(K_q|K_{q-1}|\cdots|K_0) = \tilde{\varphi}(\omega_1, \cdots, \omega_l) . \tag{4.13}$$

Then

$$d\varphi(K_q|K_{q-1}|\cdots|K_0) = 0 , \tag{4.14}$$

$$[\varphi(K_q|K_{q-1}|\cdots|K_0)] = (2\pi/\sqrt{-1})^l \varphi(\zeta) , \tag{4.15}$$

where as above $\zeta = \sum_{i=0}^{q} (-1)^i E_i$.

(4.16) Definition. Let $0 \to E_q \to E_{q-1} \to \cdots \to E_0 \to E_{-1} \to 0$ be an exact sequence of C^∞ vector-bundles on M. Denote by η_i the map from E_i to E_{i-1}. Let $D_q, D_{q-1}, \cdots, D_0, D_{-1}$ be connections for $E_q, E_{q-1}, \cdots, E_0, E_{-1}$ respectively. Then $(D_q, D_{q-1}, \cdots, D_0, D_{-1})$ is *compatible* with the exact sequence if for each $i = q, q-1, \cdots, 0$ the diagram

$$\begin{array}{ccc} C^\infty(E_i) & \longrightarrow & C^\infty(\tau^* \otimes E^i) \\ \eta_i \downarrow & & \downarrow 1 \otimes \eta_i \\ C^\infty(E_{i-1}) & \longrightarrow & C^\infty(\tau^* \otimes E_{i-1}) \end{array}$$

is commutative.

(4.17) Lemma. *Let* $0 \to E_q \to E_{q-1} \to \cdots \to E_0 \to E_{-1} \to 0$ *be an exact sequence of* C^∞ *vector bundles on* M, *and* D_{-1} *be a connection for* E_{-1}. *Then there exist connections* $D_q, D_{q-1}, \cdots, D_0$ *for* $E_q, E_{q-1}, \cdots, E_0$ *such that* $(D_q, D_{q-1}, \cdots, D_0, D_{-1})$ *is compatible with the exact sequence.*

Proof. Proceed by induction on q. If $q = 0$, the exact sequence is $0 \to E_0 \to E_{-1} \to 0$. D_{-1} then determines a unique connection D_0 for E_0 such that (D_0, D_{-1}) is compatible with the exact sequence.

Assume now that the lemma is valid for $q - 1$. Consider an exact sequence $0 \to E_q \to E_{q-1} \to \cdots \to E_0 \to E_{-1} \to 0$. Let $\eta_q(E_q)$ be the image of $\eta_q: E_q \to E_{q-1}$. Choose a C^∞ sub-vector-bundle J of E_{q-1} such that

$$E_{q-1} = J \oplus \eta_q(E_q) . \tag{4.18}$$

By the induction hypotheses there exist connections $D_{q-1}, D_{q-2}, \cdots, D_0$ for J, $E_{q-2}, \cdots, E_0$ such that

(4.19) $(D_{q-1}, D_{q-2}, \cdots, D_0, D_{-1})$ is compatible with the exact sequence $0 \to J \to E_{q-2} \to \cdots \to E_0 \to E_{-1} \to 0$.

Choose a connection D for $\eta_q(E_q)$. Let D_q be the unique connection for E_q such that

(4.20) (D_q, D) is compatible with the exact sequence $0 \to E_q \to \eta_q(E_q) \to 0$.

On $E_{q-1} = J \oplus \eta_q(E_q)$ let $D_{q-1} \oplus D$ be the direct sum connection. Thus

$$(D_{q-1} \oplus D)(s_1 + s_2) = D_{q-1}s_1 + Ds_2 , \tag{4.21}$$

whenever $s_1 \in C^\infty(J)$ and $s_2 \in C^\infty(\eta_q(E_q))$. Then $(D_q, D_{q-1} \oplus D, D_{q-2}, \cdots, D_0, D_{-1})$ is compatible with the exact sequence $0 \to E_q \to E_{q-1} \to \cdots \to E_0 \to E_1 \to 0$. This proves the lemma.

(4.22) Lemma. *Let* $0 \to E_q \to E_{q-1} \to \cdots \to E_0 \to E_{-1} \to 0$ *be an exact sequence of* C^∞ *vector bundles on* M, *and* $D_q, D_{q-1}, \cdots, D_0, D_{-1}$ *be connections for* $E_q, E_{q-1}, \cdots, E_0, E_{-1}$. *Assume that* $(D_q, D_{q-1}, \cdots, D_0, D_{-1})$ *is compatible with the exact sequence. Let* φ *be symmetric and homogeneous of degree* $l \leq n$. *Set* $K_i = K(D_i)$. *Then*

$$\varphi(K_{-1}) = \varphi(K_q | K_{q-1} | \cdots | K_0) . \tag{4.23}$$

Proof. Set $\varepsilon(i) = (-1)^i$. To prove (4.23) it suffices to show

$$\det (I + K_{-1}) = \prod_{i=0}^{q} (\det (I + K_i))^{\varepsilon(i)} . \tag{4.24}$$

To prove (4.24) proceed by induction on q. If $q = 0$, the exact sequence is $0 \to E_0 \to E_{-1} \to 0$ and (4.24) is obvious in this case.

Assume now that (4.24) is valid for $q-1$, and consider an exact sequence $0 \to E_q \to E_{q-1} \to \cdots \to E_0 \to E_{-1} \to 0$. Let $\eta_q(E_q)$ be the image of $\eta_q\colon E_q \to E_{q-1}$. Choose a C^∞ sub-vector-bundle J of E_{q-1} such that

$$E_{q-1} = J \oplus \eta_q(E_q) \,. \tag{4.25}$$

Let $\rho\colon E_{q-1} \to J$ be the projection of E_{q-1} onto J resulting from this direct sum decomposition. So we have

$$1 \otimes \rho\colon \tau^* \otimes E_{q-1} \to \tau^* \otimes J \,. \tag{4.26}$$

Define a connection ∇ for J by

$$\nabla = (1 \otimes \rho) D_{q-1} \,. \tag{4.27}$$

Then

$$\det(I + K_{q-1}) = \det(I + K_q)\det(I + K(\nabla)) \,, \tag{4.28}$$

(4.29) $(\nabla, D_{q-2}, \cdots, D_0, D_{-1})$ is compatible with the exact sequence $0 \to J \to E_{q-2} \to \cdots \to E_0 \to E_{-1} \to 0$.

The induction hypotheses and (4.29) imply

$$\det(I + K_{-1}) = \det(I + K(\nabla))^{\epsilon(q-1)} \prod_{i=0}^{q-2} (\det(I + K_i))^{\epsilon(i)} \,. \tag{4.30}$$

(4.30) and (4.28) combine to give

$$\det(I + K_{-1}) = \prod_{i=0}^{q} (\det(I + K_i))^{\epsilon(i)} \,. \tag{4.31}$$

This completes the inductive step and the proof.

(4.32) Lemma. *On M let*

$$\begin{array}{ccccccccccccc}
 & & 0 & & 0 & & & & 0 & & 0 & & \\
 & & \downarrow & & \downarrow & & & & \downarrow & & \downarrow & & \\
0 & \to & E'_q & \to & E'_{q-1} & \to & \cdots & \to & E'_0 & \to & E'_{-1} & \to & 0 \\
 & & \downarrow & & \downarrow & & & & \downarrow & & \downarrow & & \\
0 & \to & E_q & \to & E_{q-1} & \to & \cdots & \to & E_0 & \to & E_{-1} & \to & 0 \\
 & & \downarrow & & \downarrow & & & & \downarrow & & \downarrow & & \\
0 & \to & E''_q & \to & E''_{q-1} & \to & \cdots & \to & E''_0 & \to & E''_{-1} & \to & 0 \\
 & & \downarrow & & \downarrow & & & & \downarrow & & \downarrow & & \\
 & & 0 & & 0 & & & & 0 & & 0 & &
\end{array}$$

be a commutative diagram of C^∞ vector bundles in which each row and each *column are exact. Let $\iota_j(E'_j)$ be the image of the map $\iota_j\colon E'_j \to E_j$. Then there exist C^∞ sub-vector-bundles $F_q, F_{q-1}, \cdots, F_0, F_{-1}$ of $E_q, E_{q-1}, \cdots, E_0, E_{-1}$ such that*

(4.33) $$E_j = F_j \oplus \iota_j(E'_j)\ , \qquad j = q, q-1, \cdots, 0, -1\ ,$$

(4.34) *the map $\eta_j\colon E_j \to E_{j-1}$ maps F_j into F_{j-1}, $j = q, q-1, \cdots, 0$.*

Proof. Construct $F_q, F_{q-1}, \cdots, F_0, F_{-1}$ by a downward induction. First, let F_q be any C^∞ sub-vector-bundle of E_q such that

(4.35) $$E_q = F_q \oplus \iota_q(E'_q)\ .$$

Next, suppose that $F_q, F_{q-1}, \cdots, F_r$ have been constructed so that

(4.36) $$E_j = F_j \oplus \iota_j(E'_j)\ , \qquad j = q, q-1, \cdots, r\ ,$$

(4.37) $$\eta_j(F_j) \subset F_{j-1}\ , \qquad j = q, q-1, \cdots, r+1\ .$$

$\eta_r(F_r)$ is then a C^∞ sub-vector-bundle of E_{r-1}. A diagram chase shows that

(4.38) $$\eta_r(F_r) \cap \iota_{r-1}(E'_{r-1}) = \{0\}\ .$$

Hence there exists a C^∞ sub-vector-bundle F_{r-1} of E_{r-1} such that

(4.39) $$E_{r-1} = F_{r-1} \oplus \iota_{r-1}(E'_{r-1})\ ,$$

(4.40) $$\eta_r(F_r) \subset F_{r-1}\ .$$

This completes the inductive step and the proof.

(4.41) Lemma. *Let E be a C^∞ vector-bundle on M, and B a closed subset of M. On $M - B$ let D be a connection for $E|M - B$. Let Σ be a closed subset of M such that B is contained in the interior of Σ. Then on M there exists a connection $\tilde{D}$ for E such that*

(4.42) *D and $\tilde{D}$ agree on $E|M - \Sigma$.*

Proof. On M let ∇ be a connection for E. Let $\psi\colon M \to \boldsymbol{R}$ be a C^∞ function such that

(4.43) ψ vanishes on a neighborhood of B,

(4.44) $$\psi = 1 \quad \text{on} \quad M - \Sigma\ .$$

Set $\tilde{D} = \psi D + (1 - \psi)\nabla$. $\tilde{D}$ satisfies (4.42).

(4.45) Lemma. *On M let $0 \to E' \to E \to E'' \to 0$ be an exact sequence of C^∞ vector-bundles. Denote by $\iota(E')$ the image of $\iota\colon E' \to E$. Let B be a closed subset of M. On $M - B$ let F be a C^∞ sub-vector-bundle of $E|M - B$ such that*

(4.46) $$E|M - B = F \oplus \iota(E'|M - B) .$$

Let Σ be a closed subset of M such that B is contained in the interior of Σ. Then on M there exists a C^∞ sub-vector-bundle $\tilde{F}$ of E such that

(4.47) $$E = \tilde{F} \oplus \iota(E') ,$$

(4.48) $$\tilde{F}|M - \Sigma = F|M - \Sigma .$$

Proof. Denote by $\mu: E \to E''$ the map from E to E''. On $M - B$ there is a unique map $\alpha: E''|M - B \to E|M - B$ such that

(4.49) $$\alpha(E''|M - B) = F ,$$

(4.50) $$\mu\alpha = 1 .$$

On M let $\beta: E'' \to E$ be a map of C^∞ vector-bundles such that

(4.51) $$\mu\beta = 1 .$$

Let $\psi: M \to \boldsymbol{R}$ be a C^∞ function with (4.43) and (4.44) valid for ψ. On M define $\tilde{\alpha}: E'' \to E$ by

(4.52) $$\tilde{\alpha} = \psi\alpha + (1 - \psi)\beta .$$

Set $\tilde{F} = \tilde{\alpha}(E'')$. $\tilde{F}$ satisfies (4.47) and (4.48).

(4.53) Remark. Let M, X be C^∞ manifolds, E be a C^∞ vector-bundle on X, and $g: M \to X$ be a C^∞ map. Then on M there is the pull-back bundle $g^!(E)$:

(4.54) $$g^!(E)_p = E_{gp} , \qquad p \in M .$$

Let U be an open subset of X, and let $s \in C^\infty(E|U)$. Then on $g^{-1}(U)$ there is $g^!(s) \in C^\infty(g^!(E)|g^{-1}(U))$:

(4.55) $$g^!(s)p = s(gp) , \qquad p \in M .$$

If D is a connection for E, then there is the pull-back connection $g^!(D)$ for $g^!(E)$.

Let $e_1, \cdots, e_r$ be a C^∞ frame of $g^!(E)$, $\theta = \|\theta_{ij}\|$ be the connection matrix of D with respect to $e_1, \cdots, e_r$, and $\omega = \|\omega_{ij}\|$ be the connection matrix of $g^!(D)$ with respect to $g^!(e_1), \cdots, g^!(e_r)$. Then for each θ_{ij},

(4.56) $$\omega_{ij} = g^*\theta_{ij} .$$

(4.56) characterizes $g^!(D)$. Here g^* is the usual map

(4.57) $$g^*: C^\infty(\tau^*X|U) \to C^\infty(\tau^*M|g^{-1}(U)) .$$

If E' is another C^∞ vector-bundle on X, and $\eta: E' \to E$ is a map of C^∞ vector-bundles, then there is

$$(4.58) \qquad g^!(\eta): g^!(E') \to g^!(E) .$$

For $p \in M$, $g^!(\eta): g^!(E')_p \to g^!(E)_p$ is $\eta: E'_{gp} \to E_{gp}$.

Example. Suppose that M is complex-analytic, and that F is a holomorphic integrable sub-vector-bundle of T. Assume that the foliation determined by F is a fibration. Let X be the base of this fibration, and $\pi: M \to X$ the projection of M onto X. Then

$$(4.59) \qquad T/F = \pi^!(TX) .$$

On X let D be a connection of type (1,0) for TX. $\pi^!(D)$ is then a basic connection for T/F.

5. Z-sequences

As in the introduction let M be complex-analytic, and ξ an integrable subsheaf of $\underline{T}$. Let k be the leaf dimension of ξ, and S the singular set. On $M - S$, let F be the unique holomorphic sub-vector-bundle of T such that

$$(5.1) \qquad \underline{F} = \xi | M - S .$$

Here $\underline{F}$ denotes the sheaf of germs of holomorphic sections of F. On $M - S$ set

$$(5.2) \qquad \nu = T/F .$$

(5.3) Lemma. *Let W be an open subset of $M - S$. On W, let D and D' be two basic connections for $\nu | W$. Set $\tilde{W} = W \times [0,1]$. Let $\rho: \tilde{W} \to W$ and $t: \tilde{W} \to [0,1]$ be the projections. On $\tilde{W}$ define a connection ∇ for $\rho^!(\nu | W)$ by*

$$(5.4) \qquad \nabla = t\rho^!(D') + (1-t)\rho^!(D) .$$

Set $K = K(\nabla)$. Let $\varphi \in C[X_1, \cdots, X_n]$ be symmetric and homogeneous of degree l. Assume $n - k < l \leq n$. Then

$$(5.5) \qquad \varphi(K) = 0 .$$

Proof. The proof is very much like the proof of (3.28). Given $p \in W$, let W_p be an open neighborhood of p in W such that W_p is the domain of a complex-analytic coordinate system $z_1, \cdots, z_n$ with

$$(5.6) \qquad \partial/\partial z_1, \cdots, \partial/\partial z_k \in \Gamma(F | W_p)$$

Set $\tilde{W}_p = W_p \times [0,1]$. Let $A(\tilde{W}_p)$ be the ring of all C^∞ complex-valued differential forms on $\tilde{W}_p$. In $A(\tilde{W}_p)$ let $I(F, \tilde{W}_p)$ be the ideal generated by

$\rho^*(dz_{k+1}), \cdots, \rho^*(dz_n)$. This ideal has the properties:

(5.7) $$\text{If} \quad \omega \in I(F, \tilde{W}_p), \quad \text{then} \quad d\omega \in I(F, \tilde{W}_p).$$

(5.8) If $\omega_1, \cdots, \omega_{n-k+1}$ are any $n-k+1$ elements of $I(F, \tilde{W}_p)$, then $\omega_1 \wedge \cdots \wedge \omega_{n-k+1} = 0$.

On W_p let $\eta\colon T \to \nu$ be the projection of T onto ν. Let $\theta = \|\theta_{ij}\|$ and $\theta' = \|\theta'_{ij}\|$ be the connection matrices of D and D' with respect to the frame $\eta\partial/\partial z_{k+1}, \cdots, \eta\partial/\partial z_n$. Let $\omega = \|\omega_{ij}\|$ be the connection matrix of ∇ with respect to the frame $\rho^!\eta\partial/\partial z_{k+1}, \cdots, \rho^!\eta\partial/\partial z_n$. Then according to (5.4), for each ω_{ij}:

(5.9) $$\omega_{ij} = t\rho^*(\theta'_{ij}) + (1-t)\rho^*(\theta_{ij}).$$

Since D and D' are basic, (3.25) and (3.26) now imply that each ω_{ij} is in $I(F, \tilde{W}_p)$:

(5.10) $$\omega_{ij} \in I(F, \tilde{W}_p).$$

Let $\kappa = \|\kappa_{ij}\|$ be the curvature matrix of ∇ with respect to the frame $\rho^!\eta\partial/\partial z_{k+1}, \cdots, \rho^!\eta\partial/\partial z_n$. Then (5.7) and (1.13) imply that each κ_{ij} is in $I(F, \tilde{W}_p)$:

(5.11) $$\kappa_{ij} \in I(F, \tilde{W}_p).$$

Let $\sigma_1, \cdots, \sigma_n$ be the elementary symmetric functions of $X_1, \cdots, X_n$. Set $\varphi = \tilde{\varphi}(\sigma_1, \cdots, \sigma_l)$. On $\tilde{W}_p$ define differential forms $\sigma_1(\kappa), \cdots, \sigma_n(\kappa)$ by requiring:

(5.12) $$\sigma_j(\kappa) \text{ in a } 2j\text{-form on } \tilde{W}_p, \quad j = 1, \cdots, n,$$

(5.13) $$\det(I + \kappa) = 1 + \sigma_1(\kappa) + \cdots + \sigma_n(\kappa).$$

Then

(5.14) $$\varphi(K) \mid \tilde{W}_p = \tilde{\varphi}(\sigma_1(\kappa), \cdots, \sigma_l(\kappa)).$$

Since $\deg \varphi > n-k$, (5.14), (5.11) and (5.8) imply that $\varphi(K)$ vanishes on $\tilde{W}_p$. This proves the lemma.

(5.15) Definition. Let Z be a connected component of the singular set S. A Z-sequence β is a triple $\beta = (U, (E_q, E_{q-1}, \cdots, E_0), (\eta_q, \eta_{q-1}, \cdots, \eta_0))$ such that the following five conditions are satisfied:

(5.16) U is an open subset of M such that $U \cap S = Z$ and Z is a deformation retract of U.

(5.17) $E_q, E_{q-1}, \cdots, E_0$ are C^∞ complex vector-bundles on U.

(5.18) For $i = q, q-1, \cdots, 1$, η_i is a C^∞ vector-bundle map from $E_i \mid U - Z$ to $E_{i-1} \mid U - Z$.

(5.19) η_0 is a C^∞ vector-bundle map from $E_0 \mid U - Z$ to $\nu \mid U - Z$.

(5.20) On $U - Z$ the sequence

$$0 \to E_q|U - Z \to E_{q-1}|U - Z \to \cdots \to E_0|U - Z \to \nu|U - Z \to 0$$

is exact.

Remark. Note that although each E_i is a vector-bundle on all of U, η_i exists only on $U - Z$:

(5.21) $\eta_i\colon E_i|U - Z \to E_{i-1}|U - Z\,, \quad i = q, q-1, \cdots, 1\,,$

(5.22) $\eta_0\colon E|U - Z \to \nu|U - Z\,.$

(5.23) Definition. Let Z be a connected component of S. Assume that Z is compact. Let $\beta = (U, (E_q, E_{q-1}, \cdots, E_0), (\eta_q, \eta_{q-1}, \cdots, \eta_0))$ be a Z-sequence. On U let $D_q, D_{q-1}, \cdots, D_0$ be connections for $E_q, E_{q-1}, \cdots, E_0$. On $U - Z$, let D_{-1} be a connection for $\nu|U - Z$. Then $(D_q, D_{q-1}, \cdots, D_0, D_{-1})$ is *fitted to* β *if*

(5.24) D_{-1} is a basic connection for $\nu|U - Z$,

(5.25) there exists a compact subset Σ of U with Z contained in the interior of Σ such that on $U - \Sigma$, $(D_q, D_{q-1}, \cdots, D_0, D_{-1})$ is compatible with the exact sequence

$$0 \to E_q|U - \Sigma \to E_{q-1}|U - \Sigma \to \cdots \to E_0|U - \Sigma \to \nu|U - \Sigma \to 0\,.$$

(5.26) Lemma. *Let Z be a connected component of S. Assume that Z is compact. Let $\beta = (U, (E_q, E_{q-1}, \cdots, E_0), (\eta_q, \eta_{q-1}, \cdots, \eta_0))$ be a Z-sequence, and D_{-1} a basic connection for $\nu|U - Z$. Then on U there exist connections $D_q, D_{q-1}, \cdots, D_0$ for $E_q, E_{q-1}, \cdots, E_0$ such that*

(5.27) $(D_q, D_{q-1}, \cdots, D_0, D_{-1})$ *is fitted to* β.

Proof. According to Lemma (4.17) on $U - Z$ there exist connections ∇_q, $\nabla_{q-1}, \cdots, \nabla_0$ for $E_q|U - Z, E_{q-1}|U - Z, \cdots, E_0|U - Z$ such that

(5.28) $(\nabla_q, \nabla_{q-1}, \cdots, \nabla_0, D_{-1})$ is compatible with the exact sequence

$$0 \to E_q|U - Z \to E_{q-1}|U - Z \to \cdots \to E_0|U - Z \to \nu|U - Z \to 0\,.$$

Let Σ be a compact subset of U with Z contained in the interior of Σ. According to Lemma (4.41) on U there exist connections $D_q, D_{q-1}, \cdots, D_0$ for E_q, $E_{q-1}, \cdots, E_0$ such that

(5.29) D_i and ∇_i agree on $E_i|U - Z$, $i = q, q-1, \cdots, 0$.

$(D_q, D_{q-1}, \cdots, D_0, D_{-1})$ is then fitted to β.

(5.30) Remark. Note that given $\beta = (U, (E_q, E_{q-1}, \cdots, E_0), (\eta_q, \eta_{q-1}, \cdots, \eta_0))$ and given any compact subset Σ of U with Z contained in the interior of

Σ, one can then construct $D_q, D_{q-1}, \cdots, D_0, D_{-1}$ such that (5.25) is valid for these D_i and the given Σ.

(5.31) Proposition. *Let Z be a connected component of S. Assume that Z is compact. Let $\beta = (U, (E_q, E_{q-1}, \cdots, E_0), (\eta_q, \eta_{q-1}, \cdots, \eta_0))$ be a Z-sequence. Assume that $(D_q, D_{q-1}, \cdots, D_0, D_{-1})$ is fitted to β. Let $\varphi \in C[X_1, X_2, \cdots, X_n]$ be symmetric and homogeneous of degree l, where $n - k < l \leq n$. Set $K_i = K(D_i)$. On U consider the $2l$-form $\varphi(K_q | K_{q-1} | \cdots | K_0)$. Then*

$$(5.32) \qquad \varphi(K_q | K_{q-1} | \cdots | K_0) \quad \textit{has compact support.}$$

Moreover, suppose $(D'_q, D'_{q-1}, \cdots, D'_0, D'_{-1})$ is also fitted to β. Set $K'_i = K(D'_i)$. Then there exists a $2l - 1$ form ω on U such that

(5.33) *ω has compact support,*

$$(5.34) \qquad d\omega = \varphi(K'_q | K'_{q-1} | \cdots | K'_0) - \varphi(K_q | K_{q-1} | \cdots | K_0) .$$

Proof. Given $(D_q, D_{q-1}, \cdots, D_0, D_{-1})$ let Σ be as in (5.25). Then according to (4.23) on $U - \Sigma$,

$$(5.35) \qquad \varphi(K_q | K_{q-1} | \cdots | K_0) | U - \Sigma = \varphi(K_{-1}) | U - \Sigma .$$

Hence by (3.28), $\varphi(K_q | K_{q-1} | \cdots | K_0)$ vanishes on $U - \Sigma$. This proves (5.32).

To prove (5.33) and (5.34) we may assume that on $U - \Sigma$, $(D'_q, D'_{q-1}, \cdots, D'_0, D'_{-1})$ is also compatible with the exact sequence

$$(5.36) \quad 0 \to E_q | U - \Sigma \to E_{q-1} | U - \Sigma \to \cdots \to E_0 | U - \Sigma \to \nu | U - \Sigma \to 0 .$$

Define $\tilde{U}$, $\tilde{\Sigma}$, $\tilde{Z}$ by

$$(5.37) \qquad \tilde{U} = U \times [0, 1] ,$$

$$(5.38) \qquad \tilde{\Sigma} = \Sigma \times [0, 1] ,$$

$$(5.39) \qquad \tilde{Z} = Z \times [0, 1] .$$

Let $\rho: \tilde{U} \to U$ and $t: \tilde{U} \to [0, 1]$ be the projections. On $\tilde{U}$ there is the pull-back bundle $\rho^!(E_i)$, and there are the pull-back connections $\rho^!(D'_i)$ and $\rho^!(D_i)$ for $\rho^!(E_i)$. On $\tilde{U}$ define a connection ∇_i for $\rho^!(E_i)$ by

$$(5.40) \qquad \nabla_i = t\rho^!(D'_i) + (1 - t)\rho^!(D_i) , \qquad i = q, q - 1, \cdots, 0 .$$

Set $\tilde{K}_i = K(\nabla_i)$. On $\tilde{U} - \tilde{Z}$ set

$$(5.41) \qquad \tilde{\nu} = \rho^!(\nu) ,$$

$$(5.42) \qquad \nabla_{-1} = t\rho^!(D'_{-1}) + (1 - t)\rho^!(D_{-1}) ,$$

$$(5.43) \qquad \tilde{K}_{-1} = K(\nabla_{-1}) .$$

Define $i_0: U \to \tilde{U}$ and $i_1: U \to \tilde{U}$ by

(5.44) $i_0(x) = (x, 0)$, $x \in U$,

(5.45) $i_1(x) = (x, 1)$, $x \in U$,

Then (5.40) and (4.56) imply

(5.46) $i_0^*\varphi(\tilde{K}_q | \tilde{K}_{q-1} | \cdots | \tilde{K}_0) = \varphi(K_q | K_{q-1} | \cdots | K_0)$,

(5.47) $i_1^*\varphi(\tilde{K}_q | \tilde{K}_{q-1} | \cdots | \tilde{K}_0) = \varphi(K'_q | K'_{q-1} | \cdots | K'_0)$.

Hence in order to prove (5.33) and (5.34) it suffices to prove

(5.48) $\varphi(\tilde{K}_q | \tilde{K}_{q-1} | \cdots | \tilde{K}_0)$ vanishes on $\tilde{U} - \tilde{\Sigma}$.

If the exact sequence (5.36) is pulled back by $\rho^!$ to $\tilde{U} - \tilde{\Sigma}$, then on $\tilde{U} - \tilde{\Sigma}$ $(\nabla_q, \nabla_{q-1}, \cdots, \nabla_0, \nabla_{-1})$ is compatible with the pulled-back exact sequence. So according to (4.23), on $\tilde{U} - \tilde{\Sigma}$

(5.49) $\varphi(\tilde{K}_q | \tilde{K}_{q-1} | \cdots | \tilde{K}_0) | \tilde{U} - \tilde{\Sigma} = \varphi(\tilde{K}_{-1}) | \tilde{U} - \tilde{\Sigma}$.

By (5.5), $\varphi(\tilde{K}_{-1}) = 0$. This completes the proof.

Remark. Let Z be compact, and $\beta = (U, (E_q, E_{q-1}, \cdots, E_0), (\eta_q, \eta_{q-1}, \cdots, \eta_0))$ be a Z-sequence. Let $H_c^*(U; C)$ denote the cohomology of U with compact supports and coefficients C. Then there are isomorphisms:

(5.50) $H_c^j(U; C) \to H_{2n-j}(U; C) \leftarrow H_{2n-j}(Z; C)$.

The isomorphism $H_c^j(U; C) \to H_{2n-j}(U; C)$ is the usual Poincaré duality isomorphism. The isomorphism $H_{2n-j}(U; C) \leftarrow H_{2n-j}(Z; C)$ is given by the inclusion of Z in U. Recall that by (5.16), Z is a deformation retract of U. Thus a closed j-form ω on U with compact support determines an element of $H_{2n-j}(Z; C)$.

We come now to the main definition of this section.

(5.51) Definition. Let Z be a connected component of S. Assume that Z is compact. Let $\beta = (U, (E_q, E_{q-1}, \cdots, E_0), (\eta_q, \eta_{q-1}, \cdots, \eta_0))$ be a Z-sequnce. Choose connections $D_q, D_{q-1}, \cdots, D_0, D_{-1}$ such that $(D_q, D_{q-1}, \cdots, D_0, D_{-1})$ is fitted to β. Set $K_i = K(D_i)$. Let φ be symmetric and homogeneous of degree l, where $n - k < l \leq n$. Define $\mathrm{Res}_\varphi(\xi, Z, \beta) \in H_{2n-2l}(Z; C)$ to be the element of $H_{2n-2l}(Z; C)$ determined by $(\sqrt{-1}/(2\pi))^l \varphi(K_q | K_{q-1} | \cdots | K_0)$.

Remarks. (a) (5.33) and (5.34) imply that $\mathrm{Res}_\varphi(\xi, Z, \beta)$ depends only on φ, ξ, Z, and β. $\mathrm{Res}_\varphi(\xi, Z, \beta)$ does not depend on the choice of $D_q, D_{q-1}, \cdots, D_0, D_{-1}$.

(b) Since Z is a compact holomorphic subvariety of M, Z has the property:

(5.52) Let V be any open subset of M with $V \supset Z$. Then there exists an open subset V_1 of M such that $V \supset V_1 \supset Z$ and Z is a deformation retract of V_1.

(c) For $\operatorname{Res}_\varphi(\xi, Z, \beta)$ only the local structure of ξ and β near Z is relevant. Let $\beta = (U, (E_q, E_{q-1}, \cdots, E_0), (\eta_q, \eta_{q-1}, \cdots, \eta_0))$. Let W be an open subset of M with $U \supset W \supset Z$, and Z a deformation retract of W. Set $\beta | W = (W, (E_q | W, E_{q-1} | W, \cdots, E_0 | W), (\eta_q | W, \eta_{q-1} | W, \cdots \eta_0 | W))$. Then

$$\operatorname{Res}_\varphi(\xi, Z, \beta) = \operatorname{Res}_\varphi(\xi, Z, \beta | W) . \tag{5.53}$$

(5.53) can be proved by applying Remark (5.30). Choose $(D_q, D_{q-1}, \cdots, D_0, D_{-1})$ fitted to β so that the Σ of (5.25) is contained in W. $(D_q | W, D_{q-1} | W, \cdots, D_0 | W, D_{-1} | W)$ is then fitted to $\beta | W$. Hence the element of $H_{2n-l}(Z; C)$ determined by $(\sqrt{-1}/(2\pi))^l \varphi(K_q | K_{q-1} | \cdots | K_0)$ is $\operatorname{Res}_\varphi(Z, \xi, \beta)$ and is also $\operatorname{Res}_\varphi(Z, \xi, \beta | W)$.

The next proposition will be used in § 7.

(5.54) Definition. Let β and γ be two Z-sequences. Set

$$\beta = (U, (E_q, E_{q-1}, \cdots, E_0), (\eta_q, \eta_{q-1}, \cdots, \eta_0))$$

and

$$\gamma = (V, (I_s, I_{s-1}, \cdots, I_0), (\mu_s, \mu_{s-1}, \cdots, \mu_0)) .$$

Assume $s = q$. An *admissible epimorphism* or γ onto β is a pair of consisting of an open subset W of M and a diagram

$$\begin{array}{ccccccccccccc}
 & & 0 & & 0 & & & & 0 & & 0 & & \\
 & & \downarrow & & \downarrow & & & & \downarrow & & \downarrow & & \\
0 & \to & L_q & \to & L_{q-1} & \to & \cdots & \to & L_0 & \to & 0 & \to & 0 \\
 & & \downarrow & & \downarrow & & & & \downarrow & & \downarrow & & \\
0 & \to & I_q & \to & I_{q-1} & \to & \cdots & \to & I_0 & \to & \nu & \to & 0 \\
 & & \downarrow & & \downarrow & & & & \downarrow & & \downarrow & & \\
0 & \to & E_q & \to & E_{q-1} & \to & \cdots & \to & E_0 & \to & \nu & \to & 0 \\
 & & \downarrow & & \downarrow & & & & \downarrow & & \downarrow & & \\
 & & 0 & & 0 & & & & 0 & & 0 & &
\end{array} \tag{5.55}$$

of C^∞ vector-bundles such that the following six conditions are satisfied:

$$U \cap V \supset W \supset Z . \tag{5.56}$$

(5.57) Each column $0 \to L_j \to I_j \to E_j \to 0$ is defined and exact on all of W. $j = q, q-1, \cdots, 0$.

(5.58) The map $\nu \to \nu$ is the identity of $\nu | W - Z$.

(5.59) The top row $0 \to L_q \to L_{q-1} \to \cdots \to L_0 \to 0 \to 0$ is defined and exact on all of W.

(5.60) The middle row is $\gamma | W$. The bottom row is $\beta | W$.

(5.61) The diagram commutes on $W - Z$.

(5.62) Proposition. *If there is an admissible eipimorphism of γ onto β, then for all φ with $n - k < \deg \varphi \leq n$*

$$\text{Res}_\varphi (Z, \xi, \beta) = \text{Res}_\varphi (Z, \xi, \gamma) . \tag{5.63}$$

Proof. Set $\beta = (U, (E_q, E_{q-1}, \cdots, E_0), (\eta_q, \eta_{q-1}, \cdots, \eta_0))$ and $\gamma = (V, (I_q, I_{q-1}, \cdots, I_0), (\mu_q, \mu_{q-1}, \cdots, \mu_0))$. Let W be as in Definition (5.54), and consider the diagram (5.55). In view (5.52) and of (5.53) we may assume

$$U = V = W . \tag{5.64}$$

On U choose connections $\nabla_q, \nabla_{q-1}, \cdots, \nabla_0$ for $L_q, L_{q-1}, \cdots, L_0$ such that on U

(5.65) $(\nabla_q, \nabla_{q-1}, \cdots, \nabla_0)$ is compatible with the exact sequence $0 \to L_q \to L_{q-1} \to \cdots \to L_0 \to 0$.

Let D_{-1} be a basic connection for $\nu | U - Z$. On U choose connections $D_q, D_{q-1}, \cdots, D_0$ for $E_q, E_{q-1}, \cdots, E_0$ such that

(5.66) $\qquad (D_q, D_{q-1}, \cdots, D_0)$ is fitted to β.

Hence there exists a compact subset Σ of U with (5.25) valid for Σ.

Denote by $\iota_j(L_j)$ the image of $\iota_j \colon L_j \to I_j$. According to Lemma (4.32) on $U - Z$ there exist C^∞ sub-vector-bundles $F_q, F_{q-1}, \cdots, F_0$ of $I_q | U - Z$, $I_{q-1} | U - Z, \cdots, I_0 | U - Z$ such that

$$I_j | U - Z = F_j \oplus \iota_j(L_j | U - Z) , \qquad j = q, q-1, \cdots, 0 , \tag{5.67}$$

$$\mu_j(F_j) \subset F_{j-1} , \qquad j = q, q-1, \cdots, 1 . \tag{5.68}$$

According to Lemma (4.45) on U there exist C^∞ sub-vector-bundles $\tilde{F}_q, \tilde{F}_{q-1}, \cdots, \tilde{F}_0$ of $I_q, I_{q-1}, \cdots, I_0$ such that

$$I_j = \tilde{F}_j \oplus \iota_j(L_j) , \qquad j = q, q-1, \cdots, 0 , \tag{5.69}$$

$$\tilde{F}_j | U - \Sigma = F_j | U - \Sigma , \qquad j = q, q-1, \cdots, 0 . \tag{5.70}$$

On U let $\hat{D}_j$ be the unique connection for $\tilde{F}_j$ such that

(5.71) $(\hat{D}_j, D_j)$ is compatible with the exact sequence

$$0 \to \tilde{F}_j \to E_j \to 0 , \qquad j = q, q-1, \cdots, 0 .$$

On U let $\hat{\nabla}_j$ be the unique connection for $\iota_j(L_j)$ such that

(5.72) $(\nabla_j, \hat{\nabla}_j)$ is compatible with the exact sequence

$$0 \to L_j \to \iota_j(L_j) \to 0\ , \qquad j = q, q-1, \cdots, 0\ .$$

Let $\tilde{D}_j$ be the direct sum connection for I_j:

$$\tilde{D}_j = \hat{D}_j \oplus \hat{\nabla}_j\ . \tag{5.73}$$

Thus

$$\tilde{D}_j(s_1 + s_2) = \hat{D}_j s_1 + \hat{\nabla}_j s_2\ , \qquad s_1 \in C^\infty(\tilde{F}_j),\ s_2 \in C^\infty(\iota_j(L_j))\ . \tag{5.74}$$

(5.65) and (5.66) imply

$$(\tilde{D}_q, \tilde{D}_{q-1}, \cdots, \tilde{D}_0, D_{-1}) \qquad \text{is fitted to } \gamma\ . \tag{5.75}$$

Set $\tilde{K}_j = K(\tilde{D}_j)$, $K_j = K(D_j)$. (5.73) implies

$$\det(I + \tilde{K}_j) = \det(I + K_j)\det(I + K(\nabla_j))\ . \tag{5.76}$$

Set $\varepsilon(i) = (-1)^i$, (5.65) and (4.24) imply

$$\prod_{i=0}^{q} \det(I + K(\nabla_i))^{\varepsilon(i)} = 1\ . \tag{5.77}$$

Therefore

$$\prod_{i=0}^{q} (\det(I + \tilde{K}_i))^{\varepsilon(i)} = \prod_{i=0}^{q} (\det(I + K_i))^{\varepsilon(i)}\ . \tag{5.78}$$

(5.78) and (4.13) imply

$$\varphi(\tilde{K}_q \,|\, \tilde{K}_{q-1} \,|\, \cdots \,|\, \tilde{K}_0) = \varphi(K_q \,|\, K_{q-1} \,|\, \cdots \,|\, K_0)\ . \tag{5.79}$$

Due to (5.66) and (5.75), (5.63) has been proved.

6. Coherent real-analytic sheaves

In order to prove the residue existence theorem stated in the introduction, we shall have to use real-analytic sheaves. Following [1] let us state the basic facts which we need.

Let M be a complex-analytic manifold, and $n = \dim_C M$. Denote by $\mathcal{O}$ the sheaf of germs of holomorphic functions on M, and by $\mathcal{A}$ the sheaf of germs of real-analytic functions on M. Given $x \in M$, let $z_1, \cdots, z_n$ be complex-analytic coordinates defined about x with $z_i(x) = 0$. Then $\mathcal{A}_x$ is isomorphic to the ring $C\{z_1, \cdots, z_n, \bar{z}_1, \cdots, \bar{z}_n\}$ of convergent power series in z_i, $\bar{z}_i$. Any module over $\mathscr{A}_x$ has a projective resolution of length $\leq 2n$. $\mathcal{O}$ and $\mathscr{A}$ are both sheaves of rings, and there is the natural injection $\mathcal{O} \to \mathscr{A}$. If $\mathscr{F}$ is a sheaf of $\mathcal{O}$-modules, then $\mathscr{A} \otimes_{\mathcal{O}} \mathscr{F}$ is a sheaf of $\mathscr{A}$-modules.

(**6.1**) **Proposition.** *Let $\mathscr{F}$ be a coherent sheaf of $\mathcal{O}$-modules. Then $\mathscr{A} \otimes_{\mathcal{O}} \mathscr{F}$ is a coherent sheaf of $\mathscr{A}$-modules. Moreover, if $\mathscr{F}_1 \to \mathscr{F}_2 \to \mathscr{F}_3$ is an exact sequence of coherent sheaves of $\mathcal{O}$-modules, then $\mathscr{A} \otimes_{\mathcal{O}} \mathscr{F}_1 \to \mathscr{A} \otimes_{\mathcal{O}} \mathscr{F}_2 \to \mathscr{A} \otimes_{\mathcal{O}} \mathscr{F}_3$ is an exact sequence of coherent sheaves of $\mathscr{A}$-modules.*

Proof. See [1, Proposition 2.9, p. 30] and also [2, Proposition (1.5), p. 153].

(**6.2**) **Definition.** Let U be open in M. On U, let $\mathscr{F}$ be a coherent sheaf of $\mathscr{A}$-modules. A *resolution* of $\mathscr{F}$ is an exact sequence

$$0 \to H_{2n} \to H_{2n-1} \to \cdots \to H_0 \to \mathscr{F} \to 0$$

of coherent sheaves of $\mathscr{A}$-modules on U such that each H_i is locally free.

(**6.3**) **Proposition** (*Existence of resolutions*). *Let U be an open subset of M, and $\mathscr{F}$ a coherent sheaf of $\mathscr{A}$-modules on U. Let W be an open subset of U such that there is a compact B with $U \supset B \supset W$. Then on W, $\mathscr{F} \mid W$ has a resolution.*

Proof. See [1, Proposition 2.6, p. 29].

(**6.4**) **Definition.** On U, let R_1, R_2 be two resolutions of $\mathscr{F}$. A *morphism* of R_1 to R_2 is a commutative diagram

$$\begin{array}{ccccccccccc} 0 \to & H_{2n} & \to & H_{2n-1} & \to & \cdots & \to & H_0 & \to & \mathscr{F} & \to 0 \\ & \downarrow & & \downarrow & & & & \downarrow & & \downarrow & \\ 0 \to & J_{2n} & \to & J_{2n-1} & \to & \cdots & \to & J_0 & \to & \mathscr{F} & \to 0 \end{array}$$

of sheaves of $\mathscr{A}$-modules on U such that the upper row is R_1, the lower row is R_2, and the vertical arrow farthest to the right is the identity map. The morphism is said to be a morphism of R_1 *onto* R_2 if the vertical arrows are all surjections.

(**6.5**) **Proposition** (*Comparison of resolutions*). *Let U be open in M. On U, let $\mathscr{F}$ be a coherent sheaf of $\mathscr{A}$-modules, and R_1, R_2 be two resolutions of $\mathscr{F}$. Let W be an open subset of U such that there is a compact set B with $U \supset B \supset W$. Let $R_1 \mid W$, $R_2 \mid W$ be the restrictions of R_1, R_2 to W. Then on W there is a resolution R_3 of $\mathscr{F} \mid W$ such that*

(6.6) *there exists a morphism of R_3 onto $R_1 \mid W$,*

(6.7) *there exists a morphism of R_3 onto $R_2 \mid W$.*

Proof. See [7, Lemmas 13 and 14, p. 107]. In [7] these are proved on an algebraic variety using coherent algebraic sheaves. But due to [1, Corollary 2.5, p. 29] the same reasoning is valid in the real-analytic framework.

(**6.8**) **Remarks.** (a) If H is a coherent sheaf of $\mathscr{A}$-modules, then H is locally free if and only if each stalk H_x is a free $\mathscr{A}_x$-module.

(b) If E is a real-analytic vector-bundle, let $\underline{\underline{E}}$ denote the sheaf of germs of real-analytic sections of E. Then $E \rightsquigarrow \underline{\underline{E}}$ is a functor which gives an equivalence between the category of real-analytic vector-bundles on M and the category of locally free coherent sheaves of $\mathscr{A}$-modules on M.

(c) If M is compact and $\mathscr{F}$ is a coherent sheaf of $\mathscr{O}$-modules, then Propositions (6.1),(6.3), and (6.5) can be used to define the Chern classes of $\mathscr{F}$. To do this, on M let

$$0 \to H_{2n} \to H_{2n-1} \to \cdots \to H_0 \to \mathscr{A} \underset{\mathscr{O}}{\otimes} \mathscr{F} \to 0$$

be a resolution of $\mathscr{A} \otimes_{\mathscr{O}} \mathscr{F}$. Let E_i be the real-analytic vector-bundle with $\underline{\underline{E}}_i = H_i$. In $K(M)$ let ζ be the virtual bundle:

(6.9) $$\zeta = \sum_{i=0}^{2n} (-1)^i E_i \, .$$

$c_i(\mathscr{F})$ is defined by

(6.10) $$c_i(\mathscr{F}) = c_i(\zeta) \, , \qquad i = 1, \cdots, n \, .$$

It follows easily from Proposition (6.5) that ζ depends only on $\mathscr{F}$. Hence $c_i(\mathscr{F})$ is well-defined. For a detailed proof of this see [7, Lemma 11, p. 106].

More generally, let φ be symmetric and homogeneous with $\deg \varphi \leq n$. $\varphi(\mathscr{F})$ is defined by

(6.11) $$\varphi(\mathscr{F}) = \tilde{\varphi}(c_1(\mathscr{F}), \cdots, c_l(\mathscr{F})) \, ,$$

where $\varphi = \tilde{\varphi}(\sigma_1, \cdots, \sigma_l)$, and $l = \deg \varphi$.

(d) We are forced to use real-analytic sheaves because it is not known whether the propositions on existence and comparison of resolutions are true in the holomorphic category. By Proposition (6.1) a resolution in the holomorphic category, when tensored with $\mathscr{A}$, gives a resolution in the real-analytic category.

7. Proof of the residue existence theorem

As in the statement of Theorem 2 let M be complex-analytic, and ξ a full integrable sub-sheaf of $\underline{T}$. Set $Q = \underline{T}/\xi$. Let φ be symmetric and homogeneous of degree l where $n - k < l \leq n$.

(7.1) Definition. Let Z be a connected component of the singular set S. Assume that Z is compact. Choose an open subset U of M with $U \cap S = Z$ and Z a deformation retract of U such that on U there exists a resolution R:

$$0 \to \underline{\underline{E}}_{2n} \to \underline{\underline{E}}_{2n-1} \to \cdots \to \underline{\underline{E}}_0 \to \mathscr{A} \underset{\mathcal{O}}{\otimes} Q \to 0$$

of $\mathscr{A} \otimes_{\mathcal{O}} Q \mid U$.

For $i = 2n, 2n-1, \cdots, 1$ let $\eta_i \colon E_i \mid U - Z \to E_{i-1} \mid U - Z$ be the vector-bundle map which gives $\underline{\underline{E}}_i \mid U - Z \to \underline{\underline{E}}_{i-1} \mid U - Z$. Let $\eta_0 \colon E_0 \mid U - Z \to \nu \mid U - Z$ be the vector-bundle map which gives $\underline{\underline{E}}_0 \mid U - Z \to \mathscr{A} \otimes_{\mathcal{O}} Q \mid U - Z$. Then $(U, (E_{2n}, E_{2n-1}, \cdots, E_0), (\eta_{2n}, \eta_{2n-1}, \cdots, \eta_0))$ is a Z-sequence. Call this Z-sequence $\beta(R)$, and define $\operatorname{Res}_\varphi(\xi, Z)$ by

$$\operatorname{Res}_\varphi(\xi, Z) = \operatorname{Res}_\varphi(\xi, Z, \beta(R)) \,. \tag{7.2}$$

(7.3) Remarks. (a) The existence of U, R as in Definition (7.1) is implied by (5.52) and Proposition (6.3).

(b) In order for (7.2) to be legitimate it must be shown that $\operatorname{Res}_\varphi(\xi, Z, \beta(R))$ does not depend on the choice of U and R. This is implied by (5.62) and Proposition (6.5). A morphism of a resolution R_1 onto a resolution R_2 gives an admissible epimorphism of $\beta(R_1)$ onto $\beta(R_2)$.

Proof of (0.22). From (7.3)b and (5.53) it is clear that $\operatorname{Res}_\varphi(\xi, Z)$ depends only on φ and on the local structure of Q near Z. Since ξ is full, (0.18) implies that the local structure of Q near Z is determined by the local behavior of the leaves of ξ near Z. (0.22) is now evident.

Proof of (0.23). If M is compact, then on M let

$$0 \to \underline{\underline{E}}_{2n} \to \underline{\underline{E}}_{2n-1} \to \cdots \to \underline{\underline{E}}_0 \to \mathscr{A} \underset{\mathcal{O}}{\otimes} Q \to 0$$

be a resolution of $\mathscr{A} \otimes_{\mathcal{O}} Q$. Let $Z_1, \cdots, Z_r$ be the connected components of the singular set S. Choose open subsets $U_1, \cdots, U_r$ of M such that

$$U_i \cap S = Z_i \,, \qquad i = 1, \cdots, r \,, \tag{7.4}$$

$$U_i \cap U_j = \phi \qquad \text{if } i \neq j \,, \tag{7.5}$$

$$Z_i \text{ is a deformation retract of } U_i. \tag{7.6}$$

Choose compact subsets $\Sigma_1, \cdots, \Sigma_r$ of M such that

$$U_i \supset \Sigma_i \,, \qquad i = 1, \cdots, r \,, \tag{7.7}$$

$$Z_i \text{ is contained in the interior of } \Sigma_i,\ i = 1, \cdots, r. \tag{7.8}$$

Set $\Sigma = \Sigma_1 \cup \cdots \cup \Sigma_r$. On $M - S$ let D_{-1} be a basic connection for ν. On M let $D_{2n}, D_{2n-1}, \cdots, D_0$ be connections for $E_{2n}, E_{2n-1}, \cdots, E_0$ such that

(7.9) On $M - \Sigma$, $(D_{2n}, D_{2n-1}, \cdots, D_0, D_{-1})$ is compatible with the exact sequence

$$0 \to E_{2n} \mid M - \Sigma \to E_{2n-1} \mid M - \Sigma \to \cdots \to E_0 \mid M - \Sigma \to \nu \mid M - \Sigma \to 0 \,.$$

Set $K_i = K(D_i)$. According to (4.15) and (6.11),

$$[\varphi(K_{2n} | K_{2n-1} | \cdots | K_0)] = (2\pi/\sqrt{-1})^l \varphi(Q) . \tag{7.10}$$

(7.9), (4.23), and (3.28) imply

$$\varphi(K_{2n} | K_{2n-1} | \cdots | K_0) \quad \text{vanishes on } M - \Sigma . \tag{7.11}$$

Therefore $\varphi(K_{2n} | K_{2n-1} | \cdots | K_0) | U_i$ is a $2l$-form on U_i with compact support. By (7.2) the element of $H_{2n-2l}(Z; C)$ determined by $\varphi(K_{2n} | K_{2n-1} | \cdots | K_0) | U_i$ is $(2\pi/\sqrt{-1})^l \operatorname{Res}_\varphi(\xi, Z_i)$. Let ω_i be the $2l$-form on M defined by

$$\omega_i | U_i = \varphi(K_{2n} | K_{2n-1} | \cdots | K_0) | U_i , \tag{7.12}$$

$$\omega_i \text{ vanishes on } M - U_i . \tag{7.13}$$

From definition (0.21) of μ_* we then have

$$[\omega_i] = (2\pi/\sqrt{-1})^l \mu_* \operatorname{Res}_\varphi(\xi, Z_i) , \qquad i = 1, \cdots, r . \tag{7.14}$$

But

$$\varphi(K_{2n} | K_{2n-1} | \cdots | K_0) = \omega_1 + \cdots + \omega_r . \tag{7.15}$$

(7.15), (7.14), and (7.10) imply (0.23). This completes the proof of Theorem 2.

Remark. In Definition (7.1) the exact sequence of sheaves has $2n$ locally free sheaves. The next lemma asserts that $\operatorname{Res}_\varphi(\xi, Z)$ can be obtained from a suitable exact sheaf sequence of any length.

(7.16) Lemma. *Let U be an open set containing Z with $U \cap S = Z$ and Z a deformation retract of U. On U let*

$$0 \to \underline{\underline{E}}_q \to \underline{\underline{E}}_{q-1} \to \cdots \to \underline{\underline{E}}_0 \to \mathscr{A} \otimes_{\mathcal{O}} Q \to 0 \tag{7.17}$$

be an exact sequence of sheaves of $\mathscr{A}$-modules. Let β be the resulting Z-sequence. Then

$$\operatorname{Res}_\varphi(\xi, Z, \beta) = \operatorname{Res}_\varphi(\xi, Z) . \tag{7.18}$$

Proof. If $q < 2n$, add on $2n - q$ zeroes to the left of (7.17) to obtain

$$0 \to 0 \to \cdots \to 0 \to \underline{\underline{E}}_q \to \underline{\underline{E}}_{q-1} \to \cdots \to \underline{\underline{E}}_0 \to \mathscr{A} \otimes_{\mathcal{O}} Q \to 0 . \tag{7.19}$$

Let β' be the Z-sequence resulting from (7.19). β' has length $2n$ so by (7.1)

$$\operatorname{Res}_\varphi(\xi, Z) = \operatorname{Res}_\varphi(\xi, Z, \beta') . \tag{7.20}$$

But it is obvious that

$$\text{(7.21)} \qquad \operatorname{Res}_\varphi(\xi, Z, \beta) = \operatorname{Res}_\varphi(\xi, Z, \beta') .$$

This proves (7.18) when $q < 2n$.

If $q > 2n$, then by the syzygy theorem [13, Chapter VIII, Theorem 6.5′, p. 158], the kernel of $\underline{\underline{E}}_{2n-1} \to \underline{\underline{E}}_{2n-2}$ is locally free. Denote this kernel by $\underline{\underline{E}}$. Thus

$$\text{(7.22)} \qquad 0 \to \underline{\underline{E}} \to \underline{\underline{E}}_{2n-1} \to \underline{\underline{E}}_{2n-2} \to \cdots \to \underline{\underline{E}}_0 \to \mathscr{A} \underset{\mathcal{O}}{\otimes} Q \to 0$$

is a resolution of $\mathscr{A} \otimes_{\mathcal{O}} Q$. Denote this resolution by R. The map $\underline{\underline{E}}_{2n} \to \underline{\underline{E}}_{2n-1}$ gives a surjection $\underline{\underline{E}}_{2n} \to \underline{\underline{E}} \to 0$. Consider the commutative diagram:

$$\begin{array}{ccccccccccccccccccc}
0 & \to & \underline{\underline{E}}_q & \to & \underline{\underline{E}}_{q-1} & \to & \cdots & \to & \underline{\underline{E}}_{2n+1} & \to & \underline{\underline{E}}_{2n} & \to & \underline{\underline{E}}_{2n-1} & \to & \cdots & \to & \underline{\underline{E}}_0 & \to & \mathscr{A} \otimes_{\mathcal{O}} Q \to 0 \\
 & & \downarrow & & \downarrow & & & & \downarrow & & \downarrow & & \downarrow 1 & & & & \downarrow 1 & & \downarrow 1 \\
0 & \to & 0 & \to & 0 & \to & \cdots & \to & 0 & \to & \underline{\underline{E}} & \to & \underline{\underline{E}}_{2n-1} & \to & \cdots & \to & \underline{\underline{E}}_0 & \to & \mathscr{A} \otimes_{\mathcal{O}} Q \to 0
\end{array}$$

This diagram gives an admissible epimorphism of Z-sequences. Therefore by (5.63)

$$\text{(7.23)} \qquad \operatorname{Res}_\varphi(\xi, Z, \beta) = \operatorname{Res}_\varphi(\xi, Z, \beta(R)) .$$

This proves (7.18) when $q > 2n$. If $q = 2n$, then (7.18) is immediate from (7.1).

(7.24) Corollary. *Suppose that $E_q, E_{q-1}, \cdots, E_0$ are holomorphic vector-bundles on U. Let*

$$\text{(7.25)} \qquad 0 \to \underline{E}_q \to \underline{E}_{q-1} \to \cdots \to \underline{E}_0 \to Q \to 0$$

be an exact sequence of sheaves of $\mathcal{O}$-modules on U. Denote the resulting Z-sequence by β. Then

$$\text{(7.26)} \qquad \operatorname{Res}_\varphi(\xi, Z) = \operatorname{Res}_\varphi(\xi, Z, \beta) .$$

Proof. View each E_i as a real-analytic vector-bundle. According to Proposition (6.1), (7.25) gives an exact sequence

$$\text{(7.27)} \qquad 0 \to \underline{\underline{E}}_q \to \underline{\underline{E}}_{q-1} \to \cdots \to \underline{\underline{E}}_0 \to \mathscr{A} \underset{\mathcal{O}}{\otimes} Q \to 0$$

of sheaves of $\mathscr{A}$-modules on U. Hence Lemma (7.16) applies, and the corollary is proved.

8. Proof of Theorem 1

Recall the data of the theorem. M is compact and complex-analytic. L is a holomorphic line bundle on M. $\eta: L \to T$ is a holomorphic vector-bundle map.

Each zero of η is isolated. φ is symmetric and homogeneous of degree $n = \dim_C M$. Set $\xi = \eta(\underline{L})$. For $p \in \text{Zero}(\eta)$ there is the usual identification $H_0(p; C) = C$. Let $\varphi(\eta, p)$ be as in the introduction. Then due to (0.23) and (0,14), (0.10) will be implied by

$$\varphi(\eta, p) = \text{Res}_\varphi(\xi, p) . \tag{8.1}$$

The remainder of this section will be devoted to proving (8.1). The proof will have two steps:

Step 1. Replace a situation involving several vector-bundles and several connections by a much simpler situation involving only one vector-bundle and one connection.

Step 2. An explicit computation using one vector-bundle and one connection.

To begin Step 1, let W be an open subset of M. On W let X be a holomorphic section of $T|W$ such that X has no zeroes. According to (3.10) and (3.11),

$$[X, s_1 + s_2] = [X, s_1] + [X, s_2] , \tag{8.2}$$

$$[X, fs] = (Xf)s + f[X, s] , \tag{8.3}$$

whenever $s, s_1, s_2 \in C^\infty(T|W)$, and $f: W \to C$ is a C^∞ function.

If (X) denotes the sub-line-bundle of $T|W$ spanned by X, then (8.2) and (8.3) imply that there is a unique partial connection $((X) \oplus \overline{T}|W, \delta)$ for $T|W$ such that

$$i(X)\delta s = [X, s] , \qquad s \in C^\infty(T|W) , \tag{8.4}$$

$$i(\gamma)\delta s = i(\gamma)\bar{\partial} s , \qquad s \in C^\infty(T|W) , \quad \gamma \in C^\infty(\overline{T}|W) . \tag{8.5}$$

This partial connection for $T|W$ will be referred to as the partial connection for $T|W$ determined by X. Note that (8.4) implies

$$i(fX)\delta s = f[X, s] , \qquad f: W \to C . \tag{8.6}$$

(8.7) Definition. Let X be a holomorphic section of $T|W$ such that X has no zeroes. An *X-connection* for $T|W$ is a connection for $T|W$, which extends the partial connection for $T|W$ determined by X.

(8.8) Remarks. (a) A connection D for $T|W$ is an X-connection if and only if

$$i(X)Ds = [X, s] , \qquad s \in C^\infty(T|W) , \tag{8.9}$$

$$D \text{ is of type } (1,0) . \tag{8.10}$$

(b) Lemma (2.5) guarantees the existence of X-connections. Since $[X, X] = 0$ and $\bar{\partial} X = 0$, Lemma (2.11) implies that there exist X-connections D with $DX = 0$.

(8.11) Lemma. *Let X be a holomorphic section of $T \mid W$ such that X has no zeroes. Let D be an X-connection for $T \mid W$. Set $K = K(D)$. Assume* $\deg \varphi = n$. *Then*

(8.12) $$\varphi(K) = 0 \ .$$

Proof. Given $p \in W$, let W_p be an open neighborhood of p in W such that W_p is the domain of a complex-analytic coordinate system $z_1, \cdots, z_n$ with

(8.13) $$\partial/\partial z_1 = X \mid W_p \ .$$

Let $A(W_p)$ be the ring of all C^∞ complex-valued differential forms on W_p. In $A(W_p)$ let $I(X, W_p)$ be the ideal generated by $dz_2, \cdots, dz_n$. This ideal has the properties:

(8.14) If $\omega \in I(X, W_p)$, then $d\omega \in I(X, W_p)$.

(8.15) If $\omega_1, \cdots, \omega_n$ are any n elements of $I(X, W_p)$, then $\omega_1 \wedge \cdots \wedge \omega_n = 0$.

Let $\theta = \|\theta_{ij}\|$ be the connection matrix of D with respect to the frame $\partial/\partial z_1, \cdots, \partial/\partial z_n$. Then (8.9), (8.10), and (3.12) imply that each θ_{ij} is in $I(X, W_p)$:

(8.16) $$\theta_{ij} \in I(X, W_p) \ .$$

Let $\kappa = \|\kappa_{ij}\|$ be the curvature matrix of D with respect to the frame $\partial/\partial z_1, \cdots, \partial/\partial z_n$. (8.16), (8.14), and (1.13) imply that each κ_{ij} is in $I(X, W_p)$:

(8.17) $$\kappa_{ij} \in I(X, W_p) \ .$$

Since $\deg \varphi = n$, (8.17) and (8.15) imply that $\varphi(K)$ vanishes on W_p. This proves the lemma.

(8.18) Lemma. *Let X be a holomorphic section of $T \mid W$ such that X has no zeroes. Let D and D' be two X-connections for $T \mid W$. Set $\tilde{W} = W \times [0,1]$. Let $\rho \colon \tilde{W} \to W$ and $t \colon \tilde{W} \to [0,1]$ be the projections. On $\tilde{W}$ define a connection ∇ for $\rho^!(T \mid W)$ by*

(8.19) $$\nabla = t\rho^!(D') + (1 - t)\rho^!(D) \ .$$

Set $K = K(\nabla)$. Assume $\deg \varphi = n$. *Then*

(8.20) $$\varphi(K) = 0 \ .$$

Proof. Given $p \in W$, let W_p be an open neighborhood of p in W such that W_p is the domain of a complex-analytic coordinate system $z_1, \cdots, z_n$ with

(8.21) $$\partial/\partial z_1 = X \mid W_p \ .$$

Set $\tilde{W}_p = W_p \times [0,1]$. Let $A(\tilde{W}_p)$ be the ring of all C^∞ complex-valued differential forms on $\tilde{W}_p$. In $A(\tilde{W}_p)$ let $I(X, \tilde{W}_p)$ be the ideal generated by $\rho^*(dz_2)$, $\cdots, \rho^*(dz_n)$.

Let $\theta = \|\theta_{ij}\|$ and $\theta' = \|\theta'_{ij}\|$ be the connection matrices of D and D' with respect to the frame $\partial/\partial z_1, \cdots, \partial/\partial z_n$. Let $\omega = \|\omega_{ij}\|$ be the connection matrix of ∇ with respect to the frame $\rho^!(\partial/\partial z_1), \cdots, \rho^!(\partial/\partial z_n)$. Then (8.19) and (4.56) imply for each ω_{ij},

(8.22)
$$\omega_{ij} = t\rho^*(\theta'_{ij}) + (1-t)\rho^*(\theta_{ij}) .$$

Since D and D' are both X-connections, it now follows that each ω_{ij} is in $I(X, \tilde{W}_p)$:

(8.23)
$$\omega_{ij} \in I(X, \tilde{W}_p) .$$

Let $\kappa = \|\kappa_{ij}\|$ be the curvature matrix of ∇ with respect to the frame $\rho^!(\partial/\partial z_1), \cdots, \rho^!(\partial/\partial z_n)$. $I(X, \tilde{W}_p)$ is closed under d, so each κ_{ij} is in $I(X, \tilde{W}_p)$:

(8.24)
$$\kappa_{ij} \in I(X, \tilde{W}_p) .$$

The wedge product of any n elements of $I(X, \tilde{W}_p)$ is zero. Since $\deg \varphi = n$, (8.24) now implies that $\varphi(K)$ vanishes on $\tilde{W}_p$. This proves the lemma.

(8.25) Definition. Let U be an open subset of M. On U let X be a holomorphic section of $T\,|\,U$. Set $Z = \{p \in U \,|\, X(p) = 0\}$. Assume that Z is compact and connected. On U let D be a connection for $T\,|\,U$. Then D is *fitted to* X if

(8.26) there exists a compact subset Σ of U with Z contained in the interior of Σ such that on $U - \Sigma$, D is an X-connection for $T\,|\,U - \Sigma$.

Remark. Given U, X as in Definition (8.25), choose any compact subset Σ of U with Z contained in the interior of Σ. Then according to Lemma (4.41) there exists a connection D for $T\,|\,U$ such that (8.26) is valid for D and the chosen Σ.

(8.27) Lemma. *Let U be an open subset of M. On U let X be a holomorphic section of $T\,|\,U$. Assume that the zero set Z of X is compact and connected. Let D be a connection for $T\,|\,U$ such that D is fitted to X. Set $K = K(D)$. Assume* $\deg \varphi = n$. *Then*

(8.28) $\varphi(K)$ *has compact support.*

Moreover, suppose that D' is another connection for $T\,|\,U$ such that D' is also fitted to X. Set $K' = K(D')$. Then

(8.29)
$$\int_U \varphi(K) = \int_U \varphi(K') .$$

Proof. To prove (8.28), let Σ be as in (8.26). Then on $U - \Sigma$, D is an X-connection for $T|U - \Sigma$. Hence by (8.12), $\varphi(K)$ vanishes on $U - \Sigma$. This proves (8.28).

To prove (8.29) we may assume that on $U - \Sigma$, D' is also an X-connection for $T|U - \Sigma$. Let $\tilde{U} = U \times [0,1]$. Let $\rho: \tilde{U} \to U$ and $t: \tilde{U} \to [0,1]$ be the projections. On $\tilde{U}$ define a connection ∇ for $\rho^!(T|U)$ by

$$\nabla = t\rho^!(D') + (1 - t)\rho^!(D) . \tag{8.30}$$

Set $\tilde{K} = K(\nabla)$, and define $i_0, i_1: U \to \tilde{U}$ by

$$i_0(x) = (x, 0) , \qquad x \in U , \tag{8.31}$$

$$i_1(x) = (x, 1) , \qquad x \in U . \tag{8.32}$$

Then

$$\varphi(K) = i_0^*\varphi(\tilde{K}) , \tag{8.33}$$

$$\varphi(K') = i_1^*\varphi(\tilde{K}) . \tag{8.34}$$

According to (8.20), $\varphi(\tilde{K})$ vanishes on $\tilde{U} - \Sigma \times [0,1]$. Therefore

(8.35) $\quad\varphi(\tilde{K})$ has compact support.

(8.33), (8.34), and (8.35) imply that there exists a $(2n - 1)$-form ω on U with

(8.36) $\quad\omega$ has compact support,

$$d\omega = \varphi(K') - \varphi(K) . \tag{8.37}$$

(8.29) is now evident.

(8.38) Proposition. *Let U be an open subset of M. On U let X be a holomorphic section of $T|U$. Assume that the zero set Z of X is compact and connected. Let D be a connection for $T|U$ such that D is fitted to X. Set $K = K(D)$. Let ξ be the subsheaf of $T|U$ spanned by X. Assume* $\deg\varphi = n$. *Then*

$$\mathrm{Res}_\varphi(\xi, Z) = (\sqrt{-1}/(2\pi))^n \int_U \varphi(K) . \tag{8.39}$$

Proof. Due to (8.29) it will suffice to exhibit a connection D' for $T|U$ such that D' is fitted to X and such that for $K' = K(D')$ it is immediate and obvious that

$$\mathrm{Res}_\varphi(\xi, Z) = (\sqrt{-1}/(2\pi))^n \int_U \varphi(K') . \tag{8.40}$$

To construct such a D', let V be an open subset of U, and Δ be a compact subset of U with

(8.41) $$U \supset V \supset \Delta \supset Z\,,$$

(8.42) Z is contained in the interior of Δ,

(8.43) Z is a deformation retract of V.

Let D' be a connection for $T|U$ such that

(8.44) on $U - \Delta$, D' is an X-connection for $T|U - \Delta$,

(8.45) $$\text{on} \quad U - \Delta\,, \quad D'X = 0\,.$$

The existence of such a D' is implied by Lemma (2.11) and (4.41). Set $K' = K(D')$.

To verify (8.40), let (1) denote the trivial line bundle $U \times \boldsymbol{C}$. Define $\eta\colon (1) \to T|U$ by

(8.46) $$\eta(p, z) = zX(p)\,, \qquad p \in U,\ z \in \boldsymbol{C}\,.$$

Define a section s of (1) by

(8.47) $$s(p) = (p, 1)\,, \qquad p \in U\,.$$

Thus

(8.48) $$\eta s = X\,.$$

Let D_1 be the unique connection for (1) with

(8.49) $$D_1 s = 0\,.$$

On $U - Z$ set $\nu = T/\eta(1)$. Let $\mu\colon T|U - Z \to \nu$ be the projection. On $V - Z$ consider

(8.50) $$0 \to (1) \to T \to \nu \to 0\,.$$

Let β denote the Z-sequence obtained from (8.50).

For $\nu|U - \Delta$ there is a unique connection D_{-1} such that

(8.51) $$D_{-1}(\mu\gamma) = (1 \otimes \mu)D'\gamma\,,$$

whenever $\gamma \in C^\infty(T|U - \Delta)$. D_{-1} is a basic connection for $\nu|U - \Delta$. By enlarging Δ slightly and applying Lemma (4.41) we may assume that on $U - Z$ there is a basic connection $\tilde{D}_{-1}$ for ν such that

(8.52) $$D_{-1} \text{ and } \tilde{D}_{-1} \text{ agree on } \nu|U - \Delta\,.$$

Thus $(D_1, D', \tilde{D}_{-1})$ is fitted to the Z-sequence β. Set $Q = \underline{T}/\xi$. At the sheaf level (8.50) gives on V an exact sequence of sheaves of $\mathcal{O}$-modules:

(8.53) $$0 \to (\underline{1}) \to \underline{T}\,|\,V \to Q\,|\,V \to 0\ .$$

Set $K_1 = K(D_1)$. Then from (7.26) it follows directly and immediately that

(8.54) $$\operatorname{Res}_\varphi(\xi, Z) = (\sqrt{-1}/(2\pi))^n \int_U \varphi(K_1\,|\,K')\ .$$

But (8.49) implies

(8.55) $$K_1 = 0\ .$$

Therefore

(8.56) $$\varphi(K_1\,|\,K') = \varphi(K')\ .$$

(8.56) and (8.54) imply (8.40), so the proof is complete.

Remark. Definition (7.1) of $\operatorname{Res}_\varphi(\xi, Z)$ requires choosing a resolution and then choosing connections for the vector-bundles in the resolution. Thus several vector-bundles and several connections are involved. The point of (8.39) is that it replaces this complicated situation involving several vector-bundles and several connections by a much simpler situation involving only one vector-bundle and one connection.

The next proposition will use (8.39) to explicitly compute $\operatorname{Res}_\varphi(\xi, p)$ when p is an isolated zero of X. But first, a lemma which will permit an application of the Lebesgue bounded convergence theorem.

(8.57) Lemma. *Let $g\colon [0, 1) \to [0, 1)$ be a nondecreasing C^∞ function with*

(8.58) $$g(r) = r\ , \qquad \textit{for } 0 \le r \le 1/3\ ,$$

(8.59) $$g(r) = 1\ , \qquad \textit{for } 2/3 \le r < 1\ .$$

For $m = 1, 2, \cdots$ define $g_m\colon [0, 1) \to [0, 1)$ by

(8.60) $$g_m(r) = \sqrt[m]{g(r^m)}\ , \qquad r \in [0, 1)\ .$$

Then there exists a positive real number b such that for all $r \in [0, 1)$ and all $m = 1, 2, \cdots$

(8.61) $$|(dg_m/dr)(r)| < b\ .$$

Proof. Choose a real number b such that

(8.62) $$3\,|(dg/dr)(r)| < b\ , \qquad \text{for all } r \in [0, 1)\ .$$

Then (8.61) will be implied by

(8.63) $$|(dg_m/dr)(r)| \le 3\,|(dg/dr)(r^m)|\ , \qquad \text{for all } r \in [0, 1)\ .$$

On $[0, \sqrt[m]{1/3}]$ $g_m(r) = r$. So on $[0, \sqrt[m]{1/3}]$,

(8.64) $1 = |(dg_m/dr)(r)| < 3|(dg/dr)(r^m)| = 3 , \quad r \in [0, \sqrt[m]{1/3}] .$

On $[\sqrt[m]{1/3}, 1)$ differentiation of g_m gives

$$(dg_m/dr)(r) = (dg/dr)(r^m)g_m(r)r^{m-1}/g(r^m) . \tag{8.65}$$

g is nondecreasing so $g(r^m) \geq 1/3$ for all $r \in [\sqrt[m]{1/3}, 1)$. Hence (8.65) implies

$$|(dg_m/dr)(r)| \leq 3|(dg/dr)(r^m)| , \quad r \in [\sqrt[m]{1/3}, 1) . \tag{8.66}$$

(8.66) and (8.64) combine to give (8.63). The lemma is proved.

(8.67) Proposition. *Let U be an open subset of M. On U let X be a holomorphic section of $T|U$. Assume that the zero set of X consists of one point p. Let $z_1, \cdots, z_n$ be a complex-analytic coordinate system with domain U and origin p. Denote by ξ the subsheaf of $\underline{T}|U$ spanned by X. Assume* $\deg \varphi = n$. *Then*

$$\operatorname{Res}_\varphi (\xi, p) = \operatorname{Res}_p \begin{bmatrix} \varphi(A)dz_1 \cdots dz_n \\ a_1, \cdots, a_n \end{bmatrix} , \tag{8.68}$$

where $X = \sum_{i=1}^n a_i\partial/\partial z_i$ and $A = \|\partial a_i/\partial z_j\|$.

Proof. Since p is an isolated zero of the a_i, there exist positive integers $\alpha_1, \cdots, \alpha_n$ such that $z_i^{\alpha_i}$ is in the ideal generated by $a_1, \cdots, a_n$. So there exist holomorphic functions b_{ij} defined about p with

$$z_i^{\alpha_i} = \sum_{j=1}^n b_{ij}a_j . \tag{8.69}$$

Passing to a smaller U, if necessary, it may be assumed that each b_{ij} is defined on all of U. Hence (8.69) holds throughout U.

Let $z\colon U \to \boldsymbol{C}^n$ be

$$z(x) = (z_1(x), \cdots, z_n(x)) , \quad x \in U . \tag{8.70}$$

In $\boldsymbol{C}^n$ denote by B_α the set

$$B_\alpha = \left\{(\zeta_1, \cdots, \zeta_n) \in \boldsymbol{C}^n \,\middle|\, \sum_{i=1}^n (\zeta_i\bar{\zeta}_i)^{\alpha_i} < 1\right\} . \tag{8.71}$$

We may assume

$$B_\alpha \subset z(U) . \tag{8.72}$$

For if $B_\alpha \not\subset z(U)$, then replace $z_1, \cdots, z_n$ by $bz_1, \cdots, bz_n$ where b is a large positive real number. Set $w_i = bz_i$. Then

$$X = \sum_{i=1}^n (ba_i)\partial/\partial w_i , \tag{8.73}$$

$$(8.74)\qquad \|\partial(ba_i)/\partial w_j\| = \|\partial a_i/\partial z_j\| ,$$

$$(8.75)\qquad \operatorname{Res}_p \begin{bmatrix} \varphi(A)dw_1 \cdots dw_n \\ ba_1, \cdots, ba_n \end{bmatrix} = \operatorname{Res}_p \begin{bmatrix} \varphi(A)dz_1 \cdots dz_n \\ a_1, \cdots, a_n \end{bmatrix} .$$

So it is legitimate to assume that (8.72) is valid.

In U let B denote the subset

$$(8.76)\qquad B = \{x \in U \mid z(x) \in B_a\} .$$

On B define a 1-form ω by

$$(8.77)\qquad \omega = \sum_{1 \le i,j \le n} (\bar{z}_i)^{a_i} b_{ij} dz_j .$$

On B let D be the connection for $T|B$ given by

$$(8.78)\qquad D(\partial/\partial z_i) = \omega \otimes [X, \partial/\partial z_i] , \qquad i = 1, \cdots, n .$$

Set $K = K(D)$. Then (8.68) will be proved if it can be shown that

$$(8.79)\qquad \operatorname{Res}_\varphi (\xi, p) = (\sqrt{-1}/(2\pi))^n \int_B \varphi(K) ,$$

$$(8.80)\qquad (\sqrt{-1}/(2\pi))^n \int_B \varphi(K) = \operatorname{Res}_p \begin{bmatrix} \varphi(A)dz_1 \cdots dz_n \\ a_1, \cdots, a_n \end{bmatrix} .$$

To prove (8.79) construct a sequence of connections $D_1, D_2, \cdots$ for $T|B$ as follows. On B denote by $(z, z)^a$ the function

$$(8.81)\qquad (z, z)^a = \sum_{i=1}^{n} (z_i \bar{z}_i)^{a_i} .$$

Let π be the 1-form on $B - \{p\}$ defined by

$$(8.82)\qquad \pi = \omega/(z, z)^a .$$

Then on $B - \{p\}$

$$(8.83)\qquad \pi \text{ is of type } (1, 0) ,$$

$$(8.84)\qquad i(X)\pi = 1 ,$$

Note that (8.84) is implied by (8.69). For $m = 1, 2, \cdots$, let $g_m : [0, 1) \to [0, 1)$ be as in (8.60). Define $\psi_m : B \to [0, 1)$ by

$$(8.85)\qquad \psi_m(x) = g_m((z, z)^a x) , \qquad x \in B .$$

Then on B take D_m to be the connection for $T|B$ such that

(8.86) $$D_m(\partial/\partial z_i) = \psi_m\pi \otimes [X, \partial/\partial z_i]\,, \qquad i = 1, \cdots, n\,.$$

Note that near p $\psi_m\pi$ agrees with ω, so D_m is well-defined on all of B. Set $K_m = K(D_m)$. (8.83), (8.84), and (8.59) imply

(8.87) D_m is fitted to X.

So by (8.39),

(8.88) $$\mathrm{Res}_\varphi\,(\xi, p) = (\sqrt{-1}/(2\pi))^n \int_B \varphi(K_m)\,.$$

Now for all $r \in [0, 1)$,

(8.89) $$\lim_{m\to\infty} g_m(r) = r\,.$$

Hence for all $x \in B$,

(8.90) $$\lim_{m\to\infty} \psi_m(x) = (z, z)^\alpha x\,.$$

So if $x \in B$ and $v \in T_x \oplus \overline{T}_x$, then

(8.91) $$\lim_{m\to\infty} i(v)\psi_m\pi = i(v)\omega\,.$$

Moreover, if $x \in B$ and $v_1, \cdots, v_{2n} \in T_x \oplus \overline{T}_x$, then

(8.92) $$\lim_{m\to\infty} i(v_1, \cdots, v_{2n})\varphi(K_m) = i(v_1, \cdots, v_{2n})\varphi(K)\,.$$

Due to (8.61) the Lebesgue bounded convergence theorem [17, Chapter V, Theorem D, p. 110] applies to give

(8.93) $$\lim_{m\to\infty} \int_B \varphi(K_m) = \int_B \varphi(K)\,.$$

(8.93) and (8.88) imply (8.79), so (8.79) has been proved.

To prove (8.80), let θ, κ denote respectively the connection and curvature matrices of D with respect to the frame $\partial/\partial z_1, \cdots, \partial/\partial z_n$. Then

(8.94) $$\theta = -\omega A\,.$$

From (8.94) it is clear that $\theta \wedge \theta = 0$, so $\kappa = d\theta$,

(8.95) $$\kappa = -(d\omega)A + (\omega)dA\,.$$

ω is of type (1,0), so by (8.95) each entry of κ is a sum of 2-forms of type (1,1) and type (2,0). In $\varphi(\kappa)$, which is of type (n, n), the terms of type (2,0) will play no role. Set $d\omega = d'\omega + d''\omega$ where

(8.96) $d'\omega$ is of type (2,0) , .

(8.97) $d''\omega$ is of type (1,1) .

Then

(8.98) $$\varphi(\kappa) = \varphi(-(d''\omega)A) ,$$

(8.99) $$d''\omega = \sum_{i,j} \alpha_i(\bar{z}_i)^{\alpha_i - 1} d\bar{z}_i b_{ij} dz_j .$$

Since φ is homogeneous of degree n, (8.98) implies

(8.100) $$\varphi(\kappa) = (-d''\omega)^n \varphi(A) .$$

Set $\Omega = dz_1 d\bar{z}_1 \cdots dz_n d\bar{z}_n$. A straightforward calculation from (8.99) shows

(8.101) $$(-d''\omega)^n = n! \alpha_1 \cdots \alpha_n (\bar{z}_1)^{\alpha_1 - 1} \cdots (\bar{z}_n)^{\alpha_n - 1} \det \| b_{ij} \| \, \Omega .$$

Therefore

(8.102) $$\varphi(\kappa) = n! \alpha_1 \cdots \alpha_n (\bar{z}_1)^{\alpha_1 - 1} \cdots (\bar{z}_n)^{\alpha_n - 1} \varphi(A) \det \| b_{ij} \| \, \Omega .$$

Now (8.72) implies

(8.103) $$\int_B (z_1\bar{z}_1)^{\alpha_1 - 1} \cdots (z_n\bar{z}_n)^{\alpha_n - 1} \Omega = (n! \alpha_1 \cdots \alpha_n)^{-1} (2\pi/\sqrt{-1})^n .$$

Also, if $\beta = (\beta_1, \cdots, \beta_n)$ is an n-tuple of nonnegative integers with $(\beta_1, \cdots, \beta_n) \neq (\alpha_1 - 1, \cdots, \alpha_n - 1)$, then (8.72) implies

(8.104) $$\int_B \bar{z}_1^{\alpha_1 - 1} z_1^{\beta_1} \cdots \bar{z}_n^{\alpha_n - 1} z_n^{\beta_n} \Omega = 0 .$$

Expand $\varphi(A) \det \| b_{ij} \|$ in a power series in $z_1, \cdots, z_n$. Denote by λ the coefficient of $z_1^{\alpha_1 - 1} \cdots z_n^{\alpha_n - 1}$. Then (8.102)–(8.104) imply

(8.105) $$\lambda = (\sqrt{-1}/(2\pi))^n \int_B \varphi(K) .$$

If (8.105) is compared to the algorithm for computing $\operatorname{Res}_p \begin{bmatrix} \varphi(A) dz_1 \cdots dz_n \\ a_1, \cdots, a_n \end{bmatrix}$ given by (0.9), then it is evident that (8.80) has been proved.

The proof of the proposition is complete.

From (0.2) and (0.6) it is clear that (8.68) implies (8.1). Theorem 1 is proved.

9. Proof of Theorem 3

Let U be an open subset of M. On U, let ξ be a full integrable subsheaf of $\underline{T}|U$. Set $Z = \{x \in U | T_x/\xi_x$ is not a free $\mathcal{O}_x$-module$\}$. Assume that Z is compact and connected. Assume also that (0.27) and (0.28) are valid for Z. Throughout this section $\deg\varphi = n - k + 1$.

Let $Z_1, \cdots, Z_s$ be the irreducible complex-analytic components of Z of dimension $k - 1$. If $[Z_i]$ denotes the element of $H_{2k-2}(Z; C)$ given by the fundamental cycle of Z_i, then $[Z_1], \cdots, [Z_s]$ is a vector-space basis for $H_{2k-2}(Z; C)$. Hence there exist complex numbers $\lambda_1, \cdots, \lambda_s$ with

(9.1) $$\mathrm{Res}_\varphi(\xi, Z) = \sum_{i=1}^{s} \lambda_i[Z_i] \, .$$

In order to prove (0.42) we must compute $\lambda_1, \cdots, \lambda_s$.

Let V be an open subset of U such that

(9.2) V contains Z, and Z is a deformation retract of V,

(9.3) on V there is an exact sequence

$$0 \to \underline{\underline{E}}_q \to \underline{\underline{E}}_{q-1} \to \cdots \to \underline{\underline{E}}_1 \to \mathcal{A} \underset{\mathcal{O}}{\otimes} \xi \to 0$$

of sheaves of $\mathcal{A}$-modules.

On V there is the short exact sequence

(9.4) $$0 \to \mathcal{A} \underset{\mathcal{O}}{\otimes} \xi \to \underline{\underline{T}}|V \to \mathcal{A} \underset{\mathcal{O}}{\otimes} Q \to 0 \, .$$

Combining (9.3) and (9.4) gives

(9.5) $$0 \to \underline{\underline{E}}_q \to \underline{\underline{E}}_{q-1} \to \cdots \to \underline{\underline{E}}_1 \to \underline{\underline{T}}|V \to \mathcal{A} \underset{\mathcal{O}}{\otimes} Q \to 0 \, .$$

Denote by β the Z-sequence resulting from (9.5). On V, let $D_q, D_{q-1}, \cdots, D_0, D_{-1}$ be connections for $E_q, E_{q-1}, \cdots, E_1, T|V, \nu$ such that

(9.6) $$(D_q, D_{q-1}, \cdots, D_0, D_{-1}) \quad \text{is fitted to } \beta.$$

Set $K_i = K(D_i)$. As in (5.50) a closed j-form ω on Z with compact support determines an element of $H_{2n-j}(Z; C)$. (9.5) is exact, so by (7.18)

(9.7) $$(\sqrt{-1}/(2\pi))^{n-k+1}\varphi(K_q | K_{q-1} | \cdots | K_0) \quad \text{determines } \mathrm{Res}_\varphi(\xi, Z) \, .$$

Since (0.27) and (0.28) are valid for Z, Theorem (0.30) applies. Let $p \in Z_i - (Z_i \cap (Z^{(2)} \cup N))$. Let U_p be an open neighborhood of p in V such that on U_p there are defined a complex-analytic coordinate system $z_1, \cdots, z_n$ and holomorphic functions $a_k, \cdots, a_n$ as in (0.31)–(0.35). Define a holomorphic normal disc D_p by

(9.8) $$\{D_p = x \in U_p \mid z_1(x) = z_1(p), \cdots, z_{k-1}(x) = z_{k-1}(p)\} .$$

Let $i\colon D_p \to V$ be the inclusion. It may be assumed that V and U_p have been chosen so that i is proper. Hence there is the induced map of cohomology with compact supports

(9.9) $$i^*\colon H_c^*(V; C) \to H_c^*(D_p; C) .$$

Consider the homomorphism $I_p\colon H_{2k-2}(Z; C) \to C$ given by

(9.10) $$H_{2k-2}(Z; C) \cong H_c^{2n-2k+2}(V, C) \to H_c^{2n-2k+2}(D_p; C) \cong C .$$

(9.1) implies

(9.11) $$I_p(\mathrm{Res}_\varphi(\xi, Z)) = \lambda_i .$$

But then (9.7) implies

(9.12) $$\lambda_i = (\sqrt{-1}/(2\pi))^{n-k+1} \int_{D_p} i^*\varphi(K_q \mid K_{q-1} \mid \cdots \mid K_0) .$$

On D_p let A be the $(n-k+1) \times (n-k+1)$ matrix

(9.13) $$A = \|\partial a_i/\partial z_j\| , \qquad k \le i, j \le n .$$

Due to (9.12), (0.42) will be implied by

(9.14) $$(\sqrt{-1}/(2\pi))^{n-k+1} \int_{D_p} i^*\varphi(K_q \mid K_{q-1} \mid \cdots \mid K_0) = \mathrm{Res}_p \begin{bmatrix} \varphi(A)dz_k \cdots dz_n \\ a_k, \cdots, a_n \end{bmatrix} .$$

To prove (9.14), observe that D_p is itself a complex manifold with

(9.15) $$\dim_C D_p = n - k + 1 .$$

Let $T(D_p)$ denote the holomorphic tangent bundle of D_p. On D_p set

(9.16) $\underline{T}(D_p) =$ sheaf of germs of holomorphic sections of $T(D_p)$,

(9.17) $$\dot{X} = \sum_{i=k}^{n} a_i \partial/\partial z_i ,$$

(9.18) $$\dot{\xi} = \text{subsheaf of } \underline{T}(D_p) \text{ spanned by } \dot{X} .$$

Then according to (8.68),

(9.19) $$\mathrm{Res}_\varphi(\dot{\xi}, p) = \mathrm{Res}_p \begin{bmatrix} \varphi(A)dz_k \cdots dz_n \\ a_k, \cdots, a_n \end{bmatrix} .$$

So (9.14) will follow from

(9.20) $$\mathrm{Res}_\varphi(\dot{\xi}, p) = (\sqrt{-1}/(2\pi))^{n-k+1} \int_{D_p} i^*\varphi(K_q | K_{q-1} | \cdots | K_0) .$$

To prove (9.20), on D_p set

(9.21) $\dot{\mathcal{O}}$ = sheaf of germs of holomorphic functions,

(9.22) $\dot{\mathscr{A}}$ = sheaf of germs of real-analytic functions,

(9.23) $$\dot{\underline{Q}} = \underline{T}(D_p)/\dot{\xi} ,$$

(9.24) $$\dot{E}_j = i^!(E_j) , \quad j = q, q-1, \cdots, 1 ,$$

(9.25) $$\dot{T} = i^!(T) .$$

Thus $\dot{\underline{\underline{E}}}_j$ is a sheaf of $\dot{\mathscr{A}}$ modules on D_p. Now use (9.5) to construct on D_p an exact sequence

(9.26) $$0 \to \dot{\underline{\underline{E}}}_q \to \dot{\underline{\underline{E}}}_{q-1} \to \cdots \to \dot{\underline{\underline{E}}}_1 \to \dot{\underline{\underline{T}}} \to \dot{\mathscr{A}} \otimes_{\dot{\mathcal{O}}} \dot{\underline{Q}} \to 0$$

of sheaves of $\dot{\mathscr{A}}$-modules.

The sequence (9.26) is obtained by first noting that (0.35) implies

(9.27) ξ_x is a free $\mathcal{O}_x$-module for all $x \in U_p$.

Let E be the unique holomorphic vector bundle on U_p with

(9.28) $$\underline{E} = \xi | U_p .$$

Then (9.3) gives on U_p an exact sequence of vector-bundles

(9.29) $$0 \to E_q \to E_{q-1} \to \cdots \to E_1 \to E \to 0 .$$

Set $\dot{E} = i^!(E)$. Applying $i^!$ to (9.29) gives on D_p an exact sequence of vector-bundles

(9.30) $$0 \to \dot{E}_q \to \dot{E}_{q-1} \to \cdots \to \dot{E}_1 \to \dot{E} \to 0 .$$

So on D_p the sequence

(9.31) $$0 \to \dot{\underline{\underline{E}}}_q \to \dot{\underline{\underline{E}}}_{q-1} \to \cdots \to \dot{\underline{\underline{E}}}_1 \to \dot{\underline{\underline{E}}} \to 0$$

is exact.

On U_p the inclusion $\xi | U_p \subset \underline{T} | U_p$ gives a vector-bundle map $\eta : E \to T | U_p$ such that

(9.32) on U_p there is a holomorphic frame $e_1, \cdots, e_k$ of E with

$$\eta e_i = \partial/\partial z_i \text{ for } i = 1, \cdots, k-1 \text{ and } \eta e_k = \sum_{i=k}^{n} a_i \partial/\partial z_i ,$$

(9.33) $$0 \to \underline{E} \to \underline{T} \mid U_p \to Q \mid U_p \to 0 \quad \text{is exact}.$$

Restricting η to D_p gives $\dot{E} \to \dot{T}$ and thus gives a map of $\dot{\mathcal{O}}$-modules $\underline{\dot{E}} \to \underline{\dot{T}}$. On D_p let J be the holomorphic sub-vector-bundle of $\dot{T}$ spanned by $\partial/\partial z_1, \cdots, \partial/\partial z_{k-1}$. Then

(9.34) $$\dot{T} = J \oplus T(D_p) .$$

This direct sum decomposition gives a projection $\dot{T} \to T(D_p)$. Map $\underline{\dot{T}}$ to $\dot{Q}$ by

(9.35) $$\underline{\dot{T}} \to \underline{T}(D_p) \to \dot{Q} .$$

Then (9.32) and (9.23) imply

(9.36) $$0 \to \underline{\dot{E}} \to \underline{\dot{T}} \to \dot{Q} \to 0 \quad \text{is exact.}$$

Hence by (6.1)

(9.37) $$0 \to \underline{\underline{\dot{E}}} \to \underline{\underline{\dot{T}}} \to \dot{\mathscr{A}} \otimes_{\dot{\mathcal{O}}} \dot{Q} \to 0 \quad \text{is exact.}$$

Now (9.26) is the exact sequence obtained by combining (9.37) and (9.31).

On D_p let $\dot{\beta}$ be the p-sequence resulting from (9.26). Set $\dot{D}_j = i^!(D_j)$. (9.26) implies

(9.38) $$(\dot{D}_q, \dot{D}_{q-1}, \cdots, \dot{D}_0, \dot{D}_{-1}) \quad \text{is fitted to } \dot{\beta}.$$

Set $K_j = K(D_j)$. Since (9.26) is exact, (7.18) implies

(9.39) $$\operatorname{Res}_\varphi(\dot{\xi}, p) = (\sqrt{-1}/(2\pi))^{n-k+1} \int_{D_p} \varphi(\dot{K}_q \mid \dot{K}_{q-1} \mid \cdots \mid \dot{K}_0) .$$

But

(9.40) $$\varphi(\dot{K}_q \mid \dot{K}_{q-1} \mid \cdots \mid \dot{K}_0) = i^*\varphi(K_q \mid K_{q-1} \mid \cdots \mid K_0) .$$

So (9.20) has been proved. This concludes the proof of Theorem 3.

Remark. The argument of this section really verifies a very special case of the functoriality of $\operatorname{Res}_\varphi(\xi, Z)$.

10. Proof of the rigidity theorem

Let F be a holomorphic integrable sub-vector-bundle of T, $k = \dim_C F_x$, and $\nu = T/F$. If U is an open subset of M, let $A(U)$ denote the ring of all C^∞ complex-valued differential forms on U. In $A(U)$ let $I(F, U)$ be the ideal generated by all C^∞ 1-forms ω on U such that

(10.1) ω is of type $(1, 0)$,

(10.2) $i(\gamma)\omega = 0 \quad$ for every $\gamma \in C^\infty(F \mid U)$.

(10.3) Lemma. *Let D be a basic connection for ν. On U let $e_1, \cdots, e_{n-k}$ be a C^∞ frame of ν. Let $\kappa = \|\kappa_{ij}\|$ be the curvature matrix of D with respect to the frame $e_1, \cdots, e_{n-k}$. Then for each κ_{ij},*

(10.4) $$\kappa_{ij} \in I(F, U) .$$

Proof. Let $\eta: T \to T/F$ be the projection. Given $p \in U$, let U_p be an open neighborhood of p in U such that on U_p there is a complex-analytic coordinate system $z_1, \cdots, z_n$ with

(10.5) $$\partial/\partial z_1, \cdots, \partial/\partial z_k \in \Gamma(F \mid U_p) .$$

Let $\kappa' = \|\kappa'_{ij}\|$ be the curvature matrix of D with respect to the frame $\eta\partial/\partial z_{k+1}, \cdots, \eta\partial/\partial z_n$. Then according to (3.33) each κ'_{ij} is in $I(F, U)$:

(10.6) $$\kappa'_{ij} \in I(F, U) .$$

(10.4) is now implied by (1.16), and the lemma is proved.

Next, let U be open in M, and $[a, b]$ a closed interval of real numbers. Set $\tilde{U} = U \times [a, b]$, and let $\rho: \tilde{U} \to U$, $t: \tilde{U} \to [a, b]$ be the projections. For each $r \in [a, b]$ define $i_r: U \to \tilde{U}$ by

(10.7) $$i_r(x) = (x, r) , \qquad x \in U,\ r \in [a, b] .$$

(10.8) Definition. A C^∞ 1-parameter family of holomorphic foliations of U is a 1-parameter family $\{F_r\}$, $a \leq r \leq b$, such that

(10.9) for each $r \in [a, b]$, F_r is a holomorphic integrable sub-vector-bundle of $T \mid U$,

(10.10) on $\tilde{U}$ there exists a C^∞ sub-vector-bundle $\tilde{F}$ of $\rho^!(T \mid U)$ with $i_r^!(\tilde{F}) = F_r$ for each $r \in [a, b]$.

(10.11) Lemma. *Let $\{F_r\}$, $a \leq r \leq b$, be a C^∞ 1-parameter family of holomorphic foliations of U. On $\tilde{U}$ let D be a connection for $\rho^!(T)/\tilde{F}$ such that*

(10.12) *for each $r \in [a, b]$, $i_r^!(D)$ is a basic connection for T/F_r.*

Set $K = K(D)$, and assume $n - k + 1 < \deg \varphi \leq n$. Then

(10.13) $$\varphi(K) = 0 .$$

Proof. Given $p \in U$, let U_p be an open neighborhood of p in U such that

p is a deformation retract of U_p. Set $\tilde{U}_p = U_p \times [a, b]$. In $A(\tilde{U}_p)$ let I be the ideal

(10.14) $$I = \{\omega \in A(\tilde{U}_p) \mid \text{For each } r \in [a, b],\ i_r^*(\omega) \in I(F_r, U)\} .$$

On $\tilde{U}_p$ let $u_1, \cdots, u_{n-k}$ be a C^∞ frame of $\rho^!(T)/\tilde{F}$. Let $\kappa = \|\kappa_{ij}\|$ be the curvature matrix of D with respect to $u_1, \cdots, u_{n-k}$. Then (10.12), (10.4), and (4.56) imply that each κ_{ij} is in I:

(10.15) $$\kappa_{ij} \in I .$$

Hence (10.13) will follow from

(10.16) if $\omega_1, \cdots, \omega_{n-k+2}$ are any $n-k+2$ elements of I, then $\omega_1 \wedge \cdots \wedge \omega_{n-k+2} = 0$.

To prove (10.16) let $T_R\tilde{U}$ and $T_R[a, b]$ be the C^∞ tangent bundles of $\tilde{U}$ and $[a, b]$. Define $T_C\tilde{U}$, $T_C[a, b]$ by

(10.17) $$T_C\tilde{U} = C \otimes_R T_R\tilde{U} ,$$

(10.18) $$T_C[a, b] = C \otimes_R T_R[a, b] .$$

Then

(10.19) $$T_C\tilde{U} = \rho^!(T) \oplus \rho^!(\bar{T}) \oplus t^! T_C[a, b] .$$

On $\tilde{U}_p$ let $v_1, \cdots, v_{2n+1}$ be a C^∞ frame of $T_C\tilde{U}$ such that

(10.20) $$v_1, \cdots, v_k \in C^\infty(\tilde{F} \mid \tilde{U}_p) ,$$

(10.21) $$v_{k+1}, \cdots, v_n \in C^\infty(\rho^!(T) \mid \tilde{U}_p) ,$$

(10.22) $$v_{n+1}, \cdots, v_{2n} \in C^\infty(\rho^!(\bar{T}) \mid \tilde{U}_p) ,$$

(10.23) $$v_{2n+1} \in C^\infty(t^! T_C[a, b] \mid \tilde{U}_p) .$$

Let $v_1^*, \cdots, v_{2n+1}^*$ denote the dual frame for the dual bundle $(T_C\tilde{U})^* \mid \tilde{U}_p$. Then (10.14) implies that I is the ideal in $A(\tilde{U}_p)$ generated by $v_{k+1}^*, \cdots, v_n^*$ and v_{2n+1}^*. Since there are $n-k+1$ of these, (10.16) is clear. This completes the proof.

Remark. If $\{F_r\}$ is as in Lemma (10.11), then there always exist connections D for $\rho^!(T)/\tilde{F}$ such that (10.12) is valid for D. To see this, set $\tilde{\nu} = \rho^!(T)/\tilde{F}$ and $\nu_r = T/F_r$. As in Proposition (3.21) for each ν_r there is a partial connection

(10.24) $$\delta_r \colon C^\infty(\nu_r) \to C^\infty((F_r \oplus \bar{T})^* \otimes \nu_r) .$$

These δ_r fit together to give a partial connection δ for $\tilde{\nu}$:

$$(10.25)\qquad \delta: C^\infty(\mathfrak{v}) \to C^\infty((\tilde{F} \oplus \rho^!\tilde{T})^* \otimes \mathfrak{v}) .$$

A connection D for $\mathfrak{v}$ which extends this δ will satisfy (10.12).

(10.26) Definition. On U let ξ be a full integrable subsheaf of $\underline{T} \mid U$. Let $E_q, E_{q-1}, \cdots, E_1$ be real-analytic vector-bundles on U such that there is an exact sequence of sheaves of $\mathscr{A}$-modules

$$(10.27)\qquad 0 \to \underline{\underline{E}}_q \to \underline{\underline{E}}_{q-1} \to \cdots \to \underline{\underline{E}}_1 \to \mathscr{A} \underset{\mathscr{O}}{\otimes} \xi \to 0$$

on U. From (10.27) a complex

$$(10.28)\qquad 0 \to E_q \to E_{q-1} \to \cdots \to E_1 \to T \mid U$$

of real-analytic vector-bundles on U is obtained. By viewing each E_i and T as C^∞ vector-bundles, (10.28) may then be taken to be a complex of C^∞ vector-bundles on U. Any complex of C^∞ vector-bundles on U which arises in this way will be referred to as *a complex for* ξ.

Remark. Up to this point we have not precisely defined a C^∞ 1-parameter family of sheaves. This is made precise by

(10.29) Definition. A C^∞ *1-parameter family* $\{\xi_r\}$, $a \leq r \leq b$, of full integrable subsheaves of $\underline{T} \mid U$ is a 1-parameter family such that

(10.30) for each $r \in [a, b]$, ξ_r is a full integrable subsheaf of $\underline{T} \mid U$,

(10.31) on $\tilde{U} = U \times [a, b]$ there exists a complex

$$0 \to E_q \to E_{q-1} \to \cdots \to E_0 \to \rho^!(T) \to 0$$

of C^∞ vector-bundles such that for each $r \in [a, b]$,

$$0 \to i_r^!(E_q) \to i_r^!(E_{q-1}) \to \cdots \to i_r^!(E_1) \to T \mid U$$

is a complex for ξ_r.

Proof of Theorem 4. Let $\{\xi_r\}$, $a \leq r \leq b$, be a C^∞ 1-parameter family of full integrable subsheaves of $\underline{T} \mid U$. For $r \in [a, b]$, let $Z_r = \{x \in U \mid (\underline{T}/\xi_r)_x$ is not a free $\mathscr{O}_x$-module$\}$. Assume that each Z_r is compact and connected. As in (0.43) assume that there is a compact subset B of U with

$$(10.32)\qquad Z_r \subset B \qquad \text{for all } r \in [a, b] .$$

Let $i_*: H_*(Z_r; C) \to H_*(U; C)$ be the homology map induced by the inclusion of Z_r in U. If $n - k + 1 < \deg \varphi \leq n$, we then wish to prove

$$(10.33)\qquad i_* \operatorname{Res}_\varphi(\xi_a, Z_a) = i_* \operatorname{Res}_\varphi(\xi_b, Z_b) .$$

To prove (10.33) set $\tilde{U} = U \times [a, b]$, and on $\tilde{U}$ let

$$(10.34)\qquad 0 \to E_q \to E_{q-1} \to \cdots \to E_1 \to \rho^!(T)$$

be as in (10.31). On $U - Z_r$ let F_r be the unique holomorphic sub-vector-bundle of $T|U - Z_r$ such that

(10.35) $$\underline{F}_r = \xi_r | U - Z_r \, .$$

On $U - Z_r$, set $\nu_r = T/F_r$. Let V_r be an open subset of U with Z_r contained in V_r and Z_r a deformation retract of V_r. Then

(10.36) when restricted to V_r,

$$0 \to i_r^!(E_q) \to i_r^!(E_{q-1}) \to \cdots \to i_r^!(E_1) \to T \to \nu_r \to 0$$

is a Z_r-sequence.

Set $\tilde{Z} = \{(x, r) \in \tilde{U} \,|\, x \in Z_r\}$. On $\tilde{U} - \tilde{Z}$, let $\tilde{F}$ be the unique C^∞ sub-vector-bundle of $\rho^!(T)$ such that

(10.37) $$i_r^!(\tilde{F}) = F_r \qquad \text{for each } r \in [a, b] \, .$$

On $\tilde{U} - \tilde{Z}$, set $\tilde{\nu} = \rho^!(T)/\tilde{F}$. Let D_{-1} be a connection for $\tilde{\nu}$ with

(10.38) $i_r^!(D_{-1})$ is a basic connection for ν_r for each $r \in [a, b]$.

With B as in (10.32) choose a compact subset Σ of U with B contained in the interior of Σ. Set $\tilde{\Sigma} = \Sigma \times [a, b]$. On $\tilde{U}$ let $D_q, D_{q-1}, \cdots, D_1, D_0$ be connections for $E_q, E_{q-1}, \cdots, E_1, \rho^!(T)$ such that

(10.39) on $\tilde{U} - \tilde{\Sigma}$, $(D_q, D_{q-1}, \cdots, D_0, D_{-1})$ is compatible with the exact sequence

$$0 \to E_q \to E_{q-1} \to \cdots \to E_1 \to \rho^!(T) \to \tilde{\nu} \to 0 \, .$$

Set $K_i = K(D_i)$. Then according to (4.23),

(10.40) $$\text{On} \quad \tilde{U} - \tilde{\Sigma}, \varphi(K_q | K_{q-1} | \cdots | K_0) = \varphi(K_{-1}) \, .$$

Since $n - k + 1 < \deg \varphi \leq n$, (10.38) and (10.13) imply

(10.41) $$\varphi(K_{-1}) \quad \text{vanishes on} \quad \tilde{U} - \tilde{\Sigma} \, .$$

Hence

(10.42) $\varphi(K_q | K_{q-1} | \cdots | K_0)$ is a closed form on $\tilde{U}$ with compact support.

Set $D_i^r = i_r^!(D_i)$, $K_i^r = K(D_i^r)$. Let $l = \deg \varphi$. On U there is the Poincaré duality isomorphism:

(10.43) $$\alpha\colon H_{2n-2l}(U; C) \to H_c^{2l}(U; C) \, .$$

(10.36)–(10.39) imply

(10.44) for each $r \in [a, b]$, $\varphi(K_q^r | K_{q-1}^r | \cdots | K_0^r)$ is a $2l$-form on U with compact support,

(10.45) the element of $H_c^{2l}(U; C)$ given by $(\sqrt{-1}/(2\pi))^l \varphi(K_q^r | K_{q-1}^r | \cdots | K_0^r)$ is $\alpha i_* \operatorname{Res}_\varphi (\xi_r, Z_r)$.

Since α is an isomorphism, (10.33) will be proved if it can be shown that $\varphi(K_q^a | K_{q-1}^a | \cdots | K_0^a)$ and $\varphi(K_q^b | K_{q-1}^b | \cdots | K_0^b)$ give the same element of $H_c^{2l}(U)$. But (4.56) implies

(10.46) $$i_r^* \varphi(K_q | K_{q-1} | \cdots | K_0) = \varphi(K_q^r | K_{q-1}^r | \cdots | K_0^r) \qquad \text{for each } r \in [a, b] .$$

From (10.46) and (10.42) it is clear that the proof is complete.

Proof of Corollary 0.44. Let Z be as in Corollary (0.45). Choose an open subset V of U with Z contained in V and Z a deformation retract of V. Let $i_*: H_*(Z; C) \to H_*(V; C)$ be the homology map induced by the inclusion of Z in V. Then according to (0.43),

(10.47) $$i_* \operatorname{Res}_\varphi (\xi_a, Z) = i_* \operatorname{Res}_\varphi (\xi_b, Z) .$$

Since $i_*: H_*(Z; C) \to H_*(V; C)$ is an isomorphism, (10.47) implies

(10.48) $$\operatorname{Res}_\varphi (\xi_a, Z) = \operatorname{Res}_\varphi (\xi_b, Z) .$$

This proves the corollary.

11. Examples

Example 1. Let $\lambda_1, \cdots, \lambda_n$ be nonzero complex numbers. On C^n, with its usual coordinate system, let X be the homomorphic vector-field:

(11.1) $$X = \sum_{i=1}^{n} \lambda_i z_i \partial / \partial z_i .$$

The origin is the only zero of X. Let ξ be the subsheaf of $\underline{T}$ spanned by X. Assume $\deg \varphi = n$. Identify, as usual, $H_0(0, C) = C$. Then

(11.2) $$\operatorname{Res}_\varphi (\xi, 0) = \varphi(\lambda_1, \lambda_2, \cdots, \lambda_n)/(\lambda_1 \lambda_2 \cdots \lambda_n) .$$

Example 2. Let $a_0, \cdots, a_n$ be $n + 1$ distinct complex numbers. Define a holomorphic flow

(11.3) $$C \times CP^n \to CP^n$$

by

(11.4) $$(z, [z_0 : z_1 : \cdots : z_n]) \to [e^{a_0 z} z_0 : e^{a_1 z} z_1 : \cdots : e^{a_n z} z_n] .$$

Let X be the holomorphic vector field on CP^n, which generates this flow. The zeroes of X are the $n+1$ points $p_0, p_1, \cdots, p_n$ where

$$\begin{aligned} p_0 &= [1:0:0:\cdots:0]\,, \\ p_1 &= [0:1:0:\cdots:0]\,, \\ &\vdots \\ p_n &= [0:0:0:\cdots:1]\,. \end{aligned}$$

Each p_i is a non-degenerate zero of X. Let ξ be the subsheaf of $\underline{T}$ spanned by X. Identify $H_0(p_i, C) = C$, and assume $\deg \varphi = n$. Then

$$\operatorname{Res}_\varphi(\xi, p_i) = \frac{\varphi(a_0 - a_i, a_1 - a_i, \cdots, a_{i-1} - a_i, a_{i+1} - a_i, \cdots, a_n - a_i)}{(a_0 - a_i)(a_1 - a_i)\cdots(a_{i-1} - a_i)(a_{i+1} - a_i)\cdots(a_n - a_i)}\,. \tag{11.5}$$

Example 3. Fix integers k and n with $1 < k < n$. Let A be a $k \times (n+1)$ matrix of complex numbers

$$A = \begin{bmatrix} a_{10} & a_{11} & \cdots & a_{1n} \\ a_{20} & a_{21} & \cdots & a_{2n} \\ \vdots & & & \vdots \\ a_{k0} & a_{k1} & \cdots & a_{kn} \end{bmatrix}. \tag{11.6}$$

For each $i = 0, 1, \cdots, n$ denote by A_i the $k \times n$ matrix obtained by subtracting the i-th column of A from all the other columns of A

$$A_i = \begin{bmatrix} a_{10} - a_{1i} & a_{11} - a_{1i} & \cdots & a_{1n} - a_{1i} \\ a_{20} - a_{2i} & a_{21} - a_{2i} & \cdots & a_{2n} - a_{2i} \\ \vdots & \vdots & & \vdots \\ a_{k0} - a_{ki} & a_{k1} - a_{ki} & \cdots & a_{kn} - a_{ki} \end{bmatrix}. \tag{11.7}$$

Assume

(11.8) for each $i = 0, 1, \cdots, n$ all the $k \times k$ sub-matrices of A_i are non-singular.

The set of all matrices A for which (11.8) is valid is open and dense in the vector-space of all $k \times (n+1)$ matrices of complex numbers.

So given A satisfying (11.8) let V_i be the holomorphic vector field on CP^n which generates the flow

$$(z, [z_0 : z_1 : \cdots : z_n]) \to (e^{a_{i0}z}z_0 : e^{a_{i1}z}z_1 : \cdots : e^{a_{in}z}z_n)\,. \tag{11.9}$$

Let ξ be the subsheaf of $\underline{T}$ spanned by $V_1, \cdots, V_k$. ξ is integrable and full.

To describe the singular set of ξ, let $\alpha = (\alpha_1, \cdots, \alpha_{n-k+1})$ be an $(n-k+1)$-tuple of integers with

$$0 \leq \alpha_1 < \cdots < \alpha_{n-k+1} \leq n \,. \tag{11.10}$$

Define CP^{k-1}_α by

$$CP^{k-1}_\alpha = \{[z_0 : z_1 : \cdots : z_n] \in CP^n \,|\, 0 = z_{\alpha_1} = \cdots = z_{\alpha_{n-k+1}}\} \,. \tag{11.11}$$

The singular set Z of ξ is

$$Z = \bigcup CP^{k-1}_\alpha \,, \tag{11.12}$$

where the union is taken over all α satisfying (11.10). The CP^{k-1}_α are the irreducible complex-analytic components of Z. (0.27) and (0.28) are valid for Z.

If $\deg \varphi = n - k + 1$, then Theorem 3 applies. Hence according to (0.42)

$$\mathrm{Res}_\varphi\,(\xi, Z) = \sum_\alpha \#(\varphi, \xi, CP^{k-1}_\alpha)[CP^{k-1}_\alpha] \,. \tag{11.13}$$

$\#(\varphi, \xi, CP^{k-1}_\alpha)$ can be explicitly computed as follows. First, shuffle the columns of A to obtain a new matrix A_α whose first $n-k+1$ columns are the α_1-th, $\cdots, \alpha_{n-k+1}$-th columns of A. Next, form a $k \times n$ matrix B_α by subtracting the last column of A_α from all the other columns of A_α. Define complex numbers $\lambda^\alpha_1, \cdots, \lambda^\alpha_{n-k+1}$ by letting λ^α_i be the determinant of the $k \times k$ sub-matrix of B_α consisting of the i-th column of B_α and the last $k-1$ columns of B_α. Then by (0.37):

$$\#(\varphi, \xi, CP^{k-1}_\alpha) = \varphi(\lambda^\alpha_1, \cdots, \lambda^\alpha_{n-k+1}, 0, \cdots, 0)/(\lambda^\alpha_1 \cdots \lambda^\alpha_{n-k+1}) \,. \tag{11.14}$$

Combining (11.14) and (11.13) gives

$$\mathrm{Res}_\varphi\,(\xi, Z) = \sum_\alpha \varphi(\lambda^\alpha_1, \cdots, \lambda^\alpha_{n-k+1}, 0, \cdots, 0)/(\lambda^\alpha_1 \cdots \lambda^\alpha_{n-k+1})[CP^{k-1}_\alpha] \,. \tag{11.15}$$

If $n - k + 1 < \deg \varphi \leq n$, then the situation is quite different. In this case, set $l = \deg \varphi$ and let $x \in H^2(CP^n; C)$ be the element of $H^2(CP^n; C)$ dual to a hyperplane. Define a complex number $w(\varphi)$ by

$$\varphi(T) = w(\varphi)x^l \,. \tag{11.16}$$

Choose an l-tuple $\beta = (\beta_1, \cdots, \beta_l)$ of integers with

$$0 \leq \beta_1 < \cdots < \beta_l \leq n \,. \tag{11.17}$$

Define CP^{n-l}_β by

(11.18) $CP^{n-l}_\beta = \{[z_0 : z_1 : \cdots : z_n] \in CP^n \mid 0 = z_{\beta_1} = \cdots = z_{\beta_l}\}$.

Denote by $[CP^{n-l}]$ the element of $H_{2n-2l}(Z, C)$ given by the fundamental cycle of CP^{n-l}_β. $[CP^{n-l}]$ does not depend on the choice of β. Then

(11.19) $\mathrm{Res}_\varphi(\xi, Z) = w(\varphi)[CP^{n-l}]$, $\quad n - k + 1 < \deg \varphi \le n$.

Note how (11.15) and (11.19) illustrate the rigidity theorem. If A is varied, then the right-hand side of (11.15) varies, but that of (11.19) remains constant.

Example 4. Fix integers k and n with $1 < k < n$. Let Z be a compact connected complex-analytic manifold with

(11.20) $$\dim_C Z = k - 1 .$$

Set $r = n - k + 1$. Let L be a holomorphic line bundle on Z. Choose nonzero integers $n_1, \cdots, n_r$. Let M be the total space of the vector-bundle $L^{n_1} \oplus \cdots \oplus L^{n_r}$. Here L^{n_i} denotes the tensor product of L with itself n_i times. Then M is a complex manifold with

(11.21) $$\dim_C M = n .$$

Let $\pi : L^{n_1} \oplus \cdots \oplus L^{n_r} \to Z$ be the projection. The zero section of the vector-bundle gives an inclusion $Z \subset M$.

Choose a cover $\{U_\alpha\}_{\alpha \in I}$ of Z by open sets with:

(11.22) U_α is the domain of a complex-analytic coordinate system $w_1^\alpha, \cdots, w_{k-1}^\alpha$,

(11.23) on U_α there is a holomorphic section s_α of $L \mid U_\alpha$ such that s_α has no zeroes.

Let $s_\alpha^{n_i}$ denote the tensor product of s_α with itself n_i times. Then $s_\alpha^{n_1}, \cdots, s_\alpha^{n_r}$ is a holomorphic frame of $L^{n_1} \oplus \cdots \oplus L^{n_r}$ on U_α.

Set $\tilde{U}_\alpha = \pi^{-1}(U_\alpha)$. On $\tilde{U}_\alpha$ let $z_1^\alpha, \cdots, z_n^\alpha$ be the coordinate system resulting from $w_1^\alpha, \cdots, w_{k-1}^\alpha$ and $s_\alpha^{n_1}, \cdots, s_\alpha^{n_r}$. Thus if $\tilde{v} \in U_\alpha$, then

(11.24) $$z_i^\alpha(v) = w_i^\alpha(\pi v) , \quad i = 1, \cdots, k - 1 ,$$

(11.25) $$v = \sum_{i=1}^r z_{i+k-1}^\alpha(v) s_\alpha^{n_i}(\pi v) .$$

On $\tilde{U}_\alpha$ let ξ_α be the subsheaf of $\underline{T} \mid \tilde{U}_\alpha$ spanned by $\partial/\partial z_1^\alpha, \cdots, \partial/\partial z_{k-1}^\alpha$, $\sum_{i=1}^r n_i z_{i+k-1}^\alpha \partial/\partial z_{i+k-1}^\alpha$. Then

(11.26) $$\xi_\alpha \mid U_\alpha \cap U_\beta = \xi_\beta \mid U_\alpha \cap U_\beta .$$

So the $\{\xi_\alpha\}_{\alpha \in I}$ fit together to form a subsheaf ξ of $\underline{T}$, which is integrable and full. Also, (11.26) implies that ξ does not depend on the choice of the cover $\{U_\alpha\}_{\alpha \in I}$. The singular set of ξ is Z.

Set $x = c_1(L)$, so that $x \in H^2(Z; C)$. Then for any φ with $n - k + 1 \leq \deg \varphi \leq n$,

(11.27) $\mathrm{Res}_\varphi(\xi, Z)$ is the element of $H_{2n-2l}(Z, C)$ which by Poincaré duality in Z is dual to $\dfrac{\varphi(n_1, \cdots, n_r, 0, \cdots, 0)}{n_1 \cdots n_r} x^{l-r}$, $\quad r = n - k + 1$.

Example 5. Let k, n, Z be as in Example 4. Set $r = n - k + 1$. Let M be $Z \times C^r$, and $\pi_1 \colon Z \times C^r \to Z$, $\pi_2 \colon Z \times C^r \to C^r$ be the projections. Then there is the splitting

(11.28) $$T(Z \times C^r) = \pi_1^!(TZ) \oplus \pi_2^!(TC^r) .$$

Let $\lambda_1, \cdots, \lambda_r$ be nonzero complex numbers. On C^r with its usual coordinate system set

(11.29) $$X = \sum_{i=1}^{r} \lambda_i z_i \partial / \partial z_i .$$

On $Z \times C^r$ set

(11.30) $$\tilde{X} = \pi_2^!(X) .$$

Let ξ be the subsheaf of $\underline{T}$ spanned by $\tilde{X}$ and all local holomorphic sections of $\pi_1^!(TZ)$. ξ is integrable and full, and its singular set is $Z \times \{0\}$. Identify $Z \times \{0\} = Z$.

(11.31) If $\deg \varphi = r$, then $\quad \mathrm{Res}_\varphi(\xi, Z) = \dfrac{\varphi(\lambda_1, \cdots, \lambda_r, 0, \cdots, 0)}{\lambda_1 \cdots \lambda_r}[Z]$.

(11.32) If $r < \deg \varphi \leq n$, then $\quad \mathrm{Res}_\varphi(\xi, Z) = 0$.

Example 6. Fix integers d and n with $1 \leq d < n$. Let Z be a compact connected complex manifold with

(11.33) $$\dim_C Z = d .$$

Set $s = n - d$. Let $L_1, \cdots, L_s$ be holomorphic line bundles on Z, and M the total space of the vector-bundle $L_1 \oplus \cdots \oplus L_s$. Then M is a complex manifold with

(11.34) $$\dim_C M = n .$$

The zero section of the vector bundle gives an inclusion $Z \subset M$. Denote a point of M by $(u_1, \cdots, u_s)$, so that $u_i \in L_i$. Let $\lambda_1, \cdots, \lambda_s$ be nonzero complex numbers. Construct a holomorphic flow

(11.35) $$C \times M \to M$$

by

(11.36) $$(z, (u_1, \cdots, u_s)) \to (e^{\lambda_1 z}u_1, \cdots, e^{\lambda_s z}u_s) .$$

Let X be the vector-field on M, which generates this flow. Let X be the subsheaf of $\underline{T}$ spanned by X. The singular set of ξ is Z. Identify $H_0(Z, C) = C$. Assume $n = \deg \varphi$. Let $x_1, \cdots, x_d$ be the formal Chern roots of TZ. Set $y_i = c_1(L_i)$. Take the $2d$-dimensional component of $\varphi(x_1, \cdots, x_d, \lambda_1 + y_1, \cdots, \lambda_s + y_s)/[(\lambda_1 + y_1)\cdots(\lambda_s + y_s)]$. Evaluate this element of $H^{2d}(Z, C)$ on the fundamental cycle of Z. This gives

(11.37) $$\operatorname{Res}_\varphi (\xi, Z) = \frac{\varphi(x_1, \cdots, x_d, \lambda_1 + y_1, \cdots, \lambda_s + y_s)}{(\lambda_1 + y_1)\cdots(\lambda_s + y_s)}[Z] .$$

For a proof of (11.37) see [8] or [14], and also [6], [3, Theorem (8.11) and Proposition (8.13), pp. 597–599], [20], [21].

Remark. In the general problem of computing $\operatorname{Res}_\varphi (\xi, Z)$ let $Z = Z^{(1)} \supset \cdots \supset Z^{(k)}$ be as (0.26). Consider the special case when $\dim_C Z = k - 1$ and $Z^{(2)} = \phi$. It can be shown that for this special case Examples 4 and 5 above essentially solve the problem of computing $\operatorname{Res}_\varphi (\xi, Z)$.

12. On the space $B\Gamma_q^C$

In the homotopy theory of complex foliations as developed by Haefliger-Phillips-Gromov [15], [16] a complex foliation F on a manifold M determines a classifying map

(12.1) $$f_F : M \to B\Gamma_q^C .$$

Here $q = n - k$ is the codimension of F. In this section we would like to explain the relation of our residue classes $\operatorname{Res}_\varphi (\xi, Z)$ to this homotopy theory.

First recall that there is a natural map

(12.2) $$\nu : B\Gamma_q^C \to BGL(q) ,$$

which corresponds to assigning the normal bundle of a foliation. As usual $BGL(q)$ denotes the classifying space of the general linear group $GL(q, C)$. In terms of these concepts the vanishing theorem simply asserts

(12.3) $$\nu^* : H^{2j}(BGL(q); C) \to H^{2j}(B\Gamma_q^C; C) \quad \text{vanishes whenever } j > q .$$

In contrast to this, it is not difficult to show that

$$\begin{aligned}(12.4) \qquad & \nu^* : H^j(BGL(q); Z) \to H^j(B\Gamma_q^C; C) \\ & \text{is injective for all } j = 0, 1, 2, \cdots ,\end{aligned}$$

and (12.3) and (12.4) together imply

(12.5) For each $j > q$, $H_{2j-1}(B\Gamma_q^C; Z)$ is an abelian group which is not finitely generated.

See [12] for details of these first consequences of (12.3). More delicate results arise in the following manner:

Let BGL be the classifying space of the infinite general linear group. If c_1, c_2, $\cdots$ are the universal Chern classes, then

$$(12.6) \qquad H^*(BGL; C) = C[c_1, c_2, \cdots] .$$

Let $\nu_s : B\Gamma_q \to BGL$ be the composition of ν with the inclusion $BGL(q) \subset BGL$. Then (12.3) is equivalent to

$$(12.7) \qquad \nu_s^* : H^{2j}(BGL; C) \to H^{2j}(B\Gamma_q^C; C) \qquad \text{vanishes whenever } j > q.$$

We are interested in ν_s only up to homotopy type, so ν_s can be taken as an inclusion $B\Gamma_q \subset BGL$. Thus there is the pair of spaces $(BGL, B\Gamma_q)$. In this context the constructions of § 5 (e.g., Definition (5.51)) can be interpreted as lifting each $\varphi \in H^{2j}(BGL; C)$, $j > q$, to a definite and well-defined class $\hat{\varphi} \in H^{2j}(BGL, B\Gamma_q; C)$.

More precisely, let ξ be a full integrable sheaf on a complex manifold M. Set $Q = \underline{T}/\xi$. Let S be the singular set of ξ. The procedure given in § 5 and § 7 (e.g., Definition (7.1)) lifts each $\varphi(Q) \in H^{2j}(M; C)$, $j > n - k$, in a canonical fashion to a class $\hat{\varphi}(Q) \in H^{2j}(M, M - S; C)$. Now ξ determines a homotopy commutative diagram

$$(12.8) \qquad \begin{array}{ccc} M - S & \xrightarrow{f_F} & B\Gamma_q \\ \downarrow & & \downarrow \\ M & \xrightarrow[f_Q]{} & BGL \end{array}$$

where f_F classifies the foliation of $M - S$, and f_Q classifies the element of $K(M)$ given by Q.

The exactness of a resolution of Q gives on $M - S$ an exact sequence of vector-bundles

$$(12.9) \qquad 0 \to E_r \to E_{r-1} \to \cdots \to E_0 \to \nu \to 0 ,$$

which can be thought of as an explicit homotopy between the two maps of $M - S$ into BGL of (12.8). Therefore ξ defines a map of pairs:

$$f_\xi\colon (M, M - U) \to (BGL, B\Gamma_q)\ , \tag{12.10}$$

where U is a small neighborhood of S in M.

The universal liftings $\hat{\varphi} \in H^{2j}(BGL, B\Gamma_q; \boldsymbol{C})$ are now uniquely characterized by

$$f_\xi^*(\hat{\varphi}) = \hat{\varphi}(Q)\ . \tag{12.11}$$

Quite equivalently this is expressed by the formula

$$\mathrm{Res}_\varphi\,(\xi, Z) = \pi_Z f_\xi^*(\hat{\varphi})\ , \tag{12.12}$$

where

$$\pi_Z\colon H^*(M, M - S; \boldsymbol{C}) \to H_*(Z, \boldsymbol{C})$$

is induced by excision followed by Poincaré duality.

Granting (12.12) one may use the examples of § 11 to prove the following:

Proposition. *Let $d(n)$ be the dimension of $H^{2n}(BGL(n-1); \boldsymbol{C})$ over $\boldsymbol{C}$. Then there exists a surjection of abelian groups*

$$h\colon \pi_{2n-1}(B\Gamma_{n-1}^{\boldsymbol{C}}) \to \boldsymbol{C}^{d(n)}\ . \tag{12.13}$$

This result was already announced in [9] and is an easy analogue in the complex case of the recent results of Thurston [22], concerning real foliations with varying Godbillion-Vey invariants.

To prove (12.13) we first construct a homomorphism

$$\tilde{h}\colon \pi_{2n-1}(BGL, B\Gamma_{n-1}) \to \boldsymbol{C}^{d(n)} \tag{12.14}$$

in the following manner. Let $\varphi_1, \cdots, \varphi_{d(n)}$ be a basis for the symmetric polynomials in n-variables $X_1, \cdots, X_n$, which are of degree n, and lie in the ideal generated by the first $n-1$ elementary symmetric functions $\sigma_1, \cdots, \sigma_{n-1}$ of the X's. We identify the φ's with classes in $H^*(BGL)$ by interpreting the σ_i as the i-th Chern classes, and denote by $\hat{\varphi}_1, \cdots, \hat{\varphi}_{d(n)}$ the liftings of these classes to $H^*(BGL, B\Gamma_{n-1}^{\boldsymbol{C}})$. Now then $\tilde{h}$ is defined to be the evaluation of this basis on a relative class:

$$\tilde{h}(\alpha) = \{\hat{\varphi}_1(\alpha), \cdots, \hat{\varphi}_{d(n)}(\alpha)\}\ , \tag{12.15}$$

where α denotes both the element in π_{2n-1} and its image in H_{2n-1} under the Hurewicz map.

We next evaluate $\tilde{h}$ on the relative elements f_λ determined by the foliation F_λ of § 11. Recall that here $\lambda = (\lambda_1, \cdots, \lambda_n)$ is an n-tuple of nonzero complex numbers, and F_λ the foliation of $\boldsymbol{C}^n - \{0\}$ given by the vector-field $\Sigma\lambda_i z_i \partial/\partial z_i$.

According to (11.2) and (12.12) we obtain

$$\tilde{h}(f_\lambda) = \{\varphi_1(\lambda)/\sigma_n(\lambda), \cdots, \hat{\varphi}_{d(n)}(\lambda)/\sigma_n(\lambda)\} .$$

The surjectivity now follows from the folloing proposition whose simple proof is given in an Appendix:

Lemma. *The set* $A \in C^{d(n)}$ *consisting of the values* $\tilde{h}(f_\lambda)$, $\lambda \in (C - \{0\})^n$ *additively generates all of* $C^{d(n)}$.

To proceed to (12.14) consider the diagram:

$$\begin{array}{ccccccc} \pi_{2n}(BGL) & \to & \pi_{2n}(BGL, B\Gamma^C_{n-1}) & \to & \pi_{2n-1}(B\Gamma^C_{n-1}) & \to & \pi_{2n-1}(BGL) \\ & & \downarrow \tilde{h} & & & & \\ & & C^{d(n)} & & & & \end{array}$$

It is well known that $\pi_{2n-1}(BGL) = 0$, and that $\pi_{2n}(BGL) = Z$. Furthermore any decomposable element vanishes on a spherical class. Hence $\tilde{h}$ is zero on the image of π_{2n} and induces the desired surjection h: $\pi_{2n-1}(B\Gamma^C_{n-1}) \to C^{d(n)}$.

Appendix

In § 12 we needed to show that a certain subset of affine space additively generated the whole space. The general principle behind this fact is expressed in the following

Proposition. *Suppose that* $A \subset C^n$ *is a connected complex analytic subset of* C^n *of* $\dim \geq 1$, *which is not contained in any affine hyperplane of* C^n. *Then* A *generates* C^n *additively.*

Proof. Let $M \subset A$ denote the submanifold of nonsingular points in A. M will still satisfy our conditions by well known arguments. Hence it is sufficient to show that M generates C^n.

Now let span (M) denote the vector space spanned by the translates to 0 of all the tangent spaces to M. If span (M) does not equal C^n, then there is a linear form z on C^n which vanishes identically on span (M). Hence the restriction of the one form dz to M vanishes identically whence—as M is connected, M lies in a hyperplane $z =$ const. contradicting our hypothesis.

Therefore span $(M) = C^n$, and we can find a finite number of points $m_1, \cdots, m_k \in M$ whose tangent spaces already generate C^n. Now consider the map

$$M \times \cdots \times M \xrightarrow{F} C^n$$

obtained by sending a k-tuple in M to its sum. Clearly the differential of this map is onto at the point $(m_1, \cdots, m_k)$. Hence the image of F contains an open ball about $m_1 + \cdots + m_k$. But such a ball clearly generates all of C^n. q.e.d.

To apply this principle in our case, observe that the image of the map $\lambda \to \tilde{h}(f_\lambda)$ is equal to the image of a map $H: C^{n-1} \to C^{d(n)}$, which sends the $(n-1)$-tuple $\{x_i\}$ to the $d(n)$-tuple $\{m_\alpha(x)\}$, where α ranges over the multi-indexes of weight n and $m_\alpha(x)$ denotes the monomial $x_1^{\alpha_1} \cdots x_{n-1}^{\alpha_{n-1}}$. The linear independence of these monomials now clearly implies that the image of M does not lie in a hyperplane.

References

[1] M. F. Atiyah & F. Hirzebruch, *Analytic cycles on complex manifolds*, Topology **1** (1961) 25–45.

[2] ——, *The Riemann-Roch theorem for analytic embeddings*, Topology **1** (1962) 151–166.

[3] M. F. Atiyah & I. Singer, *The index of elliptic operators*. III, Ann. of Math. (2) **87** (1968) 546–604.

[4] P. Baum, *Structure of foliation singularities*, preprint.

[5] P. Baum & R. Bott, *On the zeroes of meromorphic vector fields*, Essays on Topology and Related Topics, Mémoires dédiés à Georges de Rham, Springer, Berlin, 1970, 29–47.

[6] P. Baum & J. Cheeger, *Infinitesimal isometries and Pontryagin numbers*, Topology **8** (1969) 173–193.

[7] A. Borel & J. P. Serre, *Le théorème de Riemann-Roch*, Bull. Soc. Math. France **86** (1958) 97–136.

[8] R. Bott, *A residue formula for holomorphic vector fields*, J. Differential Geometry **1** (1967) 311–330.

[9] ——, *On a topological obstruction to integrability*, Proc. Sympos. Pure Math. Vol. 16, Amer. Math. Soc., 1970, 127–132; see also *On topological obstructions to integrability*, Proc. Internat. Congress Math. (Nice, 1970), Gauthier-Villars, Paris, Vol. 1, 1971, 27–36.

[10] R. Bott & S. S. Chern, *Hermitian vector bundles and the equidistribution of the zeroes of their holomorphic sections*, Acta Math. **114** (1965) 71–112.

[11] R. Bott & A. Haefliger, *On characteristic classes of Γ-foliations*, Bull. Amer. Math. Soc. **78** (1971).

[12] R. Bott & J. Heitsch, *A remark on the integral cohomology of $B\Gamma_q$*, Topology **11** (1972) 141–146.

[13] H. Cartan & S. Eilenberg, *Homological algebra*, Princeton University Press, Princeton, 1965.

[14] C. Chen, *On the local residues of meromorphic vector fields*, preprint.

[15] A. Haefliger, *Feuilletages sur les variétés ouvertes*, Topology **9** (1970) 183–194.

[16] A. Haefliger, *Homotopy and integrability*, Proc. Summer Institute on Manifolds, Lecture Notes in Math. Vol. 197, Springer, Berlin, 1971, 133–163.

[17] P. Halmos, *Measure theory*, Van Nostrand, Princeton, 1950.

[18] R. Hartshorne, *Residues and duality*, Lecture Notes in Math. Vol. 20, Springer, Berlin, 1966.

[19] L. Illusie, to appear.

[20] J. Pasternack, *Foliations and compact Lie group actions*, Comment. Math. Helv. **46** (1971) 467–477.

[21] J. Pasternack, *Riemannian Haefliger Structures*, University of Washington preprint.

[22] W. Thurston, *Noncobordant foliations of* S^n. Bull. Amer. Math. Soc. **78** (1972) 511–514.

BROWN UNIVERSITY
HARVARD UNIVERSITY

BULLETIN OF THE
AMERICAN MATHEMATICAL SOCIETY
Volume 78, Number 6, November 1972

ON CHARACTERISTIC CLASSES OF Γ-FOLIATIONS

BY R. BOTT[1] AND A. HAEFLIGER

Communicated by Michael Altiyah, April 13, 1972

1. **Introduction.** In this note we describe and apply a general procedure for constructing characteristic classes of Γ-foliations. Our work was inspired by that of Godbillon-Vey [4] and Gelfand-Fuks [5] on the one hand, and that of Chern-Simons [3] on the other.

We know of several essentially equivalent procedures—some more in the line of [3] (see for example [2])—but the one we present here fits naturally and well nigh unavoidably into the Gelfand-Fuks theory and we have been told that Malgrange and Gelfand have also, quite independently, come upon it.

We will take the following point of view toward characteristic classes and foliations. Consider a pseudogroup Γ whose elements are diffeomorphisms of open sets in $\mathbf{R}^n$. A Γ-foliation on a smooth (that is C^∞) manifold M is by definition a maximal family F of submersions

$$f_U : U \to \mathbf{R}^n$$

of open sets U in M, such that for every $x \in U \cap V$ there exists an element $\gamma_{VU} \in \Gamma$ with $f_V = \gamma_{VU} \circ f_U$ in some vicinity of x.

Given two Γ-foliations, F on M and F' on M', a morphism from F to F' is by definition a smooth map

$$f: M \to M'$$

such that for every "local projection" $f'_U \in F'$ the composition

$$f'_U \circ f : f^{-1}U \to \mathbf{R}^n$$

is in F.

With this concept of morphism the Γ-foliations form a category $C(\Gamma)$ and we define a characteristic class of Γ-foliations with coefficients in a group $\mathbf{R}$, as a natural transformation

$$\alpha : C(\Gamma) \to H^*(\ ; \mathbf{R}).$$

Thus $\alpha(F) \in H^*(M; \mathbf{R})$ and if $f: F' \to F$ is a morphism, then

---

AMS 1970 *subject classifications*. Primary 53A10, 53C35; Secondary 53A05, 28A75.

[1] Supported in part by National Science Foundation grant GP-7952X3, while at the Institute for Advanced Study, Princeton, New Jersey.

$$\alpha(f^{-1}F) = f^*\alpha(F).$$

Alternatively, one could define this notion in terms of the classifying space $B\Gamma$—which we assume to be realized as a CW-complex [**6**].

PROPOSITION *There is a natural bijection of the* R*-valued characteristic classes on* Γ*-foliations and the cohomology group* $H^*(B\Gamma; \boldsymbol{R})$.

2. **Relation with the cohomology of formal vector-fields.** Suppose now that Γ is a transitive Lie-pseudogroup acting on $\boldsymbol{R}^n$ and let $\mathfrak{a}(\Gamma)$ denote "the Lie algebra of formal Γ vector-fields" associated to Γ. Here a vector-field defined on $U \subset \boldsymbol{R}^n$ is called a Γ vector-field if the local one parameter group which it engenders is in Γ, and $\mathfrak{a}(\Gamma)$ is defined as the inverse limit

$$\mathfrak{a}(\Gamma) = \varprojlim \mathfrak{a}^k(\Gamma)$$

of the k jets at 0 of Γ vector-fields. In the pseudogroup Γ let Γ_0 be the set of elements of Γ keeping 0 fixed and set Γ_0^k equal to the k jets of elements in Γ_0.

Then the Γ_0^k form an inverse system of Lie groups and we can find a subgroup $K \subset \varprojlim \Gamma_0^k$, whose projection on every Γ_0^k is a maximal compact subgroup for $k > 0$. This follows from the fact that the kernel of the projection $\Gamma_0^{k+1} \to \Gamma_0^k$ is a vector space for $k > 0$. The subgroup K is unique up to conjugation, and its Lie algebra k can be identified with a subalgebra of $\mathfrak{a}(\Gamma)$.

For our purposes we now need the cohomology of basic elements rel K in $\mathfrak{a}(\Gamma)$, $H(\mathfrak{a}(\Gamma); K)$, which is defined in the following manner: Let $A\{\mathfrak{a}^k(\Gamma)\}$ denote the algebra of multilinear alternate forms on $\mathfrak{a}^k(\Gamma)$, and let $A\{\mathfrak{a}(\Gamma)\}$ be the direct limit of the $A\{\mathfrak{a}^k(\Gamma)\}$. The bracket in $\mathfrak{a}(\Gamma)$ there induces a differential on $A\{\mathfrak{a}(\Gamma)\}$ in the usual way, and we write $H\{\mathfrak{a}(\Gamma)\}$ for the resulting cohomology group. The relative group $H^*(\mathfrak{a}(\Gamma); K)$ is now defined as the cohomology of the subcomplex of $A\{\mathfrak{a}(\Gamma)\}$ consisting of elements which are invariant under the natural action of K, and annihilated by all inner products with elements of k.

THEOREM 1. *Let* F *be a* Γ*-foliation on* M. *There is an algebra homomorphism*

$$\varphi: H\{\mathfrak{a}(\Gamma); K\} \to H(M; \boldsymbol{R})$$

which is a natural transformation on the category $C(\Gamma)$.

Note that $H(\mathfrak{a}(\Gamma); K)$ can be thought of as the continuous cohomology of the topological category Γ.

To construct φ one proceeds as follows:

Let $P^k(\Gamma)$ be the differentiable bundle of k jets at the origin of elements of Γ. It is a principal Γ_0^k-bundle. On the other hand, Γ acts transitively on the left on $P^k(\Gamma)$.

Denote by $A(P^\infty(\Gamma))$ the direct limit of the algebras $A(P^k(\Gamma))$ of differential forms on $P^k(\Gamma)$. The invariant forms with respect to the action of Γ constitute a differential subalgebra denoted by A_Γ.

PROPOSITION. *A_Γ is naturally isomorphic to $A(\mathfrak{a}(\Gamma))$.*

Now let F be a foliation on M. Let $P^k(F)$ be the differentiable bundle over M whose fiber over $x \in M$ is the space of k jets at x of local projections $f_U \in F$ such that $f_U(x) = 0$. This is a Γ_0^k-principal bundle. Its restriction to U is isomorphic to the inverse image by f_U of the bundle $P^k(\Gamma)$: If g^k is the k jet at 0 of an element g of Γ and $g(0) \in f_U(U)$, the isomorphism maps (x, g^k) on the k jet at x of $g^{-1}f_U$. Hence it is readily seen that the differential algebra of Γ-invariant forms on $P^k(\Gamma)$ is mapped in the algebra $A(P^k(F))$ of differential forms on $P^k(F)$.

Hence if we denote by $A(P^\infty(F))$ the direct limit of the $A(P^k(F))$, we get an injective homomorphism φ of $A(\mathfrak{a}(\Gamma))$ in $A(P^\infty(F))$ commuting with the differential.

This homomorphism is compatible with the action of K, hence induces a homomorphism on the subalgebra of K-basic elements. But the algebra $A(P^k(F); K)$ of K-basic elements in $A(P^k(F))$ is isomorphic to the algebra of differential forms on $P^k(F)/K$ which is a bundle over M with contractible fiber Γ_0^k/K. Hence $H(A(P^k(F); K))$ is isomorphic, via the de Rham theorem, to $H(M; \mathbf{R})$.

The homomorphism φ is therefore obtained as the composition

$$H(\mathfrak{a}(\Gamma); K) \to H(A(P^\infty(F); K)) = H(M; \mathbf{R}).$$

REMARK. In the same spirit, one can for instance also consider the category of differentiable fiber bundles E, with fiber a compact comogeneous space $B = G/K$, with structural group G, and a foliation transverse to the fibers.

Also let v_B be the topological Lie algebra of tangent vector fields on B (cf. [5]). Then, there is a natural homomorphism

$$H(v_B; K) \to H(E)$$

which is naturally related to the preceding one.

3. **The basic examples.** By extending the computation of Gelfand-Fuks [5] the group $H(\mathfrak{a}; K)$ can be computed in certain cases, and we will describe the results in several instances of topological interest; namely when Γ consists of the holomorphic, orientation preserving, and arbitrary diffeomorphisms of open sets in $\mathbf{R}^n$. We denote these Γ's by $\Gamma_n\mathbf{C}, \Gamma_n^+$ and Γ_n respectively.

In the following $E(u_1, u_2, \ldots)$ denotes an exterior algebra in the indicated u's and $\boldsymbol{R}[c_1, \ldots, c_n]$ denotes the polynomial ring over $\boldsymbol{R}$ in the variables $c_1, \ldots, c_q$, with degree $c_i = 2i$. We also write $\hat{\boldsymbol{R}}[c_1, \ldots, c_n]$ for the ring $\boldsymbol{R}[c_1, \ldots, c_n]$/(elements of degree $> 2q$), and denote by WO_n the differential complex:

$$WO_n = E(u_1, u_3, \ldots, u_{2k+1}) \otimes \hat{\boldsymbol{R}}[c_1, \ldots, c_n],$$

$dc_i = 0$; $du_i = c_i$ for odd i; $2k + 1$ largest odd integer $< n$. With this understood, one then has the

THEOREM 2. *The K's for Γ_n and Γ_n^+ are the orthogonal and special orthogonal groups O_n and SO_n respectively, and*

$$H^*(\mathfrak{a}(\Gamma_n); K) = H(WO_n),$$

while

$$H^*(\mathfrak{a}(\Gamma_n^+); K) = H(WO_n), \qquad n \text{ odd},$$

$$= H(WO_n)[\chi]/(\chi^2 - c_n), \qquad n \text{ even}.$$

REMARKS. (1) Every n-manifold of course carries a natural Γ_n structure, the one whose projections contain all the coordinate charts of M. Under φ the c_i's above then go into the Chern classes of the complexified tangent bundle to M, while χ goes over to the Euler class.

For a general Γ-foliation F on M, the c_i's and χ play the corresponding role for the normal bundle νF to F. The truncation of $\boldsymbol{R}[c_1, \ldots, c_n]$ thus expresses and refines the vanishing theorem of [**5**]. Finally the invariant of Godbillon-Vey is identified with the image of the class $\mu_1 c_1^{2n-1}$ under φ.

(2) The Gelfand-Fuks computations for $H\{\mathfrak{a}(\Gamma_n)\}$ can also be expressed conveniently in terms of a complex. Indeed, define W_n by

$$W_n = E(u_1, u_2, \ldots, u_n) \otimes \hat{\boldsymbol{R}}[c_1, \ldots, c_n], \qquad du_i = c_i.$$

Then $H\{\mathfrak{a}(\Gamma_n)\} \cong H(W_n)$. This group occurs naturally in our context when one seeks characteristic classes for Γ-foliations with trivialized normal bundles.

We turn next to the complex case. Here one has

THEOREM 3. *For $\Gamma_n C$ the unitary group plays the role of K, and*

$$H^*(\Gamma_n C; K) \otimes C \cong H(WU_n)$$

where WU_n is the complex $E[u_1, \ldots, u_n] \otimes \hat{\boldsymbol{R}}[c_1, \ldots, c_n] \otimes \hat{\boldsymbol{R}}[\bar{c}_1, \ldots, \bar{c}_n] \otimes \boldsymbol{C}$ with $du_k = c_k - \bar{c}_k$.

Here, of course, φ again assigns to c_i the ith Chern class of νF, while the classes of type $u_j c_1^{\alpha_1}, \ldots, c_n^{\alpha_n}$ correspond to the relative classes in the Chern-Simons theory.

4. **On the nontriviality of φ.** It seems to be very difficult, especially in

the real case, to find foliations which distinguish between the different classes defined by φ. One procedure is to extend the example of Godbillon-Vey-Roussarie, and to seek Γ-foliations on homogeneous spaces, which have nontrivial invariants after division by a suitable discrete group. From our point of view, this procedure takes the following form. Let G be a Lie group and $H \subset G$ a closed subgroup. The action of G on a vicinity of H in G/H then defines a Lie pseudogroup $\Gamma_{G,H}$ on $\boldsymbol{R}^n$, $n = \dim G/H$. For this Lie pseudogroup one, of course, finds that $\mathfrak{a}(\Gamma_{G,H}) = \mathfrak{g}$ the Lie algebra of G, while the role of K is played by the *maximal compact subgroup of H*. Consider now the inclusion $i : \Gamma_{G,H} \subset \Gamma_n$. Then our φ construction can be thought of as inducing a commutative diagram:

$$\begin{array}{ccc} H(B\Gamma_{G/H}) & \overset{i^*}{\longleftarrow} & H(B\Gamma_n) \\ \uparrow \varphi & & \uparrow \varphi \\ H(\mathfrak{g}; K) & \overset{i^*}{\longleftarrow} & H(\mathfrak{a}(\Gamma_n); O_n). \end{array}$$

Now suppose G is semisimple. Then according to the deep theory of Borel-Harish-Chandra, G contains a group D acting discretely on G/K and such that $D\backslash G/K$ is compact. From this fact and Poincaré duality, it follows that $H^*(\mathfrak{a}; K)$ injects into $H^*(D\backslash G/K)$ and this in turn easily leads to the conclusion:

PROPOSITION. *If G is semisimple, then*

$$\varphi : H(\mathfrak{g}; K) \to H(B\Gamma_{G/H})$$

is injective.

Using this principle, one obtains a weak independence result for the classes $u^1 c^\alpha$, $c^\alpha = c_1^{\alpha_1} \cdots c_n^{\alpha_n}$, $\alpha_1 + 2\alpha_2 + \cdots + n\alpha_n = n$ in $H^{2n+1}(WO_n)$. For this purpose define a minimal variety V to be the quotient of a simple semisimple complex Lie group by a maximal parabolic subgroup. (Examples are the complex Grassmannians, the quadrics and, in general, each simple group G has no more than l minimal varieties associated to it where $l = \text{rank } G$.) These varieties are compact so that if V has complex dimension n, then $c^\alpha(V)$ is well-defined.

INDEPENDENCE THEOREM. *The φ image of the classes $\mu_1 c^\alpha$ in $H^{2n+1}(WO_n)$ are as independent as the Chern numbers $c^\alpha(V)$ of the minimal varieties V of* $\dim_C V = n$.

COROLLARY. *In degree* $(2n + 1)$ *the image of* φ *has* $\dim \geqq 2$, *for* $n \geqq 2$.

These two classes correspond to the complex projective space and the quadric, which exist in every dimension. Unfortunately, they are also the only ones when n is a prime. However, these classes in turn imply that many higher dimensional ones are also nontrivial.

In the complex case the corresponding classes are much easier to detect, and there they, in fact, take on a continuum of values. Furthermore, they play a universal role for the infinitesimal Lefschetz theorem (see [2]). Precisely, one has the following:

THEOREM 4. *Let F_λ be the $\Gamma_n C$-structure defined on $C_{n+1} - 0$, by the vector-field*

$$X_\lambda = \lambda_0 z_0 \frac{\partial}{\partial z_0} + \lambda_2 z_2 \frac{\partial}{\partial z_2} + \cdots + \lambda_n z_n \frac{\partial}{\partial z_n} \qquad (\lambda_i \neq 0).$$

Then one has the formula

$$\int_{S^{2n+1}} \varphi(\mu_1 c^\alpha)[F_\lambda] = \Im c^\alpha(\lambda)/c^n(\lambda)$$

where $\Im$ denotes imaginary part and the $c_i(\lambda)$ are taken to be the ith elementary symmetric functions of the system $\lambda = \{\lambda_0, \ldots, \lambda_n\}$.

BIBLIOGRAPHY

1. R. Bott, *On a topological obstruction to integrability*, Proc. Internat. Congress Math. (Nice, 1970), vol. 1, Gauthier-Villars, Paris, 1971, pp. 27–36.

2. ———, *The Lefschetz formula and exotic characteristic classes*, Proc. Differential Geometry Conference, Rome, May 1971.

3. S. Chern and J. Simons, *Some cohomology classes in principal fiberbundles and their applications to Riemannian geometry*, Proc. Nat. Acad. Sci. U.S.A. **68** (1971), 791–794.

4. Godbillon-Vey, *Un invariant des feuilletages de codimension* 1, C. R. Acad. Sci. Paris **273** (1971), 92.

5. I. M. Gel'fand and D. B. Fuks, *Cohomologies of the Lie algebra of tangent vector fields of a smooth manifold*. I, II, Funkcional. Anal. i Priložen. **3** (1969), no. 3, 32–52; ibid. **4** (1970), 23–32. (Russian) MR **41** # 1067.

6. A. Haefliger, *Feuilletages sur les variétés ouvertes*, Topology **9** (1970), 183–194. MR **41** # 7709.

DEPARTMENT OF MATHEMATICS, HARVARD UNIVERSITY, CAMBRIDGE, MASSACHUSETTS 02138

Current address (A. Haefliger): Université de Genève, Genève, Switzerland

Lecture Notes in Mathematics

A collection of informal reports and seminars
Edited by A. Dold, Heidelberg and B. Eckmann, Zürich

279

R. Bott
S. Gitler
I. M. James

Lectures on Algebraic and Differential Topology

Delivered at the II. ELAM

Springer-Verlag
Berlin · Heidelberg · New York 1972

PREFACE

The Second Latin American School in Mathematics was held during the month of July at the Centro de Investigación del IPN, in Mexico City and was dedicated to the memory of Heinz Hopf. The School was held under the auspices of the Centro de Investigación del IPN, Organization of American States, The National Science Foundation and the Consejo Nacional de Ciencia y Tecnología.

These notes contain three of the seven series of lectures given at the School. We hope that the others will appear at some future date. We thank the participants and the lecturers for having made the School a most stimulating mathematical event.

Mexico, D.F., April 14, 1972.

Samuel Gitler
Editor

CONTENTS

LECTURES ON CHARACTERISTIC CLASSES AND FOLIATIONS

by Raoul Bott

(Notes by Lawrence Conlon)

1. Some basic notions. In this section we presume some familiarity with vector bundles and differential forms. Brief discussions of these two notions will be found in §2 and §3 respectively. We also presume (both here and throughout the notes) some familiarity with differentiable manifolds.

Let M and N be smooth manifolds of respective dimensions m and n. Let $f : M \to N$ be a smooth map. For each $x \in M$ one has the linear map

$$df_x : T_x(M) \to T_{f(x)}(N) .$$

(1.1) Definition. f as above is called an immersion if df_x is injective, $\forall\, x \in M$. f is called a submersion if df_x is surjective, $\forall\, x \in M$.

Of these two notions we are here primarily interested in submersions. By the implicit function theorem, df_x surjective implies that $f^{-1}(f(x))$ is a smooth submanifold of M of dimension $m - n$. Thus, a submersion $f : M \to N$ decomposes M into $(m-n)$-dimensional "fibers" $f^{-1}(y)$, $y \in N$. This is an example of the sort of structure we will call a foliation. The crucial property of this example is the fact that locally one can choose coordinates on M and N such that f becomes the canonical projection

$$R^n \times R^{m-n} \to R^n \quad .$$

We further notice that the vector spaces $\mathrm{Ker}(df_x)$ unite to form a smooth subbundle $\mathrm{Ker}(df) \subset T(M)$ (equivalently, a smooth $(m-n)$-dimensional distribution on M in the sense of [1, p. 10]).

(1.2) **Definition.** A smooth subbundle $E \subset T(M)$ is called integrable iff locally E is the kernel of the differential of a submersion. An integrable subbundle of $T(M)$ is also called a foliation of M.

Notice that a foliation of M gives rise to an open cover $\{U_\alpha\}$ and submersions $\{f_\alpha : U_\alpha \to R^q\}$ such that

$$\mathrm{Ker}(df_\alpha)|U_\alpha \cap U_\beta = \mathrm{Ker}(df_\beta)|U_\alpha \cap U_\beta$$

for all α,β. Furthermore, if $x \in U_\alpha \cap U_\beta$ it can be shown that there is an open neighborhood $W \subset U_\alpha \cap U_\beta$ of x such that $f_\alpha^{-1}(f_\alpha(x)) \cap W = f_\beta^{-1}(f_\beta(x)) \cap W$. Conversely, such a structure clearly determines a unique foliation of M.

We show by an example that not every subbundle $E \subset T(M)$ is integrable. Let $M = R^3 - \{0\}$ and consider the 1-form $\varphi = x\cdot dy + y\cdot dz + z\cdot dx$. φ is nowhere zero on M, hence $E = \mathrm{Ker}(\varphi)$ is a two dimensional smooth subbundle of $T(M)$. If E were integrable we could find $\{U_\alpha, f_\alpha : U_\alpha \to R^1\}$ as above such that

$$\mathrm{Ker}(df_\alpha) = \mathrm{Ker}(\varphi)|U_\alpha, \quad \forall\alpha .$$

Thus $\varphi|U_\alpha = h\cdot df_\alpha$ where $h : U_\alpha \to R$ is C^∞ and nowhere zero. On U_α this gives

$$d\varphi = dh \wedge df_\alpha = \frac{dh}{h} \wedge hdf_\alpha = \eta \wedge \varphi$$

where η is a 1-form. Thus, on each U_α

$$\varphi \wedge d\varphi = \varphi \wedge \eta \wedge \varphi = 0$$

hence $\varphi \wedge d\varphi = 0$ identically on M. But direct computation shows

$$\varphi \wedge d\varphi = (x + y + z)dx \wedge dy \wedge dz$$

contradicting the above. E, therefore, cannot be integrable.

(1.3) Theorem. (Deahna, Clebsch, Frobenius). Let $E \subset T(M)$ be a smooth subbundle described locally as the simultaneous kernel of everywhere linearly independent 1-forms $\theta_1,\ldots,\theta_q$ (i.e., for x in the domain of the θ_i's, $E_x = \{v \in T_x(M) : \theta_1(v) = \cdots = \theta_q(v) = 0\}$) . Then E is integrable iff each $d\theta_i$ belongs to the ideal generated by $\theta_1,\ldots,\theta_q$.

An equivalent formulation of this theorem is the following.

(1.3)′ Theorem. A smooth subbundle $E \subset T(M)$ is integrable iff for any two smooth sections X and Y of E, the vector field $[X,Y]$ is also a section of E.

Remark that a simple corollary of either formulation, which also follows immediately from the standard existence and uniqueness theorem for ordinary differential equations, is that any smooth 1-dimensional subbundle $E \subset T(M)$ is integrable.

Theorems (1.3) and (1.3)′, usually referred to jointly as the "Frobenius Theorem", give a complete and satisfying answer to the (essentially local) question of integrability of a subbundle $E \subset T(M)$. We next modify the integrability question to obtain an essentially global problem to which no complete answer has been given.

(1.4) If $E \subset T(M)$ is a subbundle, does there exist an integrable subbundle $E' \subset T(M)$ such that E' and E are isomorphic as vector bundles?

One of the main aims of these lectures will be to give topological obstructions to the existence of E' in terms of the characteristic classes of E.

References

1. S. Kobayashi and K. Nomizu, Foundations of Differential Geometry, Vol. I, 1963, Interscience Publishers, New York.

2. Vector bundles and Pontryagin classes

While a certain basic knowledge of vector bundles might well be presumed in these notes, we wish here to review the concept from a point of view particularly suited to our purposes.

(2.1) Definition. An n-dimensional (real) vector bundle over a space X is a continuous map $\pi : E \to X$ such that

1) $\exists$ open cover $\{U_\alpha\}_{\alpha \in A}$ of X and homeomorphisms

$\varphi_\alpha : U_\alpha \times R^n \to \pi^{-1}(U_\alpha)$, $\forall\, \alpha \in A$.

2) The diagram (where p is the standard projection)

$$\begin{array}{ccc} U_\alpha \times R^n & \xrightarrow{\varphi_\alpha} & \pi^{-1}(U_\alpha) \\ & {\scriptstyle p}\searrow \quad \swarrow{\scriptstyle \pi} & \\ & U_\alpha & \end{array}$$

is commutative, $\forall\, \alpha \in A$.

3) For any $\alpha, \beta \in A$ with $U_\alpha \cap U_\beta \neq \emptyset$ the map

$$\varphi_\beta^{-1} \circ \varphi_\alpha : (U_\alpha \cap U_\beta) \times R^n \to (U_\alpha \cap U_\beta) \times R^n$$

is of the form

$$\varphi_\beta^{-1} \circ \varphi_\alpha(x,v) = (x, g_{\beta\alpha}(x) \cdot v)$$

where $g_{\beta\alpha} : U_\alpha \cap U_\beta \to GL_n = GL(n,R)$ is continuous. Here, of course, $GL(n,R)$ denotes the group of $n \times n$ nonsingular matrices.

By a certain abuse, we often call E itself the vector bundle over X. More properly E is called the total space of the bundle.

Note that each fiber $\pi^{-1}(x)$, $x \in X$, has canonically the structure of an n-dimensional vector space over R, although $\pi^{-1}(x)$ does not usually have a canonical coordinatization as R^n. As a general principle, operations on vector spaces extend to vector bundles. Thus we may form the direct sum $E \oplus E'$ of bundles over X, the quotient E/E' of a bundle by a subbundle, etc., obtaining new bundles over X.

Remark that the maps $g_{\alpha\beta}$ satisfy the following "cocycle condition".

(2.2) On $U_\alpha \cap U_\beta \cap U_\gamma$, $g_{\alpha\beta} \cdot g_{\beta\gamma} = g_{\alpha\gamma}$.

Conversely, suppose an open cover $\{U_\alpha\}_{\alpha \in A}$ of X is given together with continuous maps

$$g_{\alpha\beta} : U_\alpha \cap U_\beta \to GL_n$$

defined for all $\alpha, \beta \in A$ such that $U_\alpha \cap U_\beta \neq \emptyset$ and, further, satisfying (2.2). On the disjoint union $\coprod_{\alpha \in A} U_\alpha \times \mathbb{R}^n$ make the identifications

$$(x,v) \sim (x, g_{\beta\alpha}(x)v)$$

$\forall\, x \in U_\alpha \cap U_\beta \neq \emptyset$, $\forall\, v \in \mathbb{R}^n$. The resulting quotient space F and the natural map $\rho : F \to X$ define a vector bundle over X.

(2.3) **Definition.** If $\pi : E \to X$ and $\rho : F \to X$ are vector bundles, a continuous map $f : E \to F$ is said to be linear if

1) $$\begin{array}{ccc} E & \xrightarrow{f} & F \\ & {\scriptstyle\pi}\searrow \quad \swarrow{\scriptstyle\rho} & \\ & X & \end{array}$$ commutes

2) f is linear on each fiber.

Furthermore, we will say that E and F are isomorphic (and write $E \simeq F$) iff there are linear maps $f : E \to F$ and $g : F \to E$ such that both $f \circ g$ and $g \circ f$ are identity maps.

(2.4) **Exercise.** Let $\pi : E \to X$ be an n-dimensional vector bundle over X. Using the definition (2.1) take an associated cocycle $\{g_{\alpha\beta}\}$ and construct a new bundle $\rho : F \to X$ as above and show that $E \simeq F$.

Let $\mathcal{H}$ be a separable real Hilbert space. Define BGL_n to be the set of n-dimensional subspaces of $\mathcal{H}$ and topologize this set in some "reasonable" way (e.g., define a metric on BGL_n by taking the distance between two n-dimensional subspaces of $\mathcal{H}$ to be the angle between these subspaces).

Let $Vect_n(X)$ denote the set of isomorphism classes of n-dimensional vector bundles over X. Also, for any two spaces X,Y let [X,Y] denote the set of homotopy classes of maps $X \to Y$.

(2.5) Theorem. If X is a paracompact space, there is a natural one-one correspondence $Vect_n(X) \leftrightarrow [X, BGL_n]$.

We indicate briefly the way in which this correspondence is defined. Given $\pi : E \to X$ of dimension n, the paracompactness of X makes it possible to find a countable open cover $\{U_i\}_{i=1}^{\infty}$ of X and a subordinate partition of unity $\{\lambda_i\}_{i=1}^{\infty}$ and a continuous $\varphi_i : E|U_i \to R^n$ mapping each fiber isomorphically onto R^n. Express $\mathcal{H}$ as a countable orthogonal direct sum of copies of R^n and let $\psi_i : R^n \to \mathcal{H}$ be the inclusion of the ith summand. Then $\varphi : E \to \mathcal{H}$ defined by

$$\varphi(e) = \sum_{i=1}^{\infty} \lambda_i(\pi(e)) \cdot \psi_i \circ \varphi_i(e)$$

is continuous and sends each fiber $\pi^{-1}(x)$ linearly and one-one onto an n-dimensional subspace of $\mathcal{H}$. Thus $g_E(x) = \varphi(\pi^{-1}(x))$ defines a continuous map $g_E : X \to BGL_n$. This construction is then shown to set up the desired correspondence. For a careful treatment of (2.5), cf. [2, Chapter III].

As an example, let M be an n-dimensional manifold and choose a smooth imbedding $M \subset R^k$. Then, for each $x \in M$, $T_x(M)$ identifies canonically with an n-dimensional subspace of $R^k \subset \mathcal{H}$. This enables us to construct $\varphi : T(M) \to \mathcal{H}$ as desired. The corresponding map

$$g_{T(M)} : M \to BGL_n$$

is called the "Gauss map". The reader who needs to increase his comfort with these ideas is invited to prove that the above construction really does well define $g_{T(M)}$ up to homotopy.

(2.6) Theorem. Modulo torsion, the cohomology ring $H^*(BGL_n)$ is isomorphic to a polynomial ring $\mathbb{Z}[p_1,\ldots,p_{[n/2]}]$ where $p_i \in H^{4i}(BGL_n)$ is canonically defined. Thus, as algebras over R,

$$H^*(BGL_n; R) \cong R[p_1,\ldots,p_{[n/2]}] \quad .$$

For a proof, cf. [1,p.375]

(2.7) Definition. If $\pi : E \to X$ is an n-dimensional vector bundle over paracompact X, let $g_E : X \to BGL_n$ be the map, unique up to homotopy, given by (2.5). Then the ith (real) Pontryagin class of E is defined to be

$$p_i(E) = g_E^*(p_i) \in H^{4i}(X;R),$$

$i = 1,\ldots,[n/2]$. Furthermore, the graded subring

$$\mathrm{Pont}^*(E) = g_E^*(H^*(BGL_n;R)) \subset H^*(X;R)$$

is called the (real) Pontryagin ring of E.

We are now in a position to state the theorem whose proof is our first main aim.

(*) Theorem. If $E \subset T(M)$ is integrable and if the quotient bundle $Q = T(M)/E$ has fiber dimension q, then $\mathrm{Pont}^k(Q) = 0$ for $k > 2q$.

This is really a <u>global</u> integrability condition. Indeed, for $\pi : E \to X$ any n-dimensional bundle, set

$$p(E) = 1 + p_1(E) + \cdots + p_{[n/2]}(E) .$$

Because of the leading term 1 this is an invertible element of the ring $H^{**}(X;R)$ of formal infinite series $a_o + a_1 + \cdots + a_r + \cdots$, $a_i \in H^i(X;R)$. If $E' \subset E$ is a subbundle, the basic "duality" formula [3]

$$p(E/E') = p(E')^{-1}p(E)$$

holds and shows that the Pontryagin classes of E/E' depend only on the isomorphism classes of E and E' and <u>not</u> on the imbedding $E' \subset E$. Thus we can reformulate (*) as follows.

(*)$'$ Theorem. If $E \subset T(M)$ is a subbundle which is isomorphic to an integrable subbundle $E' \subset T(M)$, and if $Q = T(M)/E$, $q = \dim(Q)$, then $\mathrm{Pont}^k(Q) = 0$ for $k > 2q$.

Thus we have a topological obstruction to solving the global integrability problem (1.4). For some time this was the only known topological obstruction to global integrability. However, Shulman has observed (cf. §6) that the method of proof of (*) also shows that the Massey products of elements of $\mathrm{Pont}^*(Q)$ must vanish in suitably high dimensions under the hypothesis of integrability. Examples exist in which these new obstructions do not all vanish although $\mathrm{Pont}^*(Q)$ does vanish in dimensions $> 2q$.

References

1. A. Borel and F. Hirzebruch, Characteristic classes and homogeneous spaces, II, Am. J. Math., 81 (1959), pp. 315-382.
2. D. Husemoller, Fibre Bundles, McGraw-Hill, New York, 1966.
3. J. Milnor, Lectures on Characteristic Classes, (Notes by J. Stasheff), Mimeographed notes, Princeton Univ.
4. Munkres, Elementary Differential Topology, Annals of Math. Studies, No. 54, Princeton Univ. Press, Princeton, N. J., 1963.

3. De Rham cohomology. In order to prove (*) we will use an alternative definition of the Pontryagin classes in terms of differential forms. In order to formulate this definition we need a cohomology theory based on differential forms, the de Rham cohomology $H^*_{DR}(M)$. The present section is devoted to the definition of this theory and a sketch of its basic properties.

Let Λ^*_n be the associative algebra over $\mathbb{R}$ generated by elements $1, dx_1, \ldots, dx_n$ subject to the relations

1) 1 is a multiplicative identity

2) $dx_i \cdot dx_j = - dx_j \cdot dx_i$, $\forall\, i,j$.

Remark that a consequence of 2) is the relation

$$dx_i^2 = 0, \ \forall\, i .$$

Thus any monomial $dx_{i_1} \cdots dx_{i_r} = 0$ if $r > n$. This algebra has a natural grading by the degree of a monomial:

$$\Lambda^*_n = \bigoplus_{j=0}^{n} \Lambda^j_n ,$$

where, of course, $\Lambda_n^0 = R \cdot 1 \cong R$. Each Λ_n^j is a finite dimensional vector space over R.

(3.1) Definition. Let $U \subset R^n$ be an open subset. Let $A^k(U) = \{\omega : U \to \Lambda_n^k : \omega \text{ smooth}\}$. Then $A^*(U) = \bigoplus_{k=0}^{n} A^k(U)$ with its natural graded associative algebra structure is called the algebra of differential forms on U. Each $\omega \in A^k(U)$ is called a k-form on U.

Notice that each $f \in A^0(U)$ is simply a smooth real valued function on U. In general, $\omega \in A^k(U)$ can be written uniquely as

$$\omega = \sum_{1 \leq i_1 < \cdots < i_k \leq n} f_{i_1 \cdots i_k} dx_{i_1} \cdots dx_{i_k}$$

where each $f_{i_1 \cdots i_k}$ is a smooth real valued function on U.

(3.2) Definition. The exterior differential $d : A^0(U) \to A^1(U)$ is given by the formula $df = \sum_{j=1}^{n} \frac{\partial f}{\partial x_j} dx_j$. Generally, the exterior differential $d : A^k(U) \to A^{k+1}(U)$ is given by

$$d\Big(\sum_{i_1 < \cdots < i_k} f_{i_1 \cdots i_k} dx_{i_1} \cdots dx_{i_k}\Big)$$

$$= \sum_{i_1 < \cdots < i_k} d(f_{i_1 \cdots i_k}) dx_{i_1} \cdots dx_{i_k} .$$

Remark that if $x_i : U \to R$ is the ith coordinate function, then $d(x_i) = dx_i$.

(3.3) Exercises. 1) Prove that $d^2 = 0$. Indeed, show that this is simply a fancy way of saying that, for smooth real valued functions f on U, the mixed partials satisfy $\frac{\partial^2 f}{\partial x_i \partial x_j} = \frac{\partial^2 f}{\partial x_j \partial x_i}$, $\forall\ i,j$.

2) If $\omega \in A^p(U)$, $\eta \in A^q(U)$, prove that $d(\omega\eta) = d(\omega)\eta + (-1)^p \omega d(\eta)$.

(3.4) Definition. If $U \subset R^n$ and $V \subset R^m$ are open sets, $F : U \to V$ a smooth map, then define

$$F^* : A^0(V) \to A^0(U)$$

by $F^*(f) = f \circ F$. Define

$$F^*(dx_i) = d(F^*(x_i))$$

and extend this to a unique homomorphism of graded algebras

$$F^* : A^*(V) \to A^*(U).$$

Remark that $F^* \circ d = d \circ F^*$.

We would like to extend all of the above notions to any n-dimensional manifold M. Using local coordinate charts on M, one can define a smooth k-form on a coordinate neighborhood $U_\alpha \subset M$. ω_α is said to agree with ω_β if the smooth change of coordinates on $U_\alpha \cap U_\beta$ transforms ω_α into ω_β. If all of the k-forms ω_α agree, they are said to define a (smooth) k-form $\omega \in A^k(M)$. By the above it is clear that

$$d : A^k(M) \to A^{k+1}(M)$$

is still well-defined and that $d^2 = 0$.

(3.5) Definition. Let $Z^k_{DR}(M) = \{\omega \in A^k(M) : d\omega = 0\}$ and $B^k_{DR}(M) = \{\omega \in A^k(M) : \omega = d\eta$, some $\eta \in A^{k-1}(M)\}$. The de Rham cohomology of M is defined to be

$$H^*_{DR}(M) = \bigoplus_{k=0}^{n} H^k_{DR}(M)$$

where

$$H^k_{DR}(M) = \frac{Z^k_{DR}(M)}{B^k_{DR}(M)} \quad .$$

Notice that a smooth map $F : M \to N$ induces $F^* : A^*(N) \to A^*(M)$ with $F^* \circ d = d \circ F^*$, hence induces

$$F^* : H^*_{DR}(N) \to H^*_{DR}(M).$$

Clearly, $(id)^* = id$ and $(F \circ G)^* = G^* \circ F^*$, so de Rham cohomology defines a contravariant functor from the category of smooth manifolds and smooth maps to the category of graded $\mathbb{R}$-algebras and graded $\mathbb{R}$-algebra homomorphisms.

A basic property of de Rham cohomology is its homotopy invariance. This will be a consequence of the following lemma.

(3.6) **Lemma.** Let $\pi : M \times \mathbb{R} \to M$ be the standard projection. Then $\pi^* : H^*_{DR}(M) \to H^*_{DR}(M \times \mathbb{R})$ is bijective.

Proof. Let $i_0 : M \to M \times \mathbb{R}$ be defined by $i_0(x) = (x,0)$. Then $\pi \circ i_0 = id$, so $i_0^* \circ \pi^* = id : H^*_{DR}(M) \to H^*_{DR}(M)$. We have to show that $\pi^* \circ i_0^*$ is also an identity map.

Every $\omega \in A^k(M \times \mathbb{R})$ is a linear combination of k-forms of the following two types:

a) $\pi^*(\varphi) \cdot f(x,t)dt$, $\varphi \in A^{k-1}(M)$

b) $\pi^*(\varphi) \cdot f(x,t)$, $\varphi \in A^k(M)$.

Define $S : A^k(M \times \mathbb{R}) \to A^{k-1}(M \times \mathbb{R})$ by

$$S(\pi^*(\varphi) \cdot f(x,t)) = 0$$

$$S(\pi^*(\varphi) \cdot f(x,t)dt) = \pi^*(\varphi) \cdot g$$

where

$$g(x,t) = \int_0^t f(x,u)du \ .$$

This is clearly well defined and a straightforward computation gives

$$(-1)^{k-1}(d \circ S - S \circ d) = id - \pi^* \circ i_0^*$$

on $A^k(M \times R)$. Thus $id - \pi^* \circ i_0^*$ maps Ker(d) into Im(d), hence induces the zero homomorphism in cohomology. This proves the assertion.

q.e.d.

An obvious induction proves the following important corollary.

(3.7) Corollary. (Poincaré Lemma).

$$H^k_{DR}(R^q) = H^k_{DR}(\text{point}) = \begin{cases} R, & k = 0 \\ 0, & k > 0 \end{cases} .$$

This corollary is usually formulated differently. Indeed, consider the sequence

$$0 \rightarrow R \xrightarrow{i} A^0(R^q) \xrightarrow{d} A^1(R^q) \xrightarrow{d} \cdots \xrightarrow{d} A^q(R^q) \rightarrow 0 \ .$$

Here, if $a \in R$ then $i(a)$ is the constant function a on R^q. Evidently

$$Im(i) = Ker\{d: A^0(R^q) \rightarrow A^1(R^q)\} \ .$$

Then, (3.7) is entirely equivalent to the assertion that this sequence is exact.

(3.8) Corollary. If $f_0, f_1 : M \rightarrow N$ are smooth and are smoothly homotopic, then $f_0^* = f_1^* : H^*_{DR}(N) \rightarrow H^*_{DR}(M)$.

Proof. Let $i_0, i_1 : M \to M \times R$ be defined by

$$i_0(x) = (x,0)$$

$$i_1(x) = (x,1).$$

If $F : M \times I \to N$ is the smooth homotopy between f_0 and f_1 it can be smoothly extended to $F : M \times R \to N$ and

$$f_0 = F \circ i_0$$

$$f_1 = F \circ i_1 \quad .$$

Thus

$$f_0^* = i_0^* \circ F^*$$

$$f_1^* = i_1^* \circ F^*$$

and it will be enough to show $i_0^* = i_1^*$ in cohomology. Consulting the proof of (3.6) we see that $i_0^* = (\pi^*)^{-1}$ and that the same argument shows $i_1^* = (\pi^*)^{-1}$. Thus $i_0^* = i_1^*$. q.e.d.

Closely related to the above discussion is an important operation in de Rham cohomology called "integration along the fiber". We describe it briefly.

If M is oriented and n-dimensional, and if $A_0^n(M)$ denotes the space of n-forms with compact support, then each $\omega \in A_0^n(M)$ has a well defined integral $\int_M \omega \in R$ (cf.[3,p.118]). If the orientation of M is reversed, this integral receives the opposite sign. If M has boundary $\partial M \neq \phi$, all of our definitions make sense and one has the following basic theorem.

(3.9) Theorem. (Stokes) Let $i : \partial M \to M$ be the inclusion. Then $\int_M d\omega = \int_{\partial M} i^* \omega$, $\forall \omega \in A_0^{n-1}(M)$.

Let $\pi : E \to M$ be a smooth fiber bundle with fiber a compact manifold F. We assume that $\partial M \neq \emptyset$ and note that, if $\partial F \neq \emptyset$, then $\partial E \neq \emptyset$ and π restricts to a smooth fiber bundle $\pi^{\partial} : \partial E \to M$ with fiber ∂F. Let $e = \dim(E)$, $f = \dim(F)$.

(3.10) Theorem. For $\pi : E \to M$ as above, there is a canonical homomorphism

$$\pi_* : A^r(E) \to A^{r-f}(M), \quad r \geq 0 ,$$

which is zero for $r < f$ and satisfies

$$\pi_* \circ d + (-1)^{f+1} d \circ \pi_* = \pi_*^{\partial} \circ i^*$$

for $r \geq f$. Here, if $\partial E = \emptyset$, the right hand side is interpreted as 0.

Indeed, one can show the existence of a unique π_* satisfying the equation

$$\int_E \psi \cdot \pi^*(\varphi) = \int_M \pi_*(\psi) \cdot \varphi$$

for all $\psi \in A^r(E)$, $\varphi \in A_0^{e-r}(M)$. Then, for $\omega \in A^r(E)$, $\varphi \in A_0^{e-r-1}(M)$, we have $\int_M (\pi_* d\omega + (-1)^{f+1} d\pi_*\omega) \cdot \varphi = \int_M \pi_*(d\omega) \cdot \varphi$

$$+ (-1)^{f+1} \int_M \{d(\pi_*(\omega) \cdot \varphi) - (-1)^{r-f} \pi_*(\omega) \cdot d\varphi\}$$

$$= \int_M \pi_*(d\omega) \cdot \varphi + (-1)^r \int_M \pi_*(\omega) \cdot d\varphi$$

$$= \int_E d\omega \cdot \pi^*(\varphi) + (-1)^r \int_E \omega \cdot \pi^*(d\varphi)$$

$$= \int_E d(\omega \cdot \pi^*(\varphi)) = \int_{\partial E} i^*(\omega) \cdot \pi^{\partial *}(\varphi)$$

$= \int_M \pi_*^{\partial} i^*(\omega) \cdot \varphi$. This proves the desired identity.

If $\partial E = \emptyset$, (3.10 says that π_* either commutes or anticommutes with d, hence, that it induces a homomorphism

$$\pi_* : H_{DR}^r(E) \to H_{DR}^{r-f}(M) .$$

This homomorphism can also be produced in the standard cohomology of fiber bundles by a spectral sequence technique (cf. [1, §8]).

Returning to the study of the de Rham cohomology functor, we remark first that $H^*_{DR}(M)$ has the structure of a graded algebra over $\mathbb{R}$. Indeed, the relation

$$d(\omega \cdot \eta) = d(\omega) \cdot \eta + (-1)^{\deg(\omega)}\omega \cdot d\eta ,$$

assigned in (3.3) as an exercise, implies that $Z^*_{DR}(M)$ is a subring of $A^*(M)$ and that $B^*_{DR}(M)$ is an ideal in $Z^*_{DR}(M)$. Thus $H^*_{DR}(M)$ has a graded algebra structure inherited from that of $A^*(M)$.

Let $\check{H}^*(M;\mathbb{R})$ denote the standard Čech cohomology of M with coefficients $\mathbb{R}$. This is a graded algebra under the cup product. The following theorem is of fundamental importance for relating topological invariants of manifolds to their differential geometric properties.

(3.11) Theórem (de Rham) There is a canonical isomorphism of graded algebras

$$\theta_M : H^*_{DR}(M) \longrightarrow \check{H}^*(M;\mathbb{R}).$$

Furthermore, if $f : M \to N$ is smooth, then the diagram

$$\begin{array}{ccc} H^*_{DR}(M) & \xrightarrow{\theta_M} & \check{H}^*(M;\mathbb{R}) \\ \uparrow f^* & & \uparrow f^* \\ H^*_{DR}(N) & \xrightarrow{\theta_N} & \check{H}^*(N;\mathbb{R}) \end{array}$$

is commutative.

We sketch a proof of (3.11) due to André Weil [4]. Let U be an open cover of M and for each pair (p,q) of nonnegative integers, set

$$K^{p,q}(U) = \check{C}^p(U;A^q) ,$$

the pth Čech cochain module with values in the q-forms. Precisely, each $c \in K^{p,q}(U)$ is a function which to each ordered $(p+1)$-tuple $(U_{\alpha_0},\ldots,U_{\alpha_p})$ of elements of U assigns a q-form

$$c_{\alpha_0\alpha_1\ldots\alpha_p} \in A^q(U_{\alpha_0} \cap \ldots \cap U_{\alpha_p})$$

(interpreted as 0 if $U_{\alpha_0} \cap \ldots \cap U_{\alpha_p} = \emptyset$). The Čech coboundary

$$\delta : K^{p,q}(U) \rightarrow K^{p+1,q}(U)$$

is given by the formula

$$\{\delta(c)\}_{\alpha_0\ldots\alpha_{p+1}} = \sum_{i=0}^{p+1} (-1)^i \varphi_{\alpha_i}(c_{\alpha_0\ldots\hat{\alpha}_i\ldots\alpha_{p+1}})$$

where

$$\varphi_{\alpha_i} : A^q(U_{\alpha_0} \cap \ldots \cap \hat{U}_{\alpha_i} \cap \ldots \cap U_{\alpha_{p+1}}) \rightarrow A^p(U_{\alpha_0} \cap \ldots \cap U_{\alpha_{p+1}})$$

is the restriction map. As usual, $\delta^2 = 0$. The de Rham coboundary

$$d : K^{p,q}(U) \rightarrow K^{p,q+1}(U)$$

is given by

$$\{d(c)\}_{\alpha_0\ldots\alpha_p} = d(c_{\alpha_0\ldots\alpha_p}) .$$

Clearly $d\delta = \delta d$ on $K^{**}(U)$. Set

$$D' = \delta$$
$$D'' = (-1)^p d \quad (\text{on } K^{p,q}(U))$$
$$D = D' + D''$$

and remark that $D^2 = 0$. Set

$$K^n(U) = \bigoplus_{p+q=n} K^{p,q}(U)$$

and remark that

$$D : K^n(U) \to K^{n+1}(U).$$

One can also define a multiplication

$$K^{p,q}(U) \times K^{r,s}(U) \to K^{p+r,q+s}(U)$$

by

$$(\omega\eta)_{\alpha_0 \ldots \alpha_{p+r}} = (-1)^{qr}\, \omega_{\alpha_0 \ldots \alpha_p} \cdot \eta_{\alpha_p \ldots \alpha_{p+r}}$$

on their common domain. This also defines a multiplication

$$K^n(U) \times K^m(U) \to K^{n+m}(U) .$$

Note that if $\omega \in K^{p,q}$, $\eta \in K^{r,s}$, then

$$D'(\omega\eta) = \delta((-1^{rq}\omega \cdot \eta)$$

where $\cdot$ indicates exterior multiplication of forms, hence

$$D'(\omega\eta) = (-1)^{rq} \{\delta(\omega) \cdot \eta + (-1)^p \omega \cdot \delta(\eta)\}$$

$$= (-1)^{rq}(-1)^{rq} D'(\omega)\eta + (-1)^{p+rq}(-1)^{q(r+1)}\omega\, D'(\eta)$$

$$= D'(\omega)\eta + (-1)^{p+q}\omega\, D'(\eta) .$$

Likewise,

$$D''(\omega\eta) = (-1)^{p+r} d((-1)^{rq}\omega \cdot \eta)$$

$$= (-1)^{p+r+rq}\{d(\omega) \cdot \eta + (-1)^q \omega \cdot d\eta\}$$

$$= (-1)^{p+r+rq+p+(q+1)r} D''(\omega)\eta + (-1)^{p+q+r+rq+r+rq}\omega\, D''(\eta)$$

$$= D''(\omega)\eta + (-1)^{p+q}\omega\, D''(\eta) .$$

Thus $D : K^*(U) \to K^*(U)$ is an antiderivation.

Note that the kernel of

$$D' : K^{0,q}(U) \to K^{1,q}(U)$$

is exactly $A^q(M)$ and that the kernel of

$$D'' : K^{p,0}(U) \to K^{p,1}(U)$$

is eactly $\check{C}^p(U;R)$. We define the homomorphisms

$$\alpha : A^*(M) \to K^*(U)$$

$$\beta : \check{C}^*(U;R) \to K^*(U)$$

to be the inclusion maps. These are homomorphisms of cochain complexes and homomorphisms of graded algebras.

(3.12) Definition. U is called a simple covering if every finite nonempty intersection of elements of U is a contractible set.

(3.13) Lemma. Every open cover of M admits a refinement which is a simple covering.

Proof. Put a Riemannian metric on M and choose the refinement such that each element is a geodesically convex set (cf. [2,p.39]). Any finite nonempty intersection is then again geodesically convex, hence contractible.

q.e.d.

Since we are going to pass to the limit over open coverings, (3.13) allows us to assume that U is a simple covering.

Consider the first quadrant array

$$
\begin{array}{ccccccccccc}
\vdots & & \vdots & & \vdots & & & & & & \\
\uparrow d & & \uparrow D'' & & \cdot & & & & \uparrow D'' & & \\
A^q(M) & \xrightarrow{\alpha} & K^{0q}(U) & \xrightarrow{D'} & \cdot & & \cdots & \xrightarrow{D'} & K^{pq}(U) & \xrightarrow{D'} & \cdots \\
\uparrow d & & \uparrow D'' & & \cdot & & & & \uparrow D'' & & \\
\vdots & & \vdots & & \vdots & & & & \vdots & & \\
\uparrow d & & \uparrow D'' & & \cdot & & & & \cdot & & \\
A^1(M) & \xrightarrow{\alpha} & K^{01}(U) & \xrightarrow{D'} & \cdot & & & & \cdot & & \\
\uparrow d & & \uparrow D'' & & \uparrow D'' & & & & \uparrow D'' & & \\
A^0(M) & \xrightarrow{\alpha} & K^{00}(U) & \xrightarrow{D'} & K^{10}(U) & \xrightarrow{D'} & \cdots & \xrightarrow{D'} & K^{p0}(U) & \xrightarrow{D'} & \cdots \\
 & & \uparrow \beta & & \uparrow \beta & & & & \uparrow \beta & & \\
 & & \check{C}^0(U;R) & \xrightarrow[\delta]{} & \check{C}^1(U;R) & \xrightarrow[\delta]{} & \cdots & \xrightarrow[\delta]{} & \check{C}^p(U;R) & \xrightarrow[\delta]{} & \cdots
\end{array}
$$

(3.14) **Lemma.** In the above diagram, the rows are exact. If U is a simple covering the columns are also exact.

Proof. Let $\lambda = \{\lambda_\alpha\}$ be a smooth partition of unity subordinate to the open cover $U = \{U_\alpha\}$. If $c \in \check{C}^p(U;A^q)$, define $L(c) \in \check{C}^{p-1}(U;A^q)$ by

$$\{L(c)\}_{\alpha_0 \ldots \alpha_p} = \sum_\alpha \lambda_\alpha \cdot c_{\alpha\alpha_0 \ldots \alpha_{p-1}}$$

where each $\lambda_\alpha \cdot c_{\alpha\alpha_0 \ldots \alpha_{p-1}}$ is interpreted in the obvious way as an element of $A^q(U_{\alpha_0} \cap \ldots \cap U_{\alpha_{p-1}})$. If $D'(c) = 0$, then a straightforward computation gives $c = D'L(c)$, which proves exactness of the rows. If U is simple, then the columns are exact by (3.7). q.e.d.

Assume in all that follows that U is simple.

(3.15) **Lemma.** In cohomology, $\alpha^* : H^*_{DR}(M) \to H^*(K^*,D)$ and

$\beta^* : H^*(U;R) \to H^*(K^*,D)$ are one-one.

The proof of this is a trivial exercise.

(3.16) Lemma. In cohomology; α^* and β^* are onto.

Proof. Let $z \in K^r(U)$ such that $D(z) = 0$. Write

$$z = \sum_{p=0}^{n} z_p , \qquad z_p \in K^{p,n-p}(U) .$$

Since $D(z) = 0$, we have in particular that $D''(z_0) = 0$. By exactness of columns there is $u \in K^{0,n-1}$ such that $D''(u) = z_0$. Then $z - D(u)$ has component zero in $K^{0,n}(U)$. Repeating this process we finally produce a D-cocycle w lying in $K^{n,0}(U)$ cohomologous to z. But $D(w) = 0$ implies that

$$D'(w) = 0 \text{ and } D''(w) = 0 ,$$

hence $w = \beta(v)$ where $v \in \check{C}^n(U;R)$ is a δ-cocycle. This shows β^* onto. An entirely parallel argument using exactness of the rows proves α^* onto.

q.e.d.

(3.11) follows easily. Indeed

$$\alpha^* : H^*_{DR}(M) \to H^*(K^*(U),D)$$

is an isomorphism for every simple covering U, hence the right hand side has already attained its limit value. Since

$$\beta^* : \check{H}^*(U;R) \to H^*(K^*(U),D)$$

is also an isomorphism, the left hand side of this relation has likewise reached its limit value $\check{H}^*(M;R)$. Thus, set $\theta_M = (\beta^*)^{-1} \circ \alpha^*$, the desired isomorphism of graded algebras. We leave it to the reader to check the naturality of θ_M relative to smooth maps.

References

1. A. Borel and F. Hirzebruch, Characteristic classes and homogeneous spaces, I, Am. J. Math., 80 (1958), pp. 458-538.
2. S. Helgason, Differential Geometry and Symmetric Spaces, Academic Press, New York, 1962.
3. S. Sternberg, Lectures on Differential Geometry, Prentice-Hall, Englewood Cliffs, N. J., 1964.
4. A. Weil, Sur les théorèmes de de Rham, Comment. Math. Helv., 26 (1952), pp. 119-145.

4. Connections on vector bundles.

Let $\pi : E \to M$ be a smooth vector bundle and let $\Gamma(E)$ denote the set of smooth cross sections. That is, the elements $s \in \Gamma(E)$ are smooth functions

$$s : M \to E$$

such that $\pi \circ s = \text{id}$. Clearly $\Gamma(E)$ is a vector space over $\mathbb{R}$ and a module over the algebra $\mathfrak{F}$ of all smooth real valued functions on M. Let $\mathfrak{X}$ denote $\Gamma(T(M))$.

(4.1) Definition. A connection on E is an $\mathbb{R}$-bilinear map

$$\nabla : \mathfrak{X} \times \Gamma(E) \to \Gamma(E)$$

written $\nabla(X,s) = \nabla_X(s)$, such that

1) $\nabla_X(fs) = X(f)s + f\,\nabla_X(s)$

2) $\nabla_{fX}(s) = f\,\nabla_X(s)$

for all $X \in \mathfrak{X}$, $s \in \Gamma(E)$, $f \in \mathfrak{F}$.

While the motivation of this definition may not at first be evident, a little reflection shows that ∇_X is a kind of directional derivative which differentiates smooth sections of E in X-directions. Indeed, property 2) in the definition shows that at any point of M, $\nabla_X(s)$ depends (for fixed s) only on the value of X at that point. Thus if $X_0 \in T_{x_0}(M)$, ∇_{X_0} is a well defined operator transforming smooth sections of E defined near x_0 into vectors in the fiber $\pi^{-1}(x_0)$. Property 1) says that this is a first order differential operator.

(4.2) Lemma. There exists a connection ∇ on E.

Proof. Let $\{U_\alpha\}$ be an open cover of M such that for each α there is a smooth frame $s_\alpha = (s_\alpha^1,\ldots,s_\alpha^q)$ defined on U_α . Then each $s_\alpha^i \in \Gamma(E|U_\alpha)$ and for each $x \in U_\alpha$, $\{s_\alpha^1(x),\ldots,s_\alpha^q(x)\}$ is a basis for $\pi^{-1}(x)$. Let $\{\lambda_\alpha\}$ be a smooth partition of unity subordinate to this cover.

Over U_α define a connection ∇^α on $E|U_\alpha$ by

$$\nabla_X^\alpha(s_\alpha^i) = 0, \quad \forall\, i,\ \forall\, X ,$$

which extends to arbitrary sections s by the property 1) in (4.1). $\nabla = \sum_\alpha \lambda_\alpha \nabla^\alpha$ then makes sense and satisfies (4.1). q.e.d.

We may consider ∇ as an $\mathfrak{F}$-linear map

$$\nabla : \mathfrak{X} \to \mathrm{Hom}_R(\Gamma(E),\Gamma(E)) .$$

Both of these $\mathfrak{F}$-modules are also Lie algebras over R. Generally, ∇ is <u>not</u> a homomorphism of Lie algebras and the measure of its failure to be such is an important geometric quantity.

(4.3) **Definition.** The curvature k of the connection ∇ is the function

$$k : \mathfrak{X} \times \mathfrak{X} \rightarrow \mathrm{Hom}_R(\Gamma(E), \Gamma(E))$$

given by

$$k(X,Y) = [\nabla_X, \nabla_Y] - \nabla_{[X,Y]}$$

$$= \nabla_X \nabla_Y - \nabla_Y \nabla_X - \nabla_{[X,Y]}.$$

Remark that $k(X,Y) = -k(Y,X)$.

(4.4) **Lemma.** If $f,g,h \in \mathfrak{F}$, $X,Y \in \mathfrak{X}$, and $s \in \Gamma(E)$, then $k(fX,gY)(hs) = fgh \cdot k(X,Y)(s)$.

The proof is a direct verification which we leave to the reader. As a consequence, consider a smooth frame $(s^1_\alpha, \ldots, s^q_\alpha)$ for $E|U_\alpha$ and note that there is a $q \times q$ matrix k^α of 2-forms, $k^\alpha = (k^\alpha_{ij})$ such that

$$k(X,Y)(s^i_\alpha) = \sum_{j=1}^{q} k^\alpha_{ji}(X,Y) s^j_\alpha .$$

Here, of course, we use the identification of $A^q(M)$ with the set of alternating $\mathfrak{F}$-multilinear q-forms on $\mathfrak{X}$ which is standard in differential geometry. The fact that $k^\alpha_{ji}(X,Y) = -k^\alpha_{ji}(Y,X)$ together with $\mathfrak{F}$-bilinearity then identifies k^α_{ji} as an element of $A^2(U_\alpha)$.

Referring to the definition (2.1) of vector bundles, we note that, on $U_\alpha \cap U_\beta$, k^α and k^β are related by

$$k^\alpha = g_{\alpha\beta} k^\beta g^{-1}_{\alpha\beta} .$$

The proof of this crucial formula is an elementary exercise using (4.4).

We also note that on U_α there is a $q \times q$ matrix θ^α of 1-forms, $\theta^\alpha = (\theta^\alpha_{ij})$, such that

$$\nabla_X(s^i_\alpha) = \sum_{j=1}^{q} \theta^\alpha_{ij}(X) s^i_\alpha .$$

The following basic structural equation will be found proven in various books on differential geometry (e.g., cf.[1, p.118]).

(4.5) Lemma. $d\theta^{\alpha} - \theta^{\alpha} \cdot \theta^{\alpha} = k^{\alpha}$.

Here, of course, the matrix product $\theta^{\alpha} \cdot \theta^{\alpha}$ involves the usual multiplication of 1-forms in the entries.

As a final remark about connections, we define the notion of parallel translation. If $\sigma : [0,1] \to M$ is a smooth curve, one defines the notion of a "section of E along σ" as follows.

(4.6) Definition. A section of E along σ is a smooth curve $s : [0,1] \to E$ such that $\pi \circ s = \sigma$.

If s is a section of E along σ, the connection ∇ defines the "covariant derivative of s along σ", a new section $\frac{Ds}{dt}$ of E along σ. If σ is regular (hence locally one-one) it is an easy exercise to extend s locally to a section of E and show that $\nabla_{\dot\sigma(t)}(s)$ does not depend on this extension, hence defines $\frac{Ds}{dt}$. To handle general σ, proceed as in [2, §8] .

(4.7) Definition. A section s of E along σ is parallel iff $\frac{Ds}{dt} \equiv 0$.

Given $e_0 \in T_{\sigma(0)}(M)$, the parallel translate of e_0 along σ will be the unique section e of E along σ satisfying

1) $e(0) = e_0$

2) e is parallel.

The existence and uniqueness of e is an elementary consequence of the theory of ordinary differential equations.

References

1. S. Kobayashi and K. Nomizu, Foundations of Differential Geometry, Vol. I Interscience Publishers, New York. N. Y., 1963.

2. J. Milnor, Morse Theory, Princeton University Press, Princeton, N. J., 1963.

5. The Pontryagin classes. We suppose given a smooth vector bundle

$$\pi : E \to M$$

and a connection ∇ on E. Let k be the curvature of ∇. Let gl_q denote the Lie algebra of $GL_q = GL(q,R)$. Thus gl_q is the set of all $q \times q$ real matrices, while GL_q is the set of nonsingular $q \times q$ real matrices.

(5.1) Definition. A polynomial function

$$\varphi : gl_q \to R$$

is called invariant if

$$\varphi(m) = \varphi(gmg^{-1}), \forall g \in GL_q, \forall m \in gl_q \quad .$$

Since sums and products of invariant polynomials are invariant, these polynomials form an algebra over R.

There are certain basic invariant polynomials $\Sigma_0, \Sigma_1, \Sigma_2, \ldots$ which are defined as follows.

$$\Sigma_p(m) = \text{trace } (m^p).$$

The following purely algebraic theorem will be of use.

(5.2) Theorem. The algebra of invariant polynomials has $\{\Sigma_i\}_{i=0}^{\infty}$ as a set of generators.

For a proof of (5.2), cf. the Appendix.

Now recall the curvature matrices k^{α} of 2-forms defined on U_{α}. If φ is an invariant polynomial of degree r, then $\varphi(k^{\alpha}) \in A^{2r}(U_{\alpha})$. Furthermore, on $U_{\alpha} \cap U_{\beta}$,

$$\varphi(k^{\alpha}) = \varphi(g_{\alpha\beta} k^{\beta} g_{\alpha\beta}^{-1}) = \varphi(k^{\beta}).$$

Thus these 2r-forms fit together to give a well defined element $\varphi(k) \in A^{2r}(M)$. Note that this form depends only on the curvature k, not on choices of local trivializations of E.

(5.3) Lemma. $d(\varphi(k)) = 0$.

Proof. Let $x_0 \in M$ be any point, U_{α} an open neighborhood of x_0 which trivializes E and on which there is a smooth spherical coordinate system centered at x_0. Let s_0 be any frame for $\pi^{-1}(x_0)$ and spread this out to a smooth frame s_{α} over U_{α} by parallel translations along the radial curves of the spherical coordinate system. Then, for all $X \in T_{x_0}(M)$, $\theta^{\alpha}(X) = 0$. (i.e., every θ^{α}_{ij} is a 1-form which vanishes at x_0). On U_{α},

$$d(\varphi(k)) = d(\varphi(k^{\alpha})) = d(\varphi(d\theta^{\alpha} - \theta^{\alpha} \cdot \theta^{\alpha})) .$$

But $\varphi(d\theta^{\alpha} - \theta^{\alpha} \cdot \theta^{\alpha})$ is a sum of monomials of forms in which either two of the forms vanish at x_0 or all of the forms are $d\theta^{\alpha}_{ij}$'s. Thus

$$\{d(\varphi(k))\}_{x_0} = 0 .$$

But x_0 was an arbitrary point of M, hence $d(\varphi(k)) = 0$. q.e.d.

(5.4) Proposition. $[\varphi(k)] \in H^{2r}_{DR}(M)$ is independent of the choice of connection.

Proof. Let ∇^0 and ∇^1 be two connections on E. Let k^0 and k^1 be the corresponding curvatures. In a natural way,

$$\pi \times id : E \times R \longrightarrow M \times R$$

defines a smooth vector bundle over $M \times R$ with the same fiber dimension as E. Define a connection $\tilde{\nabla}$ on this bundle as follows. On sections s which are constant in R - directions, let $\tilde{\nabla}_{\partial/\partial t}(s) = 0$. If $X \in T_{(x,t)}(M \times \{t\})$, define

$$\tilde{\nabla}_X(s) = (1-t)\nabla^0_X(s) + t\,\nabla^1_X(s) \quad .$$

Every section is a function-linear combination of sections constant in R-directions, hence $\tilde{\nabla}$ extends to all sections by use of property 1) in (4.1). It is easy to check that $\tilde{\nabla}$ is a connection. Let $\tilde{k}$ be its curvature and consider $[\varphi(\tilde{k})] \in H^{2r}_{DR}(M \times R)$. Let i_0 and $i_1 : M \longrightarrow M \times R$ be as usual. Then $i_0^* = i_1^*$ in cohomology by the proof of (3.6), and

$$i_0^*[\varphi(\tilde{k})] = [\varphi(k^0)]$$

$$i_1^*[\varphi(\tilde{k})] = [\varphi(k^1)] \ .$$

q.e.d.

(5.5) **Definition.** If φ is an invariant polynomial of degree r, $\varphi(E) = [\varphi(k)] \in H^{2r}_{DR}(M)$ is called the Pontryagin class corresponding to φ. The set of all such Pontryagin classes is the graded subalgebra

$$\mathrm{Pont}^*(E) \subset H^*_{DR}(M)$$

called the Pontryagin algebra of E.

(5.6) **Proposition.** $\mathrm{Pont}^j(E) = 0$ if j is not divisible by 4.

Proof. From the definition it is clear that this is 0 if j is not divisible by 2. To obtain the finer result, we put a smoothly varying positive definite inner product $<\,,\,>$ on the fibers of E. This is easily done locally and then pieced together with a smooth partition of unity. Next we want a connection ∇ on E such that

$$\nabla_X < s_1,s_2 > = < \nabla_X s_1,s_2 > + < s_1,\nabla_X s_2 > ,$$

$\forall X \in \mathfrak{X}$, $\forall s_1, s_2 \in \Gamma(E)$. On U_α choose a smooth orthonormal frame s_α. This is easy to do by applying Gram-Schmidt orthonormalization to any smooth frame. Then define ∇^α on U_α by requiring

$$\nabla^\alpha_X(s^i_\alpha) = 0, \quad i = 1,\dots,q.$$

It is easy to check that ∇^α has the required property. Piece these local connections together by a smooth partition of unity and obtain ∇ as desired.

The curvature k of this connection has the important property

$$< k(X,Y)s_1,s_2 > = < s_1,-k(X,Y)s_2 > .$$

The proof of this is left as an elementary exercise. One immediate consequence is that, if on U_α we choose s_α to be a smooth orthonormal frame, then the matrix k^α is antisymmetric. Thus also, if $r > 0$ is odd, then $(k^\alpha)^r$ is antisymmetric, hence

$$\Sigma_r(k^\alpha) = \text{trace } ((k^\alpha)^r) = 0.$$

Thus $\Sigma_r(k) = 0$. But by (5.2) every invariant φ of odd degree is a linear combination of polynomials of the form $\Sigma_{i_1}\Sigma_{i_2}\cdots\Sigma_{i_m}$ in which some of the indices i_j are odd. Thus, $\deg(\varphi)$ odd implies $\varphi(k) = 0$. q.e.d.

This proposition enables us to define the total Pontryagin class of E to be

$$p(E) = 1 + p_1(E) + \cdots + p_{[q/2]}(E)$$

$$= [\det(I + \frac{\sqrt{-1}}{2\pi} k)]$$

where each $p_j(E) \in H^{4j}_{DR}(M)$. No imaginary components arise because of (5.6). These classes $1, p_1(E),\dots,p_{[q/2]}(E)$ actually generate the whole algebra

$\text{Pont}^*(E)$.

Note that this definition gives an easy proof of the formula $p(E \oplus E') = p(E) \cdot p(E')$.

For the case of complex vector bundles, Chern classes can be defined in complex de Rham cohomology $H^*_{DR}(M;C)$ by exactly the same techniques (using complex smooth p-forms $A^p_C(M)$). If E has complex dimension n, one obtains $c(E) = 1 + c_1(E) + \cdots + c_n(E)$, $c_i(E) \in H^{2i}_{DR}(M;C)$, and again $c(E \oplus E') = c(E) \cdot c(E')$.

For complex analytic bundles on a complex analytic manifold, holomorphic connections do not generally exist, although they can be defined locally. As a consequence, if one insists on using these local holomorphic connections there is generally no way of producing forms in the de Rham complex representing Chern classes, but it is possible to define in this way representative cocycles in the double complex $\check{C}^*(U; A^*_C(M)) = K^*(U)$. We sketch this construction.

Let $U = \{U_\alpha\}$ be a simple covering of M such that each U_α is a trivializing neighborhood for the complex analytic vector bundle E. One defines a holomorphic connection ∇^α on $E|U_\alpha$ by defining it as trivial on a holomorphic frame field. One then obtains a curvature form k^α on U_α and, if φ is an invariant polynomial of degree r on $g\ell(q,C)$, $\varphi(k^\alpha) \in A^{2r}_C(U_\alpha)$. Thus we define $\varphi^0 \in \check{C}^0(U;A^{2r}_C)$ by

$$\varphi^0_\alpha = \varphi(k^\alpha).$$

On $U_\alpha \cap U_\beta$ we have the two connections ∇^α and ∇^β . Thus on $(U_\alpha \cap U_\beta) \times C$ put the connection $(1-z)\nabla^\alpha + z\nabla^\beta$ defined analogously to the $\tilde{\nabla}$ in the proof of (5.4). Let $k^{\alpha\beta}$ be the corresponding curvature form, $\varphi(k^{\alpha\beta}) \in A^{2r}((U_\alpha \cap U_\beta) \times C)$. Let Δ^p denote the standard p-simplex.

Then

$$(U_\alpha \cap U_\beta) \times \Delta^1 \subset (U_\alpha \cap U_\beta) \times K \subset (U_\alpha \cap U_\beta) \times C$$

and the projection

$$\pi^{\Delta^1} : (U_\alpha \cap U_\beta) \times \Delta^1 \to U_\alpha \cap U_\beta$$

defines (via integration along the fiber)

$$\pi_*^{\Delta^1} : \check{C}^1(U;A_C^{2r}) \to \check{C}^1(U;A_C^{2r-1}).$$

Define $\qquad \varphi^1 \in \check{C}^1(U;A_C^{2r-1})$ by

$$\varphi^1_{\alpha\beta} = \pi_*^{\Delta^1}(\varphi(k^{\alpha\beta})).$$

On $U_\alpha \cap U_\beta \cap U_\gamma$ we work with three connections and the convex combination of these over $(U_\alpha \cap U_\beta \cap U_\gamma) \times C^2$ and we use

$$\pi^{\Delta^2} : (U_\alpha \cap U_\beta \cap U_\gamma) \times \Delta^2 \to U_\alpha \cap U_\beta \cap U_\gamma$$

to define $\qquad \varphi^2 \in \check{C}^2(U;A_C^{2r-2})$ by

$$\varphi^2_{\alpha\beta\gamma} = \pi_*^{\Delta^2}(\varphi(k^{\alpha\beta\gamma})) .$$

Continuing in this way, we produce

$$\bar{\varphi} = (\epsilon_0\varphi^0, \epsilon_1\varphi^1, \ldots, \epsilon_n\varphi^n) \in K^{2r}(U).$$

Here each $\epsilon_i = (-1)^{[(i+1)/2]}$. This makes $\bar{\varphi}$ a cocycle because of the relation (3.10)

$$\pi_*^{\Delta^p} \circ d + (-1)^{p+1} d \circ \pi_*^{\Delta^p} = \pi_*^{\partial\Delta^p} \circ i^* = \sum_{j=0}^{p} (-1)^j \pi_*^{\Delta^{p-1}(j)}$$

where $\Delta^{p-1}(j)$ is the jth face of Δ^p. The last equality is by the combinatorial version of Stoke's Theorem [1, p. 109]. Since $d\varphi(k^{\alpha_0\ldots\alpha_p}) = 0$, the above relation guarantees that

$$d(\varphi^0) = 0$$

$$d(\varphi^p) = (-1)^{p+1}\delta(\varphi^{p-1}), \quad p > 0 .$$

Thus $\overline{\varphi}$ is a cocycle and we obtain the Chern class

$$\varphi(E) = [\overline{\varphi}] \in H^{2r}(K^*(U)) = H^{2r}_{DR}(M;C).$$

We might further note that, by the definition of integration along the fiber, the cochains φ^p will be 0 for $p > r$ (i.e., below the diagonal in $K^{**}(U)$).

References

1. S. Sternberg, Lectures on Differential Geometry, Prentice-Hall, Englewood Cliffs, N. J., 1964.

6. Characteristic classes and integrability

We here suppose that E is an integrable subbundle of $T(M)$. Let $Q = T(M)/E$ with fiber dimension q. Thus we assume the hypotheses of (*) in § 2 and must prove

$$\text{Pont}^k(Q) = 0, \quad k > 2q .$$

Since E is integrable, (1.3)′ implies that all $X,Y \in \Gamma(E)$ must have $[X,Y] \in \Gamma(E)$. If $Z \in \Gamma(Q)$, then $Z = \pi(\widetilde{Z})$, some $\widetilde{Z} \in \mathfrak{X}$, where $\pi : T(M) \rightarrow Q$ is the canonical projection. $\widetilde{Z}$ is well defined modulo elements of $\Gamma(E)$. Thus for $X \in \Gamma(E)$ and $Z \in \Gamma(Q)$,

$$\nabla_X(Z) = \pi[X,\widetilde{Z}]$$

is well defined. This is clearly $\mathbb{R}$-bilinear as a map

$$\nabla : \Gamma(E) \times \Gamma(Q) \longrightarrow \Gamma(Q)$$

and satisfies

1) $\nabla_X(fZ) = X(f)Z + f\nabla_X(Z)$,

2) $\nabla_{fX}(Z) = f\,\nabla_X(Z)$,

as is easily verified. This satisfies the definition of a connection on Q except that the variable X is restricted to range over $\Gamma(E) \subset \mathfrak{X}$. In order to complete ∇ to a connection, we use a Riemannian metric on $T(M)$ to split this bundle into the direct sum of E and the orthogonal complement bundle to E in $T(M)$. This complement is isomorphic to Q, so by the choice of Riemannian metric we have obtained an isomorphism

$$T(M) \cong E \oplus Q .$$

Let $\bar{\nabla}$ be any connection on Q. For $X \in \mathfrak{X} = \Gamma(E) \oplus \Gamma(Q)$, write $X = X_E + X_Q$. Then define

$$\nabla_X(Z) = \nabla_{X_E}(Z) + \bar{\nabla}_{X_Q}(Z),$$

$\forall\, Z \in \Gamma(Q)$. Thus if $X \in \Gamma(E) \subset \mathfrak{X}$, $X_Q = 0$ and we obtain the previous formula for $\nabla_X(Z)$. It is trivial to check that this defines a connection on Q.

(6.1) **Definition.** A basic connection ∇ on Q is one such that

$$\nabla_X(Z) = \pi[X,\tilde{Z}],$$

$\forall\, X \in \Gamma(E)$, where $\tilde{Z} \in \mathfrak{X}$ is such that $\pi(\tilde{Z}) = Z$.

We have proven

(6.2) **Lemma.** Under the assumption that E is integrable, there exists a basic connection on Q.

(6.3) Lemma. Let ∇ be a basic connection on Q, k the curvature of ∇. Then $k(X,X') \equiv 0$, $\forall\ X,X' \in \Gamma(E)$.

Proof. Let $Z \in \Gamma(Q)$ and $\tilde{Z} \in \mathfrak{X}$ with $\pi(\tilde{Z}) = Z$. Then

$$k(X,X')(Z) = \nabla_X\nabla_{X'}(Z) - \nabla_{X'}\nabla_X(Z) - \nabla_{[X,X']}(Z)$$

$$= \nabla_X(\pi[X',\tilde{Z}]) - \nabla_{X'}(\pi[X,\tilde{Z}]) - \pi[[X,X'],\tilde{Z}] .$$

But we can choose

$$\widetilde{\pi[X,\tilde{Z}]} = [X,\tilde{Z}]$$

$$\widetilde{\pi[X',\tilde{Z}]} = [X',\tilde{Z}]$$

so

$$k(X,X')(Z) = \pi[X,[X',\tilde{Z}]] - \pi[X',[X,\tilde{Z}]] - \pi[[X,X'],\tilde{Z}] = \pi(0) = 0$$

by the Jacobi identity. q.e.d.

(6.4) Lemma. Let $U_\alpha \subset M$ be a simultaneously trivializing neighborhood for Q and E, s_α a smooth frame for Q over U_α. Let $I_\alpha(E)$ be the ideal in $A^*(U_\alpha)$ generated by those 1-forms which vanish on $\Gamma(E|U_\alpha)$. Let k^α be the curvature matrix associated to the frame s_α by a basic connection. Then each $k^\alpha_{ij} \in I_\alpha(E)$.

Proof. Over U_α, E can be described as the set of tangent vectors on which certain 1 forms $\theta_1,\ldots,\theta_q$ vanish, these 1-forms being linearly independent at each point of U_α. In particular, $I_\alpha(E)$ is generated by $\theta_1,\ldots,\theta_q$. Complete these to a basis of 1-forms by $\theta_{q+1},\ldots,\theta_n$. These last restrict to a basis of E^*_p , $\forall\ p \in U_\alpha$. Consider a nontrivial form

$$\omega = \sum_{q+1<i<j} g_{ij}\theta_i \cdot \theta_j .$$

Clearly there are $X, X' \in \Gamma(E|U_\alpha)$ such that

$$\omega(X, X') \not\equiv 0.$$

By (6.3) it follows that each $k^\alpha_{ij} \in I_\alpha(E)$. q.e.d.

We now prove Theorem (*) of §2. Elements of $I_\alpha(E)$ are typically of the form

$$\omega = \sum_{i=0}^{q} \omega_i \cdot \theta_i , \quad \omega_i \in A^*(U_\alpha) .$$

Thus $I_\alpha(E)^{q+1} = 0$. If φ is an invariant polynomial of degree $> q$, then (6.4) and this remark imply

$$\varphi(k^\alpha) = 0 .$$

Thus $\varphi(k) = 0$ for $\deg(\varphi) > q$ and (*) follows immediately by our definition of the Pontryagin algebra.

As remarked in §2, this gives the global integrability theorem (*)′. By exactly parallel reasoning one also obtains the following global holomorphic analogue.

(*)″ Theorem. If M is a complex analytic manifold, $T(M)$ the holomorphic tangent bundle, $E \subset T(M)$ a complex subbundle which is isomorphic to a holomorphic integrable subbundle $E' \subset T(M)$, $Q = T(M)/E$, and $q = \dim_{\mathbb{C}}(Q)$, then $\mathrm{Chern}^k(Q) = 0$ for $k > 2q$.

Using this theorem, we exhibit the first known counterexample to global integrability. Let CP^n denote the complex n-dimensional projective space, $T = T(CP^n)$ its holomorphic tangent bundle.

(6.5) Theorem. If n is odd, then T contains a holomorphic subbundle of complex codimension one. If, furthermore, $n > 1$, no holomorphic subbundle of T with codimension one is integrable.

Proof. Let $[n+1]$ denote the trivial bundle $CP^n \times C^{n+1}$ over CP^n. Each $x \in CP^n$ is a one dimensional subspace of C^{n+1}, hence one can define

$$S = \{(x,v) \in [n+1]: v \in x\}$$

$$Q = \{(x,v) \in [n+1]: v \perp x\} ,$$

holomorphic bundles of respective complex dimensions 1 and n, $[n+1] = S \oplus Q$. It is rather well known that there is a canonical isomorphism

$$T \cong \mathrm{Hom}(S,Q)$$

of holomorphic bundles. Indeed, given $\varphi \in \mathrm{Hom}(S_x,Q_x)$, $x \in CP^n$, choose any nonzero $(x,v) \in S_x$, and define

$$\sigma : U \to CP^n$$

on $U = \{z \in C : |z| < 1\}$, by letting $\sigma(z)$ be the one dimensional subspace of C^{n+1} containing $(1-z)v + z\varphi(v)$. Then σ is a holomorphic curve with $\sigma(0) = x$. The holomorphic tangent vector to CP^n at x determined by this curve will be labeled $\nu(\varphi)$. In this way one obtains the desired isomorphism

$$\nu : \mathrm{Hom}(S,Q) \to T(CP^n).$$

Let n be odd, choose a basis $v_1,\ldots,v_{n+1}$ of $(C^{n+1})^*$, and form the nondegenerate antisymmetric bilinear form

$$\omega = v_1 \wedge v_2 + v_3 \wedge v_4 + \cdots + v_n \wedge v_{n+1} .$$

Let H denote the dual bundle S^* and let H^2 be the symmetric square of H. Thus each $\eta \in H_x^2$ is simply a homogeneous function of degree 2 from $S_x \to C$. H^2 is a line bundle. Define

$$\omega_x : \mathrm{Hom}(S_x, Q_x) \to H_x^2$$

by

$$\omega_x(\varphi)(v) = \omega(v,\varphi(v)), \quad \forall\, v \in S_x .$$

Since ω is nondegenerate and $\varphi(v)$ ranges over the orthogonal complement of $v \neq 0$, ω_x is a surjection, $\forall\, x \in CP^n$, hence defines a holomorphic bundle surjection

$$\omega_* : T = \mathrm{Hom}(S,Q) \rightarrow H^2 .$$

$\mathrm{Ker}(\omega_*)$ is a holomorphic subbundle of T of complex codimension one.

Finally, suppose that $n > 1$ and let $E \subset T$ be holomorphic of codimension one. If E is integrable,

$$\{c_1(T/E)\}^2 \in \mathrm{Chern}^4(T/E) = 0$$

by (*)''. Since $H^*(CP^n;C)$ is well known to be generated by 1 and by $\mu = c_1(S^*) \in H^2(CP^n;C)$ with the single relation $\mu^n = 0$, the above together with $n > 1$ implies that $c_1(T/E) = 0$. But

$$c(T) = c(E) \cdot c(T/E)$$

then implies $c_n(T) = 0$. On the other hand

$$T \oplus [1] = \mathrm{Hom}(S,Q \oplus S) = \mathrm{Hom}(S,[n+1]) = S^* \oplus \cdots \oplus S^* \text{ (n+1 times)},$$

so

$$c(T) = c(T \oplus [1]) = c(S^*)^{n+1} = (1 + \mu)^{n+1}$$

shows that $c_n(T) \neq 0$. This contradiction proves that E cannot be integrable. q.e.d.

As remarked in § 2, the proof of (*) has really proved something much stronger. We indicate an idea, due to Shulman, which exploits this stronger result.

Consider cohomology classes $[\alpha]$, $[\beta]$, $[\gamma]$ (in any suitable cohomology theory) such that

$$[\alpha]\cdot[\beta] = 0$$

$$[\beta]\cdot[\gamma] = 0 \quad .$$

Then, at the cocycle level,

$$\alpha\cdot\beta = \delta(x)$$

$$\beta\cdot\gamma = (-1)^{\deg(\alpha)}\delta(y)$$

so that $\alpha\cdot\beta\cdot\gamma$ is cohomologous to 0 in two different ways:

$$\alpha\cdot\beta\cdot\gamma = \delta(x\cdot\gamma)$$

$$\alpha\cdot\beta\cdot\gamma = \delta(\alpha\cdot y) \quad .$$

Thus $x\cdot\gamma - \alpha\cdot y$ is a cocycle and we set

$$< [\alpha],\ [\beta],\ [\gamma] > = [x\cdot\gamma - \alpha\cdot y] \in H^* \quad .$$

This is not well defined independently of the various choices, but it is easy to check that the indeterminacy is the ideal $I^*([\alpha],[\gamma])$ in H^* generated by $[\alpha]$ and $[\gamma]$. Thus $<[\alpha],[\beta],[\gamma]>$ is well defined as an element of $H^*/I^*([\alpha],[\gamma])$. This is called the Massey triple product of $[\alpha],[\beta],[\gamma]$. Higher order Massey products can also be defined [1].

The following is immediate from our definitions.

(6.6) Lemma. If $[\alpha]\cdot[\beta] = 0 = [\beta]\cdot[\gamma]$ and if the representative cocycles can so be chosen that $\alpha\cdot\beta = 0 = \beta\cdot\gamma$, then $<[\alpha],[\beta],[\gamma]> = 0$.

(6.7) Theorem. (Shulman) If E is integrable and $a,b,c \in \mathrm{Pont}^*(Q)$ are such that $\deg(a) + \deg(b) > 2q$ and $\deg(b) + \deg(c) > 2q$, then $< a,b,c >$ is defined and $= 0$.

Proof. Let k be the curvature of a basic connection and let $\varphi_1,\varphi_2,\varphi_3$ be invariant polynomials such that $a = [\varphi_1(k)]$, $b = [\varphi_2(k)]$, $c = [\varphi_3(k)]$. Then $a\cdot b = [(\varphi_1\varphi_2)(k)] = 0$ and $b\cdot c = [(\varphi_2\varphi_3)(k)] = 0$, so $< a,b,c >$ is defined. But actually,

$$\varphi_1(k) \cdot \varphi_2(k) = 0$$

$$\varphi_2(k) \cdot \varphi_3(k) = 0 ,$$

so by (6.6), $< a,b,c > = 0$. q.e.d.

For maximum usefulness of (6.7), it should be noted that the isomorphism between de Rham cohomology and Čech cohomology preserves Massey products. This follows directly from the fact that

$$\alpha : A^*(M) \to K^*(U)$$
$$\beta : \check{C}^*(U;R) \to K^*(U)$$

are ring homomorphisms.

We note also that similar treatments are possible for higher order Massey products. Using (6.7), Shulman has exhibited bundles which satisfy (*) but are not integrable.

References

1. W. S. Massey, Some higher order cohomology operations, Symposium Internacional de Topología Algebraica, Univ. Nac. Autonoma de Mexico and UNESCO, Mexico City, 1958.

7. Haefliger structures and the functor Γ_q. We assume here some acquaintance with the basic concepts of sheaves (cf. [1], [2]).

Over R^q construct the sheaf $\mathcal{D}$ of germs of local diffeomorphisms of $R^q \to R^q$. That is, if $x \in R^q$, the stalk $\mathcal{D}_x$ is the set of germs at x of diffeomorphisms of open neighborhoods of x onto open sets of R^q. If $x,y \in R^q$ we adopt the notation $\mathcal{D}_{(x,y)}$ for the set $\{\gamma \in \mathcal{D}_x : \gamma(x) = y\}$.

(7.1) **Definition.** A Haefliger cocycle on a topological space X (also called a Γ_q-cocycle on X) consists of the following data:

1) An open cover $\{U_\alpha\}_{\alpha \in A}$ of X.

2) For each $\alpha \in A$, a continuous map $f_\alpha : U_\alpha \to R^q$ (the germ of which at $x \in U_\alpha$ will be denoted f_α^x).

3) For each $x \in U_\alpha \cap U_\beta$ a germ $\gamma_{\alpha\beta}^x \in \mathcal{D}_{(f_\beta(x), f_\alpha(x))}$ such that

a) The assignment $x \to \gamma_{\alpha\beta}^x$ defines a continuous map $U_\alpha \cap U_\beta \to \mathcal{D}$

b) $f_\alpha^x = \gamma_{\alpha\beta}^x \circ f_\beta^x$

c) $\gamma_{\alpha\beta}^x \circ \gamma_{\beta\delta}^x = \gamma_{\alpha\delta}^x$.

(7.2) **Definition.** Γ_q-cocycles $c = \{U_\alpha, f_\alpha, \gamma^x{}_{\alpha\beta}\}_{\alpha,\beta \in A}$ and $c' = \{U_\lambda, f_\lambda, \gamma_{\lambda\mu}^x\}_{\lambda,\mu \in B}$ are called equivalent if $\exists$ cocycle corresponding to the covering $\{U_\zeta\}_{\zeta \in C}$, C = disjoint union of A and B, which restricts (in the obvious sense) to c on $\{U_\alpha\}_{\alpha \in A}$ and to c' on $\{U_\lambda\}_{\lambda \in B}$.

(7.3) **Exercise.** Show that the above relation of equivalence is an equivalence relation. [Hint. If $\{U_a, f_a, \gamma_{ab}^x\}_{a,b \in A \cup B}$ and $\{U_s, f_s, \gamma_{st}^x\}_{s,t \in B \cup C}$ restrict to the same Γ_q-cocycle on $\{U_\lambda\}_{\lambda \in B}$, then for $a \in A$, $s \in C$, $x \in U_a \cap U_s$, find $\lambda \in B$ with $x \in U_\lambda$ and try to define $\gamma_{sa}^x = \gamma_{s\lambda}^x \circ \gamma_{\lambda a}^x$. Show this independent of λ].

(7.4) **Definition.** A Haefliger structure (or Γ_q-structure) on X is an equivalence class of Haefliger cocycles on X. The set of all such structures on X is denoted $H^1(X;\Gamma_q)$.

Of course, the principal example of a Haefliger structure which we have in mind is a foliation. Indeed, an integrable subbundle $E \subset T(M)$ of codimension q allows us to cover M with coordinate patches U_α and to find submersions

$$f_\alpha : U_\alpha \to R^q$$

such that $E|U_\alpha = \mathrm{Ker}(df_\alpha)$. The fact that the maps f_α are submersions together with the fact that $\mathrm{Ker}(df_\alpha)$ and $\mathrm{Ker}(df_\beta)$ agree on $U_\alpha \cap U_\beta$ implies that for each $x \in U_\alpha \cap U_\beta$ there is a unique $\gamma^x_{\alpha\beta} \in \mathcal{D}_{(f_\beta(x), f_\alpha(x))}$ such that $\gamma^x_{\alpha\beta} \circ f^x_\beta = f^x_\alpha$. This uniqueness makes it easy to verify the various properties of a Haefliger cocycle. Finally, two cocycles obtained from the same foliation are clearly equivalent, hence the foliation defines a unique element of $H^1(M;\Gamma_q)$.

One defines the normal bundle Q of a Haefliger structure by the GL_q-cocycle $g_{\alpha\beta}(x) = d(\gamma^x_{\alpha\beta})$. Equivalent Γ_q-cocycles give equivalent GL_q-cocycles, hence the normal bundle of a Γ_q structure is uniquely determined up to isomorphism. For the case of a foliation E, Q is just $T(M)/E$.

Remark that a continuous map $g : Y \to X$ can be used to pull back a Γ_q-cocycle on X to one on Y. Indeed, $\{U_\alpha, f_\alpha, \gamma^x_{\alpha\beta}\}_{\alpha,\beta \in A}$ pulls back to $\{\bar{U}_\alpha, \bar{f}_\alpha, \bar{\gamma}^y_{\alpha\beta}\}_{\alpha,\beta \in A}$ where

$$\bar{U}_\alpha = g^{-1}(U_\alpha)$$

$$\bar{f}_\alpha = f_\alpha \circ g$$

$$\bar{\gamma}^y_{\alpha\beta} = \gamma^{g(y)}_{\alpha\beta} .$$

This pull-back by g respects equivalence, hence defines

$$g^* : H^1(X;\Gamma_q) \to H^1(Y;\Gamma_q).$$

This makes $H^1(;\Gamma_q)$ into a contravariant functor from the category of topological spaces to the category of sets.

It is an unpleasant fact of life that this functor is not homotopy invariant. As an example, consider the maps

$$g_0 : R^q \to R^q , \quad g_0(x) = 0, \quad \forall\, x \in R^q$$

$$g_1 : R^q \to R^q , \quad g_1(x) = x, \quad \forall\, x \in R^q .$$

These maps are homotopic. On the image space R^q consider the Γ_q-cocycle c with just one element $U_1 = R^q$ in the open cover, just one $f_1 = id : U_1 \to R^q$, and every γ_{11}^x = germ of the identity diffeomorphism. We claim that $g_0^*(c)$ is not equivalent to $g_1^*(c)$. To see this clearly we introduce a definition and a lemma.

(7.5) Definition. If $c = \{U_\alpha, f_\alpha, \gamma_{\alpha\beta}^x\}_{\alpha,\beta \in A}$ is a Γ_q-cocycle on X, the level set of c through $x \in U_\alpha$ is defined to be $X_\alpha^x = \{y \in U_\alpha : f_\alpha(y) = f_\alpha(x)\}$.

(7.6) Lemma. If $\{U_\alpha, f_\alpha, \gamma_{\alpha\beta}^x\}_{\alpha,\beta \in A}$ and $\{U_\lambda, f_\lambda, \gamma_{\lambda\mu}^x\}_{\lambda,\mu \in B}$ are equivalent Γ_q-cocyles on X, then for any $x \in U_\alpha \cap U_\lambda$, $\alpha \in A$, $\lambda \in B$, there is an open neighborhood W of x such that $X_\alpha^x \cap W = X_\lambda^x \cap W$.

The proof of (7.6) is an elementary application of definitions. Returning to our example, $g_0^*(c)$ and $g_1^*(c)$ cannot be equivalent since the only level set for $g_0^*(c)$ is R^q while the level sets for $g_1^*(c)$ are the single points of R^q.

In order to obtain a homotopy invariant functor, we impose a further equivalence relation on $H^1(X;\Gamma_q)$.

(7.7) Definition. If $\alpha, \alpha' \in H^1(X;\Gamma_q)$ we say that α and α' are homotopic and write $\alpha \simeq \alpha'$ if and only if $\exists \quad \lambda \in H^1(X \times I; \Gamma_q)$ such that $\alpha = i_0^*(\lambda)$, $\alpha' = i_1^*(\lambda)$.

Here, of course, $i_0, i_1 : X \to X \times I$ are the usual face maps.

(7.8) Exercises. 1) Prove that homotopy is an equivalence relation on $H^1(X;\Gamma_q)$.

2) If $f : X \to Y$ is continuous, prove that $f^* : H^1(Y;\Gamma_q) \to H^1(X;\Gamma_q)$ preserves the relation of homotopy.

3) If $f, g : X \to Y$ are homotopic maps, $\alpha \in H^1(Y;\Gamma_q)$, then $f^*(\alpha) \simeq g^*(\alpha)$.

(7.9) Definition. $\Gamma_q(X)$ = set of homotopy classes in $H^1(X;\Gamma_q)$.

By 2) and 3) of (7.8) we see that $\Gamma_q(\ )$ is a homotopy invariant contravariant functor.

Haefliger [4], [5] has shown that the functor Γ_q is representable. That is, one can construct a space $B\Gamma_q$ such that, for "reasonable" spaces X, there is a canonical one-one correspondence between the set $\Gamma_q(X)$ and the set of homotopy classes of maps $[X, B\Gamma_q]$. Furthermore, this correspondence is natural with respect to the set maps induced by continuous maps, so we have an equivalence of functors

$$\Gamma_q(\ ) \cong [\ \ , B\Gamma_q]$$

on a reasonable category of spaces. There are three ways of showing this.

1) Γ_q is shown to be representable by checking the axioms of Brown [2] for representable functors. This works for CW-complexes X.

2) One can mimic Milnor's construction of the classifying space of a topological group [6]. This works for all paracompact spaces X.

3) One can mimic the "abstract nonsense" approach to classifying spaces due to Graeme Segal [7].

The first approach is written up in [5] and the second in [4]. We will take a look at the third approach in the next section.

If M is a smooth manifold, a smooth Γ_q-cocycle is defined in the obvious way, and an element of $H^1(M;\Gamma_q)$ is said to be a smooth Haefliger structure if it is represented by a smooth Γ_q-cocycle. One has generalizations of the theorems of § 6 to these smooth structures.

(7.10) Theorem. If Q is the normal bundle to a smooth Γ_q-structure γ on M, then $\mathrm{Pont}^k(Q) = 0$ for $k > 2q$. If $a, b, c \in \mathrm{Pont}^*(Q)$ with $\deg(a) + \deg(b) > 2q$, $\deg(b) + \deg(c) > 2q$, then $< a,b,c > = 0$.

The proof proceeds by analogy with the proofs in § 6. Over $U_\alpha \subset M$ let ∇^α be the connection obtained by pulling back the standard connection on R^q by f_α. Patch these together with a partition of unity and prove that the curvature matrices k^α for the resulting ∇ have $k^\alpha_{ij} \in I_\alpha(\gamma)$, the ideal in $A^*(U_\alpha)$ generated by those 1-forms which locally are pull-backs via f_α of 1-forms on R^q. Clearly $I_\alpha(\gamma)^{q+1} = 0$, so $\varphi(k) = 0$ for any invariant φ with $\deg(\varphi) > q$.

In §10, we generalize $\mathrm{Pont}^k(Q) = 0$, $k > 2q$, to the normal bundle of any Γ_q-structure on any reasonable space X.

References

1. G. Bredon, Sheaf Theory, McGraw-Hill, New York, 1967.
2. E. H. Brown, Abstract homotopy theory, Trans. Amer. Math. Soc. 119 (1965), pp. 79-85.

3. R. Godement, Topologie Algebrique et Theorie des Faisceaux, Hermann, Paris, 1958.

4. A. Haefliger, Homotopy and integrability, Lecture Notes in Mathematics, No. 197, Springer-Verlag, New York, 1971, pp. 133-163.

5. A. Haefliger, Feuilletages sur les variétés ouvertes, Topology 9 (1970), pp. 183-194.

6. D. Husemoller, Fibre Bundles, McGraw-Hill, New York, 1966.

7. G. Segal, Classifying spaces and spectral sequences, Institut des Hautes Etudes Scientifiques, Publications Mathematiques, No. 34 (1968), pp. 105-112.

8. Topological categories and classifying spaces. The basic reference for this section is [1].

Recall the notion of a category C. We write $C = \mathfrak{O} \cup \mathfrak{M}$ where $\mathfrak{O}$ is the class of objects of C and $\mathfrak{M}$ is the class of morphisms of C described as follows. For each pair $X,Y \in \mathfrak{O}$, $\mathrm{Hom}(X,Y)$ is a set and

$$\mathfrak{M} = \coprod_{(X,Y) \in \mathfrak{O} \times \mathfrak{O}} \mathrm{Hom}(X,Y),$$

a disjoint union. The basic axioms are:

1) If $X,Y,Z \in \mathfrak{O}$, there is a map (called composition)

$$\mathrm{Hom}(Y,Z) \times \mathrm{Hom}(X,Y) \to \mathrm{Hom}(X,Z)$$

(written $(f,g) \mapsto f \circ g$).

2) Composition is associative.

3) For each $X \in \mathfrak{O}$, $\exists\ 1_X \in \mathrm{Hom}(X,X)$ such that $f \circ 1_X = f$ and $1_X \circ g = g$ whenever these compositions are defined.

Typical examples are the category of sets and set mappings, the category of topological spaces and continuous mappings, the category of groups and homomorphisms, etc. Not all categories, however, need be so enormous. Indeed, in order to avoid logical difficulties in what follows, we consider only "small" categories (ones in which $\mathfrak{O}$ and $\mathfrak{M}$ are <u>sets</u>).

As an example of a small category, let G be a group and define the category whose only object is G with Hom(G,G) = G, and define composition to be the group operation.

As another example, let X be a topological space, $\{U_\alpha\}_{\alpha \in A}$ an open cover of X. Define a category whose objects are the finite subsets $\Sigma \subset A$ such that

$$U_\Sigma = \bigcap_{\alpha \in \Sigma} U_\alpha \neq \phi$$

and let

$$\mathrm{Hom}(\Sigma,\Sigma') = \begin{cases} \text{inclusion map } U_\Sigma \hookrightarrow U_{\Sigma'}, \text{ if defined} \\ \phi \quad \text{otherwise} \end{cases} .$$

Composition is composition in the usual sense.

(8.1) Definition. A topological category $C = \mathfrak{O} \cup \mathfrak{M}$ is a small category such that $\mathfrak{O}$ and $\mathfrak{M}$ are topological spaces and the following maps are continuous.

1) $\mathfrak{M} \to \mathfrak{O} \times \mathfrak{O}$ defined by $f \mapsto (X,Y)$, $\forall f \in \mathrm{Hom}(X,Y)$.

2) Composition $\mathfrak{M} * \mathfrak{M} \to \mathfrak{M}$, where $\mathfrak{M} * \mathfrak{M} = \{(f,g) \in \mathfrak{M} \times \mathfrak{M} : f \circ g$ is defined$\}$.

3) $\mathfrak{O} \to \mathfrak{M}$ defined by $X \mapsto 1_X$.

Remark that any small category can be regarded as a topological category with the discrete topology on $\mathfrak{O}$ and $\mathfrak{M}$.

As a first example, let G be a topological group and again form the category with one object G and Hom(G,G) = G. This clearly satisfies (8.1).

As a slightly less trivial example, let X be a topological space,

$U = \{U_\alpha\}_{\alpha \in A}$ an open cover of X, and form the topological category X_U whose objects are the pairs (Σ,x) where $\Sigma \subset A$ is as in our earlier example and $x \in U_\Sigma$. $\mathrm{Hom}((\Sigma,x),(\Sigma',x'))$ is empty if either $x \neq x'$, or $U_\Sigma \not\subset U_{\Sigma'}$. Otherwise it consists of the single inclusion $(U_\Sigma,x) \to (U_{\Sigma'},x)$. Topologize $\mathfrak{O}$ by taking as basic neighborhoods sets of the form $\{(\Sigma,x) : \Sigma \text{ fixed}, x \in W \subset U_\Sigma, W \text{ open in } U_\Sigma\}$. Similarly, topologize $\mathfrak{M}$ by basic neighborhoods of the form $\{(U_\Sigma,x) \hookrightarrow (U_{\Sigma'},x) : \Sigma \text{ and } \Sigma' \text{ fixed}, x \in W \subset U_\Sigma, W \text{ open in } U_\Sigma\}$.

As a third example, we construct the topological category Γ_q which is pertinent to the study of Haefliger structures. Let $\mathfrak{O} = R^q$ with the usual topology. Given $x, y \in R^q$, let $\mathrm{Hom}(x,y) = \mathfrak{D}_{(x,y)}$ as in § 7. Thus $\mathfrak{M} = \mathfrak{D}$ with the sheaf topology. It is easy to check that composition $\mathfrak{D} * \mathfrak{D} \to \mathfrak{D}$ is continuous. The map $x \to 1_x$ (the germ at x of the identity diffeomorphism) defines a natural continuous section $R^q \to \mathfrak{D}$. Finally, let

$$\iota : \mathfrak{D} \to \mathfrak{D}$$

be the homeomorphism which to each germ γ of a local diffeomorphism g assigns the germ γ^{-1} of g^{-1}. If

$$\pi : \mathfrak{D} \to R^q$$

is the sheaf projection, then the map $\mathfrak{M} \to \mathfrak{O} \times \mathfrak{O}$ in (8.1) becomes

$$\mathfrak{D} \to R^q \times R^q$$

$$\gamma \to (\pi(\gamma), \pi \circ \iota(\gamma)) \quad ,$$

hence is continuous.

We next take up the construction of the classifying space of a topological category $C = \mathfrak{O} \cup \mathfrak{M}$. Let $A_0C = \mathfrak{O}$ and, if $n > 0$ is an integer, let A_nC denote the set of all finite sequences

$$X_0 \xrightarrow[f_1]{} X_1 \xrightarrow[f_2]{} \cdots \xrightarrow[f_n]{} X_n$$

where $X_i \in \mathfrak{G}$ and $f_i \in \mathrm{Hom}(X_{i-1}, X_i)$. A_0C has the topology of $\mathfrak{G}$ and A_nC is topologized as a subset of $\mathfrak{m} \times \mathfrak{m} \times \cdots \times \mathfrak{m}$. Define functions

$$\partial_i : A_nC \to A_{n-1}C \ , \quad 0 \leq i \leq n$$

$$s_i : A_nC \to A_{n+1}C \ , \quad 0 \leq i \leq n$$

by

$$\partial_0(X_0 \xrightarrow[f_1]{} X_1 \xrightarrow[f_2]{} \cdots \xrightarrow[f_n]{} X_n) = X_1 \xrightarrow[f_2]{} \cdots \xrightarrow[f_n]{} X_n$$

$$\partial_n(X_0 \xrightarrow[f_1]{} X_1 \xrightarrow[f_2]{} \cdots \xrightarrow[f_n]{} X_n) = X_0 \xrightarrow[f_1]{} \cdots \xrightarrow[f_{n-1}]{} X_{n-1}$$

$$\partial_i(X_0 \xrightarrow[f_1]{} \cdots \xrightarrow[f_n]{} X_n) = X_0 \xrightarrow[f_1]{} \cdots \to X_{i-1} \xrightarrow[f_{i+1} \circ f_i]{} X_{i+1} \to \cdots \xrightarrow[f_n]{} X_n$$

$$s_i(X_0 \xrightarrow[f_1]{} \cdots \xrightarrow[f_n]{} X_n) = X_0 \xrightarrow[f_1]{} \cdots \xrightarrow[f_i]{} X_i \xrightarrow[1_{X_i}]{} X_i \xrightarrow[f_{i+1}]{} \cdots \xrightarrow[f_n]{} X_n \ .$$

Let Δ_n denote the standard n-simplex with its usual topology, $E_0, \ldots, E_n$ the vertices, and define functions

$$\epsilon_i : \Delta_{n-1} \to \Delta_n \ , \quad 0 \leq i \leq n$$

$$p_i : \Delta_{n+1} \to \Delta_n \ , \quad 0 \leq i \leq n$$

by the linear extensions of

$$\epsilon_i(E_j) = \begin{cases} E_j & , \ 0 \leq j \leq i-1 \\ E_{j+1}, & i \leq j \leq n-1 \end{cases}$$

$$p_i(E_j) = \begin{cases} E_j & , \ 0 \leq j \leq i \\ E_{j-1}, & i+1 \leq j \leq n+1 \ . \end{cases}$$

Then on the disjoint union of the topological spaces $A_nC \times \Delta_n$, $0 \leq n < \infty$, make the identifications

$$(a, \epsilon_i(t)) \sim (\partial_i(a), t)$$

$$(s_i(a), t) \sim (a, p_i(t)).$$

The resulting quotient space is designated by BC and is called the classifying space of C.

(8.2) Definition. Let $C = \mathfrak{O} \cup \mathfrak{m}$ and $C' = \mathfrak{O}' \cup \mathfrak{m}'$ be categories. A function $F : C \to C'$ is called a (covariant) functor if

(1) $F(\mathfrak{O}) \subset \mathfrak{O}'$, $F(\mathfrak{m}) \subset \mathfrak{m}'$

(2) $\varphi \in \mathrm{Hom}(X,Y) \Rightarrow F(\varphi) \in \mathrm{Hom}(F(X),F(Y))$.

(3) $F(\varphi \circ \psi) = F(\varphi) \circ F(\psi)$ whenever the compositions are defined.

(4) $F(1_X) = 1_{F(X)}$, $\forall\, X \in \mathfrak{O}$.

If C and C' are topological categories, a continuous functor $F : C \to C'$ is a functor such that $F : \mathfrak{O} \to \mathfrak{O}'$ and $F : \mathfrak{m} \to \mathfrak{m}'$ are both continuous.

Clearly a continuous functor $F : C \to C'$ induces a continuous map

$$BF : BC \to BC'$$

and

$$B(F \circ G) = BF \circ BG$$

$$B(1_C) = 1_{BC}$$

where $1_C : C \to C$ is the identity functor. B is a functor from the category of topological categories and continuous functors to the category of topological spaces and continuous maps.

We consider two important examples. Let X be a space, $U = \{U_\alpha\}_{\alpha \in A}$ an open cover of X, and form the topological category X_U defined above. The topological group $GL_q = GL(q,R)$ also defines a topological category as described earlier, and we consider a continuous functor

$$F : X_U \to GL_q .$$

For $\alpha, \beta \in A$, $x \in U_\alpha \cap U_\beta$, there are unique morphisms

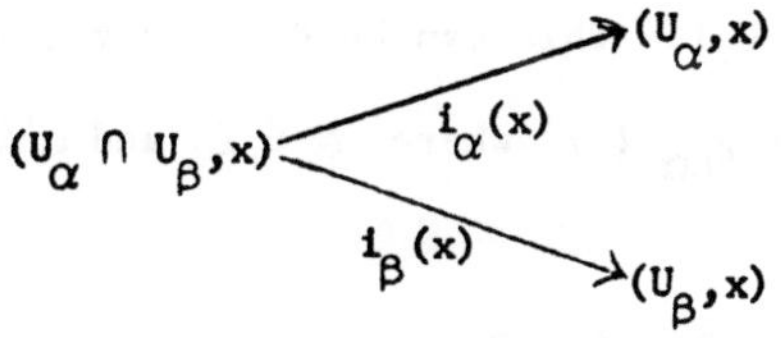

in X_U. Define

$$g_{\alpha\beta} : U_\alpha \cap U_\beta \to GL_q$$

by

$$(8.3) \qquad g_{\alpha\beta}(x) = F(i_\alpha(x)) \circ F(i_\beta(x))^{-1} .$$

By the continuity of F, $g_{\alpha\beta}$ is continuous.

(8.4) Lemma. $\{g_{\alpha\beta}\}_{\alpha,\beta\in A}$ satisfies the cocycle condition (2.2).

Proof. If $U_\alpha \cap U_\beta \cap U_\gamma \neq \emptyset$, then for any $x \in U_\alpha \cap U_\beta \cap U_\gamma$ we have the commutative diagram

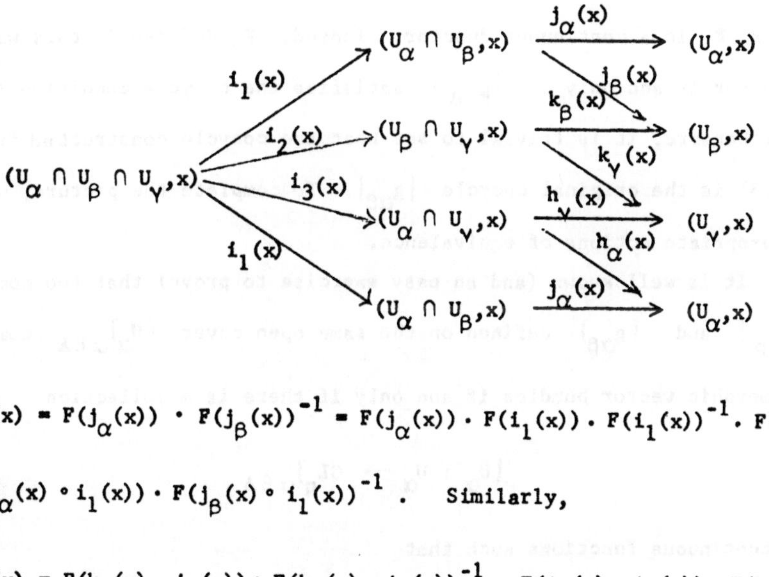

$g_{\alpha\beta}(x) = F(j_\alpha(x)) \cdot F(j_\beta(x))^{-1} = F(j_\alpha(x)) \cdot F(i_1(x)) \cdot F(i_1(x))^{-1} \cdot F(j_\beta(x))^{-1}$

$= F(j_\alpha(x) \circ i_1(x)) \cdot F(j_\beta(x) \circ i_1(x))^{-1}$. Similarly,

$g_{\beta\gamma}(x) = F(k_\beta(x) \circ i_2(x)) \cdot F(k_\gamma(x) \circ i_2(x))^{-1} = F(j_\beta(x) \circ i_1(x)) \cdot F(k_\gamma(x) \circ i_2(x))^{-1}$.

Thus $\quad g_{\alpha\beta}(x) g_{\beta\gamma}(x) = F(j_\alpha(x) \circ i_1(x)) \cdot F(k_\gamma(x) \circ i_2(x))^{-1}$

$= F(h_\alpha(x) \circ i_3(x)) \cdot F(h_\gamma(x) \circ i_3(x))^{-1} = F(h_\alpha(x)) \cdot F(h_\gamma(x))^{-1} = g_{\alpha\gamma}(x)$. q.e.d.

Conversely, let a GL_q-cocycle $\{g_{\alpha\beta}\}_{\alpha,\beta\in A}$ be given. Extend this to a cocycle $\{g_{\Sigma\Sigma'}\}$ over $\{U_\Sigma\}$. This can be done, e.g., by linearly ordering A and defining $g_{\Sigma\Sigma'}(x) = g_{\alpha\alpha'}(x)$ where $\alpha \in \Sigma$ and $\alpha' \in \Sigma'$ are the maximal elements. Then define

$$F : X_U \to GL_q$$

by

$$F((U_\Sigma,x) \to (U_{\Sigma'},x)) = g_{\Sigma'\Sigma}(x).$$

This is continuous and

$$F(1_{(U_\Sigma,x)}) = 1_{GL_q}$$

$$F([(U_{\Sigma'},x) \to (U_{\Sigma''},x)] \circ [(U_\Sigma,x) \to (U_{\Sigma'}(x))])$$

$$= F((U_\Sigma,x) \to (U_{\Sigma''},x)) = g_{\Sigma''\Sigma}(x)$$

$$= g_{\Sigma''\Sigma'}(x)g_{\Sigma'\Sigma}(x) = F((U_{\Sigma'},x) \to (U_{\Sigma''},x))F((U_\Sigma,x) \to (U_{\Sigma'},x)) \quad .$$

Thus F is a continuous functor. Indeed, F defined in this way is a functor if and only if $\{g_{\alpha\beta}\}$ satisfies the cocycle condition (2.2). Furthermore, it is trivial to see that the cocycle constructed from F by (8.3) is the original cocycle $\{g_{\alpha\beta}\}$. To complete the picture, we need appropriate notions of equivalence.

It is well known (and an easy exercise to prove) that two cocycles $\{g_{\alpha\beta}\}$ and $\{g'_{\alpha\beta}\}$ defined on the same open cover $\{U_\alpha\}_{\alpha\in A}$ come from isomorphic vector bundles if and only if there is a collection

$$\{\theta_\alpha : U_\alpha \to GL_q\}_{\alpha\in A}$$

of continuous functions such that

$$g'_{\alpha\beta}(x) = \theta_\alpha(x)g_{\alpha\beta}(x)\theta_\beta(x)^{-1} ,$$

$\forall\, x \in U_\alpha \cap U_\beta$, $\forall\, \alpha,\beta \in A$. We call such cocycles equivalent (or cohomologous) and denote the set of equivalence classes by $H^1(U;GL_q)$.

(8.5) Definition. If $F, H : C \to C'$ are continuous functors, a natural transformation $\theta : F \to H$ is a continuous map

$$\theta : \mathfrak{O} \to \mathfrak{M}'$$

(where $\mathfrak{O}$ denotes the set of objects in C and $\mathfrak{M}'$ the set of morphisms in C') such that

$$\theta(L) \in \mathrm{Hom}(F(L),H(L)), \ \forall\, L \in \mathfrak{O} \ ,$$

and for any $\alpha \in \mathrm{Hom}(L,L')$, the diagram

$$\begin{array}{ccc} F(L) & \xrightarrow{F(\alpha)} & F(L') \\ \Big\downarrow{\scriptstyle \theta(L)} & & \Big\downarrow{\scriptstyle \theta(L')} \\ H(L) & \xrightarrow{H(\alpha)} & H(L') \end{array}$$

is commutative, $\forall\, L,L' \in \mathfrak{O}$. F and H are said to be isomorphic (written $F \cong H$) if there are natural transformations

$$\theta : F \to H$$

$$\varphi : H \to F$$

such that $\theta(L) \circ \varphi(L) = 1_{H(L)}$ and $\varphi(L) \circ \theta(L) = 1_{F(L)}$, $\forall\, L \in \mathfrak{O}$. The set of isomorphism classes of continuous functors $C \to C'$ will be denoted $\pi[C,C']$.

(8.6) Theorem. The construction (8.3) sets up a one-one correspondence between $H^1(U;GL_q)$ and $\pi[X_U,GL_q]$.

Proof. Given $F_1, F_2 : X_U \to GL_q$, suppose $\theta : F_2 \to F_1$ is an isomorphism. Let $\{g^1_{\alpha\beta}\}$ and $\{g^2_{\alpha\beta}\}$ be the cocycles defined via (8.3) by F_1 and F_2 respectively. Write $\theta_\alpha(x) = \theta(U_\alpha,x)$. Then, for $x \in U_\alpha \cap U_\beta$,

$$g^1_{\alpha\beta}(x) = F_1(i_\alpha(x))F_1(i_\beta(x))^{-1}$$

$$= \theta_\alpha(x)F_2(i_\alpha(x))\theta(U_\alpha \cap U_\beta,x)^{-1}\theta(U_\alpha \cap U_\beta,x)F_2(i_\beta(x))^{-1}\theta_\beta(x)^{-1}$$

$$= \theta_\alpha(x)g^2_{\alpha\beta}(x)\theta_\beta(x)^{-1} \ .$$

Thus $\pi[X_U,GL_q]$ is mapped in a well defined way into $H^1(U;GL_q)$.

If F_1 and F_2 give rise to equivalent cocycles $g^1_{\alpha\beta} = \theta_\alpha g^2_{\alpha\beta}\theta^{-1}_\beta$, then, as in the remarks following the proof of (8.4), extend them to cocycles $\{g^1_{\Sigma\Sigma'}\}$ and $\{g^2_{\Sigma\Sigma'}\}$ over the cover $\{U_\Sigma\}_{\Sigma \text{ finite} \subset A}$, and note that these cocycles are still equivalent, hence obtain $g^1_{\Sigma\Sigma'} = \theta_\Sigma g^2_{\Sigma\Sigma'} \theta^{-1}_{\Sigma'}$. Define

$$\theta(U_\Sigma,x) : F_2(U_\Sigma,x) \to F_1(U_\Sigma,x)$$

by

$$\theta(U_\Sigma,x) = \theta_\Sigma(x) .$$

This is readily seen to be an isomorphism of continuous functors, so $\pi[X_U,GL_q] \to H^1(U;GL_q)$ is one-one. It is onto by the remarks following the proof of (8.4). q.e.d.

Thus, $\pi[X_U,GL_q]$ is in natural one-one correspondence with the set of isomorphism classes of q-dimensional vector bundles over X for which each $U_\alpha \in U$ is a trivializing neighborhood. Thus, if each U_α is contractible then $\pi[X_U,GL_q]$ is identified with the set of isomorphism classes of q-dimensional vector bundles over X.

For numerable coverings U (i.e., open coverings which admit a subordinate partition of unity) the space BX_U has canonically the homotopy type of X [1, Prop. (4.1)]. Any continuous functor $F : X_U \to GL_q$ defines a continuous map $BF : BX_U \to BGL_q$, hence defines a unique homotopy class

$$\widetilde{BF} \in [X,BGL_q] .$$

By [1, Prop. (2.1)], $F_1 \cong F_2$ implies $\widetilde{BF}_1 = \widetilde{BF}_2$. Thus BGL_q appears to be a good candidate up to homotopy for the classifying space for vector

bundles described in a quite different way in §2. [1,p.107] sketches t
relation between this construction and the Milnor construction.

With the above discussion as motivation, consider the topological ca
Γ_q and form the space $B\Gamma_q$. Mimicry of the above discussion with suita
adjustments shows that giving a Haefliger cocycle amounts to giving a
continuous functor

$$F : X_U \to \Gamma_q ,$$

hence (again assuming U numerable) the Haefliger cocycle will determi
an element

$$\widetilde{BF} \in [X, B\Gamma_q].$$

Thus $B\Gamma_q$ is an excellent candidate for the Γ_q-classifying space discus
in § 7. We will not attempt a more detailed treatment here.

Remark. There is a continuous functor

$$\nu : \Gamma_q \to GL_q$$

defined by

$$\nu(\gamma^x) = d\gamma^x ,$$

the Jacobian at x of any local diffeomorphism whose germ is
$\gamma^x \in \mathrm{Hom}(x,y)$. This gives rise to a very important continuous map

$$B\nu : B\Gamma_q \to BGL_q .$$

Indeed, in any homotopy commutative triangle

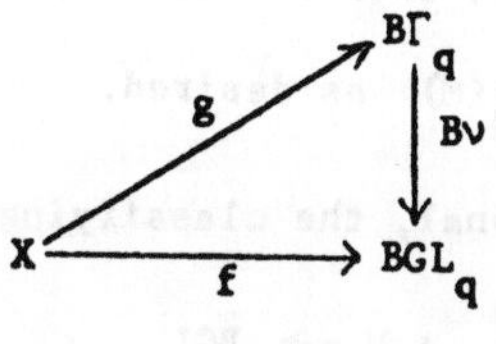

g is the classifying map of an element of $\Gamma_q(X)$ whose normal bundle is
classified by f.

References

1. G. Segal, Classifying spaces and spectral sequences, Institut des Hautes Études Scientifiques, Publications Mathematiques, No. 34 (1968), pp. 105-112.

9. Integrable homotopy and $Fol_q(M)$. For M a manifold, we define an equivalence relation called "integrable homotopy" between codimension q foliations of M and so obtain the set $Fol_q(M)$ of integrable homotopy classes. If M is an open manifold, an application due to Haefliger [1] of the Phillips-Gromov theorem will show how to describe $Fol_q(M)$ by means of the space $B\Gamma_q$.

(9.1) Definition. Let E_0 and E_1 be foliations of codimension q on M (i.e., E_0 and E_1 are integrable subbundles of $T(M)$ of codimension q). An integrable homotopy between these foliations is a subbundle $\tilde{E} \subset T(M \times R)$ of codimension q such that

1) $\tilde{E}$ is integrable.

2) $\tilde{E}_{(x,t)} + T_{(x,t)}(M \times \{t\}) = T_{(x,t)}(M \times R)$

(i.e., $\tilde{E}$ is transversal to each slice $M \times \{t\}$).

3) $\tilde{E}_{(x,0)} \cap T_{(x,0)}(M \times \{0\}) = E_{0x}$ and $\tilde{E}_{(x,1)} \cap T_{(x,1)}(M \times 1) = E_{1x}$.

By standard techniques this is shown to be an equivalence relation, so we obtain the set $Fol_q(M)$ as desired.

If M is n-dimensional, the classifying map

$$g_M : M \to BGL_n$$

for the tangent bundle is called the Gauss map. If M has a codimension q foliation E, the tangent bundle splits

$$T(M) = Q \oplus E$$

and so there is a homotopy lift of the Gauss map

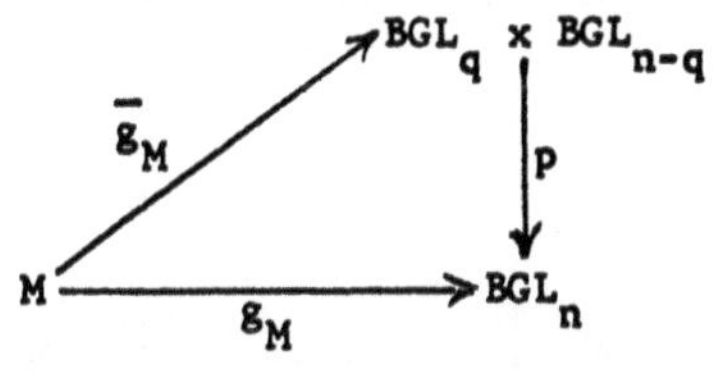

Since Q is the normal bundle of a foliation, $\bar{g}_M$ admits a homotopy lift

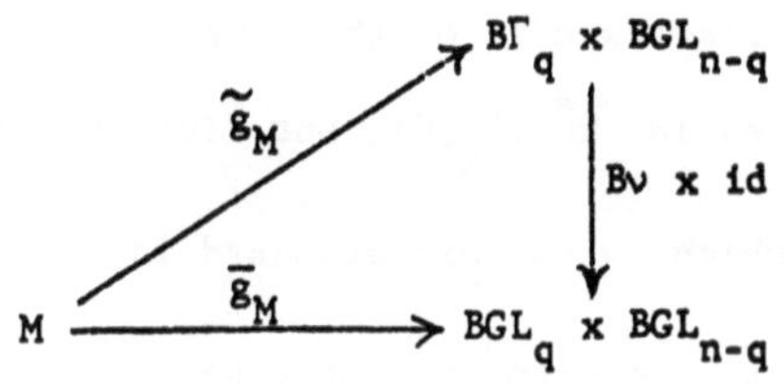

Set $\rho = p \circ (B\nu \times id)$.

(9.2) Theorem. (Phillips, Gromov, Haefliger). If M is an open manifold (i.e., $\partial M = \phi$ and M has no compact component) the above construction sets up a one-one correspondence between $Fol_q(M)$ and the set of homotopy classes of homotopy lifts $\tilde{g}_M$ of g_M :

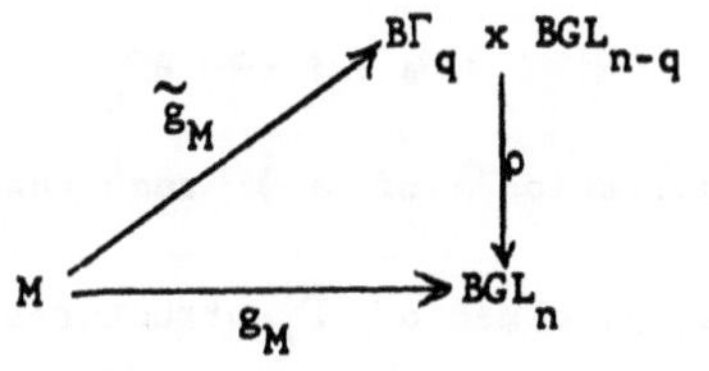

The proof of this theorem is quite difficult. It depends on the Phillips-Gromov theorem, of which an account by Poenaru will be found in [2]. The theorem in the form given here is then established by Haefliger in [1].

References

1. A. Haefliger, Feuilletages sur les variétés ouvertes, Topology 9 (1970), pp. 183-194.
2. V. Poenaru, Homotopy theory and differentiable singularities, Lecture Notes in Mathematics, No. 197, Springer-Verlag, N. Y., 1971, pp. 106-132.

10. The topology of $B\Gamma_q$. In this section we sketch in fair detail some results on the algebraic topological invariants of $B\Gamma_q$. The most novel of these results will be the construction of certain "exotic" characteristic classes in $H^*(B\Gamma_q;\mathbb{C})$, but first we discuss some facts about the map $B\nu$ which by now are standard in the theory of Γ_q-structures. The first of these is an elementary consequence of (*).

(10.1) Theorem. $B\nu^* : H^k(BGL_q;\mathbb{R}) \to H^k(B\Gamma_q;\mathbb{R})$ is zero if $k > 2q$.

Proof. We use singular cohomology. If the assertion is false, choose $\psi \in H^k(BGL_q;\mathbb{R})$, $k > 2q$, such that $\varphi = B\nu^*(\psi) \neq 0$. Then there is a homology class $\sigma \in H_k(BGL_q;\mathbb{R})$ such that $\varphi(\sigma) \neq 0$. By the definition of singular homology, there is a finite polyhedron P, a homology class $\sigma' \in H_k(P;\mathbb{R})$, and a continuous map

$$s : P \to B\Gamma_q$$

(the "geometric realization" of σ) such that $s_*(\sigma') = \sigma$. s determines a homotopy class of Γ_q-structures on P and by [1, p.188] we can thicken P to an open manifold M having the same homotopy type as P such that

$$s : M \to B\Gamma_q$$

corresponds to a foliation of M. Furthermore, by construction

$s^*(\varphi) \neq 0$ in $H^k(M;\mathbb{R})$. Let

$$f = B\nu \circ s : M \to BGL_q \ ,$$

Q the corresponding bundle on M. Q is the normal bundle to our foliation of M, hence $\mathrm{Pont}^k(Q) = 0$. But

$$0 \neq s^*(\varphi) = s^* B\nu^*(\psi) = f^*(\psi) \in \mathrm{Pont}^k(Q) \ ,$$

a contradiction. q.e.d.

(10.2) Corollary. Let $\gamma \in \Gamma_q(X)$, Q the normal bundle of γ. Then $\mathrm{Pont}^k(Q) = 0$, $k > 2q$.

This is the generalization of (*) promised at the end of §7. It is an immediate corollary of (10.1).

As is well known, from the point of view of homotopy theory, every continuous map is a fibration. Thus let $F\Gamma_q$ denote the homotopy theoretic fiber of

$$B\nu : B\Gamma_q \to BGL_q \ .$$

We sketch the proof of the following theorem from [1].

(10.3) Theorem. $\pi_i(F\Gamma_q) = 0$, $0 \leq i \leq q$. Consequently, $B\nu^* : \pi_i(B\Gamma_q) \to \pi_i(BGL_q)$ is an isomorphism, $0 \leq i \leq q$, and a surjection for $i = q + 1$.

If $i < q$, (10.3) is deduced from (9.2) as follows. $S^i \times R^{q-i}$ is parallelizable, hence for the Gauss map

$$g : S^i \times R^{q-i} \to BGL_q$$

we may take a constant map. Then by (9.2) the homotopy classes of lifts

$$\tilde{g} : S^i \times R^{q-i} \to B\Gamma_q$$

correspond one-one to $\mathrm{Fol}_q(S^i \times R^{q-i})$ and furthermore, g being constant,

$\tilde{g}$ factors through $F\Gamma_q \subset B\Gamma_q$ as do all covering homotopies. This says that

$$\pi_i(F\Gamma_q) = \mathrm{Fol}_q(S^i \times R^{q-i}) ,$$

$0 \leq i < q$. But for any q-dimensional manifold M, it is clear that $\mathrm{Fol}_q(M)$ has only one element, so $\pi_i(F\Gamma_q) = 0$, $0 \leq i < q$.

For the case $i = q$, the proof proceeds differently. One must consider any continuous map $f : S^q \to F\Gamma_q$ and the element $\gamma \in \Gamma_q(S^q)$ corresponding to

$$S^q \xrightarrow{f} F\Gamma_q \hookrightarrow B\Gamma_q .$$

Since $B\nu \circ f$ is constant, γ has trivial normal bundle. γ can be represented by a foliation on a suitable open neighborhood of S^q in R^{q+1}. The submersion theorem [3] of Phillips then allows (cf. [1, p.192]) γ to be extended to a Γ_q-structure on D^{q+1}. That is, one finds $\omega \in \Gamma_q(D^{q+1})$ such that

$$\gamma = i^*(\omega)$$

where

$$i : S^q \to D^{q+1}$$

is the inclusion. ω has trivial normal bundle since every vector bundle over a contractible space is trivial. Thus $f : S^q \to F\Gamma_q$ extends to a map $\tilde{f} : D^{q+1} \to F\Gamma_q$, which proves $\pi_q(F\Gamma_q) = 0$.

Next we take up the exotic or "secondary" characteristic classes of foliations, the existence of which will be seen to be another consequence of the proof of (*). [Redactor's note. At the time these lectures were given the theory of these characteristic classes had not come into proper focus. Indeed, only the class $\omega(E)$ (cf.(10.7)) was discussed in the lectures and the existence of a foliation E with $\omega(E) \neq 0$ was left as an open question. Later in the summer a Comptes Rendus note by Godbillon-Vey appeared and furnished an example (due to Roussarie -- cf. (10.8)) of such an E. Much of what follows was developed in a seminar conducted by Milnor and Bott in the late summer 1971 at La Jolla and later, during the fall of 1971, at the Institute for Advanced Study in Princeton.]

We begin with an example. Let $E \subset T(M)$ be a foliation of codimension one. Assume that the normal bundle Q is trivial. Then we can write $E = \mathrm{Ker}$ where $\theta \in A^1(M)$ is nowhere zero. Furthermore, by (1.3) we can write

$$d\theta = \theta \cdot \eta$$

for suitable $\eta \in A^1(M)$.

The form θ is a trivialization of Q^*, hence under a choice of decomposition

$$\mathfrak{X} = \Gamma(E) \oplus \Gamma(Q)$$

there is a unique $Z \in \Gamma(Q)$ such that

$$\theta(Z) \equiv 1.$$

Relative to this global section of Q any connection ∇ on Q has a corresponding connection form φ such that

$$\nabla_X(Z) = \varphi(X)Z, \ \forall \ X \in \mathfrak{X} .$$

(10.4) **Lemma.** $\eta \in A^1(M)$ satisfies $d\theta = \theta \cdot \eta$ iff η is the connection form relative to Z of a basic connection on Q.

Proof. Let φ be the connection form relative to Z of a basic connection. Then $\varphi' \in A^1(M)$ is another such iff $\varphi | \Gamma(E) = \varphi' | \Gamma(E)$. Let $X \in \Gamma(E)$. Then, if $d\theta = \theta \cdot \eta$,

$$-\frac{1}{2}\eta(X) = (\theta \cdot \eta)(X,Z) = d\theta(X,Z)$$

$$= -\frac{1}{2}\theta([X,Z]) + \frac{1}{2}X(\theta(Z)) - \frac{1}{2}Z(\theta(X))$$

$$= -\frac{1}{2}\theta([X,Z]) = -\frac{1}{2}\theta(\varphi(X)Z) = -\frac{1}{2}\varphi(X),$$

where we have used the standard formula [2,p.36] for the exterior derivative of a 1-form. Thus $\eta|\Gamma(E) = \varphi|\Gamma(E)$, so η is the form relative to Z of a basic connection.

For the converse, we have by the above that

$$d\theta(X,Z) = -\frac{1}{2}\varphi(X) = (\theta\cdot\varphi)(X,Z)$$

whenever $X \in \Gamma(E)$. If $X, Y \in \Gamma(E)$, then $[X,Y] \in \Gamma(E)$, so the formula for $d\theta$ shows $d\theta(X,Y) = 0$. Finally, $d\theta(Z,Z) = 0$, so

$$d\theta = \theta\cdot\varphi.$$

q.e.d.

(10.5) Lemma. Let η be as in (10.4). Then $\eta\cdot d\eta$ is a closed form. Furthermore, $[\eta\cdot d\eta] \in H^3_{DR}(M)$ is independent of the choice of η.

Proof. $d\eta = k$, the curvature form of a basic connection, by the structure equation (4.5). Thus

$$d(\eta\cdot d\eta) = d\eta\cdot d\eta = k^2 .$$

By the proof of (*), $k^2 = 0$.

If $d\theta = \theta\cdot\eta = \theta\cdot\eta'$, then $\eta' = \eta + f\theta$ for a suitable smooth function f. Thus

$$\eta'\cdot d\eta' = \eta\cdot d\eta + \eta\cdot d(f\theta) + f\theta\cdot d\eta + f\theta\cdot d(f\theta).$$

But

$$f\theta\cdot d(f\theta) = f\theta\cdot df\cdot\theta + f^2\theta\cdot d\theta$$

$$= f^2\theta\cdot d\theta = f^2\theta\cdot\theta\cdot\eta = 0$$

and

$$\theta\cdot d\eta = \theta\cdot d\eta - \theta\cdot\eta^2 = d(-\theta\cdot\eta) = -d^2(\theta) = 0 ,$$

so,

$$\eta'\cdot d\eta' = \eta\cdot d\eta + \eta\cdot d(f\theta) - f\theta\cdot d\eta = \eta\cdot d\eta - d(\eta\cdot f\theta) .$$

q.e.d.

(10.6) Exercise. Show that $[\eta \cdot d\eta]$ is also independent of the choice of trivialization θ of Q^*.

(10.7) Definition. If $E \subset T(M)$ is a foliation of codimension one having trivial normal bundle, then $\omega(E) = [\eta \cdot d\eta] \in H^3_{DR}(M)$.

Remark that the exotic characteristic class $\omega(E)$ is defined because of the vanishing of k^2. By the above, it depends only on the foliation E (indeed, by standard homotopy arguments, only on the class of E in $Fol_1(M)$). The fact that it need not always be zero is established by an example due to Roussarie which we now describe.

Let $SL(2,\mathbb{R})$ denote the subgroup of GL_2 consisting of the elements of determinant 1. It is well known that this group admits discrete subgroups G such that the right coset space

$$M = G\backslash SL(2,\mathbb{R})$$

is a compact manifold. Since $SL(2,\mathbb{R})$ has dimension three, so does M.

(10.8) Theorem. (Roussarie) M as above admits a codimension one foliation $E \subset T(M)$ such that $\omega(E) \neq 0$.

Proof. Let $K \subset SL(2,\mathbb{R})$ be the two dimensional subgroup of matrices

$$\begin{pmatrix} a & 0 \\ b & c \end{pmatrix}, \quad a\,c = 1 .$$

Thus $SL(2,\mathbb{R})$ is foliated by the left cosets of K, a foliation invariant under the action of G from the left with trivial normal bundle also invariant under G. The Lie algebra of left invariant vector fields on $SL(2,\mathbb{R})$ identifies naturally with the Lie algebra of real matrices

$$\begin{pmatrix} x & y \\ z & w \end{pmatrix}, \quad x + w = 0 .$$

The Lie algebra of left invariant vector fields tangent to the above foliation corresponds to the subalgebra of the above set of matrices in which $y = 0$. Thus $\Gamma(E)$ is spanned by two left invariant vector fields X and Y corresponding to matrices as follows

$$X \leftrightarrow \begin{pmatrix} 1 & 0 \\ 0 & -1 \end{pmatrix}, \quad Y \leftrightarrow \begin{pmatrix} 0 & 0 \\ 1 & 0 \end{pmatrix}$$

while there is a trivializing section $Z \in \Gamma(Q)$ such that

$$Z \leftrightarrow \begin{pmatrix} 0 & 1 \\ 0 & 0 \end{pmatrix} .$$

The brackets of these fields are found by taking the commutators of the corresponding matrices. Thus

$$[X,Y] = -2Y$$

$$[X,Z] = 2Z$$

$$[Y,Z] = -X .$$

Let X^*, Y^*, Z^* be the dual basis of 1-forms. Then Z^* is the θ of our previous discussion and, by the formula for $d\theta$,

$$d\theta(X,Y) = -\frac{1}{2} Z^*([X,Y]) = 0$$

$$d\theta(X,Z) = -\frac{1}{2} Z^*([X,Z]) = -1$$

$$d\theta(Y,Z) = -\frac{1}{2} Z^*([Y,Z]) = 0$$

since the terms $\frac{1}{2} X(Z^*(Y))$, $\frac{1}{2} Y(Z^*(X))$, etc. all vanish. Thus

$$d\theta = -X^* \cdot Z^* = \theta \cdot X^* ,$$

so the η of (10.4) is X^* . Computations exactly like the above show

$$d\eta = \frac{1}{2} Y^* \cdot Z^*$$

and so

$$\eta \cdot d\eta = \frac{1}{2} X^* \cdot Y^* \cdot Z^* .$$

This is a volume element (i.e., a nowhere zero 3-form) on $SL(2,\mathbb{R})$ since X^*, Y^*, Z^* are everywhere linearly independent. All of these constructions and definitions are invariant under the left action of G, hence carry over to M. On M the $\eta \cdot d\eta$ corresponding to the foliation is also a volume element. Since M is compact, it is a well known fact that $[\eta \cdot d\eta] \in H^3_{DR}(M)$ is nonzero. q.e.d.

As a first step toward generalizing the class $\omega(E)$, we show how it can be obtained by comparing two connections on Q. Indeed, quite generally let Q be any smooth q-dimensional vector bundle over an m-dimensional manifold M and let $\nabla^0, \nabla^1, \ldots, \nabla^n$ be connections on Q.

Let $\nabla^{0,1,\ldots,n}$ be the connection on the bundle

$$Q \times \mathbb{R}^n \to M \times \mathbb{R}^n$$

obtained by affine combinations

$$(1-a_1 - \cdots - a_n)\, \nabla^0 + a_1 \nabla^1 + \cdots + a_n \nabla^n$$

as in § 5. Let $I(gl_q)$ denote the ring of invariant polynomials on gl_q (cf. (5.1)) and define

$$\lambda(\nabla^0, \nabla^1, \ldots, \nabla^n) : I(gl_q) \to A^*_{\mathbb{C}}(M)$$

as follows. For the case $n = 0$, let

$$\lambda(\nabla^0)(\varphi) = \left(\frac{\sqrt{-1}}{2\pi}\right)^{\deg(\varphi)} \varphi(k^0)$$

for each $\varphi \in I(gl_q)$, where k^0 is the curvature of ∇^0. For the general case, set

$$\lambda(\nabla^0, \ldots, \nabla^n)(\varphi) = (-1)^{[n/2]}\, \pi_*(\lambda(\nabla^{0,\ldots,n})(\varphi)|M \times \Delta^n)$$

where

$$\pi_* : A^*(M \times \Delta^n) \to A^*(M)$$

is integration along the fiber (cf. (3.10)).

Note that for $n = 0$ this is a ring homomorphism. Also note that $\lambda(\nabla^0,\ldots,\nabla^n)(\varphi)$ is real for $\deg(\varphi)$ even and pure imaginary for $\deg(\varphi)$ odd. Finally, remark that $\lambda(\nabla^0,\ldots,\nabla^n)$ is antisymmetric in the $n+1$ entries.

As in § 5, we use the combinatorial version of Stoke's Theorem and the relation (3.10) together with the fact that $\lambda(\nabla^{0,\ldots,n})(\varphi)$ is a closed form (cf. (5.3)) to verify the formula

$$(\#)\qquad d(\lambda(\nabla^0,\ldots,\nabla^n)(\varphi)) = \sum_{i=0}^{n} (-1)^i \lambda(\nabla^0,\ldots,\hat{\nabla}^i,\ldots,\nabla^n)(\varphi) .$$

Indeed, the affine combination of the connections ∇^i was so set up that

$$\pi_*^{\partial}(\lambda(\nabla^{0,\ldots,n})(\varphi)|M \times \partial\Delta^n) = \sum_{i=0}^{n} (-1)^i \pi_*(\lambda(\nabla^{0,\ldots,\hat{i},\ldots,n})(\varphi)|M \times \Delta^{n-1}) ,$$

hence the formula(3.10)

$$(\pi_* \circ d + (-1)^{n+1} d \circ \pi_*)(\lambda(\nabla^{0,\ldots,n})(\varphi)|M \times \Delta^n) = \pi_*^{\partial}(\lambda(\nabla^{0,\ldots,n})(\varphi)|M \times \partial\Delta^n)$$

gives

$$\begin{aligned} d(\lambda(\nabla^0,\ldots,\nabla^n)(\varphi)) &= (-1)^{[n/2]} d\{\pi_*(\lambda(\nabla^{0,\ldots,n})(\varphi)|M \times \Delta^n)\} \\ &= (-1)^{n+1+[n/2]} \pi_*^{\partial}(\lambda(\nabla^{0,\ldots,n})(\varphi)|M \times \partial\Delta^n) \\ &= (-1)^{n+1+[n/2]+[(n-1)/2]} \sum_{i=0}^{n} (-1)^i \lambda(\nabla^0,\ldots,\hat{\nabla}^i,\ldots,\nabla^n)(\varphi) . \end{aligned}$$

But

$$n + 1 + [n/2] + [(n-1)/2] = 2n$$

so we obtain (#).

Returning to $\omega(E)$, let $E \subset T(M)$ be a foliation of codimension one with trivial normal bundle Q. Let $Z \in \Gamma(Q) \subset \mathfrak{X}$ be a trivializing section, θ the corresponding 1-form, $d\theta = \theta \cdot \eta$, and let ∇^1 be the basic connection on Q for which

$$\nabla^1_X(Z) = \eta(X)Z, \ \forall\, X \in \mathfrak{X}.$$

Let ∇^0 be the connection on Q for which

$$\nabla^0_X(Z) = 0, \ \forall\, X \in \mathfrak{X}.$$

Finally, note that $I(\mathfrak{gl}_1) = \mathbb{R}[x]$.

(10.9) Lemma. The following three formulas hold.

1) $\lambda(\nabla^1)(x) = (\frac{\sqrt{-1}}{2\pi})d\eta$

2) $\lambda(\nabla^0,\nabla^1)(x) = (\frac{\sqrt{-1}}{2\pi})\eta$

3) $\lambda(\nabla^0,\nabla^1)(x^2) = (\frac{-1}{4\pi^2})\eta \cdot d\eta$.

Proof. Since $d\eta$ is the curvature of ∇^1, the first formula is immediate. If $\pi : M \times \mathbb{R} \to M$ is the standard projection, then a trivial computation shows that the connection form of $\nabla^{0,1}$ is

$$\psi = t\,\pi^*(\eta).$$

Thus the curvature form is

$$k^{0,1} = d\psi = dt \cdot \pi^*(\eta) + t \cdot \pi^*(d\eta).$$

The square of this 2-form is

$$2t\, dt \cdot \pi^*(\eta \cdot d\eta) + t^2\, \pi^*(d\eta)^2.$$

Plugging these into our formula gives 2) and 3). q.e.d.

(10.10) Corollary. The class $\omega(E)$ is $-4\pi^2$ times the cohomology class of the closed 3-form

$$\lambda(\nabla^0,\nabla^1)(x^2) = \lambda(\nabla^0,\nabla^1)(x) \cdot \lambda(\nabla^1)(x).$$

Noting that ∇^0 is a Riemannian connection (it preserves the metric in which $\|Z\| = 1$) we use (10.10) to motivate a general definition of secondary characteristic classes for foliations. These will all be obtained by comparing a Riemannian connection and a basic connection on the normal bundle Q and will be independent of the choices of connections.

(10.11) Theorem. $I(gl_q) = \mathbb{R}[c_1, c_2, \ldots, c_q]$ where the invariant polynomials c_i are defined by

$$\det(I + tA) = 1 + \sum_{i=1}^{q} t^i c_i(A) \quad .$$

For a proof, cf. the appendix.

If $E \subset T(M)$ is a codimension q foliation, $Q = T(M)/E$, and if ∇^1 is a basic connection on Q, then by (*)

$$\lambda(\nabla^1) : \mathbb{R}[c_1, \ldots, c_q] \to A_C^*(M)$$

annihilates all elements of degree $> q$.

Thus, if

$$\mathbb{R}_q[c_1, \ldots, c_q] = \mathbb{R}[c_1, \ldots, c_q]/\{\varphi : \deg(\varphi) > q\}$$

it follows that $\lambda(\nabla^1)$ defines

$$\lambda(\nabla^1) : \mathbb{R}_q[c_1, \ldots, c_q] \to A_C^*(M) \, .$$

On the other hand, if ∇^0 is a Riemannian connection on Q, then the argument for (5.6) shows that

$$\lambda(\nabla^0) : \mathbb{R}[c_1, \ldots, c_q] \to A_C^*(M)$$

annihilates elements of odd degree. In particular,

$$\lambda(\nabla^0)(c_{2i-1}) = 0, \quad 1 \le 2i-1 \le q \quad .$$

Thus, application of (#) shows

(10.12) $$d\{\lambda(\nabla^0,\nabla^1)(c_{2i-1})\} = \lambda(\nabla^1)(c_{2i-1}).$$

These relations enable us to build a cochain complex WO_q and a homomorphism of cochain complexes

$$\lambda_E : WO_q \to A_C^*(M) .$$

As a graded algebra,

$$WO_q = R_q[c_1,\ldots,c_q] \otimes E(h_1,h_3,\ldots,h_\ell)$$

where ℓ is the largest odd integer $\leq q$, each c_i is assigned degree $2i$, and each h_i is assigned degree $2i-1$. Here, of course, $E(h_1,\ldots,h_\ell)$ designates the exterior algebra over R generated by the elements h_i. A unique antiderivation of degree 1

$$d : WO_q \to WO_q$$

is defined by requiring

$$d(c_i) = 0, \quad 1 \leq i \leq q$$

$$d(h_i) = c_i, \quad i = 1,3,\ldots,\ell .$$

Clearly $d^2 = 0$. The unique homomorphism of graded R-algebras

$$\lambda_E : WO_q \to A_C^*(M)$$

is defined by requiring

$$\lambda_E(c_i) = \lambda(\nabla^1)(c_i), \quad 1 \leq i \leq q ,$$

$$\lambda_E(h_i) = \lambda(\nabla^0,\nabla^1)(c_i), \quad i = 1,3,\ldots,\ell .$$

By (10.12) and the fact that $\lambda(\nabla^1)(c_i)$ is a closed form,

$$\lambda_E \circ d = d \circ \lambda_E .$$

Thus, in cohomology λ_E induces a homomorphism of graded $\mathbb{R}$-algebras

$$\lambda_E^* : H^*(WO_q) \rightarrow H^*_{DR}(M;\mathbb{C}).$$

λ_E depends, of course, on the choices of ∇^0 and ∇^1, but λ_E^* does not.

(10.13) Proposition. λ_E^* depends only on the foliation E, not on the choices of Riemannian connection ∇^0 and basic connection ∇^1.

Proof. If ∇^1 and $\bar{\nabla}^1$ are basic connections, then $(1-t)\nabla^1 + t\bar{\nabla}^1$ is also a basic connection, $0 \leq t \leq 1$. Similarly the set of Riemannian metrics on Q is a convex set, so any two Riemannian connections are readily seen to be homotopic through Riemannian connections. Thus, as in the proof of (5.4), the assertion follows by the homotopy invariance of $H^*_{DR}(M;\mathbb{C})$. q.e.d.

A notion of homotopy between codimension q foliations E and E' on M which is weaker than (9.1) is obtained simply by requiring a codimension q foliation F on $M \times \mathbb{R}$ such that $M \times \{0\}$ and $M \times \{1\}$ are transverse to F and $i_0^*(F) = E$, $i_1^*(F) = E'$. Then, for purely homotopy theoretic reasons, we obtain

(10.14) Proposition. λ_E^* depends only on the homotopy class of E.

The homomorphism $\lambda_E^* : H^*(WO_q) \rightarrow H^*_{DR}(M;\mathbb{C})$ has a certain naturality property. Indeed, if $N \xrightarrow{i} M$ is a submanifold transversal to the leaves of the foliation E, there is induced a codimension q foliation $E' = i^*(E)$ on N. The normal bundle of E' is just $Q|N = i^*(Q)$ and both Riemannian and basic connections on Q restrict to Riemannian and basic connections respectively on $i^*(Q)$.

(10.15) Proposition. Under the above hypotheses, the diagram

$$\begin{array}{ccc} H^*(WO_q) & \xrightarrow{\lambda_E^*} & H^*_{DR}(M;C) \\ & \searrow^{\lambda_{E'}^*} & \downarrow{i^*} \\ & & H^*_{DR}(N;C) \end{array}$$

is commutative.

As a consequence of this naturality we can define a canonical homomorphism

$$\lambda^* : H^*(WO_q) \to H^*(B\Gamma_q;C)$$

where we use singular cohomology on $B\Gamma_q$. Indeed, if $\alpha \in H^r(WO_q)$, we can determine

$$\lambda^*(\alpha) \in H^r(B\Gamma_q;C) = \mathrm{Hom}_C(H_r(B\Gamma_q;C),C)$$

by showing how to evaluate it on homology classes $[\sigma] \in H_r(B\Gamma_q;C)$. As in the proof of (10.1), there is an open manifold M and a codimension q foliation E of M, the classifying map of which

$$s : M \to B\Gamma_q$$

satisfies

$$[\sigma] = s_*[\sigma']$$

for suitable $[\sigma'] \in H_r(M;C)$. Indeed, M is a neighborhood of a polyhedron P in a suitable Euclidean space such that P is a deformation retract of M and

$$s : P \to B\Gamma_q$$

is the geometric realization of the cycle σ. Then define

$$\lambda^*(\alpha)[\sigma] = \lambda_E^*(\alpha)[\sigma'] .$$

Here, of course, we assume the de Rham theorem for singular cohomology [4].

In order to see that $\lambda^*(\alpha)[\sigma]$ is well defined, let R be another polyhedron with map

$$u : R \longrightarrow B\Gamma_q$$

which is the geometric realization of a cycle $\bar{\sigma} = u_\#(\sigma'')$ homologous to σ. Let $R \subset N$ be a deformation retract, N an open manifold with codimension q foliation F classified by the extension of u to N. One then builds a polyhedron K and a map

$$v : K \longrightarrow B\Gamma_q$$

such that P and R are subpolyhedra and $s = v|P$, $u = v|K$, and such that the cycles σ' and σ'' are homologous in K. Let $K \subset W$ as usual, W an open manifold with codimension q foliation G classified by v. Without loss of generality, we suppose M and N open submanifolds of W, the foliations E and F respective restrictions of G, the maps s and u respective restrictions of v. Thus the naturality (10.15) gives a commutative diagram

$$\begin{array}{ccc} & \overset{\lambda_E^*}{\nearrow} & H^r_{DR}(M;C) \\ & & \uparrow i^* \\ H^r(WO_q) & \xrightarrow{\lambda_G^*} & H^r_{DR}(W;C) \\ & & \downarrow j^* \\ & \underset{\lambda_F^*}{\searrow} & H^r_{DR}(N;C) \end{array}$$

where i and j are the inclusions. Since $i_*[\sigma'] = j_*[\sigma'']$ in $H_r(W;C)$, we have

$$\lambda_E^*(\alpha)[\sigma'] = i^*\lambda_G^*(\alpha)[\sigma']$$

$$= \lambda_G^*(\alpha)\{i_*[\sigma']\} = \lambda_G^*(\alpha)\{j_*[\sigma'']\}$$

$$= j^*\lambda_G^*(\alpha)[\sigma''] = \lambda_F^*(\alpha)[\sigma''] .$$

(10.16) Theorem. There is a canonical homomorphism

$$\lambda^* : H^*(WO_q) \longrightarrow H^*(B\Gamma_q;C)$$

with the property that, for any foliation $E \subset T(M)$ with classifying map $g : M \longrightarrow B\Gamma_q$, the diagram

$$\begin{array}{ccc} H^*(WO_q) & \xrightarrow{\lambda^*} & H^*(B\Gamma_q;C) \\ & \searrow^{\lambda_E^*} & \downarrow{g^*} \\ & & H^*(M;C) \end{array}$$

is commutative.

$Im(\lambda^*)$ is, then, the desired set of universal exotic characteristic classes for foliations (actually, for Γ_q-structures).

The extent to which $\lambda^* \neq 0$ is of fundamental interest, and for this Roussarie's example plays a decisive role. In WO_1, the elements 1, c_1, h_1, and c_1h_1 are an R-linear basis, and the only cycles are the linear combinations of 1, c_1, and c_1h_1. $c_1 = d(h_1)$, hence

$$H^r(WO_1) = \begin{cases} R\ , & r = 0,3 \\ 0\ , & \text{otherwise} \end{cases}$$

and the generator in $H^3(WO_1)$ is $[c_1h_1]$. By (10.9), if $E \subset T(M)$ is a codimension one foliation with trivial Q, then

$$\omega(E) = (-4\pi^2)\lambda_E^*[c_1h_1] .$$

By (10.8) we obtain

(10.17) Theorem. (Roussarie) $\lambda^*[c_1h_1] \neq 0$ in $H^3(B\Gamma_1;C)$.

Remark that, in defining $\lambda^*[c_1h_1]$ we used not only the codimension one hypothesis, but restricted ourselves to foliations with trivial normal bundle. The classifying map

$$g : M \to B\Gamma_1$$

for such a foliation factors through $F\Gamma_1$, hence the above technique provides a nontrivial element in $H^3(F\Gamma_1;C)$. With this clue we can modify the definition of λ^* to obtain a canonical homomorphism

$$\lambda^* : H^*(W_q) \to H^*(F\Gamma_q;C)$$

for a suitable cochain complex W_q. Indeed, considering only codimension q foliations $E \subset T(M)$ with trivial normal bundle, we replace the general Riemannian connection ∇^0 in the above construction with a <u>flat</u> connection ∇^0 (i.e., one for which Q admits independent global sections $s_1,\dots,s_q$ with $\nabla^0_X(s_i) = 0$, $\forall X \in \mathfrak{X}$). Then $\lambda(\nabla^0)(c_i) = 0$ for all i, so we define

$$W_q = R_q[c_1,\dots,c_q] \otimes E(h_1,h_2,\dots,h_q)$$

with $\deg(c_i) = 2i$, $\deg(h_i) = 2i-1$, and $d(c_i) = 0$, $d(h_i) = c_i$. The homomorphism

$$\lambda_E : W_q \to A^*_C(M)$$

is defined by

$$\lambda_E(c_i) = \lambda(\nabla^1)(c_i)$$

$$\lambda_E(h_i) = \lambda(\nabla^0,\nabla^1)(c_i)$$

for ∇^0 a flat connection, ∇^1 a basic connection. One obtains the desired homomorphism of $H^*(W_q)$ into $H^*(F\Gamma_q;C)$ by mimicry of the previous techniques.

In conclusion, we mention an application of the analogue of $\omega(E)$ in the theory of holomorphic foliations. Here we study a fibration

$$F\Gamma_q C \to B\Gamma_q C \xrightarrow{B\nu} BGL_q C$$

where $\Gamma_q C$ is the topological category with C^q as the space of objects and with morphisms the sheaf of germs of local holomorphic diffeomorphisms of C^q. The holomorphic analogue of $\omega(E)$ gives a universal class

$$\omega \in H^3(F\Gamma_1 C;C) .$$

Let $M = C^2 - \{0\}$, an open complex analytic manifold having the homotopy type of S^3. Then the set of homotopy classes of maps $[M,F\Gamma_1 C]$ canonically identifies with $\pi_3(F\Gamma_1 C)$. Let $\sigma \in H_3(M;C)$ be the homology class represented by the unit sphere $S^3 \subset M$. The class ω then defines a map

$$\omega_* : \pi_3(F\Gamma_1 C) \to C$$

by

$$\omega_*([f]) = \{f^*(\omega)\}(\sigma)$$

for each $[f] \in [M,F\Gamma_1 C]$.

On M consider the nowhere zero holomorphic 1-form

$$\theta_{\alpha\beta} = \alpha z_2 dz_1 + \beta z_1 dz_2$$

where α and β are nonzero complex constants. $\mathrm{Ker}(\theta_{\alpha\beta})$ is a one dimensional holomorphic subbundle of the holomorphic tangent bundle $T(M)$, hence is integrable, and the foliation has complex codimension one with trivial complex normal bundle. From the identity

$$d\theta_{\alpha\beta} = (\beta-\alpha)\, dz_1 \cdot dz_2$$

$$= \theta_{\alpha\beta} \cdot \{ (\frac{\beta-\alpha}{|\alpha z_2|^2+|\beta z_1|^2}) (\bar{\alpha}\bar{z}_2 dz_2 - \bar{\beta}\bar{z}_1 dz_1) \}$$

$$= \theta_{\alpha\beta} \cdot \eta_{\alpha\beta}$$

we obtain the connection form $\eta_{\alpha\beta}$ for a suitable complex C^∞ basic connection. Since $\eta_{\alpha\beta}$ is of the form $f\psi$ where

$$f = \frac{\beta-\alpha}{|\alpha z_2|^2+|\beta z_1|^2}$$

$$\psi = \bar{\alpha}\bar{z}_2 dz_2 - \bar{\beta}\bar{z}_1 dz_1 \quad ,$$

we obtain

$$\eta_{\alpha\beta} \cdot d\eta_{\alpha\beta} = f^2\psi \cdot d\psi = \overline{\alpha\beta}\, f^2\varphi \cdot dz_1 \cdot dz_2$$

where

$$\varphi = \bar{z}_1 d\bar{z}_2 - \bar{z}_2 d\bar{z}_1 \quad .$$

Making the change of variables

$$u = \alpha z_2$$

$$v = \beta z_1$$

we obtain

$$\eta_{\alpha\beta} \cdot d\eta_{\alpha\beta} = (\frac{\beta}{\alpha} + \frac{\alpha}{\beta} - 2) \left\{ \frac{(\bar{v}\, d\bar{u} - \bar{u}\, d\bar{v}) \cdot dv \cdot du}{(|u|^2 + |v|^2)^2} \right\} \quad .$$

The unit sphere relative to the (u,v)-coordinates is homotopic to the unit sphere relative to the (z_1,z_2)-coordinates, hence evaluation of $[\eta_{\alpha\beta} \cdot d\eta_{\alpha\beta}]$ on the corresponding hemology class becomes

$$\int_S \eta_{\alpha\beta} \cdot d\eta_{\alpha\beta} = (\frac{\beta}{\alpha} + \frac{\alpha}{\beta} - 2) \int_{S^3} (\bar{v}\, d\bar{u} - \bar{u}\, d\bar{v}) \cdot dv \cdot du \quad .$$

Letting

$$u = x_1 + ix_2$$

$$v = x_3 + ix_4$$

one shows that on S^3 the integrand on the right is a real nowhere zero 3-form, hence

$$\int_{S^3} \eta_{\alpha\beta} \cdot d\eta_{\alpha\beta} = c\left(\frac{\beta}{\alpha} + \frac{\alpha}{\beta} - 2 \right)$$

where c is a nonzero real constant independent of α and β. By suitable choices of α and β this expression can be made to assume any complex value γ. Indeed, set $\alpha = 1$ and solve the resulting equation

$$c\beta^2 - (\gamma + 2c)\beta + c = 0$$

for nonzero β. If

$$f_{\alpha\beta} : M \to F\Gamma_1 C$$

corresponds to the foliation in question, then $\omega_*([f_{\alpha\beta}])$ assumes any complex value γ by suitable choices of α and β.

(10.18) Theorem. $\omega_* : \pi_3(F\Gamma_1 C) \to C$ is a surjection.

References

1. A. Haefliger, Feuilletages sur les variétés ouvertes, Topology 9 (1970), pp. 183-194.
2. S. Kobayashi and K. Nomizu, Foundations of Differential Geometry, Vol. I, 1963, Interscience Publishers, New York.
3. A. Phillips, Submersions of open manifolds, Topology 6 (1967), 171-206.
4. A. Weil, Sur les théorèmes de de Rham, Comment. Math. Helv., 26 (1952), pp. 119-145.

Appendix A. We discuss the algebra of invariant polynomials. Denote by F the field $\mathbb{R}$ or $\mathbb{C}$ and by $I_q(F)$ the algebra of polynomials on $gl(q,F)$ invariant under conjugation by $GL(q,F)$. Let $S_q(F)$ denote the algebra of symmetric polynomials in q variables over F, and let $\sigma_i \in S_q(F)$ be the ith elementary symmetric function. It is well known that

$$S_q(F) = F[\sigma_1,\sigma_2,\ldots,\sigma_q] ,$$

a polynomial algebra (cf. [2,p. 177]).

Given $\varphi \in I_q(F)$, the restriction of φ to the set of diagonal matrices produces a symmetric polynomial in the diagonal entries. Indeed, any permutation of these diagonal entries can be produced via conjugation by suitable elements of $GL(n,F)$. This defines a canonical homomorphism of algebras

$$\rho : I_q(F) \to S_q(F) .$$

Lemma A. ρ is surjective.

Proof. Recall the elements $c_i \in I_q(F)$ defined by the formula

$$\det(I + tA) = 1 + \sum_{i=1}^{q} t^i c_i(A) .$$

The restriction of c_i to the set of diagonal matrices is simply the ith elementary symmetric function. That is, $\rho(c_i) = \sigma_i$, $1 \leq i \leq q$, so ρ is surjective. q.e.d.

Lemma B. $\rho : I_q(\mathbb{C}) \to S_q(\mathbb{C})$ is injective.

Proof. Let $\Delta_q(\mathbb{C})$ denote the vector space of upper triangular $q \times q$ matrices over $\mathbb{C}$. Since a matrix is diagonalizable iff its minimal polynomial is a product of distinct linear factors [1,p. 175] it follows that the subset $\Delta_q^*(C) \subset \Delta_q(C)$ consisting of those matrices whose diagonal entries are pairwise distinct is a set of diagonalizable matrices. Clearly $\Delta_q^*(C)$ is dense in $\Delta_q(C)$, hence the subset of all diagonalizable matrices in $\Delta_q(C)$ is dense.

It is an elementary fact that every matrix in $\mathfrak{gl}(q,C)$ is conjugate under $GL(q,C)$ to an upper triangular matrix. Indeed, if A is a linear transformation of C^q, the algebraic closure of C guarantees the existence of a nonzero eigenvector $v_1 \in C^q$ for A. A defines a linear transformation of $V_1 = C^q/\text{span}(v_1)$, hence has a nonzero eigenvector $\bar{v}_2 \in V_1$. This gives $v_2 \in C^q$ linearly independent of v_1 such that $A(v_2) \in \text{span}(v_1,v_2)$. Proceeding inductively, produce a basis $v_1,v_2,\ldots,v_q$ of C^q such that $A(v_i) \in \text{span}(v_1,\ldots,v_i)$, $1 \leq i \leq q$. Relative to this basis A is represented by an upper triangular matrix.

By these remarks, the set of all $q \times q$ diagonalizable matrices is dense in $\mathfrak{gl}(q,C)$. If $\varphi \in I_q(C)$ and $\rho(\varphi) = 0$, it follows that φ vanishes on every diagonalizable matrix. By continuity and denseness, $\varphi = 0$. q.e.d.

Lemma C. $\rho : I_q(R) \longrightarrow S_q(R)$ is injective.

Proof. For purely formal reasons, every $\varphi \in I_q(R)$ is also invariant when interpreted as a polynomial on $\mathfrak{gl}(q,C)$. This gives an inclusion of $I_q(R)$ as a real subalgebra of $I_q(C)$. Likewise, $S_q(R)$ is a real subalgebra of $S_q(C)$. The diagram

$$\begin{array}{ccc} I_q(R) & \subset & I_q(C) \\ \rho \downarrow & & \downarrow \rho \\ S_q(R) & \subset & S_q(C) \end{array}$$

is commutative, so Lemma B gives the assertion. q.e.d.

These lemmas establish (10.11).

Finally, recall the elements $\Sigma_i \in I_q(F)$ defined by

$$\Sigma_i(A) = \operatorname{trace}(A^i) .$$

Then

$$\rho(\Sigma_i)(x_1,\ldots,x_q) = x_1^i + x_2^i + \cdots + x_q^i .$$

Lemma D. The elements $\{\Sigma_1,\Sigma_2,\ldots,\Sigma_r,\ldots\}$ generate $I_q(F)$.

Proof. Since ρ is an isomorphism, it is enough to prove that the symmetric polynomials $\{\rho(\Sigma_1),\rho(\Sigma_2),\ldots,\rho(\Sigma_r),\ldots\}$ generate $S_q(F)$. This will follow by induction on k from the classical formula (in which we use the convention $\sigma_0 = 1$)

$$0 = k\sigma_k + \sum_{i=1}^{k} (-1)^i \sigma_{k-i}\, \rho(\Sigma_i), \quad 1 \leq k \leq q .$$

In order to prove this formula, first define

$$f = 1 + \sigma_1 t + \cdots + \sigma_q t^q \in S_q(F)[t] .$$

That is,

$$f = (1 + x_1 t)(1 + x_2 t)\cdots(1 + x_q t).$$

Then, for purely formal reasons,

$$\begin{aligned} -t\left(\frac{df}{dt}\right)/f &= -t\frac{d}{dt}(\log f(t)) \\ &= -tx_1/(1 + x_1 t) - \cdots - tx_q/(1 + x_q t) \\ &= \sum_{i\geq 1} (x_1^i + \cdots + x_q^i)(-t)^i = \sum_{i\geq 1} (-1)^i \rho(\Sigma_i)t^i . \end{aligned}$$

Multiplying both sides by f gives

$$\left(\sum_{j\geq 0}\sigma_j t^j\right)\left(\sum_{i\geq 1} (-1)^i \rho(\Sigma_i)t^i\right) = -t\sum_{k\geq 0} k\sigma_k t^{k-1}$$

and so

$$\sum_{k\geq 1}\{(\sum_{i+j=k}(-1)^i\sigma_j\,\rho(\Sigma_i)) + k\sigma_k\}t^k = 0 .$$

Here, of course, we also use the convention $\sigma_j = 0$ if $j > q$. q.e.d.

References

1. K. Hoffman and R. Kunze, Linear Algebra, Prentice-Hall, Englewood Cliffs, N. J., 1961.

2. D. Husemoller, Fibre Bundles, McGraw-Hill, New York, 1966.

Appendix B - Construction of BC

by J. Stasheff

The history of classifying spaces for groups is extensive [9]. Here we will study several important examples and compare alternative forms of the construction.

If G is a topological group, we can form the associated category $\mathcal{G}$ with one object $*$ and $Mor(*,*) = G$, composition being given by $g \circ g' = g'g$, to conform with established usage. Segal's $B\mathcal{G}$ is then exactly BG as defined by Milgram [5] or, in an obscure exposition, by Stasheff [10, p. 289]. If $G = S^0$, S^1 or S^3 with the usual multiplication then BG is homeomorphic (respectively) to RP^∞, CP^∞ or HP^∞. If S^0, S^1 or S^3 is given the minimal cell structure, the induced cell structure on the projective space is the standard one. The identifications corresponding to the degeneracies s_i serve to achieve this minimality of the cell decomposition, but for our purposes they get in the way. The construction can be reworked as follows:

Recall that $A_0C = ObC$ and $A_nC \subset Mor\, C \times \dots \times Mor\, C$ is the subset consisting of n-tuples $(f_1,\dots,f_n)$ such that $f_{i+1} \circ f_i$ is defined. Let $\sigma = \sigma^n$ denote an $(n+1)$-tuple of integers $0 \leq i_0 \leq i_1 \leq \dots \leq i_n$, and Δ_σ the simplex with vertices in σ. Define

$$\mathcal{B}C = \coprod_{n,\sigma^n} \Delta_\sigma \times A_n/\sim$$

where the equivalence relation $\sim$ is defined by the face relations:

if $t_{i_k} = 0$,

$$(t_{i_0},\dots,t_{i_n}, f_1,\dots,f_n) \sim (t_{i_0},\dots,\hat{t}_{i_k},\dots,f_{k+1} \circ f_k,\dots).$$

This $\mathcal{B}$ is, like Segal's B, a functor from topological categories to topological spaces.

There is a map $\mathcal{B}C \to BC$ (as defined by Segal or on p. 61) which collapses the degenerate simplices, where any $A_n C$ coordinate is the identity. This is the standard reduction of a simplicial object to the normalized form, as it appears in singular theory or the Eilenberg-Zilber theorem [3, p. 236]. Just as there, it is a homotopy equivalence in this case [10, p. 289].

If G is a topological group and $\mathcal{G}$ is the associated category, we have $A_n\mathcal{G} = G^n$. The space $\mathcal{B}\mathcal{G}$ is identical with "BG" as defined by Dold and Lashof [2]. The point of their construction was that it did not use the inverses in G and so went through for a topological monoid (=associative H-space). For example, if ΩX is the (Moore) space of loops on a connected CW-complex X, then $\mathcal{B}\Omega X$ has the homotopy type of X. Closer to our interest, if H(F) is the monoid of auto-homotopy - equivalences of F to itself, $\mathcal{B}$ H(F) classifies Hurewicz fibrations with fibre F over CW-base spaces [1,7].

For a group G, we also have Milnor's construction [6] which we will denote $\mathbb{B}G$. For comparison with $\mathcal{B}\mathcal{G}$, we note that $\mathbb{B}G$ can be defined by first taking

$$EG = \coprod_{n,\sigma^n} \Delta_\sigma \times G \times G^n/\approx$$

where the equivalence relation is: if $t_{i_k} = 0$,

$$(t_{i_0},\dots,t_{i_n},g_0,g_1,\dots,g_n) \approx (\dots,\hat{t}_{i_k},\dots,\hat{g}_k,\dots) .$$

Then define $\mathbb{B}G$ as the quotient by the action $g(g_0,\dots,g_n) = (gg_0,\dots,gg_n)$. (One can profitably think of homogeneous coordinates in projective geometry.).

The comparison between $\mathcal{B}\mathcal{G}$ and $\mathbb{B}G$ can be made quite explicit in terms of the isomorphism $\phi: G^n \to G^n$ given by

$$(g_1, g_2, \ldots, g_n) \to (g_1, g_1 g_2, \ldots, g_1 \cdots g_n) .$$

We then map $\mathcal{B}\mathcal{G} \to \mathbb{B}G$ by

$$(t_{i_0}, \ldots, t_{i_n}, g_1, \ldots, g_n) \mapsto (t_{i_0}, \ldots, t_{i_n}, e, \phi(g_1, \ldots, g_n)).$$

The map respects the identifications as we indicate on $\Delta^2 \times G^2$:

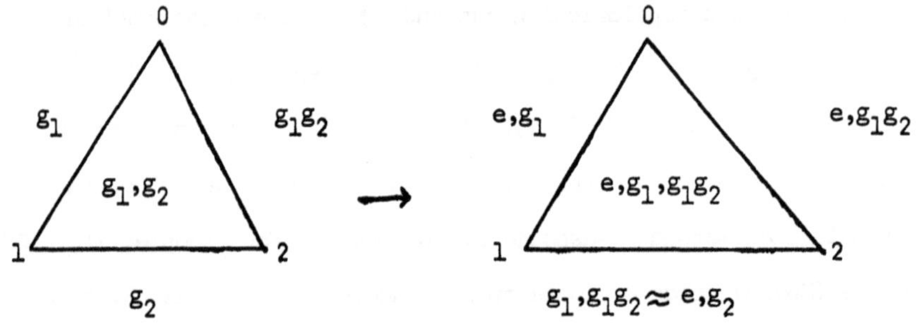

This isomorphism is the one which relates the standard inhomogeneous and homogeneous resolutions of a (abstract) group. The inverse to ϕ depends on the existence of inverses in G.

The variety of constructions available can be summarized as follows:

	homogeneous	inhomogeneous
normalized	unpublished	Milgram Stasheff Segal
unnormalized	Milnor	Dold-Lashof

There are also a variety of topologies, but these agree up to homotopy [8].

For a discrete group G, these various constructions are realizations of the various standard complexes for computing $H_*(G)$. Indeed $B\mathcal{G}$ (or $\mathcal{B}\mathcal{G}$ or $\mathbb{B}G$) is a $K(G,1)$. For any topological group G, the construction $\mathcal{B}\mathcal{G}$ is a realization of the bar construction $\bar{B}$ in homological algebra [3] in the sense that if G is a CW-complex with cellular multiplication (e.g. discrete) then, using cellular chains [5, 9],

$$C_* \mathcal{B}\mathcal{G} \simeq \bar{B}C_*G \quad .$$

MacLane has shown in general that B can be regarded as the tensor product of two functors [4].

In the case of $G = GL_q(\mathbb{F})$, $\mathbb{F} = R$, C or H to recognize $BGL_q(\mathbb{F})$ as the usual Grassmanian up to homotopy, it is perhaps easiest to use a homotopy characterization in terms of a universal bundle [6] or in terms of classification of G-structures as we do in Appendix C.

References

1. A. Dold, Halbexakte Homotopiefunktoren, Lecture Notes in Math., no. 12, Springer-Verlag, Berlin and New York, 1966.

2. A. Dold and R. Lashof, Principal quasi-fibrations and fibre homotopy equivalence of bundles, Illinois J. Math. 3 (1959), 285-305.

3. S. MacLane, Homology, Die Grundlehren der math. Wissenschaften, Band 114, Academic Press, New York; Springer-Verlag, Berlin, 1963.

4. ______, Milgram's classifying space as a tensor product of functors, Steenrod Conference, Lecture Notes in Math., no.168, Springer-Verlag, Berlin and New York.

5. R. Milgram, The bar construction and abelian H-spaces, Illinois J. Math. 11 (1967), 242-250.

6. J.Milnor, Construction of universal bundles. II, Ann. of Math. (2) 63 (1956), 430-436.

7. J.D. Stasheff, A classification theorem for fibre spaces, Topology 2 (1963), 239-246.

8. __________, Associated fibre spaces, Michigan Math. J. 15 (1968), 457-470.

9. __________, H-spaces and classifying spaces, Proc. Symp. Pure Math. 22, AMS, 1971.

10. __________, Homotopy associativity of H-spaces. I , II, Trans. Amer. Math. Soc. 108 (1963), 275-312.

Appendix C - Classification via $\mathcal{B}\Gamma_q$

by J. Stasheff

The approach to classification in § 8 uses the classifying space BX_U associated to an open cover $U = \{U_\alpha\}$. We can construct a space $\mathcal{B}U$ without passing through a category as follows: Order the index set A and let σ now denote a subset $\{i_0 < i_1 < \ldots < i_n\}$. Define $\mathcal{B}U$ as the quotient of $\coprod \Delta_\sigma \times U_\sigma$ by the equivalence relation: if $t_{i_k} = 0$,

$$(t_{i_0},\ldots,t_{i_n}, x \in U_\sigma) \sim (\ldots,\hat{t}_{i_k},\ldots, x \in U_{\sigma'}) \text{ where } \sigma' = (i_0,\ldots,\hat{i}_k,\ldots).$$

There is an obvious map $r\colon \mathcal{B}U \to X$ induced by $\Delta_\sigma \times U_\sigma \to U_\sigma$. If U has a subordinate locally finite partition of unity $\{\lambda_\alpha\}$, we can construct $\lambda : X \to \mathcal{B}U$ by choosing for each x, $\sigma(x) = \{\alpha \mid \lambda_\alpha(x) \neq 0\}$ and then defining $\lambda(x) = (\lambda_{\alpha_0}(x),\ldots,\lambda_{\alpha_n}(x), x \in U_{\sigma(x)})$. Clearly $r\lambda = 1$ and it is easy to construct a linear homotopy: $\lambda r \simeq 1$.

The construction and λ look as follows:

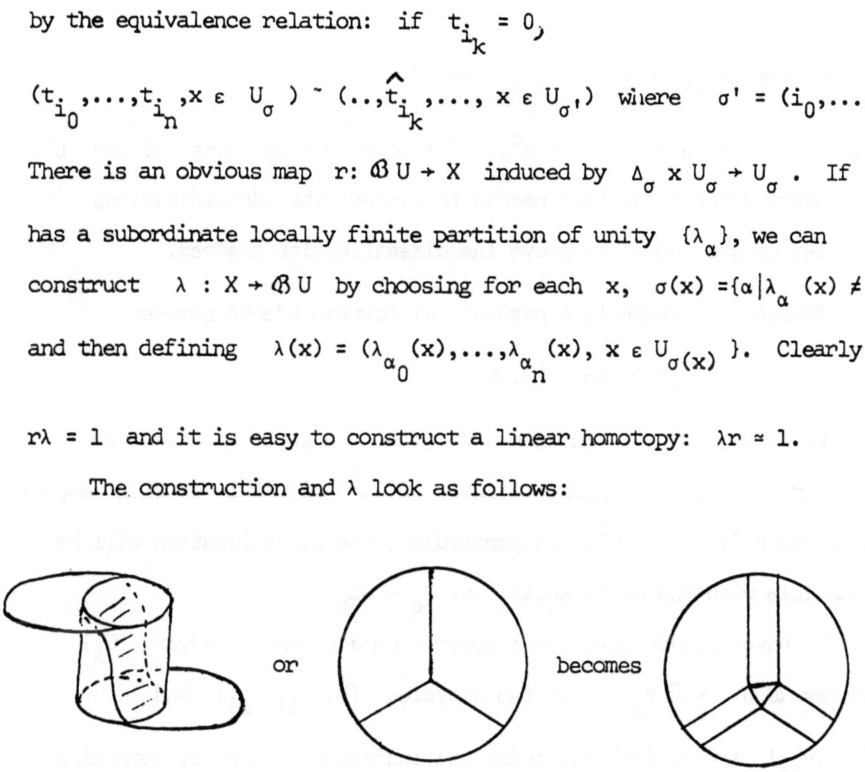

By contrast, Segal avoids choosing an ordering essentially by considering the first barycentric subdivision (which has a canonical local ordering). In the second case above this looks like:

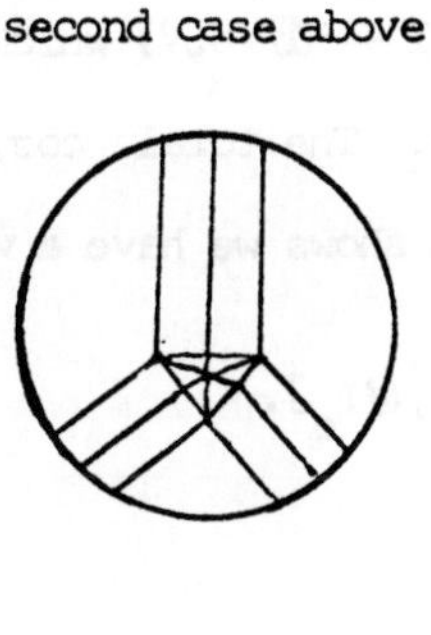

As mentioned in §8, a cocycle $\{U, \gamma_\alpha, \gamma_{\alpha\beta}\}$ induces $BX_U \to B\Gamma_q$. The same holds for $\mathcal{B}U$ if the covering is indexed by the non-negative integers. This is the case for any numerable cover, i.e. one with a subordinate locally finite partition of unity. The definition is even simpler:

$$\Delta_\sigma \times U_\sigma \to \Delta_\sigma \times A_n \Gamma_q$$

is induced by

$$x \longmapsto (\gamma_{i_1 i_0}(x), \ldots, \gamma_{i_n i_{n-1}}(x)) .$$

(For $n = 0$, we use $\gamma_i : U_i \to R^q$.) The cocycle conditions a) and b) are exactly the conditions needed to respect the identifications.

Let us use $\mathcal{B}U$ to prove the classification theorem.

Theorem D. There is a natural 1-1 correspondence between

$$\Gamma_q(X) \text{ and } [X, \mathcal{B}\Gamma_q].$$

We will prove this in such a way that the proof carries over, verbatim, to any topological category C and naturally so with respect to functors $\mathcal{F}: C \to C^1$. In particular, the classification will be compatible with the differential $d: \Gamma_q \to GL_q$.

We have already seen how a cocycle on a numerable cover $\{U_i\}$ induces $\mathcal{B}U \to \mathcal{B}\Gamma_q$. If two cocycles $\{U, \gamma_i, \gamma_{ij}\}$ and $\{V, v_k, v_{k\ell}\}$ are equivalent, order the indices of $U \cup V$ by decreeing those of U to be less than those of V. If the partition of unity for U gives $\lambda: X \to \mathcal{B}U$ and for V we have $\mu: X \to \mathcal{B}V$, use $t\lambda + (1-t)\mu$ to obtain $X \times I \to \mathcal{B}(U \cup V)$ which restricts to λ at one end and μ at the other. The total cocycle assumed on $U\ V$ induces $\mathcal{B}(U \cup V) \to \mathcal{B}\Gamma_q$ and shows we have a well defined map:

$$\psi: H^1(X; \Gamma_q) \to [X, \mathcal{B}\Gamma_q].$$

The relation of homotopy on cocycles induces a homotopy between corresponding classifying maps so , ψ factors:

$$\psi: \Gamma_q(X) \to [X, \mathcal{B}\Gamma_q].$$

We will prove ψ is a bijection by constructing a universal cocycle on $\mathcal{B}\Gamma_q$. We specify the covering in terms of the functions $t_i:\Delta_\sigma \to I$ if $i \in \sigma$, extended to be 0 on Δ_σ if $i \notin \sigma$. Notice these induce well-defined continuous maps $t_i:\mathcal{B}\Gamma_q \to I$.

We set $U_i = t_i^{-1}(0,1]$. Trivially the t_i are a partition of unity, but only point finite in general, not locally finite. Following [2 or 3] we shrink gradually to obtain a locally finite partition of unity subordinate to $\{ U_i \} = U$.

For $x \in \mathcal{B}\Gamma_q$, let $W_i(s,x) = \max(0, t_i(x) - s \sum_{j<i} t_j(x))$ and let $v_i(s,x) = W_i(s,x)/ \sum_0^\infty W_j(s,x)$. Note the division is in fact by a finite sum of non-zero terms, hence $v_i:I \times \mathcal{B}\Gamma_q \to I$ is continuous. For $s = 0$, we have $v_i(0, \ ) = t_i$ while at $s = 1$ we have a locally finite partition of unity since, if $\sum_0^n t_j(X) = 1$, then there exists a neighborhood N of x such that $\sum_0^n t_j(y) > 1/2$ for $y \in N$ and hence $v_i(i,y) \leq \max(0,a)$ where $a < 0$ for $i > n$.

Now we are ready to construct our universal example $\gamma \in H^1(\mathcal{B}\Gamma_q;\Gamma_q)$.

Define $\gamma_i:U_i \to R^q$ by

$$\gamma_i(t_{i_0},\dots,t_{i_n},f_1,\dots,f_n) = \text{source } f_{i+1} \quad \text{for } i < n$$

$$= \text{target } f_i \quad \text{for } i > 0$$

except that for $n = 0$ let $\gamma_0(1,x) = x \in R^q$.

Similarly define, for $i > j$,

$$\gamma_{ij}(t_{i_0},\ldots,t_{i_n},f_1,\ldots,f_n) = f_i \circ \ldots \circ f_{j+1}\ .$$

The condition $\gamma_{ij}\gamma_{jk} = \gamma_{ik}$ is immediate if we define $\gamma_{ij} = \gamma_{ji}^{-1}$ for $i < j$ and γ_{ii} to be the appropriate identity. Let γ denote the class of $\{U, \gamma_i, \gamma_{ij}\}$ in $H^1(\mathcal{B}\Gamma_q;\Gamma_q)$ or in the set of structures $\Gamma_q(\mathcal{B}\Gamma_q)$.

Given $f\colon X \to \mathcal{B}\Gamma_q$, we have $f^*(\gamma)$ in $\Gamma_q(X)$ and can apply the natural transformation ψ. To see that $\psi(f^*\gamma) = [f]$, we look directly at the definition. The class $f^*(\gamma)$ is represented on the covering $\{f^{-1}(U_i)\}$ by $\gamma_i \circ f\colon f^{-1}(U_i) \to R^q$ and $\gamma_{ij} \circ f\colon f^{-1}(U_i) \cap f^{-1}(U_j) \to \Gamma_q$. As a partition of unity for $\{f^{-1}(U_i)\} = U$, we choose $u_i(x) = v_i(1,f(x))$

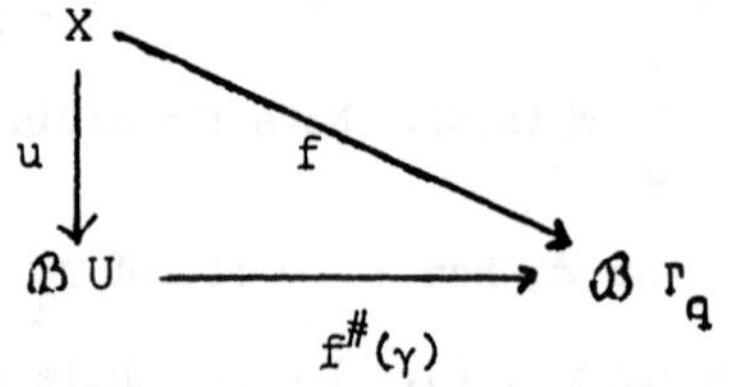

We have, for $x \in U_{\sigma(x)}$, that $f(x)$ can be represented as $(t_{i_0} f(x),\ldots,f_1(x),\ldots,f_n(x))$, so we have $x \to (u_{i_0}(x),\ldots,x) \longrightarrow (u_{i_0}(x),\ldots, \gamma_{i_1 i_0} f(x),\ldots)$ where $\gamma_{i_{j+1} i_j} f(x) = f_{j+1}(x)$. The homotopies $v_i(s,f(x))$ thus define a homotopy from f to $\mathcal{B}f^{\#}(\gamma) \circ u$. This shows $\psi\colon H^1(X;\Gamma_q) \to [X, \mathcal{B}\Gamma_q]$ is onto.

On the other hand, let $v = \{V, v_i, v_{ij}\}$ be a cocycle representative of $H^1(X;\Gamma_q)$. If $\lambda_i\colon V_i \to I$ is a subordinate locally finite partition of unity, we have $((\mathcal{B}v)\circ\lambda)^{-1}(U_i) = \{x \mid \lambda_i(x) \neq 0\}$, which is a subset of V_i. On that subset, the pull back of the universal γ_i

$H^1(X;C)$ or $C(X)$ signifies unless only invertible morphisms in C are used. However our universal structure involved γ_{ij} only for $i > j$, so our proof does go through in this generality.

Theorem E. Let C be a topological category and X a paracompact space. There is a 1-1 correspondence between $\iota(X)$ and $[X, \mathcal{B}C]$.

Indeed, Milnor has pointed out that $\mathcal{B}C$ can be described as "the space of cocycles with values in C." That is, let $\mathcal{B}C$ as a set consist of pairs $(\vec{t}, \{f_{ij}\})$ where $\vec{t} = (t_0, \ldots, t_i, \ldots)$ is a sequence of numbers $t_i \geq 0$, almost all 0 such that $\Sigma t_i = 1$ while $f_{ij} \in \text{Mor } C$ (ObC if $i = j$) runs over all pairs $i \geq j$ such that $t_i \neq 0$, $t_j \neq 0$ satisfying the relations $f_{ij} \in \text{Mor}(f_{jj}, f_{ii})$ and $f_{ij} \circ f_{jk} = f_{ik}$ for $i > j > k$. To topologize $\mathcal{B}C$ so as to be homeomorphic to $\mathcal{B}C$ as previously defined, use the direct limit of the quotient topologies of the maps $\Delta_\sigma \times A_n C \to \mathcal{B}C$ by $(t_{i_0}, \ldots, f_1, \ldots, f_n) \to (\vec{t}, \{f_{ij}\})$ where $\vec{t}$ is zero except for t_{i_j} and $f_{i_j i_k} = f_j \circ \ldots \circ f_{k+1}$.

Notice the proof is functorial in C, i.e. if $\mathcal{F}: C \to D$ is a continuous functor between topological categories, it induces $\mathcal{B}C \to \mathcal{B}D$ and we have a commutative diagram

$$\begin{array}{ccc} C(X) & \leftrightarrow & [X, \mathcal{B}C] \\ \downarrow & & \downarrow \\ D(X) & \leftrightarrow & [X, \mathcal{B}D] \end{array}$$

In particular this holds for the differential $d: \Gamma_q \to GL_q$ which is a homomorphism (= functor).

If we specialize to a group, we can say more.

Theorem F. Let G be a topological group. For paracompact X, there are natural 1-1 correspondences between the set of equivalence classes of G-bundles over X and $H^1(X;G)$ and $[X, \mathcal{B}\mathcal{G}]$.

The first correspondence is well-known: a bundle can be defined over X from the disjoint union $\coprod U_\alpha \times G$ by identification over

is given as follows: $\lambda_i(x)$ appears as some barycentric coordinate of $\mathcal{B}\, v o \lambda(x)$ and $v_{ij}(x)$ for some j appears as the corresponding Γ_q-coordinate so we have

$$x \longmapsto (\ldots, \lambda_i(x), \ldots, v_{ij}(x), \ldots) \overset{\gamma_i}{\longmapsto} \text{target } v_{ij} = v_i(x)$$

and similarly the pull back of the universal γ_{ij} agrees with v_{ij} where $\lambda_i(x) \neq 0 \neq \lambda_j(x)$. Thus if $h\colon X \times I \to \mathcal{B}\Gamma_q$ is a homotopy from f_0 to f_1, then $h^*(\gamma)$ restricts as desired to $f_0^*(\gamma)$ and $f_1^*(\gamma)$, showing $\psi : \Gamma_q(X) \to [X, \mathcal{B}\Gamma_q]$ is one-to-one.

Our proofs works in much greater generality. The discussion of Γ_q-structures or a footnote in Segal [4] indicates how to define $H^1(X;C)$ for any topological category C. In terms of an open covering $U = \{U_\alpha\}$, over an ordered index set, a cocycle consists of continuous maps

$$\gamma_\alpha : U_\alpha \to \mathrm{Ob}C$$
$$\gamma_{\alpha\beta} : U_\alpha \cap U_\beta \to \mathrm{Mor}\, C \quad \text{for } \alpha > \beta$$

such that a) $\gamma_{\alpha\beta}(x) \in \mathrm{Hom}(\gamma_\beta(x), \gamma_\alpha(x))$ and b) $\gamma_{\alpha\beta}\gamma_{\beta\delta} = \gamma_{\alpha\delta}$ on $U_\alpha \cap U_\beta \cap U_\delta$ for $\alpha > \beta > \delta$. Two such cocycles $\{U, \gamma_\alpha, \gamma_{\alpha\beta}\}$ and $\{V, \nu_\alpha, \nu_{\alpha\beta}\}$ are cohomologous if they are restrictions of a cocycle on $U \cup V$ with either the indices of U preceding those of V or vice versa. The equivalence classes form $H^1(X;C)$. The set $C(X)$ of C-structures on X is the quotient of $H^1(X;C)$ by the homotopy relation as above. Since our definition depends on the ordering of the covering, it is far from clear what

$U_\alpha \cap U_\beta$ via $\gamma_{\alpha\beta}$. The second correspondence implies one between $H^1(X;G)$ and $G(X)$. That is $H^1(X;G)$ already satisfies the homotopy axiom in this case. This in turn follows from the basic fact of bundle theory: A bundle $p: E \to X \times I$ is equivalent to $p_t \times 1: E_t \times I \to X \times I$ where $p_t = p|p^{-1}(X \times t)$. The example of §7 shows this breaks down for Γ_q-structures even if X is an interval. It is the comparison of the γ_α-part of the cocycle that fails here but not for groups where there is only one object and the γ_α are unique.

The strong result in terms of bundles suggests one further extension of this type of classification, namely to fibrations in the sense of the covering homotopy property rather than local triviality. Here the homotopy is built in, but we lack strict inverses since the appropriate notion of equivalence is fibre homotopy equivalence. Indeed, Dold shows local triviality is often present up to homotopy.

Theorem [2]. Let X be paracompact, weakly locally contractible. Then $p:E \to X$ has the WCHP if and only if the map p is locally fibre homotopy trivial, i.e. for some covering $\{U_\alpha\}$, there exist mutually inverse fibre homotopy equivalences

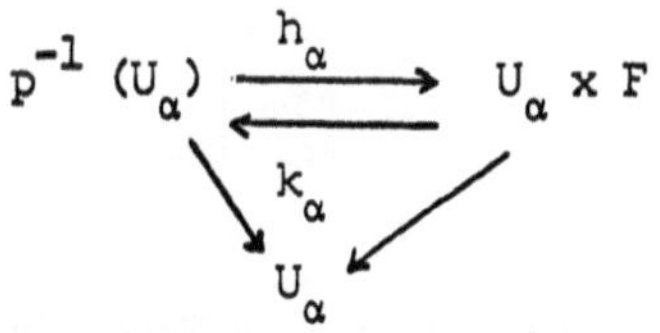

If we attempt to define a cocycle as before with $H(F) =$ {homotopy equivalence $F \to F$} taking the role of G, we would obtain $(x, \gamma_{\alpha\beta}(x)y) = h_\alpha k_\beta(x,y)$ for $x \in U_\alpha \cap U_\beta$, but the cocycle condition breaks down for we have $h_\alpha k_\beta h_\beta k_\delta \simeq h_\alpha k_\delta$. It turns out that a specific homotopy $\gamma_{\alpha\beta\delta}: I \times U_\alpha \cap U_\beta \cap U_\delta \to H(F)$ is relevant to classifying the fibration. Wirth [7] has carried out a full program along these lines and finds higher homotopies $I^{n-1} \times U_\sigma \to H(F)$ are also important. One way of summarizing his result is the following: We have seen that the cocylce condition corresponds to an associated map being a functor or, in pseudogroup terminology, a homomorphism. With topological categories or topological pseudogroups, it makes sense to talk of functors "up to strong homotopy" or s(trongly) h(omotopy) m(ultiplicative) maps [6]. Associated to these conditions, we can define a "cocycle up to strong homotopy." Similarly the equivalence or cobounding relation can with effort be generalized up to homotopy so as to define $H^1_\pi(X,C)$ as a limit over all covers of X.

Theorem G. [Wirth, 7]. There is a 1-1 correspondence between fibre homotopy equivalence classes of fibrations (with the WCHP) over paracompact, weakly locally contractible X with fibres of the homotopy type of F and the set $H^1_\pi(X; H(F))$.

Since shm maps induce maps of classifying spaces, so do functors up to strong homotopy. Thus either of the above sets is in 1-1 correspondence with $[X, \mathcal{B}H(F)]$, as was known for the first mentioned set [1, 5].

References

1. A. Dold, Halbexakte Homotopiefunktorem, Lecture Notes in Math, No. 12, Springer-Verlag, Berlin and New York, 1966.

2. ______, Partitions of unity in the theory of fibrations, Ann. of Math. (2) 78 (1963), 223-228.

3. D. Husemoller, Fibre Bundles, McGraw-Hill, New York, 1966.

4. G. Segal, Classifying spaces and spectral sequences, Inst. Hautes Études Sci. Publ. Math., No. 34 (1968), 105-112.

5. J.D. Stasheff, A classification theorem for fibre spaces, Topology 2 (1963), 239-246.

6. ____________, H-spaces from a homotopy point of view, Lecture Notes in Math, 161, Springer-Verlag, Berlin and New York, 1970.

7. J. Wirth, Fibre spaces and the higher homotopy cocycle relations, Thesis, Notre Dame, Ind., 1964.

Some Remarks on Continuous Cohomology

By

Raoul Bott*)

Reprinted from
Monifolds-Tokyo 1973
Proceedings of the International Conference
on Manifolds and Related Topics in Topology
Tokyo, 1973
University of Tokyo Press, 1975

Some Remarks on Continuous Cohomology

Raoul Bott*)

This is in the nature of a report on work in progress by A. Haefliger and myself on questions of continuous cohomology in various geometric contexts, and its implications for foliation theory. I will further discuss a related formula of Hochschild and myself which is also essentially contained from a slightly different point of view in independent work of Kamber and Tondeur [6].

Continuous cohomology arises naturally whenever the coefficient group of a cohomology theory is topologised: One may then consider the subcomplex of continuous cochains of the theory and study its homology. Notable examples of this situation are:

a) The continuous cohomology of a Lie group in the sense of Hu and Van Est,
b) The continuous cohomology of the Lie algebra of vector fields on a manifold in the sense of Gelfand-Fuks, and
c) The continuous cohomology of Haefliger's category Γ encountered in foliation theory.

We hope to show that there are interesting relations between all of these, but (to set the stage) consider first the example furnished by the singular theory with values in $\boldsymbol{R}$. Here we single out the subcomplex $S_c^*(X; \boldsymbol{R})$ of the singular $\boldsymbol{R}$-valued cochains $S^*(X; \boldsymbol{R})$ on X, whose values vary continuously when the chains on X are moved continuously. Clearly the inclusion $S_c^* \subset S^*$ induces a homomorphism

$$H_c^*(X; \boldsymbol{R}) \to H^*(X; \boldsymbol{R})$$

which for *reasonable spaces* is of course an isomorphism. Similarly, one can speak of smooth cohomology for manifolds and in a sense the de Rham theorem may be interpreted as asserting that the "smooth cohomology of a manifold" equals its ordinary cohomology. In short, nothing of great interest occurs in the "absolute" setting. On the other hand, the "relative" form of this notion does give rise to interesting examples such as a) and c) mentioned above.

In the relative theory one envisions a map

$$X^\delta \to X,$$

which is often given simply by the identity map so that X^δ is X in some coarser topology, and then studies the cochains on X^δ which are continuous relative to the topology of X. This complex—denoted by $S_c^*(X^\delta/X)$—then lies between $S_c^*(X)$ and $S_c^*(X^\delta)$, and its cohomology therefore mediates between the cohomology of the other two. Thus one has natural maps:

$$(1) \qquad H_c^*(X^\delta) \xleftarrow{\alpha} H_c^*(X^\delta/X) \xleftarrow{\beta} H_c^*(X),$$

and in practice the groups $H_c(X^\delta/X)$ and $H_c^*(X) \cong H^*(X)$ turn out to be computable while $H_c^*(X^\delta) = H^*(X^\delta)$ is not.

As a first example we consider the case of a Lie group G, in its usual topology, and

*) Part of the research reported herein was supported by NSF Grant #GP43613X.

let G^δ denote G in the discrete topology. The identity map then induces a continuous map

$$G^\delta \to G,$$

and hence a corresponding map

$$BG^\delta \to BG,$$

of the classifying spaces, so that the relative groups $H_c^*(BG^\delta/BG; \boldsymbol{R})$ are well-defined.

Now $H^*(BG^\delta; \boldsymbol{R})$ is of course the Hopf-Eilenberg-MacLane cohomology of the abstract group G^δ, and as such can be computed from the Eilenberg MacLane complex $\mathscr{F}^*(G)$ for G. Recall that this complex takes the form:

$$\mathscr{F}^*(G): \mathscr{F}(G) \xrightarrow{\delta} \mathscr{F}(G \times G) \xrightarrow{\delta} \mathscr{F}(G \times G \times G) \longrightarrow \cdots, \tag{2}$$

with $\mathscr{F}(X)$ denoting the space of real-valued functions on X and δ the usual boundary operator. Clearly then, this complex also has a natural *continuous* subcomplex $\mathscr{F}_c^*(G)$ given by the continuous functions, and the cohomology of this complex was already investigated by Van Est [11] in the 1950's. In particular, he showed that the cohomology of $\mathscr{F}_c^*(G)$, which we denote by $H_{\text{cve}}(G; \boldsymbol{R})$, is computable in terms of the relative Lie-algebra cohomology $H^*(\mathfrak{g}; K)$ of the Lie algebra $\mathfrak{g}$ of G modulo a maximal compact subgroup $K \subset G$.

Recall here that $H^*(\mathfrak{g}; K)$ is defined as the cohomology of the complex of G-invariant differential forms on G/K. Thus,

$$H^*(\mathfrak{g}; K) = H\{\text{Inv}_G \cdot A(G/K)\},$$

where A denotes the de Rham functor of differential forms.

In any case, the theorem of Van Est asserts that

$$H_{\text{cve}}(G; \boldsymbol{R}) \equiv H(\mathscr{F}_c^*(G)) \cong H^*(\mathfrak{g}; K). \tag{3}$$

On the other hand it is not difficult to show that the complex $\mathscr{F}_c^*(G)$ also computes the relative group $H^*(BG^\delta/BG; \boldsymbol{R})$ so that in our context the Van Est theorem gives the isomorphism

$$H_c^*(BG^\delta/BG; \boldsymbol{R}) \cong H_{\text{cve}}(G; \boldsymbol{R}) \cong H^*(\mathfrak{g}; K). \tag{4}$$

Composing this isomorphism with α, one obtains classes in $H^*(BG^\delta)$ which play an important role in the theory of discrete arithmetic subgroups D of G; in fact, the most important results in this subject usually assert that the restriction of these "continuous" classes to $H^*(BD)$ generates $H^*(BD)$ up to certain dimensions.

Now in foliation theory one encounters an "infinite" analogue of this situation in which the Lie group G is replaced by an object which might best be referred to as a "Lie-category".

By definition, such an object is a category $\mathscr{C}$ whose objects $o(\mathscr{C})$ and morphisms $m(\mathscr{C})$ are endowed with a C^∞-structure such that the defining maps of the category $\mathscr{C}$:

$$o(\mathscr{C}) \leftrightarrows m(\mathscr{C}) \quad \text{and} \quad m(\mathscr{C}) \underset{o(\mathscr{C})}{\times} m(\mathscr{C}) \to m(\mathscr{C})$$

are all smooth. Here the dotted line sends an object of $\mathscr{C}$ into its identity map and the other two maps on the left send a morphism to its source and target, respectively. For the arrow on the right $m(\mathscr{C}) \underset{o(\mathscr{C})}{\times} m(\mathscr{C})$ denotes the subset of composable morphisms in $m(\mathscr{C}) \times m(\mathscr{C})$ (which we assume to be smooth) and the arrow is there given by composition.

Let me first give a "finite" example of such a Lie category, to illustrate that these objects do arise quite naturally in geometry. For this purpose let M be a C^∞-manifold and let G be a Lie group which acts smoothly on M, say on the left. The pair (M, G) then defines a Lie-category $\mathscr{C} = \mathscr{C}(M, G)$ as follows. The objects of $\mathscr{C}$ are simply the points of M, with their C^∞ structure, while the morphisms of $\mathscr{C}$ are given by the manifold $M \times G$ in its natural C^∞ structure, with

$$\operatorname{Hom}(p, q) = \{(p, g) \mid gp = q\}; \qquad p \in M, g \in G.$$

Note that here $\mathscr{C}(pt, G)$ yields back the Lie group G, now considered as a special type of Lie-category. Observe further that this example also admits a natural analogue $\mathscr{C}^\delta$ of the G^δ encountered earlier. Indeed, to obtain $\mathscr{C}^\delta$ we simply leave the objects of $\mathscr{C}(M, G)$ as before, but now give the morphisms the C^∞ structure of $M \times G^\delta$. (By and large our C^∞ manifolds need not be separable or Hausdorff–for that matter.) In any case, one should again have a suitable notion of continuous cohomology $H_c^*(B\mathscr{C}^\delta / B\mathscr{C})$. To define this concept I must recall a construction that seems to crop up more and more in various branches of algebraic topology, and which I learned from G. Segal, D. Quillen, and, in retrospect, long ago from Arnold Shapiro. The construction in question assigns to every "topological category" $\mathscr{C}$ an explicit classifying space $B\mathscr{C}$, in a sense generalizing the Milnor construction for the classifying space of a topological group.

For this purpose one first associates to $\mathscr{C}$ the "semisimplicial space":

$$N\mathscr{C} : o(\mathscr{C}) \leftleftarrows m_1(\mathscr{C}) \overset{\leftarrow}{\leftleftarrows} m_2(\mathscr{C}) \overset{\leftarrow}{\underset{\leftarrow}{\leftleftarrows}} m_3(\mathscr{C}) \cdots$$

whose $m_k(\mathscr{C})$ are those k-tuples of morphisms $(\gamma_1, \ldots, \gamma_k)$ in $m(\mathscr{C})$ with $\gamma_i \circ \gamma_{i+1}$ well-defined, and the maps in question are obtained by either dropping γ_1 and γ_k (these are the "extremal maps") or by composing adjacent ones. To obtain $B\mathscr{C}$ one realizes $N\mathscr{C}$ in the usual manner: Namely, one forms the space $B\mathscr{C} = |N\mathscr{C}|$:

$$B\mathscr{C} = o(\mathscr{C}) \cup m_1(\mathscr{C}) \times \Delta^1 \cup m_2(\mathscr{C}) \times \Delta^2 \cup \cdots,$$

where Δ^k is the standard k-simplex and $m_k(\mathscr{C}) \times \Delta^k$ is attached to the earlier material via the boundary maps in $N\mathscr{C}$. (See [1][9] for details.) Now then, this being granted, we can return to our Lie categories $\mathscr{C}$ and $\mathscr{C}^\delta$, consider the natural functor $\mathscr{C}^\delta \to \mathscr{C}$, the induced map

$$B\mathscr{C}^\delta \to B\mathscr{C},$$

and so finally arrive at the continuous cohomology $H_c^*(B\mathscr{C}^\delta / B\mathscr{C})$.

As I mentioned earlier, the foliation theory of Haefliger leads one to an "infinite" example of this situation in which the space $B\mathscr{C}^\delta$ becomes geometrically significant as the classifying space of foliations. Precisely, the appropriate categories for this theory, to be denoted by Γ_q and J_q, are the following ones:

The objects $o(\Gamma_q)$ of Γ_q are the points of Euclidean q-space $\boldsymbol{R}^q$, while the morphisms from x to y in Γ_q are taken to be the germs of diffeomorphisms of a neighborhood of x to a neighborhood of y taking x to y. Furthermore $o(\Gamma_q) = \boldsymbol{R}^q$ is given the usual topology and C^∞-structure, while $m(\Gamma_q)$ is given the *sheaf topology*. That is, if $\mathscr{F}^{(q)}$ denotes the sheaf (over R^q) of germ of smooth maps from R^q to R^q, then $m(\Gamma_q)$ is contained in $\mathscr{F}^{(q)}$ as the invertible germs and $m(\Gamma_q)$ is given the induced topology. Hence in particular the maps $o(\Gamma_q)\leftrightarrows m(\Gamma_q)$ are etale, whence $m(\Gamma_q)$ inherits a C^∞-structure as a q-dimensional (*non-Hausdorff*) manifold, and our axioms for a Lie category are easily seen to be valid.

Note however, that in many ways Γ_q behaves quite differently from our earlier example. For instance, the spaces $m_k(\Gamma_q)$ are also seen to be etale over R^q and all q-dimensional! We will see later how this fact alone suffices to prove the "vanishing theorem" for the characteristic classes of foliation [1].

The category Γ_q plays the role of $\mathscr{C}^\delta$ in our framework. The analogue of $\mathscr{C}$ is the following "infinite Lie category" J_q.

For each $k\geq 0$ let J_q^k denote the Lie category whose objects are the points of R^q, and whose morphisms are the k-jets of diffeomorphisms of R^q. The composition in J_q^k is of course induced from Γ_q, and we have natural smooth maps (functors)

$$\Gamma_q \to J_q^k$$

sending a germ to its k-jet.

We also have natural projections of J_q^{k+1} on J_q^k and the inverse limit of these categories is by definition the category J_q:

$$J_q = \varprojlim_k J_q^k .$$

The C^∞ structure on J_q is now defined naturally in at least the following weak sense, that a map f of a manifold M into J_q is called smooth if and only if all the projections $\pi_2 \circ f$ of f:

$$M \xrightarrow{f} J_q \xrightarrow{\pi_k} J_q^k$$

are smooth. Now then, with this understood, we have the natural map

$$\Gamma_q \to J_q$$

and the corresponding continuous cohomology $H_c(B\Gamma_q/BJ_q)$. Further, in this context the Van Est theorem, essentially due to Haefliger, has the following analogue:

Theorem. *Let $\mathfrak{a}_q$ denote the Lie algebra of formal vector-fields in q-variables, and let $H_c(\mathfrak{a}_q, O_q)$ denote the relative cohomology of $\mathfrak{a}_q$ modulo the orthogonal group, in the sense of Gelfand-Fuks. Then*

$$H_c(B\Gamma_q/BJ_q) \cong H_c^*(\mathfrak{a}_q ; O_q) . \tag{5}$$

I have written $H_c(\mathfrak{a}_q, O_q)$ for the Gelfand-Fuks cohomology, because their theory is also really a *continuous* theory. Indeed the cochain complex for this group consists precisely of the alternating forms on $\mathfrak{a}_q/SO_q$ which are continuous relative to the completion topology in the formal power series ring. In any case the R-modules $H^*(\mathfrak{a}_q; O_q)$ are finite dimensional, and can be explicitly computed by pushing the work of Gelfand-Fuks a little, and the result can be expressed as follows:

Write $\hat{R}[c_1, \ldots, c_q]$ for the polynomial ring in variables c_i of degree $2i$ modulo the ideal of elements of degree $>2q$. Also write $E(u_1, u_3, \ldots, u_{2k+1})$ for the exterior algebra on the u's with $2k+1$ the largest odd integer less than q. Finally denote by WO_q the differential algebra:

$$\hat{R}[c_1, \ldots, c_q] \otimes E(u_1, \ldots, u_{2k+1})$$

with $dc_i = 0$, $du_i = c_i$.

Then one has the formula:

$$H^*(\mathfrak{a}_q; O_q) = H^*(WO_q). \tag{6}$$

There is also a much more detailed description of the actual generators of this group due to Vey, and of course similar results for various other Γ's associated to other transitive pseudo-groups, see [2], [3] for instance.

But returning to Γ_q itself, consider the map (functor)

$$\mathfrak{D}: J_q \to GL_q(\boldsymbol{R})$$

sending a jet of $m(J_q)$ to its linear part. It is then not hard to see that $\mathfrak{D}$ induces an isomorphism

$$\mathfrak{D}^*: H^*(BGL_q(\boldsymbol{R})) \cong H^*(BJ_q)$$

and that the continuous or even C^∞-cohomology of BJ_q and BGL_q are equal to the ordinary cohomology: Thus $H_c^* = H^*$ in both cases. Hence one obtains the diagram:

$$\begin{array}{ccccc} H^*(B\Gamma_q) & \longleftarrow & H_c^*(B\Gamma_q/BJ_q) & \longleftarrow & H_c^*(BJ_q) \\ & \nwarrow^{\alpha'} & \wr\wr & & \wr\wr \\ & & H^*(WO_q) & \xleftarrow{\beta'} & H^*(BGL_q\boldsymbol{R}) \end{array}$$

in which β' sends the Pontrjagin classes p_k of $H^*(BGL_q; \boldsymbol{R})$ to the c_{2k} in WO_q. Furthermore $\alpha' \circ \beta'$ corresponds to

$$\nu^*: H^*(BGL_q\boldsymbol{R}) \to H^*(B\Gamma_q), \quad \nu: \Gamma_q \to J_q \to GL_q(\boldsymbol{R}). \tag{7}$$

Now the vanishing theorem, alluded to earlier, is precisely that ν^* vanishes in dimensions $>2q$. On the other hand, in view of the definition of $\hat{R}[c_1, \ldots, c_q]$ that follows immediately from (7). Thus the homomorphisms

$$\alpha': H^*(WO_q) \to H^*(B\Gamma_q)$$

not only gives the vanishing phenomenon but also produces new potential characteristic classes in $H^*(B\Gamma_q)$, the so-called exotic classes of $B\Gamma_q$. Godbillon-Vey-Rousarie [5] were the first to show that some of these can be non-trivial but this is, on the whole, a much deeper part of the theory, and the most exciting examples of this type have definitely been described by Thurston. He showed that the class of u_1c_1 in $H^3(WO_1)$ can take on arbitrary real numbers on spherical cycles of $B\Gamma_1$, see [10], and recently he has "varied" the classes of $u_1c_1^2$ similarly. The most satisfying state of affairs would of course be if:

1) $$\alpha': H^*(WO_q) \to H^*(B\Gamma_q)$$

were injective, and

2) The image of α' generated $H^*(B\Gamma_q)$ modulo operations induced by discontinuous automorphisms of $\boldsymbol{R}$ as an abelian group.

Let me now turn to a brief account of how André Haefliger and I think of these Van Est type theorems at present.

First observe the following quite general fact, which seems to be known to the experts, but for which I really know no reference in the literature. For example, it is not in G. Segal's paper [9]. The assertion is simply this. Suppose

$$X: X_0 \leftleftarrows X_1 \overset{\leftarrow}{\underset{\leftarrow}{\leftarrow}} X_2 \cdots$$

is a semi-simplicial space, and $|X|$ is a geometric realization. Then consider the double complex

$$S^*X = \sum S^q(X_p)$$

given by the singular q cochains on X_p, with the differential operators d (of the singular theory) and δ of the semisimplicial structure. Then if $d \pm \delta = D$ denotes the total operator of S^*X, one has a natural isomorphism

(8) $$H_D(S^*X) \cong H^*(|X|) .$$

Now clearly it follows from this that if, say

$$N: N_0 \leftleftarrows N_1 \overset{\leftarrow}{\underset{\leftarrow}{\leftarrow}} N_2 \cdots$$

is a semisimplicial manifold, then we can form the double complex

$$AN = \sum A^q(N_p)$$

where A^q denotes the q-forms, and define the de Rham theory H_{DR} of $|N|$ to be the cohomology of AN relative to the total operator "$D = d + \delta$":

$$H_{DR}(|N|) = H_D(AN) .$$

This de Rham theory then naturally maps into the continuous (or smooth) cohomology of $|N|$, and in good situations one has isomorphisms

(9) $$H^*_{DR}(|N|) \cong H^*_c(|N|) \cong H^*(N) .$$

Notably this is the case if the de Rham theorem holds for each N_p:

(10) $$H_{DR}(N_p) \cong H(N_p) .$$

The importance of (9) is that it makes de Rham techniques available where they are not traditionally—such as in $B\mathscr{C}$, for example, when $\mathscr{C}$ is a Lie category or even a Lie group.

An instance quite unrelated to our present purpose where this technique also proved very effective is in the recent definition of Hodge structure on a singular variety of Deligne. This example was pointed out to me only a few days ago by D. Mumford. Roughly speaking, Deligne replaces a singular variety X by semisimplicial nonsingular Y, in the sense that

$$|Y| \cong X ,$$

and using the Hodge filtration on AY one can then induce one on X.

In our context let us see what this extended de Rham theorem (9) yields for the classical case of a Lie group G. Here NG takes the form

$$NG : pt \leftleftarrows G \leftleftarrows G \times G \leftleftarrows G \times G \times G \cdots$$

and as (10) is clearly valid:

$$H_{DR}(|NG|) \cong H(|NG|) = H(BG).$$

Now if one filters ANG relative to q and p one obtains two spectral sequences converging to $H(BG)$. The one associated to q then easily yields that $H^*(BG)$ = Bar construction on $H^*(G)$ = Polynomials on primitive space of $H^*(G)$. On the other hand, filtering by p, leads to an E_q^1 term which is simply the cohomology relative to δ of the complex ANG. Note that δ preserves q so that what has to be computed is the cohomology of the complex:

$$A^qNG : A^q(pt) \xrightarrow{\delta} A^q(G) \xrightarrow{\delta} A^q(G \times G) \longrightarrow \cdots \quad \text{for each } q.$$

For $q=0$ this is precisely the Eilenberg MacLane complex of G, but with *smooth* cochains. Hence for $q=0$ one obtains the smooth-(continuous)-cohomology of G à la Van Est:

$$H_\delta(A^0NG) = H_{\text{cve}}(G; \boldsymbol{R}).$$

Now this formula has the following generalization as was shown by Hochschild and myself, see [2], and which is also essentially contained in the work of Kamber-Tondeur [6].

Theorem. *Let* $\mathfrak{g}$ *denote the Lie algebra of G and let* $S^q\mathfrak{g}^*$ *denote the q-th symmetric power of its dual, considered as a G module under the adjoint action. Then*

(11) $$H_\delta^p(A^qNG) = H_{\text{cve}}^{p-q}(G; S^q\mathfrak{g}^*).$$

Note in particular that in view of the well-known results that

(12) $$H^0_{\text{cve}}(G; M) = \text{Inv}_G \cdot M$$
$$H^i_{\text{cve}}(G; M) = 0 \qquad \text{for } i > 0 \text{ if } G \text{ is compact.}$$

The edge homomorphism of (11) defines a natural map

(13) $$\text{Inv}_G \cdot S^q\mathfrak{g}^* \to H^{2q}(BG; \boldsymbol{R}),$$

which is an isomorphism when G is compact. This homomorphism is of course the Chern-Weil homomorphism in a new setting, derived directly from the extended de Rham theorem.

In this framework, the vanishing theorem alluded to earlier also becomes quite transparent. Indeed let

$$\nu : \Gamma_q \to GL_q$$

be the natural map. Clearly ν induces smooth map

$$N\Gamma_q \xrightarrow{N\nu} NGL_q.$$

Hence on the de Rham level,

$$(N\nu)^{\#} : ANGL_q \to AN\Gamma_q$$

induces the map $\nu^* : H^*(BGL_q) \to H^*(B\Gamma_q)$. On the other hand, the semisimplicial manifold $N\Gamma_q$ has dimension q! Hence $(N\nu)^{\#}$ factors through the complex

$$A_q NGL_q = ANGL_q / \sum_{p>q} A^p NGL_q .$$

Now the fact that for $r>q$ the Chern-Weil homomorphism maps into 0 in this quotient complex shows that ν^* vanishes in dim$>2q$. Q.E.D.

Note that in this context the exotic classes also make their appearance and possibly in the simplest of all ways. Namely in view of the vanishing, $(N\nu)^{\#}$ induces a map

$$H_D(A_q NGL_q) \to H^*(B\Gamma_q),$$

and it can be shown that

$$H_D^*(A_q NGL_q) \cong H^*(WO_q) .$$

Unfortunately, this rather attractive approach to the exotic classes which has also been worked out in great detail in [6], only yields the "linear exotic" classes and fails for more subtle Lie categories.

Let me now close with a few indications on how this technique leads to the Van Est theorem. Suppose then that G is a Lie group, and K a maximal compact subgroup of G. Then, as is well-known, G/K will be contractible. Next let $\mathscr{C}$ be the Lie category given by the left action of $\mathscr{C}$ on G/K. Thus:

(14) $$N\mathscr{C} : G/K \leftleftarrows G/K \times G \;\overset{\leftarrow}{\underset{\leftarrow}{\leftarrow}}\; G/K \times G \times G \cdots .$$

Also let $\mathscr{C}^\delta$ be the discrete version of $\mathscr{C}$, so that

(15) $$N\mathscr{C}^\delta : G/K \leftleftarrows G/K \times G^\delta \;\overset{\leftarrow}{\underset{\leftarrow}{\leftarrow}}\; G/K \times G^\delta \times G^\delta \cdots$$

Now there is an obvious square of categories

$$\begin{array}{ccc} \mathscr{C}^\delta & \longrightarrow & \mathscr{C} \\ \downarrow & & \downarrow \\ G^\delta & \longrightarrow & G \end{array}$$

where the vertical functors simply map G/K to a point, and this construction induces a map of semisimplicial manifolds

$$\begin{array}{ccc} N\mathscr{C}^\delta & \overset{\simeq}{\longrightarrow} & N\mathscr{C} \\ \downarrow & & \downarrow \\ NG^\delta & \overset{\simeq}{\longrightarrow} & NG \end{array}$$

with the vertical maps smooth fiberings with fiber G/K. Hence the vertical maps induce isomorphisms of cohomology, and in fact of all three types, namely:

(16) $$H^*(B\mathscr{C}^\delta) \cong H^*(BG^\delta), \; H_{DR}(N\mathscr{C}) \cong H_{DR}^*(NG)$$
$$H_c^*(B\mathscr{C}^\delta / B\mathscr{C}) \cong H_c^*(BG^\delta / BG) .$$

We next consider the de Rham complex $AN\mathscr{C}$ and its image in $AN\mathscr{C}^{\delta}$. Clearly the cohomology of $AN\mathscr{C}^{\delta}$ maps into the continuous cohomology of $B\mathscr{C}^{\delta}/B\mathscr{C}$. But, i^* applied on the 0's level simply yields $A(G/K)$ and the kernel of the semisimplicial δ in this position:

$$A(G/K) \underset{\delta_1}{\overset{\delta_0}{\rightrightarrows}} A(G/K \times G^{\delta})$$

is given precisely by the left invariant forms on G/K. The isomorphisms (12) therefore lead first of all to a natural map

(17) $$H(\mathfrak{g}, K) = H\{\mathrm{Inv}_G A(G/K)\} \to H_c(BG^{\delta}/BG)\,.$$

On the other hand, using the fact that $N\mathscr{C}$ can also be written as the quotient of the acyclic semisimplicial manifold

(18) $$G \leftleftarrows G \times G \leftleftarrows G \times G \times G \cdots$$

by the semisimplicial manifold

$$K \leftleftarrows K \leftleftarrows K \cdots$$

included diagonally, and using an average, over the compact K, of the contracting homotopy in (18) one arrives at the desired result that (17) is in fact an isomorphism.

In the infinite case of Γ_q, an analogous procedure works. We replace Γ_q by the category $\mathscr{C}_q^{\delta}$ obtained from the natural action of Γ_q on a "manifold" Γ_q/O_q. This analogue of G/K is obtained in the following manner:

Let $J_{q,0}$ denote the ∞ jets of diffeomorphisms of $\boldsymbol{R}^q$ with source 0. Then Γ_q acts on J_q on the left, and $O(q)$, the rotations of $\boldsymbol{R}_q$, on the right. Further, these commute and we define Γ_q/O_q by:

$$\Gamma_q/O_q = J_q/O(q)\,.$$

This is a contractible infinite dimensional manifold, and just as in the finite dimensional case, one deduces that there is a natural isomorphism

$$H(\mathfrak{a}_q; O_q) = H\{\mathrm{Inv}_{\Gamma_q} A(\Gamma_q/O)\} \xrightarrow{\sim} H_c(\Gamma_q/J_q)\,.$$

I will not try to discuss the details of these steps here, as my thinking at least is very fluid on these matters at present. It seems difficult to steer a course between excessive abstract nonsense and a much too specialized point of view.

References

[1] Bott, R., Lectures on characteristic classes and foliations, Lectures on Algebraic and Differential Topology, delivered at the II. ELAM, 1–74, Lecture Notes in Math., **279**, Springer, Berlin, 1973.

[2] ———, On the Chern-Weil homomorphism, to appear in Advances in Math.

[3] Bott, R. and Haefliger, A., On the characteristic classes of Γ-foliations, Bull. Amer. Soc., **78** (1972), 1039–1044.

[4] Godbillon, C., Cohomologies d'algèbre de Lie de champs de vecteurs formels, Séminaire Bourbaki, Lecture Notes in Math., **421**, 1–19, Springer, Berlin, 1972.

[5] Godbillon, C. and Vey, J., Un invariant des feuilletages de codimension 1, C. R. Acad. Sci. Paris Sér. A, **273** (1971), 92.
[6] Kamber, F. W. and Tondeur, Ph., Algèbres de Weil semi-simpliciales, C. R. Acad. Sci. Paris Sér. A., **276** (1973), 1177–1179.
[7] ———, Characteristic invariants of foliated bundles, to appear.
[8] Gelfand, M. and Fuks, D. B., Cohomologies of the Lie algebra of tangent vector fields of a smooth manifold I, II, Funkcional Anal. i Priložen, **3** (1969), 32–52; ibid., **4** (1970), 23–32 (Russian), MR. 41 (1971) 1067.
[9] Segal, G., Classifying spaces and spectral sequences, Inst. Hautes Études Sci. Publ. Math. **34** (1968), 105–112.
[10] Thurston, W., Noncobordant foliations of S^3, Bull. Amer. Math. Soc., **78** (1972), 511–514.
[11] Van Est, Group cohomology and Lie algebra cohomology in Lie groups I, II, Nederl. Akad. Wetensch. Proc. Ser. A, **56** (1953), 484–492, 493–504.

Department of Mathematics
Harvard University
Cambridge, Mass. 02138
U. S. A.

Reprinted from ADVANCES IN MATHEMATICS
All Rights Reserved by Academic Press, New York and London
Vol. 11, No. 3, December 1973

On the Chern–Weil Homomorphism and the Continuous Cohomology of Lie-Groups

R. BOTT*

Department of Mathematics, Harvard University, Cambridge, Massachusetts 02138

1. INTRODUCTION

This note describes a formula found in collaboration with G. Hochschild during the spring of 1972 at Berkeley, California. Therefore I would like first to thank him for all he taught me during those months, and, second, to chide him for steadfastly refusing to share the blame with me as a joint author. Since that time independent work of Kamber Tondeur has also appeared [6], which covers much of this material but from a rather different point of view. (Thus Theorem 2.10 of their Manuscripta paper is essentially equivalent to our Theorem 1.)

The aim of our discussion was really to explain the Chern–Weil homomorphism to topologists, and I hope that in this respect we have succeeded. The present point of view also fits very well into the general framework of foliations and the continuous cohomology encountered there, as will be seen in a joint paper now in progress with A. Haefliger. However, in this note I will stick to the finite-dimensional case, and essentially start from scratch. To explain our formula recall first of all that the Chern–Weil homomorphism links the cohomology of the classifying space BG of a Lie-group, G, to the invariant forms on the Lie algebra $\mathfrak{g}$ of G. More precisely, if $\mathfrak{g}^*$ denotes the dual of $\mathfrak{g}$ as a G-module under the adjoint action, and $S\mathfrak{g}^*$ the symmetric algebra on $\mathfrak{g}^*$ in its induced G-module structure, then the Chern–Weil construction defines a homomorphism from the invariants $\mathrm{Inv}_G(S\mathfrak{g}^*)$ of $S\mathfrak{g}^*$ to $H^*(BG) = H^*(BG; \mathbb{R})$

$$\varphi: \mathrm{Inv}_G(S\mathfrak{g}^*) \to H^*(BG). \tag{1.1}$$

The construction of this φ was inspired by differential geometric considerations and generalizes earlier constructions for the Pontryagin

* This research was partially supported by a National Science Foundation grant GP-31359-X-1.

and Chern classes from the curvature of a connection. Partly because of these geometric antecedents this homomorphism φ has also been treated with some apprehension by topologists in general. In a recent thesis [7], Shulman constructs φ more in the spirit of the algebraic topology and this note answers a question which arises naturally in his framework.

To motivate our main result let me review here the very elementary point of view of "characteristic classes in terms of transition functions" which I have advocated for some time now (see [1], [2]) and which is also the starting point of Schulman's thesis.

In this framework we think of a principal G-bundle ξ as being described by its 1-cocycle $g = \{g_{UV}\}$ of smooth transition functions

$$g_{UV}: U \cap V \to G$$

defined on the overlaps of open sets of a covering $\mathfrak{U}$ of the base manifold M of ξ, and we seek characteristic classes of ξ in the Cěch–DeRham complex," $\check{C}(\mathfrak{U}; A)$, of M *relative to* $\mathfrak{U}$. Thus $\check{C}(\mathfrak{U}; A) = \Sigma_{p,q} C^p(\mathfrak{U}; A^q)$ where $C^p(\mathfrak{U}; A^q)$ are the Cěch p-cochains relative to the cover $\mathfrak{U}$, in the presheaf of q-forms on M. This complex has two differential operators, the d of the DeRham theory, and the δ of the Cěch theory, which combine with a suitable sign, to yield a total differential $D = d \pm \delta$. The homology $H\check{C}(\mathfrak{U}; A)$, of $\check{C}(\mathfrak{U}; A)$ relative to this D figures already in A. Weil's 1951 proof of DeRham's theorem, the important point being that for every open cover $\mathfrak{U}$ the natural inclusion of $A(M) \subset \check{C}^0(\mathfrak{U}; A)$ induces an isomorphism of the DeRham groups of M with $H\check{C}(\mathfrak{U}; A)$:

$$HAM \simeq H\check{C}(\mathfrak{U}; A). \tag{1.2}$$

This granted it suggests itself to construct characteristic classes for ξ, via a cocycle g for ξ over $\mathfrak{U}$, in $\check{C}(\mathfrak{U}; A)$ by the following procedure: Once and for all we choose differential forms $\varphi = \{\varphi_p\}$, $p = 0, 1$, with

$$\varphi_p \in A(G \times \cdots \times G) \qquad p \text{ factors.}$$

Next, given any cocycle g, of ξ valid on $\mathfrak{U}$, and any nontrivial intersection

$$U_\sigma = U_0 \cap \cdots \cap U_p$$

of $(p + 1)$ open sets in $\mathfrak{U}$, let

$$g_\sigma: U_\sigma \to G \times \cdots \times G \qquad (p \text{ factors})$$

be the product map

$$g_\sigma = g_{U_0U_1} \times g_{U_1U_2} \times \cdots \times g_{U_{p-1}U_p}.$$

Then for each σ, $g_\sigma{}^*\varphi_p$ is a well defined form on U_σ, and the collection of these forms clearly defines a cochain φ

$$g^*\varphi \in \check{C}(\mathfrak{U}, A).$$

One is next led to ask when $g^*\varphi$ is *universally closed* in $\check{C}(\mathfrak{U}, A)$ (relative to D).

For this purpose consider the naturally bigraded complex $\mathcal{O}\!\mathcal{L}(G) = \Sigma\mathcal{O}\!\mathcal{L}^{p,q}(G)$ with

$$\mathcal{O}\!\mathcal{L}^{p,q}(G) = A^q(G \times \cdots \times G) \qquad p \text{ factors}, \tag{1.3}$$

so that we may then think of φ as an element of $\mathcal{O}\!\mathcal{L}G$.

Now of course $\mathcal{O}\!\mathcal{L}G$ has the differential d of DeRham theory, but can also be equipped with a differential operator δ

$$\delta\colon \mathcal{O}\!\mathcal{L}^{q,p}(G) \to \mathcal{O}\!\mathcal{L}^{q,p+1}(G)$$

which commutes with d and is compatible with the Cěch δ of $\check{C}(\mathfrak{U}; A)$ under g^*. In short then, the complex $\mathcal{O}\!\mathcal{L}(G)$ also has a natural *total operator* D defined on it and the cochain $g^*\varphi$ *will be universally* closed, *if and only if* φ *is* D-closed in $\mathcal{O}\!\mathcal{L}(G)$.

At this stage one is therefore naturally led to consider the homology of $\mathcal{O}\!\mathcal{L}(G)$ relative to D as furnishing one with the simplest characteristic classes of G-bundles, and to compute $H\mathcal{O}\!\mathcal{L}(G)$ explicitly, there are the standard methods of treating double complexes. Namely, one filters $\mathcal{O}\!\mathcal{L}(G)$ either by p, or by q, and considers the resulting spectral sequences.

If one filters by $\leqslant p$, the resulting spectral sequence obviously has the E_1-term

$$E_1^{p,q} = H^q(\underbrace{G \times \cdots \times G}_{p})$$

and the d_1, induced by δ, in $A(G)$, is then seen to yield the cobar construction on the coalgebra $H^*(G)$. Furthermore, it then follows without trouble that the E_2 term of this sequence is the symmetric algebra generated by the *primitive* elements $\mathscr{P}(G)$ of $H(G)$, considered in $H^{1,*}(G)$ and that all the further differential operators vanish. All this is done by Schulman in [7], and clearly leads to the evaluation of $H\mathcal{O}\!\mathcal{L}G$:

$$H\mathcal{O}\!\mathcal{L}(G) \cong S\mathscr{P}(G). \tag{1.4}$$

This is of course the "correct result" in the sense that the characteristic classes predicted by the classification theorems are given by the cohomology of the classifying space BG of G, and this cohomology is, over the reals, well known to be a polynomial algebra on the primitive elements of $H^*(G)$. Furthermore, this is *not* coincidence; indeed the spectral sequence obtained above is very naturally related to the sequences obtained from any of the standard procedures for constructing BG, see [7] for details.

Note finally that up to this point the precise nature of the q-form functor A^q, is nowhere brought out essentially. For instance if we interpret $A^q(U)$ as singular cochains on U, (1.4) would still be valid by the same argument.

Now our main observation is that the special nature of the DeRham functor A^q, can be exploited to yield an explicit evaluation of the cohomology of $\mathfrak{A}(G)$ relative to δ which in turn leads to a generalization of the Chern–Weil homomorphism. Precisely we have:

THEOREM 1. *Let G be a Lie-group and consider the δ-closed subcomplex, $\mathfrak{A}^qG$, of $\mathfrak{A}G$ determined by the q forms. Thus*

$$\mathfrak{A}^qG\colon A^q(p) \xrightarrow{\delta} A^q(G) \xrightarrow{\delta} A^q(G \times G) \cdots .$$

Then the δ-homology of $\mathfrak{A}^qG$ is given by

$$H_\delta{}^p\mathfrak{A}^qG = H_c^{p-q}(G; S^q\mathfrak{g}^*), \tag{1.5}$$

where $\mathfrak{g}$ is the Lie algebra of G considered as a G-module under the adjoint action, $S^q\mathfrak{g}^*$ denotes the q-th symmetric power, and the subscript c denotes the "smooth" or equivalently "continuous" cohomology of G with values in $S^q\mathfrak{g}^*$.

Remarks. (a) That (1.5) is connected to the Chern–Weil homomorphism is clear once one recalls the first facts concerning the continuous cohomology. These are that:

$$H_c{}^0(G; W) = \mathrm{Inv}_G\, W, \tag{1.6}$$

and that

$$H_c{}^i(G, W) = 0, \qquad \text{for} \quad i > 0, \tag{1.7}$$

if G is compact.

Thus by (1.5)

$$H_\delta{}^p\mathfrak{A}^pG = \mathrm{Inv}_G\, S^p\mathfrak{g}^*,$$

whence, as there is no cohomology above the diagonal, the "edge-homomorphism" of (1.5) induces a natural arrow,

$$\Phi\colon \mathrm{Inv}_G(S\mathfrak{g}^*) \to H\mathfrak{A}G$$

which, under the identification $H^*\mathfrak{A}G \simeq H^*BG$ goes over into the Chern–Weil homomorphism. Further by (1.7) Φ is *an isomorphism* when G is compact. The formula (1.5) therefore easily yields the classical properties of φ, and at the same time clarifies the situation in the non-compact case. Further, our procedure involves no "curvature" arguments. Rather, one encounters the homomorphism Φ as a direct consequence of the DeRham functor A^q.

(b) On the debit side it should be pointed out that Theorem 1—although we could not find it explicitly in the literature— is not really new, but in some sense a specialization of theorems of Illusie outlined in [2], on Chern classes in the algebraic category. However, as this paper is highly inaccessible to most geometers and topologists, and as continuous cohomology does not really fit into the algebraic theory a direct derivation of this theorem, such as we bring below, might be of use in any case.

There are also some very recent notes by Kamber and Tondeur [4], which are closely related to (1.5) via the Van Est Theorem.

The proof of our theorem, which we bring in the next section, fits very naturally into the Dold-Puppe Yoga of nonadditive functors [3] and especially their notion of semisimplicial derivation of a functor F defined on the category of modules. Indeed, the shift from the *exterior algebra* of the DeRham theory to the *symmetric algebra* of the Chern–Weil homomorphism will be seen to correspond precisely to the following fact.

PROPOSITION 1.8. *In characteristic* 0 *the first derived* (*in the sense of Dold-Puppe*) *of the exterior power* Λ^q, *is given by the q-th syspension of* S^q, *the q-th symmetric power.*

For completeness sake we will review some of the notions from scratch in the next section, before proving this proposition. The proof of Theorem 1 is then brought in Section 3, with the aid of the Hochschild–Mostow characterization of the continuous cohomology of G.

2. $\mathfrak{A}G$ AS A SEMI-SIMPLICIAL OBJECT

The simplest description of the complex $\mathfrak{A}G$ described in the introduction is in semi-simplicial terms. Indeed, in that framework,

$\mathcal{A}G$ is seen to be derived from a *semi-simplicial homogeneous space* NG, associated to G, by first applying the DeRham functor to obtain a *cosemisimplicial module* ANG, and thereafter the functor k converting such objects into cochain complexes. Thus one has the factorization

$$\mathcal{A}G = k \circ A \circ NG. \tag{2.1}$$

Precisely, let Ord denote the category of finite linearly ordered sets, with order preserving maps, and denote as is usual, the ordered set $\{0 < 1 < \cdots < n\}$ by $\mathbf{n}$. A semisimplicial set, space, manifold, module, group, etc., is then by definition a contravariant functor from Ord to the category of sets, spaces, manifolds, groups, modules, etc. Covariant functors of this type are unfortunately called cosemisimplicial sets, spaces, etc.

Note that if X is an object in one of these categories, then X defines *two* "trivial" semisimplicial objects of that same category which we denote by X again, and PX respectively.

The semisimplicial object X is simply defined by:

$$X(\mathbf{n}) = X, \qquad \text{all maps, } \to\text{, the Identity,} \tag{2.2}$$

while PX is defined in terms of the underlying sets $\mathbf{n}^\natural$ and $X^\natural$, of $\mathbf{n} \in \text{Ord}$ and X by:

$$PX(\mathbf{n}) = \text{Hom}(\mathbf{n}^\natural, X^\natural). \tag{2.3}$$

Less pedantically, this of course means that

$$PX(\mathbf{n}) = X \times \cdots \times X \qquad (n + 1 \text{ copies}).$$

Note also that if X is a manifold, group, etc. PX is a semisimplicial manifold, group, etc., in a natural manner.

Note further that the diagonal maps $\Delta\colon X \to X \times \cdots \times X$ combine to define a natural morphism

$$X \xrightarrow{\Delta} PX. \tag{2.4}$$

Now suppose G is a group. Then $PG(\mathbf{n})$ is a group for each $\mathbf{n}$, and $\Delta\colon G(\mathbf{n}) \to PG(\mathbf{n})$ is the diagonal inclusion. We may therefore define a new functor NG by setting

$$NG(\mathbf{n}) = PG(\mathbf{n})/G(\mathbf{n}), \tag{2.5}$$

and if G is a Lie-group, NG is clearly a semisimplicial manifold. The DeRham functors A^q, are therefore well defined on $NG(\mathbf{n})$, and the composition $A^q \circ NG$, defines the cosemisimplicial object:

$$A^q \circ NG\colon \text{Ord} \to \mathbb{R}\text{-modules}. \tag{2.6}$$

At this stage one needs only explain the functor k to make sense of the right-hand side of (2.1). For this purpose let M: Ord $\to \mathcal{M}$ be any covariant functor with values in modules. Then kM denotes the cochain complex with the module $M(\mathbf{n})$ in dimension n, and differential operator

$$\delta\colon M(\mathbf{n}) \to M(\mathbf{n}+1)$$

defined by:

$$\delta = \sum (-1)^i M(\epsilon_i) \tag{2.7}$$

where $\epsilon_i\colon \mathbf{n} \to \mathbf{n}+1$, $i = 0, \ldots n+1$ is the unique monotone map whose image does not include i. With this understood we have the following straightforward proposition which we leave to the reader.

PROPOSITION 2.8. *The formula* (2.1) *is valid. That is the cochain complex* $k \circ A^q \circ NG$ *can be identified with the complex* $\mathfrak{A}^q G$.

To proceed further we will need to decompose $A^q \circ NG$ a good deal more. In particular we will want to bring out clearly the role of the exterior power in this functor and to do this we will need the following two auxiliary constructions.

Given two cosemisimplicial modules M and N, their product $M \times N$ is defined to be the functor of the same sort given by

$$(M \times N)(\mathbf{n}) = M(\mathbf{n}) \otimes N(\mathbf{n}). \tag{2.9}$$

Also given a module M one has constructions CM and ΣM, analogous to the earlier PG and NG constructions. Explicitly let us first agree to identify a module M with the trivial functor

$$M\colon \text{Ord} \to \mathcal{M}$$

which assigns M to all the objects $\mathbf{n}$ of Ord and sends all maps into 1. Next we take the forgetful functor $\mathbf{n} \to \mathbf{n}^\natural$ and convert it to the covariant functor

$$C\mathbb{Z}\colon \text{Ord} \to \mathcal{M}$$

by setting

$$C\mathbb{Z}(\mathbf{n}) = \mathbb{Z}(n^\natural). \tag{2.10}$$

Here $\mathbb{Z}$(Set) denotes the free group generated by the set. One now has a natural arrow

$$C\mathbb{Z} \xrightarrow{r} \mathbb{Z} \longrightarrow 0$$

defined by

$$r(\mathbf{n}) \Big(\sum_{\alpha=0,\ldots n} a_\alpha \otimes \alpha \Big) = \sum a_\alpha, \qquad a_\alpha \in \mathbb{Z},$$

The kernel of r therefore defines a new functor $\Sigma\mathbb{Z}$ which we call the "suspension of $\mathbb{Z}$" and which naturally fits into the exact sequence:

$$0 \to \Sigma\mathbb{Z} \to C\mathbb{Z} \to \mathbb{Z} \to 0.$$

More generally if M is any covariant functor from Ord to modules, we define its cone and suspension CM and ΣM, by

$$\begin{aligned} CM &= C\mathbb{Z} \otimes M \\ \Sigma M &= \Sigma\mathbb{Z} \otimes M, \end{aligned} \qquad \text{(tensor over } \mathbb{Z}) \tag{2.11}$$

and one has the natural exact sequence:

$$0 \to \Sigma M \to CM \to M \to 0. \tag{2.12}$$

The reason for this terminology is of course that $H(kC\mathbb{Z})$ and hence $H(kCM)$ vanishes, whence if we abbreviate $H(kM)$ to $H(M)$:

$$H(\Sigma M) = M \qquad \text{(in dimension 1)},$$

that is, $H^i(\Sigma M) = 0$ for $i \neq 1$ and $H^1(\Sigma M) = M$ for any module M.

With this notation understood, we now have the following *basic decomposition* of $A^q NG$:

Decomposition Lemma. *Let G be a Lie-group, and let $\mathfrak{g}^*$ be the $\mathbb{R}$-module of left-invariant 1-forms on G. We make $\mathfrak{g}^*$ into a G-module under the action induced by right multiplication, make $A^0 \circ PG$ into a G-module valued functor through the action of G on $PG(\mathbf{n})$ given by right diagonal multiplication, and let G act trivially on $\Sigma\mathbb{Z}$.*

So understood, the functor $A^q \circ NG$ *has the decomposition*:

$$A^q \circ NG \cong \mathrm{Inv}_G \{A^0 \circ PG \times \Lambda^q (\Sigma\mathbb{Z} \times \mathfrak{g}^*)\}. \tag{2.13}$$

This formidable expression is actually just a very pedantic description of how the forms on the base of a fibering

$$E \xrightarrow{\pi} B,$$

appear in the complex of forms on the total space when both the tangent space to E and the normal space to the fibers of π admit trivializations,

Indeed let $\nu(\pi)$ be the subbundle of the cotangent bundle $T_E{}^*$ to E, whose elements restrict to 0 on the fibers of π. Then under π^*, $A^q(B)$ is clearly mapped isomorphically onto the space $\Gamma_F(\Lambda^q\nu^*)$ of those C^∞ sections of $\Lambda^q\nu^*$ which are constant on each fiber relative to the trivialization of ν^*/F induced by π^*

$$\pi^*: A^q(B) \cong \Gamma_F(\Lambda^q\nu^*).$$

Now suppose in addition that ν^* admits a global trivialization

$$\alpha: \nu^* \to W \tag{2.15}$$

with a fixed vector space W. Such an α identifies $\Gamma(\nu^*)$ with the W-valued functions on E, that is, with $A^0(E) \otimes W$ and $\Gamma_F(\nu^*)$ then becomes identified with a certain submodule of $A^0(E) \otimes W$. In our situation we apply this principle to the projection

$$PG(\mathbf{n}) \xrightarrow{\pi} NG(\mathbf{n})$$

and the trivialization of $\nu^*(\pi)$ induced by the left invariant forms on G. Indeed the inclusion

$$\iota: \mathfrak{g}^* \to A^1G,$$

extends to an arrow

$$\iota(\mathbf{n}): C\mathfrak{g}^*(\mathbf{n}) \to A^1PG(\mathbf{n}) \tag{2.16}$$

by the formula

$$\iota(\mathbf{n}) \sum_{k=0}^{n} a_k \otimes x_k = \sum a_k \cdot \pi_k{}^* x_k$$

where the π_k: $PG(\mathbf{n}) \to G$ are the natural projections. Now because of the left invariance of the forms the image of any element in $C\mathfrak{g}^*$ under $\iota(\mathbf{n})$ restricts to the *same* left invariant form on *every fiber* of π. That is, under the map

$$G \to PG(\mathbf{n})$$

given by $g \to (p_0 \cdot g \times p_1 \cdot g \times \cdots \times p_n \cdot g)$, $p_i \in G$, the element $\sum a_\alpha \pi_\alpha{}^* x_\alpha$ maps to $\sum a_\alpha x_\alpha \in \mathfrak{g}^*$. Hence under $\iota(\mathbf{n})$ the kernel $\sum \mathfrak{g}^*(\mathbf{n})$ (as defined in (2.12)) maps into $\nu^*(\pi)$, and in fact *induces a trivialization* of $\nu^*(\pi)$, because the image of $C\mathfrak{g}^*(\mathbf{n})$ under $\iota(\mathbf{n})$ spans the cotangent space of $PG(\mathbf{n})$ at every point.

In the resulting identification of $\Gamma(\nu^*)$ with $A^0(G) \otimes \sum \mathfrak{g}^*(\mathbf{n})$ it is now clear that the image of A^1NG is precisely given by the G-invariant elements, and more generally

$$A^qNG(\mathbf{n}) \simeq \mathrm{Inv}_G\,\{A^0PG(\mathbf{n}) \otimes \Lambda^q\Sigma\mathfrak{g}^*(\mathbf{n})\}$$

as was to be shown.

3. The Proof of Theorem 1

The decomposition formula of the last section clearly yields the result

$$\mathfrak{A}^qG = \mathrm{Inv}_G\,[k\,\{A^0PG \times \Lambda^q\Sigma\mathfrak{g}^*\}]. \qquad 3.1)$$

Now consider the cochain complex

$$\mathscr{C}^qG = k\,(A^0PG \times \Lambda^q\Sigma\mathfrak{g}^*), \qquad (3.2)$$

and recall the Cartier extension of the Eilenberg–Zilber Theorem [3] which asserts that for any two cosemisimplicial modules M and N, the cochain complexes $k(M \times N)$ and $k(M) \otimes k(N)$ are naturally homotopic:

$$k(M \times N) \sim k(M) \otimes k(N). \qquad (3.3)$$

Hence, in our characteristic 0 case,

$$H(\mathscr{C}^qG) \simeq H(A^0PG) \otimes H\,(\Lambda^q\Sigma\mathfrak{g}^*). \qquad (3.4)$$

We now assert that

LEMMA 3.1. *$H(A^0PG) = \mathbb{R}$ (in* dim 0) *while*

$$H(\Lambda^q \Sigma\mathfrak{g}^*) = S^q\mathfrak{g}^* \qquad (\textit{in } \dim q). \tag{3.5}$$

The first assertion is well known and corresponds to the contractability of the total space in the bar construction. The second one which is equivalent to Proposition 1.8 of the Introduction, is also rather clear from topological considerations and most probably known to experts. In any case we will bring proofs in the next section. But assuming this lemma, we can proceed to the proof of Theorem 1 as follows.

Clearly (3.4) and the lemma imply that

$$H\mathscr{C}^qG = S^q\mathfrak{g}^* \qquad (\text{in } \dim q).$$

On the other hand one checks directly that qua-G-modules the cochain complex $\mathscr{C}$ is *continuously injective* in the sense of [5].

Equation (3.5) therefore asserts that up to a shift of q to the right; $\mathscr{C}^qG$ is a *continuously injective resolution* of $S^q\mathfrak{g}^*$. Hence by the very definition of continuous cohomology, à la Hochschild–Mostow,

$$H(\mathfrak{A}^qG) = H_c(G; S^q\mathfrak{g}^*)$$

with a shift in dimension of q; that is

$$H^p(\mathfrak{A}^qG) = H_c^{p-q}(G; S^q\mathfrak{g}^*),$$

as was to be shown.

4. SEMISIMPLICIAL SUSPENSION AND THE PROOF OF LEMMA 3.1

The first assertion of our lemma, namely that

$$H(A^0 \circ PG) \simeq \mathbb{R} \qquad (\text{in } \dim 0),$$

of course amounts to the classic cone construction, and holds for any object X. That is, given an X in one of the categories discussed, then

$$H(A^0 \circ PX) = \mathbb{R} \qquad (\text{in } \dim 0) \tag{4.1}$$

for "any" interpretation of A^0 as arbitrary, continuous, or smooth functions on the objects. Indeed one merely chooses a point $e \in X$, and in terms of it defines a contracting homotopy on kA^0PX by the formula:

$$\lambda\varphi(x_0, \ldots, x_{p-1}) = \varphi(e, x_0, \ldots, x_{p-1}).$$

The formula (3.6) is less trivial, but as we will show follows directly from Eilenberg–Zilber–Cartier and the analog of (4.1) for a semi-simplicial module M, that is, the formula:

$$H(CM) = 0. \tag{4.2}$$

To see (4.2) one may use the following construction due to Dold and Puppe. Given any cosemisimplicial module M, they associate to M a cochain complex $k^\flat M$ which is much smaller than kM *but has the same cohomology*, by setting

$$k^\flat M_n = M(\mathbf{n}) \Big/ \Big\{ \sum_{0<i} M(\epsilon_i)\, M(\mathbf{n}-1) \Big\}$$

and letting the differential operator of $k^\flat M$ be the one induced by $\delta_0 = M(\epsilon_0)$ on $k^\flat M$ under the natural projection:

$$kM \xrightarrow{\flat} k^\flat M.$$

A powerful computational device of [3] is the proposition that $\flat$ induces an isomorphism:

$$H(kM) \simeq H(k^\flat M). \tag{4.3}$$

Now consider the object CM for M a fixed module: This object "starts" with

$$M \rightrightarrows M \oplus M \rightrightarrows\!\!\!\!\!\rightarrow M \oplus M \oplus M \quad \text{etc.}$$

whence $k^\flat CM$ is easily seen to reduce to

$$M \xrightarrow{\delta_0} M \longrightarrow 0 \longrightarrow 0$$

with δ_0 the identity. Q.E.D.

Thus kCM is an "augmented cone" with no homology in any dimension, whence it follows directly from the exact sequence

$$0 \to \Sigma M \to CM \to M \to 0,$$

that

$$H(k\,\Sigma M) = M \qquad \text{(in dimension 1).} \tag{4.4}$$

Now suppose F is any functor from modules to modules. In [3] Dold and Puppe define the semisimplicial suspension or derivation LF, of F, which is a functor from semisimplicial modules to graded modules, obtained by first syspending and then taking homology. On a fixed module M (or equivalently on the trivial cosemisimplicial object defined by M) LF can therefore be defined explicitly by

$$LF(M) = H(F \circ \Sigma M), \tag{4.5}$$

and the assertion (3.5) of Lemma 3.1 is equivalent to the assertion that

$$L\Lambda^q(M) = S^q(M) \qquad \text{(in dim } q\text{).} \tag{4.6}$$

We will prove (4.6) in a slightly more general setting which explains the operation L on all functors F, obtained by "symmetrization" from the functor

$$T_q\colon M \to M \otimes \cdots \otimes M \qquad (q \text{ copies}). \tag{4.7}$$

First of all observe that the theorem of Eilenberg–Zilber–Cartier immediately evaluates LT_q. Indeed

$$T_q(M) = M \times \cdots \times M \qquad (q \text{ copies})$$

for any cosemisimplicial object. Hence

$$T_q\,\Sigma M = \Sigma M \times \cdots \times \Sigma M \qquad \text{for any module } M.$$

It therefore follows from Eilenberg–Zilber–Cartier plus characteristic 0 that

$$\begin{aligned} LT_q(M) = H(kT_q\,\Sigma M) &= H(k\,\Sigma M) \otimes \cdots \otimes H(k\,\Sigma M) \\ &= M \otimes \cdots \otimes M \qquad \text{(in dim } q\text{).} \end{aligned} \tag{4.8}$$

Thus for T_q, the operation L simply "moves T_q up by q steps":

$$LT_q \cong T_q \qquad \text{(in dim } q\text{).} \tag{4.9}$$

Now this formula admits a refinement which in addition clears up how, at least in characteristic 0, the subfunctors of T_q, such as Λ^q, S_q, etc. behave under L. To formulate it we first "promote" T_q to a functor from $\mathbb{R}$-modules to modules over the group ring $\mathbb{R}\mathbf{S}_q$ of the symmetric group on q letters.

The action of $\mathbb{R}\mathbf{S}_q$ on T_q is of course given by the permutation of the factors. We also denote by $\bar{T}_q$ the functor $M \to M \otimes \cdots \otimes M$ (q factors) on which $\mathbb{R}\mathbf{S}_q$ acts by twisting the operation of $\mathbf{S}_q$ on T_qM with the determinant character:

$$\sigma \to \det \sigma$$

Thus in $\bar{T}_q(M)$,

$$\sigma(m_1 \otimes \cdots \otimes m_q) = \det \sigma(m_{\sigma^{-1}(1)} \otimes m_{\sigma^{-1}(2)} \otimes \cdots \otimes m_{\sigma^{-1}(q)}).$$

An immediate consequence of $E - Z - C$ is now the

Proposition 4.10. *Considered as functors from $\mathbb{R}$-modules to $\mathbb{R}\mathbf{S}_q$-modules, the q-th powers T_q and $\bar{T}_q$ are "interchanged" by L. That is,*

$$LT_q(M) = \bar{T}_qM \qquad \text{(in dim } q)$$

$$L\bar{T}_q(M) = T_qM \qquad \text{(in dim } q).$$

Indeed this is just a consequence of the well known fact that under an equivalence

$$k(X \times X) \sim k(X) \otimes k(X)$$

the interchange of factors on the left is expressed, on the right, by the chain map $u^p \otimes v^q \to (-1)^{pq} v^q \otimes u^p$; u and v in dim p and q, respectively. Now in our situation $k(\Sigma M)$ has homology only in dim 1, whence all transpositions pick up the sign -1. Q.E.D.

Consider now the *subfunctors* of T_q given by symmetrization with respect to a central idempotent e in $\mathbb{R}\mathbf{S}_q$. That is, given e with $e^2 = 1$, $e\sigma = \sigma e$ all $\sigma \in \mathbf{S}_q$ define

$$F_e(M) = e \cdot T_qM.$$

We clearly have arrows $F_e(M) \underset{i}{\overset{e}{\leftrightarrows}} T_q(M)$ given by the inclusion and the projection e respectively, which induce a direct sum decomposition

$$HkF_e \sum M \leftrightarrows HkT_q \sum M = \bar{T}_q(M) \qquad \text{(in dim } q). \tag{4.11}$$

Hence the left-hand side of (4.11) is given by $e\bar{T}_q M$ (in dim q). Thus one has the formula:

$$LF_e(M) \cong e\bar{T}_q M \qquad \text{(in dim } q\text{)}. \tag{4.12}$$

On the purely module level this can also be expressed as follows:

The group ring $\mathbb{R}\mathbf{S}_q$ admits an automorphism denoted by —, given by

$$\bar{\sigma} = \det \sigma \cdot \sigma,$$

and $\bar{T}_q M$ is clearly obtained from $T_q M$ via this automorphism. Hence qua vector spaces, we have

$$e \cdot \bar{T}_q M = \bar{e} T_q M, \tag{4.13}$$

so that on this level we can write (4.12) in the form

$$LF_e(M) \cong F_{\bar{e}}(M) \qquad \text{(in dim } q\text{)}. \tag{4.14}$$

Finally, as the bar clearly interchanges alternation with symmetrization, this yields the desired formula

$$L\Lambda^q \simeq S^q \qquad \text{(in dim } q\text{)}.$$

Also note that in the language of Young diagrams (it can be shown that) the bar corresponds to exchanging the rows with the columns in a diagram.

References

1. R. Bott, "Lectures on Characteristic Classes and Foliations" (Notes by Lawrence Conlon), pp. 1–94, Springer-Verlag, Berlin, 1972.
2. Paul Baum and R. Bott, "On the Zeroes of Meromorphic Vector-Fields," pp. 1–47, Springer-Verlag, 1970.
3. A. Dold and D. Puppe, Homologie Nicht-Addiliver Funktoren, Anwendungen, *Ann. Inst. Fourier (Grenoble)* 11 (1961), 201–312.
4. L. Illusie, Complexe cotangent relatif d'un faisceau d'algebras, *C. R. Acad. Sc., Paris* **268** (1969), 323–326.
5. G. Hochschild and G. D. Mostow, Cohomology of Lie groups, *Illinois J. Math.* **6** (1962), 367–401.
6. F. W. Kamber and Ph. Tondeur, "Cohomologie des algèbres de Weil relatives tronquées," "Algèbres de Weil semi-simpliciales," "Homomorphisme caractéristique d'un fibré principal feuilleté," Comptes Rendu Acad. Sci. Paris, t. 276 (5 Fevrier 1973, 2 Mai 1973, 21 Mai 1973), also, "Invariants of foliated bundles" to appear in Manuscripta Mathematica.
7. H. Shulman, "On Characteristic Classes," Ph.D. Thesis, University of California, Berkeley, 1972.

Printed by the St Catherine Press Ltd., Tempelhof 37, Bruges, Belgium.

Notes on Gel'fand–Fuks Cohomology and Characteristic Classes

by

M. Mostow and J. Perchik

These notes were prepared in response to
a series of lectures given by
Raoul Bott
at the Eleventh Holiday Symposium
at New Mexico State University,
Las Cruces, December 1973,
sponsored by the NSF.

Preface

These notes make up the written record of the Eleventh Holiday Symposium at New Mexico State University in Las Cruces, New Mexico on 27–31 December 1973. Professor Raoul Bott of Harvard University was the principal lecturer and the notes from his lectures comprise the first (and major) portion of these Proceedings. In addition to Professor Bott's lectures, there were sessions for contributed papers and versions of these talks ranging from abstracts to actual texts comprise the balance of the Proceedings. There is also appended a list of participants.

Financial support for this Symposium was supplied in part by New Mexico State University and to a large extent by the National Science Foundation. To these institutions the Symposium Committee wishes to record its thanks. Since all Symposium activities (including board and lodging of the participants) were held at the Holy Cross Retreat of the Franciscan Friars, the Symposium Committee also thanks them for their warm hospitality.

Robert J. Wisner
Warren M. Krueger
New Mexico State University
May 1975

Foreword

The following notes were written by Mark Mostow and Jim Perchik in response to my lectures at the 1973 Holiday Symposium in Las Cruces, New Mexico.

My aim during this highly concentrated Symposium – two lectures a day for five days by the main speaker as well as two or three more by interested participants – was to explain to nonexperts some of the phenomena which I associate with the work of Gel'fand–Fuks on the Lie algebra of vector fields of a smooth manifold.

In particular I tried to outline: (1) a "new" proof of the finite dimensionality of the G.F. cohomology along lines devised by G. Segal and myself; (2) the relationship of this cohomology to characteristic classes of various sorts, and especially the ones encountered in Foliation theory à la Haefliger.[†]

My lectures were mainly informal with theorems illustrated by example, rather than proved in detail, and this of course makes note taking in the strong sense very difficult. Rather, Jim and Mark have each worked out a section of the material according to their own tastes and have supplied proofs, background material and precise references where they most felt the need for them.

By and large I am very pleased with what they have wrought and I hope that nonexperts in some parts of the material will also find this most unBourbaki-like account of a piece of mathematics refreshing.

Since these lectures were delivered the field has moved forward considerably: At this date, I have in hand independent outlines of the proof of my main conjecture (pg. 175), one by G. Segal and one by A. Haefliger. Quite recently, P. Trauber and Stasheff–Anderson have announced an affirmative solution. In the opposite direction I should point out that during the write up of these notes Jim Perchik pointed out some technical difficulties in our original argument to the G.F. spectral sequence, but this was cleared up by G. Segal after an anxious S.O.S. from us! Also, Mark Mostow pointed out that much greater care has to be taken in the geometric realization of simplicial spaces than I originally did and he has explained and clarified these matters in his write up.

Granting the truth of the conjecture – and I still haven't truly digested any of the versions – the most challenging next step in this general area would seem to me to be a proof that $F\Gamma_q$ (that is $B\overline{\Gamma}_q$) is $2q$ connected. (Thurston shows it to be $(q+1)$ connected.) Also of course, the case of Hamiltonian manifolds is still quite unexplored, and in particular, the finite dimensionality in that case is completely untouched.

Raoul Bott
October 9, 1974

[†] I acknowledge with thanks that part of this research was supported by NSF grant GP4361.

Contents

Notes on Gel'fand–Fuks Cohomology and Characteristic Classes

M. Mostow and J. Perchik

1. Bundles, classifying spaces, and characteristic classes

We regard a bundle as a twisted product. (See [41] and [64] for more details on bundles.)

Definition. Let X, Y, and F be topological spaces. Let G be a topological group which is a subgroup of $\mathrm{Aut}(F) = \{\text{homeomorphisms of } F\}$ (i.e., G acts continuously on F on the left). Let $\pi : Y \to X$ be a continuous surjection such that F is homeomorphic to $\pi^{-1}(x)$ for all $x \in X$. Then the combination $\{Y, \pi, X\}$ is a *bundle* (also a *G-bundle*) with *base space* X, *total space* Y, *fiber* F, and *structural group* G, if it satisfies the following local triviality and gluing conditions:

1) There exists an open covering $\{U_a\}$ of X and maps (called *horizontal projections*)

$$f_a : \pi^{-1}(U_a) \longrightarrow F$$

such that for all indices a, the map

$$\varphi_a = (\pi, f_a) : \pi^{-1}(U_a) \longrightarrow U_a \times F$$

is a homeomorphism (see diagram).

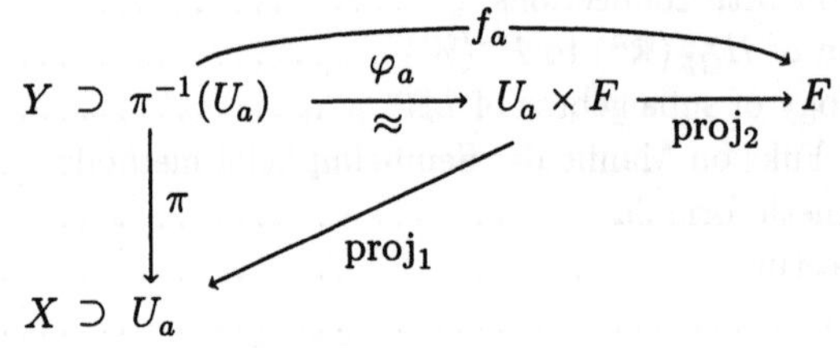

2) For all pairs of indices a, b, there is a continuous map

$$g_{ba} : U_a \cap U_b \longrightarrow G$$

(called a *transition function*) which relates f_a and f_b. Specifically, for all y in $\pi^{-1}(U_a \cap U_b)$, if $x = \pi(y)$, then

$$f_b(y) = g_{ba}(x) \cdot f_a(y)$$

Note. By abuse of notation one often refers to the total space Y as the bundle.

Definition. Let all notation be as above, and suppose $F = G$. Let the equation

$$f_b(y) = g_{ba}(x) \cdot f_a(y)$$

(in the definition of bundle) denote multiplication in G. Then the bundle is a *principal G-bundle.* The multiplication map

$$F \times G \longrightarrow F$$

defines a *right* action of G on Y which preserves each fiber. Specifically, in φ_a coordinates, for $g \in G$ and $y \in \pi^{-1}(U_a)$,

$$(\pi(y), f_a(y)) \cdot g = (\pi(y), f_a(y) \cdot g),$$

and this definition is independent of coordinates.

Definition. A bundle with fiber $F = \mathbb{R}^n$ or $\mathbb{C}^n$ and structure group $G = GL(n, \mathbb{R})$ or $GL(n, \mathbb{C})$, resp., acting in the usual way on F, is called a *vector bundle* or an *n-plane bundle.*

The transition maps g_{ba} describe the twisting of a bundle. For example, a bundle is *trivial,* i.e., a product $X \times F$, if and only if the maps g_{ba} can all be taken to be

$$g_{ba} : U_a \cap U_b \longrightarrow id \in G.$$

One observes that the maps g_{ba} satisfy a compatibility condition, called the *cocycle condition,* on triple intersections. Namely,

$$g_{ab}(x) \cdot g_{bc}(x) = g_{ac}(x) \quad \text{for all} \quad x \in U_a \cap U_b \cap U_c.$$

In fact, the maps $\{g_{ab}\}$ contain all the information necessary to reconstruct the bundle, if we are also given the action of G on the fiber F; and given any set of maps $\{g_{ab}\}$ satisfying the cocycle condition (such a set $\{g_{ab}\}$ is called a *cocycle on X with values in G*), one can construct a bundle having $\{g_{ab}\}$ as transition functions. One simply takes the disjoint union

$$\coprod_a U_a \times F$$

and divides by the relations

$$(x, f) \in U_a \times F \sim (x, g_{ba}(x) \cdot f) \in U_b \times F.$$

A *bundle map* is a commutative diagram

$$\begin{array}{ccc} Y' & \longrightarrow & Y \\ \pi' \downarrow & & \downarrow \pi \\ X' & \longrightarrow & X \end{array}$$

where $\pi : Y \to X$ and $\pi' : Y' \to X'$ are G-bundles with the same fiber F, and

$$Y' \longrightarrow Y$$

restricted to any fiber is a homeomorphism belonging to G. Two bundles with the same base space X are *isomorphic* if there is a bundle map

$$\begin{array}{ccc} Y' & \longrightarrow & Y \\ \pi' \downarrow & & \downarrow \pi \\ X' & \xrightarrow{=} & X \end{array}$$

In terms of cocycles, the condition for two bundles over X to be isomorphic is the *coboundary condition.* Namely, if there exist maps,

$$h_a : U_a \longrightarrow G$$

such that

$$h_b(x) \cdot g_{ba}(x) = g'_{ba}(x) \cdot h_a(x) \quad \text{for all} \quad x \in U_a \cap U_b,$$

then the cocycles $\{g_{ba}\}$ and $\{g'_{ba}\}$ define isomorphic bundles.

One defines $H^1(X;G)$ to be the set of isomorphism classes of G-bundles over X. Roughly speaking, $H^1(X;G)$ is the set of cocycles $\{\{g_{ab}\}\}$ modulo the set of coboundaries $\{\{h_a\}\}$. In case G is the whole group $\mathrm{Aut}(F)$, $H^1(X;G)$ is the set of isomorphism classes of fiber bundles over X with fiber F.

Now G-bundles pull back under continuous maps. Specifically, if

$$\pi : Y \longrightarrow X$$

is a G-bundle with cocycle $\{g_{ab}\}$ relative to the cover $\{U_a\}$ of X, and if

$$f : X' \longrightarrow X$$

is a continuous map, then by defining $f^{-1}(Y) = Y'$ to be

$$X' \, x_{f,\pi} \, Y,$$

the fiber product of X' and Y [41], we have a commutative diagram

$$\begin{array}{ccc} f^{-1}(Y) = Y' & \longrightarrow & Y \\ \pi' \downarrow & & \downarrow \pi \\ X' & \underset{f}{\longrightarrow} & X. \end{array}$$

$\pi' : Y' \to X'$ is a G-bundle (denoted $f^{-1}(Y)$, or $f^{-1}(\xi)$ if ξ = the bundle $\{Y, \pi, X\}$), defined relative to the open cover $\{f^{-1}(U_a)\}$ of X' by the cocycle $\{g_{ab} \circ f\}$. Thus the above diagram is a G-bundle map.

Let Q now be the functor from the category *Spaces* (= *Topological spaces*) to *Sets* which takes a space X to

$$Q(X) = H^1(X;G) = \{\text{isomorphism classes of } G\text{-bundles over } X\},$$

and a map $f : X' \to X$ into the map

$$Q(f) : Q(X) \to Q(X')$$

defined by

$$Q(f)(\overline{Y}) = \overline{f^{-1}(Y)},$$

where $\pi : Y \to X$ is a bundle and $\overline{Y}$ its isomorphism class.

Q is a contravariant functor from *Spaces* to *Sets*, and is in fact *homotopy invariant*, i.e., if f and g are homotopic maps from X' to X, then

$$Q(f) = Q(g) \quad \text{as maps:} \quad Q(X) \to Q(X').$$

This follows from the standard results of bundle theory [41].

Let us now fix a G-bundle

$$\pi : E \longrightarrow Z.$$

Let

$$[X, Z] = \{\text{homotopy classes of maps:} \quad X \to Z\}.$$

Then $[\cdot, Z]$ is a homotopy invariant contravariant functor from *Spaces* to *Sets*. By the homotopy invariance of the functor Q, there is a natural transformation of functors from $[\cdot, z]$ to Q. Namely, to each space X, we assign the map

$$[X, Z] \longrightarrow Q(X)$$

given by

$$(f : X \longrightarrow Z) \mapsto \overline{f^{-1}(E)}.$$

If the map $[X, Z] \to Q(X)$ is an isomorphism for each X, we say that

$$[\cdot, Z] \longrightarrow Q$$

is an *isomorphism of functors*, and that Z is a *classifying space* for the functor Q (in our case, Z is a classifying space for G-bundles, but the definitions are valid for *any* homotopy invariant contravariant functor Q : *Spaces*→ *Sets*). [11]

There are several ways of finding a classifying space Z[11,6,41,64]. By its universal classifying property, Z is unique up to homotopy type if it exists.

Definition. BG will always denote any classifying space Z for G-bundles. If $\pi : E \to BG$ is a principal G-bundle, then E will be denoted EG.

$$\pi : EG \longrightarrow BG$$

is called a *universal principal G-bundle.*

The following characterization of the classifying spaces is due to Steenrod ([64], p. 102).

Theorem 1.1. *Let G be a topological subgroup, and let $\pi : E \to Z$ be a principal G-bundle. Then Z is a classifying space for G-bundles on finite simplicial complexes if and only if $\pi_i(E) = 0$ for all i.*

We have deliberately been rather vague about what the fiber of any particular G-bundle is, because we are most interested in its cocycle $\{g_{ab}\}$. This vagueness is justified by the following. As we saw above, given any cocycle $\{g_{ab}\}$ we can construct a bundle with any given fiber F (where F is a space upon which G acts on the left), having $\{g_{ab}\}$ as its cocycle. Alternatively, suppose we have a *principal* G-bundle

$$\pi : Y \longrightarrow X$$

and want to find a bundle with the same cocycle $\{g_{ab}\}$ but with fiber F.

Exercise. Show that such a bundle is given by

$$Y \times_G F \longrightarrow X,$$

where $Y \times_G F = Y \times F$ modulo the action of G; i.e.,

$$(yg, f) \sim (y, gf) \quad \text{for} \quad y \in Y,\ f \in F,\ g \in G.$$

In particular,

$$EG \times_G F \longrightarrow BG$$

is a *universal G-bundle with fiber F*; notice that the base space BG does not depend on F.

In general, to represent any functor Q via a classifying space, one must restrict the category of spaces, for example to finite polyhedra or paracompact spaces. Later on we shall represent functors such as

$$Q = \{\text{homotopy classes of foliations on manifolds}\}$$

and

$$Q = (\text{Gel'fand–Fuks cohomology of a manifold})$$

in terms of appropriate classifying spaces (though the classification of Gel'fand–Fuks cohomology will be different from the type of classification being discussed now).

Assuming we have found a classifying space Z for any functor Q (for example $\{G\text{-bundles}\}$), the obstructions to trivializing or extending a *Q-structure* (i.e., an element of $Q(X)$, where X is some space), can be studied by standard obstruction theory [64]. Specifically, let

$$h \in Q(X)$$

be a Q-structure on a space X. Then h is classified by the homotopy class of a map

$$f_h : X \longrightarrow Z.$$

h is a *trivial Q-structure* if and only if f_h is homotopic to a trivial map. For example, in the case

$$Q = \{\text{isomorphism classes of } G\text{-bundles}\}$$

and $Z = BG$, as before, we have the diagram

$$\begin{array}{ccc} f_h^{-1}(EG) & \longrightarrow & EG \\ \downarrow & & \downarrow \\ X & \xrightarrow{f_h} & BG = Z \end{array}$$

and the bundle $h = (f_h^{-1}(EG) \to X)$ is *trivial* (i.e., a product G-bundle $X \times G$) if and only if f_h is a nullhomotopic map.

The obstructions to homotoping f_h to a trivial map lie in

$$H^q(X; \pi_q(Z)), \quad q = 0, 1, 2, \dots ;$$

the "universal obstructions" lie in $\pi_q(Z)$, $q = 0, 1, \dots$. One can regard the elements of $H^*(Z)$ as another measurement of the nontriviality of the functor Q, since if

$$s \in H^*(Z),\ f_h : X \to Z \quad \text{classifies } h \in Q(X), \text{ and } f_h^* s \neq 0 \text{ in } H^*(X),$$

then f_h is not nullhomotopic, so h is not a trivial Q-structure on X.

Whether or not a functor

$$Q : \mathit{Spaces} \longrightarrow \mathit{Sets}$$

has a classifying space Z, we define a *characteristic class* $s = \{s_h\}$ of the functor Q to be a choice

$$s_h \in H^*(X) \qquad \text{for each } h \in Q(X) \text{ for each space } X,$$

such that if

$$f : X' \longrightarrow X$$

and

$$h' \in Q(X') \quad \text{and} \quad h \in Q(X)$$

are such that

$$h' = Q(f)(h),$$

then

$$s_{h'} = f^* s_h,$$

where

$$Q(f) : Q(X) \longrightarrow Q(X')$$

and

$$f^* : H^*(X) \longrightarrow H^*(X').$$

In other words, a characteristic class is a choice of cohomology classes natural with respect to morphisms of Q-structures [53]. For example, a characteristic class for G-bundles is a choice of

$$s_h \in H^*(X)$$

for each G-bundle

$$h = (\pi : Y \longrightarrow X),$$

such that given a G-bundle map

$$h'\left\{\begin{array}{ccc} Y' & \longrightarrow & Y \\ \pi'\downarrow & & \downarrow\pi \\ X' & \underset{f}{\longrightarrow} & X \end{array}\right\} h,$$

$$f^* s_h = s_{h'}, \in H^*(X').$$

If the functor Q does have a classifying space Z, let s be an element of $H^*(Z)$. If for each $h \in Q(X)$ with classifying map

$$f_h : X \longrightarrow Z$$

we assign

$$s_h \underset{def.}{=} f_h^*(s) \in H^*(X),$$

then $\{s_h\}$ is a characteristic class for Q. For example, if $Q = \{G\text{-bundle isom. classes}\}$ and

$$s \in H^*(BG),$$

then for each bundle $h = (Y \to X)$, classified by

$$h\left\{\begin{array}{ccc} Y & \longrightarrow & EG \\ \downarrow & & \downarrow \\ X & \underset{f_h}{\longrightarrow} & BG \end{array}\right.$$

we have

$$s_h = f_h^*(s) \in H^*(X).$$

Exercise. For any homotopy invariant contravariant functor Q: *Spaces* $\to$ *Sets* with classifying space Z, the map of sets

$$H^*(Z) \longrightarrow \{\text{characteristic classes of } Q\}$$

defined by

$$s \longrightarrow \{s_h\}$$

is a 1-1 correspondence.

Note. Because of this result, we shall usually let s designate the characteristic class $\{s_h\}$.

Examples of classifying spaces and characteristic classes for G-bundles

Example 1. Let G be a discrete group, i.e., any group with the discrete topology imposed on its underlying set. Let $K = K(G,1)$ be an Eilenberg–MacLane space, i.e., a CW complex with

$$\pi_1(K) = G \qquad \pi_i(K) = 0,\ i \neq 1.$$

Such spaces always exist [20]. Let $\widetilde{K}$ be the universal covering space of K. Then

$$\pi_i(\widetilde{K}) = 0 \quad \text{for all} \quad i,$$

so $\widetilde{K}$ is contractible. G acts freely and discretely on $\widetilde{K}$, and

$$K = \widetilde{K}/G.$$

Exercise. $\widetilde{K} \to K$ is a G-bundle.

Thus by Theorem 1.1,

$$BG = K, \quad EG = \widetilde{K}.$$

In this case $\pi_i(BG) = G$ for $i = 1$ and 0 otherwise, but $H^*(BG)$ is hard to compute for most discrete groups G.

Example 1a. Let $G = \mathbb{Z}_2$. Then $BG = \mathbb{R}P_\infty = S^\infty/\mathbb{Z}_2 =$ infinite dimensional real projective space.

Example 1b. $G = \mathbb{Z}$. Then $BG = S^1 = \mathbb{R}/\mathbb{Z}$. Since

$$H^i(S^1) = \begin{cases} \mathbb{Z} & \text{for } i = 0, 1 \\ 0 & \text{otherwise,} \end{cases}$$

$\mathbb{Z}$-bundles have characteristic classes only in dimensions 0 and 1.

Example 1c. G non-Abelian and discrete. Here BG is hard to find explicitly.

Example 2. $G = \pi_1(S)$, where S is a Riemann surface of genus > 1. Then

$$BG = S.$$

Existence of BG for all Lie groups G

For all Lie groups G, in particular for $G = GL(n, \mathbb{R}), GL(n, \mathbb{C}), O(n), U(n)$, and other matrix groups, BG exists ([3], also cf. infra §6). In contrast to the discrete case, here $H^*(BG)$ is relatively easy to compute, while $\pi_i(BG)$ is hard to find.

By Theorem 1.1 a principal G bundle $E \to Z$ is universal if and only if E is contractible. Thus a strategy to find BG is to look for a contractible space E upon which G acts freely, let

$$Z = E/G,$$

and then prove that $E \to Z$ is a bundle, which proves that $Z = BG$. (The hard part is to show that Z has open cover $\{U_a\}$ which trivializes the bundle. This can usually be done for Lie groups using topological arguments.)

Example 1. Embed G as a subgroup of $GL(n)$ for suitably large n, via a representation of G. Let

$$\begin{aligned} E &= \{n\text{-frames in } \mathbb{R}^\infty\} \\ &= \{(v_1, \ldots, v_n) | v_i \in \mathbb{R}^\infty \\ &\quad \text{and } v_1, \ldots, v_n \text{are linearly independent}\}, \end{aligned}$$

where

$$\mathbb{R}^\infty = \{(a_1, a_2, \ldots) | a_i \in \mathbb{R} \quad \text{and the sequence is eventually} \quad 0\}.$$

Exercise. Prove E is contractible [41].

Now $GL(n, \mathbb{R})$ acts freely on E (via multiplication of an n-tuple of vectors by an $n \times n$ matrix), so $G \subset GL(n)$ acts freely on E. Thus we can take

$$BG = E/G.$$

Definition. Let E be as in Example 1, and let $G = GL(n, \mathbb{R})$. Then E/G is called the *Grassmannian* space of n-planes in $\mathbb{R}^\infty$.

Example 2. One constructs Grassmannians from quotients $U(n + m)/U(m)$ for large m.

2. Semi-simplicial objects

There is a very general kind of construction, called a semi-simplicial object, which will appear again and again in these notes. (See [21], [56], and [50] for general references.) Philosophically, one can regard a semi-simplicial space as a generalization of a space, or a semi-simplicial module as a generalization of a module, in much the same way that a chain or cochain complex is a generalization of a module. In fact, one can even construct a semi-simplicial space which is a "free resolution" of a manifold, in the sense that a chain complex is a free resolution of a module. This will be used later to prove the finite dimensionality of the Gel'fand–Fuks cohomology of a compact manifold.

Semi-simplicial objects can be defined most concisely using the language of category theory [56]. Let *Ord* be the category whose objects are finite sets ordered by a transitive $\leq$ relation, and whose morphisms are order-preserving set maps. The objects of *Ord* are called 0,1,2, etc. where $0 = \{0\} = \cdot$, $1 = \{0, 1\} =:$, $2 = \{0, 1, 2\} = \vdots$. The ordering of the set n is $0 \leq 1 \leq \ldots \leq n$. Of all the morphisms in *Ord*, we focus attention on the injections from n to $n+1$, which we call ∂_i or *face maps*, and the surjections from $n+1$ to n which we call ϵ_i or *degeneracy maps.* (In fact, all morphisms in *Ord* are compositions of face and degeneracy maps.)

The category *Ord* is represented by the picture

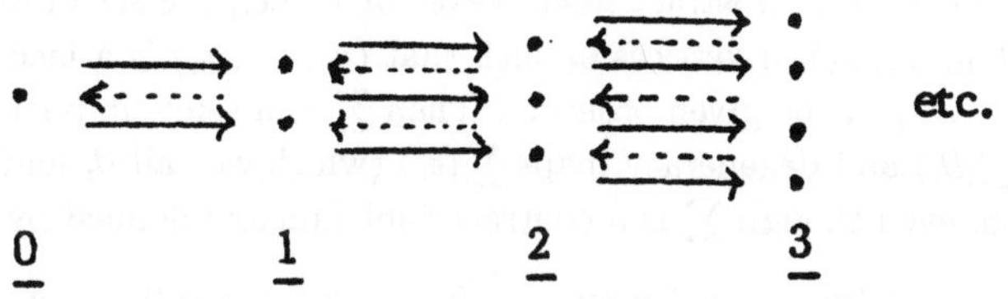

In this picture, the sets of vertically placed dots are the objects 0,1,2,... of *Ord.* The solid arrows are the morphisms ∂_i and the dashed arrows are the morphisms ϵ_i. These morphisms are defined by the following:

$$\partial_i : n \to n+1, \quad i = 0, 1, \ldots, n+1$$

$$\partial_i(j) = \begin{cases} j & \text{if } j < i \\ j+1 & \text{if } j \geq 1 \end{cases} \qquad \text{for } j \in n = \{0, 1, \ldots, n\}.$$

That is, ∂_i maps n to $n+1$ injectively, preserving order and omitting $i \in n+1$ from its range.

$$\epsilon_i : n \longrightarrow n-1, \quad i = 0, \ldots, n-1$$

$$\epsilon_i(j) = \begin{cases} j & \text{if } \ j \leq i \\ j-1 & \text{if } \ j > i \end{cases} \qquad \text{if } j \in n = \{0, 1, \ldots, n\}.$$

That is, ϵ_i maps n to $n-1$ surjectively, preserving order and mapping i and $i+1 \in n$ to $i \in n-1$.

Observe that there are $n+2$ face maps ∂_i from n to $n+1$, and n degeneracy maps ϵ_i from n to $n-1$.

Now let C be any category, for example *Sets, Topological spaces, C^∞ manifolds, Modules.* Let A be a contravariant functor from *Ord* to C. Then A is called a semi-simplicial object (abbreviated s.s. object) in C, i.e., a semi-simplicial space, manifold, module, etc. If B is a covariant functor from *Ord* to C, then B *co-semi-simplicial* (co-s.s.) space, module, etc. (The nomenclature reflects the fact that semi-simplicial objects were studied before co-semi-simplicial objects.) Explicitly, let $A(n)$ be the object in C associated to the object n in *Ord.* Then for each face map $\partial_i : n \to n+1$, there is a map $A(\partial_i) : A(n+1) \to A(n)$. Designating $A(\partial_i)$ by a solid arrow as before, we write the semi-simplicial object A as

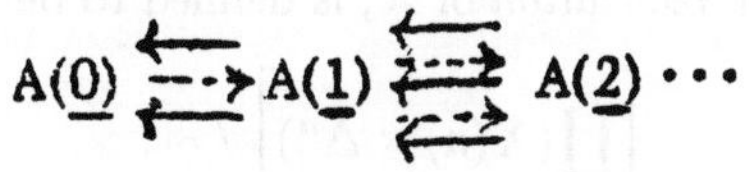

The co-semi-simplicial object B is written

$$B(\underline{0}) \;\begin{smallmatrix}\longrightarrow\\ \dashleftarrow\\ \longrightarrow\end{smallmatrix}\; B(\underline{1}) \;\begin{smallmatrix}\longrightarrow\\ \dashleftarrow\\ \longrightarrow\\ \dashleftarrow\\ \longrightarrow\end{smallmatrix}\; B(\underline{2}) \;\cdots\quad .$$

Example. Given a polyhedron P with an ordering on its vertices, let $\sum(n)$ be the set of abstract n-simplexes of P, i.e., the set of ordered $(n+1)$-tuples $(v_0, \dots, v_n)$ of vertices v_i such that $(v_0, \dots, v_n)$ is a face of P and $v_0 \leq v_1 \leq \dots \leq v_n$ in the given ordering. Then $\sum$ is a semi-simplicial set with face maps $\sum(\partial_i)$ and degeneracy maps $\sum(\epsilon_i)$ (which we call ∂_i and ϵ_i by abuse of notation, even though $\sum$ is a contravariant functor) defined by

$$\partial_i(v_0, \dots, v_n) = (v_0, \dots, \hat{v}_i, \dots, v_n), \quad i = 0, \dots, n$$
$$\epsilon_i(v_0, \dots, v_n) = (v_0, \dots, v_i, v_i, \dots, v_n), \quad i = 0, \dots, n.$$

Note that the union of the images of all the ϵ_i's is exactly the set of degenerate simplexes, i.e. simplexes with a vertex repeated.

Now there are three important general constructions done on semi-simplicial objects. The first of these is geometric realization of a semi-simplicial space.

Let X be any semi-simplicial space, i.e.

$$X : \mathit{Ord} \longrightarrow \mathit{Top.\ spaces}.$$

(As a special case we may allow X to be an s.s. set, since we may make any set into a topological space by giving it the discrete topology). We write X as

$$X(\underline{0}) \;\begin{smallmatrix}\longleftarrow\\ \dashrightarrow\\ \longleftarrow\end{smallmatrix}\; X(\underline{1}) \;\begin{smallmatrix}\longleftarrow\\ \dashrightarrow\\ \longleftarrow\\ \dashrightarrow\\ \longleftarrow\end{smallmatrix}\; X(\underline{2}) \;\cdots\qquad .$$

Define a co-semi-simplicial space $\Delta =$

$$\Delta^0 \;\begin{smallmatrix}\longrightarrow\\ \dashleftarrow\\ \longrightarrow\end{smallmatrix}\; \Delta^1 \;\begin{smallmatrix}\longrightarrow\\ \dashleftarrow\\ \longrightarrow\\ \dashleftarrow\\ \longrightarrow\end{smallmatrix}\; \Delta^2 \;\cdots$$

where Δ^n is the Euclidean n-simplex, and

$$\partial_i = \Delta(\partial_i) : \Delta^{n-1} \longrightarrow \Delta^n$$

and

$$\epsilon_i = \Delta(\epsilon_i) : \Delta^n \longrightarrow \Delta^{n-1}$$

are the standard face and degeneracy maps of singular homology theory. In barycentric coordinates,

$$\Delta(\partial_i)(t_0, \dots, t_{n-1}) = (t_0, \dots, t_{i-1}, 0, t_i, \dots, t_{n-1})$$

and

$$\Delta(\epsilon_i)(t_0, \dots, t_n) = (t_0, \dots, t_i + t_{i+1}, \dots, t_n).$$

Now $|X|$, the *geometric realization* of X, is defined to be

$$\left[\coprod_n (X(n) \times \Delta^n)\right] / \sim,$$

the quotient space of the disjoint union (topological sum) of the spaces

$$X(n) \times \Delta^n$$

under the relations

$$(x, \Delta(f)(t)) \sim (X(f)(x), t),$$

or in abuse of notation

$$(x, f(t)) \sim (f(x), t)$$

for $x \in X(m)$, $t \in \Delta^n$, and $f : n \longrightarrow m$ any morphism in *Ord.* (These relations are generated by the relations with $f = \partial_i$ or $f = \epsilon_i$.)

Example. Let P be a polyhedron with $\sum$ its associated abstract simplicial complex as before. Then

$$|\Sigma| = \coprod_n \Sigma(n) \times \Delta^n$$

modulo the relations

$$[(v_0, \ldots, v_n), (t_0, \ldots, t_{i-1}, 0, t_{i+1}, \ldots, t_n)] \\ \sim [(v_0, \ldots, \hat{v}_i, \ldots, v_n), (t_0, \ldots, t_{i-1}, t_{i+1}, \ldots, t_n)]$$

and

$$(v_0, \ldots, v_i, v_i, \ldots, v_n), (t_0, \ldots, t_{n+1})] \\ \sim [(v_0, \ldots, v_i, \ldots, v_n), (t_0, \ldots, t_i + t_{i+1}, \ldots, t_{n+1})].$$

Writing $[(v_0, \ldots, v_n), (t_0, \ldots, t_n)]$ as $t_0 v_0 + \ldots + t_n v_n$, we recognize $|\Sigma|$ as the usual geometric realization of Σ, namely $|\Sigma| = P$.

The second construction that arises with semi-simplicial objects is making a chain complex out of a semi-simplicial module (or a cochain complex out of a co-semi-simplicial module). Let M be a semi-simplicial module, $M =$

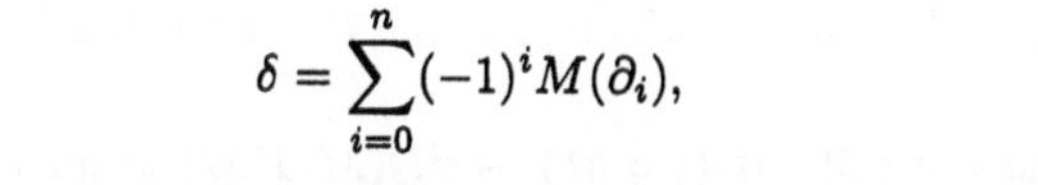

Define a boundary map

$$\delta : M(n) \longrightarrow M(n-1)$$

by taking the alternating sum of the face maps from $M(n)$ to $M(n-1)$, i.e.

$$\delta = \sum_{i=0}^{n} (-1)^i M(\partial_i),$$

where $\partial_i : n - 1 \longrightarrow n$. The fact that M is a functor implies that $\delta^2 = 0$ so that

$$M(0) \underset{\delta}{\leftarrow} M(1) \underset{\delta}{\leftarrow} M(2) \cdots$$

is a chain complex, sometimes denoted $k \circ M$; i.e. *k:Semi-simplicial modules⟶ Chain complexes* is the functor which takes alternating sums of face maps. Similarly, if N is a co-semi-simplicial module, then $k \circ N$ is a cochain complex.

Note. By abuse of notation, we shall often write $H_\delta(M(^*))$ or $H_\delta(M)$ to mean $H_\delta(ko)$ the homology of M, or refer to a semi-simplicial module as a chain complex.

In computing the homology of the chain complex $k \circ M$ or of the cochain complex $k \circ N$, it is often useful to work with a normalized complex in which the degenerate elements have been quotiented out. More precisely, one proceeds as follows. (See [18] for details and proofs.)

Let $M =$

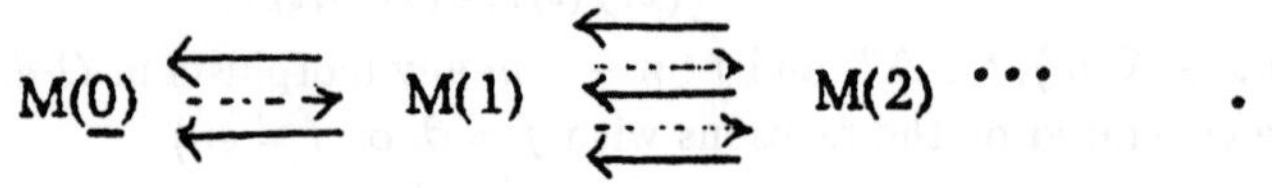

be an s.s. module. For each module M_n, we define the submodules

$$M'_n = \bigcap_{i=1}^{n} \ker(\partial_i : M_n \to M_{n-1}) \quad \text{(N.B. } \partial_0 \text{ is omitted)}$$

$$DM_n = \sum_{i=0}^{n-1} \text{image} \quad (\epsilon_i : M_{n-1} \to M_n) \quad \text{(all } \epsilon_i \text{ are included).}$$

Exercise. $\{M'_n\}$ is a subcomplex of $k \circ M$ (It is called the *normal subcomplex*). The boundary map $\sum_{i=0}^{n}(-1)^i\partial_i$ reduces to ∂_0 on this subcomplex. $\{DM_n, \sum_{i=0}^{n}(-1)^i\partial_i\}$ is a subcomplex of $k \circ M$ (called the *subcomplex of degenerate elements*).

Theorem 2.1. (Normalization Theorem of Eilenberg–MacLane. See [18], pp. 218–225)

1. $k \circ M \cong M' \oplus DM$, i.e.

$$M_n \cong M'_n \oplus DM_n \ \forall n.$$

2. *The inclusion $M' \to k \circ M$ of the normal subcomplex into the full complex is a homotopy equivalence of chain complexes. DM is null-homotopic.*

Definition. M/DM is called the *normalized* (chain) *complex* of M.

Corollary 2.1.1. *The composition $M' \to kM \to M/DM$ is an isomorphism between the normal subcomplex and the normalized complex.*

Corollary 2.1.2. *$H_\delta(k \circ M) \to H_\delta(M/DM)$ is an isomorphism, where H_δ denotes homology of chain complexes under the boundary map $\delta = \sum_{i=0}^{n}(-1)^i\partial_i$.*

Note. Analogous results exist for a co-semi-simplicial module N. In this case we define the degenerate complex by $DN = N/\bigcap_i \ker \epsilon_i$ and the normal complex by $N' = N/\sum_{i>0} \text{image } \partial_i$. ([18], pp. 283–290).

3. Spectral sequences and double complexes

We recall some basic notions from spectral sequences as they apply to computing the (co) homology of a double complex [61,58]. Assume we have a double complex $A = \oplus_{p,q} A^{p,q}$ of R-modules (R is any ring).

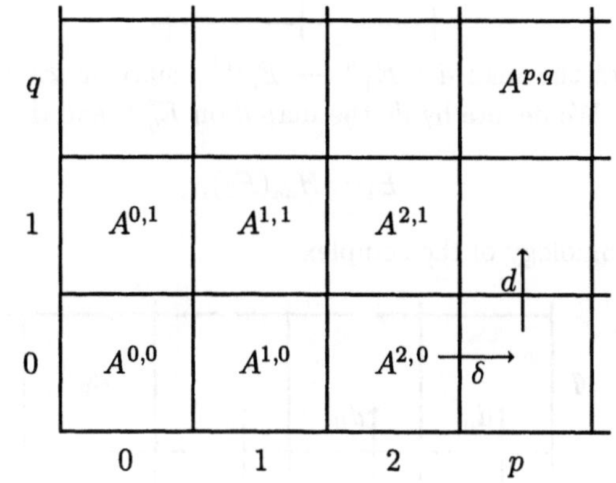

A is bigraded by p and q, and graded by $p+q$. Let there be given boundary maps δ and d on A such that $\delta : A^{p,q} \to A^{p+1,q}, d : A^{p,q} \to A^{p,q+1}$, and $d\delta = \delta d$. (One could also consider the cases $\delta : A^{p,q} \to A^{p-1,q}$, etc.) We assume that for each fixed $n, A^{p,n-p}$ is 0 for all but a finite number of p. This is true, for example, if $A^{p,q}$ is a first quadrant complex as in the diagram, i.e., if $A^{p,q} = 0$ if p or q is negative. We define a boundary map

$$D = d \pm \delta = d + (-1)^q \delta$$

on A. Then A is a cochain complex with respect to D, and we seek $H_D(A)$.

To find $H_D(A)$ using spectral sequences, we filter A as

$$A = F_0 \supset F_1 \supset F_2 \supset \cdots,$$

where

$$F_i = \oplus_{p \geq i} A^{p,q}.$$

The associated graded module $\oplus_p F_p/F_{p+1}$ of the filtration is called E_0. E_0 is bigraded by the filter degree p and the complementary degree q, and graded by $p+q$. In our case

$$E_0^{p,q} \cong A^{p,q}.$$

Thus the original picture of A above is also a picture of E_0. Now the map D on $A^{p,q}$ is $d \pm \delta$,

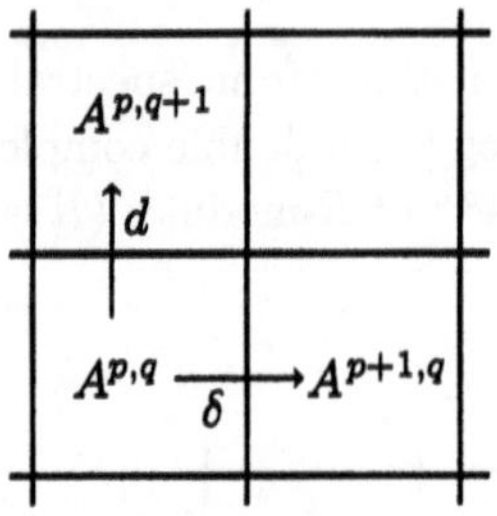

and this induces the map $d : E_0^{p,q} \to E_0^{p,q+1}$ since in $E_0^{p,q}$ we are working modulo $A^{p+1,q}$. We denote by d_0 the map d on $E_0^{p,q}$, and define

$$E_1 = H_{d_0}(E_0),$$

i.e., E_1 is the homology of the complex

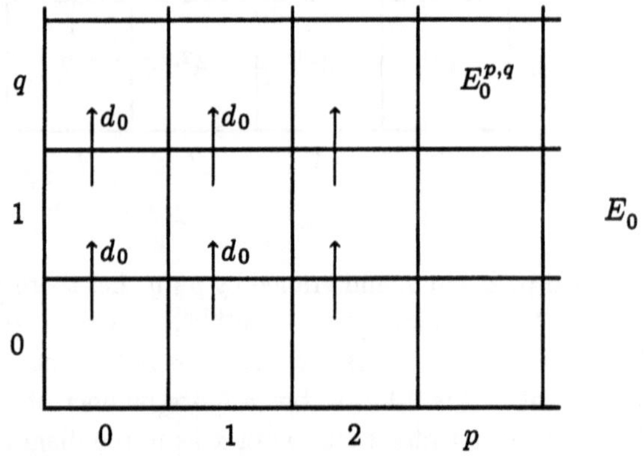

under the maps d_0. Hence

$$E_1 = H_d(A).$$

D induces a map $d_1 = \delta$ on E_1:

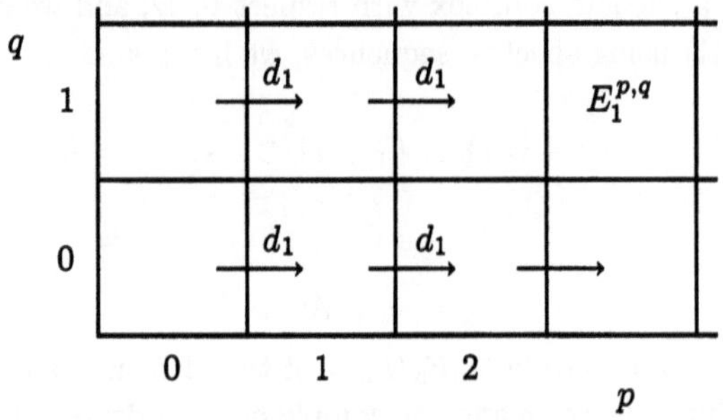

We define

$$E_2 = H_{d_1}(E_1) = H_\delta(H_d(A)).$$

What does it mean to say $\varphi \in E_2^{p,q}$?

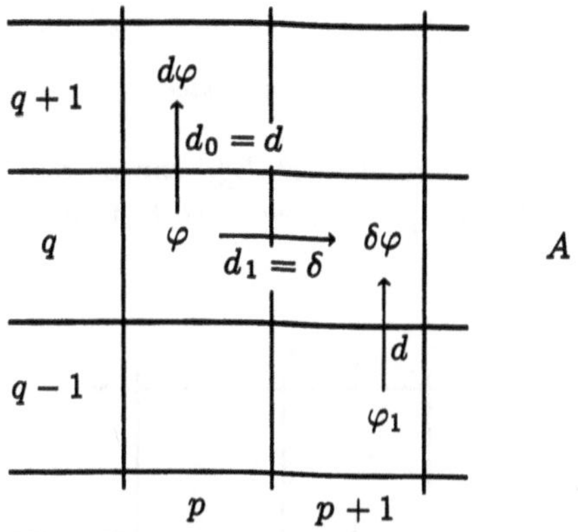

We can consider $\varphi \in A^{p,q}$ with the following properties. First, $0 = d_0\varphi = d\varphi$. Second $0 = d_1\varphi = \delta\varphi$ in $E_1^{p+1,q}$. This means that $\delta\varphi$ is a boundary in $E_0^{p+1,q}$ with respect to d_0, i.e., that there exists φ_1 in $A^{p+1,q-1}$ such that $d\varphi_1 = \delta\varphi$. Thus $D(\varphi \pm \varphi_1)$ has entry 0 in $A^{p,q+1}$ and in $A^{p+1,q}$. Thus $E_2^{p,q}$ consists of elements of $A^{p,q}$ which can be continued downward one step towards being D-cocycles. Actually $E_2^{p,q}$ is the quotient of these partial cocycles by partial coboundaries.

The spectral sequence continues:

$$H_{d_2}(E_2) = E_3, \ldots, H_{d_n}(E_n) = E_{n+1},$$

where $d_2, d_3, \ldots$ are maps induced by D. The maps d_i point in the following directions:

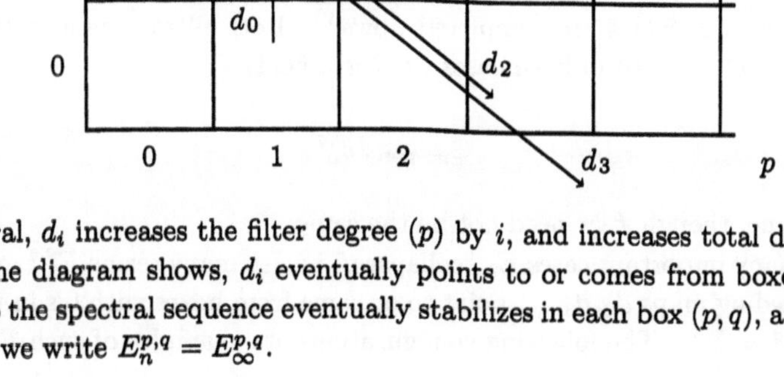

In general, d_i increases the filter degree (p) by i, and increases total degree by 1. As the diagram shows, d_i eventually points to or comes from boxes which are 0, so the spectral sequence eventually stabilizes in each box (p, q), and when it does, we write $E_n^{p,q} = E_\infty^{p,q}$.

The result of spectral sequence theory is that

> $\oplus_{p+q=n} E_\infty^{p,q}$ *is the associated graded module of* $H_D^n(A)$
> *which is isomorphic as a vector space to* $H_D^n(A)$ *if* R *is a field.*

We note that any non-zero element $E_\infty^{p,q}$ can be represented by an element φ of $A^{p,q}$ which can be completed downward to form a D-cocycle which is not a D-coboundary.

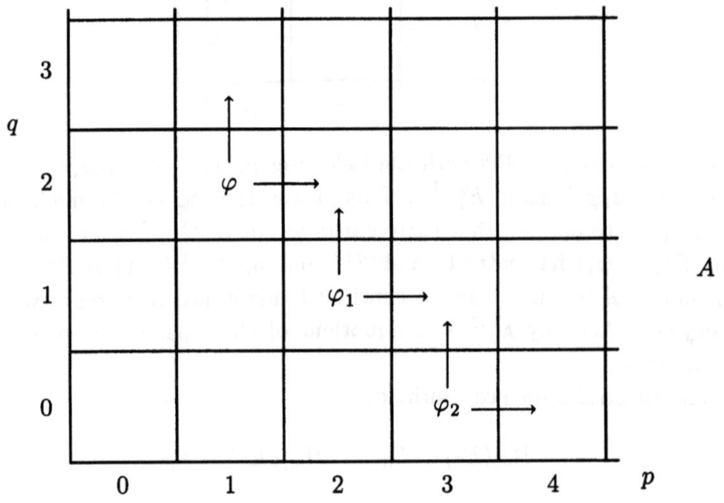

For example, if $\varphi \in A^{1,2}$ is non-zero in $E_\infty^{1,2}$, then there exist φ_1, φ_2 as shown so that $D(\varphi + \varphi_1 + \varphi_2) = 0$ and $(\varphi + \varphi_1 + \varphi_2)$ is not a D-coboundary.

Recall that filtering by p gave $E_1 = H_d(A), E_2 = H_\delta(H_d(A))$ and completed D-cocycles downward. Similarly, we could filter A by q. This would give a spectral sequence with

$$E_1 = H_\delta(A),\ E_2 = H_d(H_\delta(A)),$$

and in which cocycles are completed upward. Both filtrations give the same answer for $H_D(A)$, so in both cases (if R is a field)

$$\oplus_{p+q=n} E_\infty^{p,q}$$

is the same, though $E_\infty^{p,q}$ need not be the same.

In many important cases E_2 "collapses," i.e., so many terms $E_2^{p,q}$ are 0 that all succeeding maps $d_2, d_3, \ldots$ point to or come from boxes with 0's in them, so that $E_\infty^{p,q} = E_2^{p,q}$. The following configurations are examples of such E_2 terms:

(blank boxes denote 0, x's denote non-zero entries)

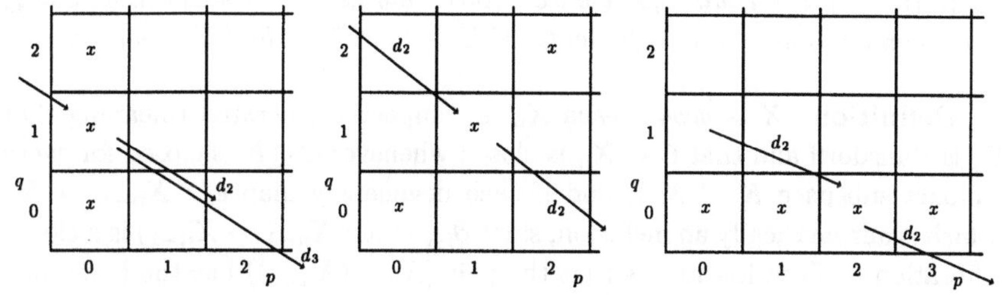

(The d_2 maps are drawn assuming filtering was by p, but $E_\infty = E_2$ also if filtering had been by q.)

4. Cohomology of a semi-simplicial space

Let X: *Ord* $\to$ *Spaces* be a s.s space, and let $|X|$ be its geometric realization. Let S_q: *Spaces* $\to$ *Groups* be the functor which assigns to each space its group of singular q-chains with coefficients in $\mathbb{Z}$. Then

$$S_qX : Ord \longrightarrow Groups$$

is a contravariant functor, hence an s.s. group. Applying the alternating sum functor k of Section 2, we make S_qX into a chain complex:

$$S_q(X)(0) \xleftarrow{\delta} S_qX(1) \xleftarrow{\delta} \cdots$$

At the same time we have the singular boundary maps $d : S_qX(\underline{n}) \to S_{q-1}X(\underline{n})$. d and δ commute, so there is a double complex S_*X:

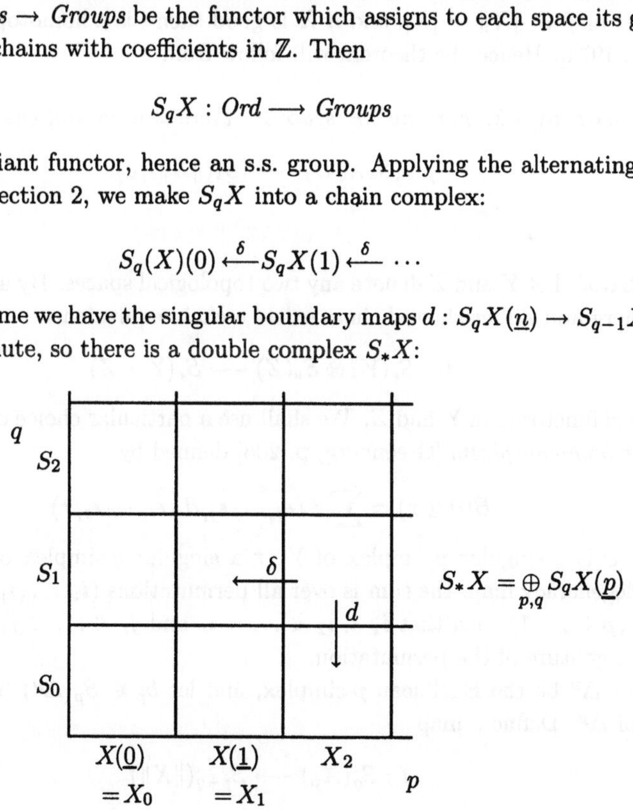

Let $D = d + (-1)^q\delta$, and grade S_*X by $p + q$. Then (S_*X, D) is a chain complex, so $H_D(S_*X)$ is defined.

Theorem 4.1 (Moore). *There is a natural chain map $f : (S_*X, D) \to (S_*(|X|), d)$. f is a chain equivalence (hence induces isomorphism in homology and cohomology with any coefficients) if X is "good" in the following sense:*

Definition. X is *good* if each X_n is compactly generated (meaning that X_n is Hausdorff and that $C \subset X_n$ is closed whenever $C \cap K$ is closed for every compact subspace $K \subset X_n$), and if each degeneracy map $\epsilon_i : X_{n-1} \to X_n$ (which is automatically an inclusion, since $\partial_i \epsilon_i = id : X_{n-1} \to X_{n-1}$) is a closed cofibration (cofibration means that the pair $(X_n, \epsilon_i(X_{n-1}))$ has the homotopy extension property with respect to any space).

Proof. We define the unnormalized geometric realization of X, denoted $\|X\|$, to be the space

$$\coprod_n X_n \times \Delta^n / \sim,$$

where $\sim$ means to mod out only by face maps ∂_i (rather than by face *and* degeneracy maps as in the construction of $|X|$ (see Section 2)). There is a natural map $i : \|X\| \to |X|$, and if X is good then i is a homotopy equivalence [Segal, 1974]. Hence the theorem will follow from

Theorem 4.2. *For any s.s. space X, there is a natural chain equivalence*

$$f : (S_*X, D) \longrightarrow (S_*(\|X\|), d).$$

Proof. Let Y and Z denote any two topological spaces. By acyclic models, Eilenberg and Zilber showed there exists a chain equivalence

$$B : S_*(Y) \otimes S_*(Z) \longrightarrow S_*(Y \times Z)$$

which is functorial in Y and Z. We shall use a particular choice of B called the *shuffle homomorphism* [Greenberg, p. 208] defined by

$$B(\partial \otimes \tau) = \sum \pm(\epsilon_{j_q} \cdots \epsilon_{j_1} \partial,\ \epsilon_{i_p} \cdots \epsilon_{i_1} \tau)$$

where ∂ is a singular p-simplex of Y, τ a singular q-simplex of Z, ϵ_k is the k-th degeneracy map, the sum is over all permutations $(i_1, \ldots, i_p, j_1, \ldots, j_p)$ of $(0, \ldots, p+q-1)$ such that $i_1 < i_2 < \ldots < i_p$ and $j_1 < \ldots < j_q$, and the sign is the signature of the permutation.

Let Δ^p be the Euclidean p-simplex, and let $\partial_p \in S_p(\Delta^p)$ be the identity map of Δ^p. Define a map

$$f : S_q(X_p) \longrightarrow S_{p+q}(\|X\|)$$

by the composition

$$S_q(X_p) \xrightarrow{\otimes \partial_p} S_q(X_p) \otimes S_p(\Delta^p) \xrightarrow{B} S_{p+q}(X_p \times \Delta^p) \xrightarrow[\pi_*]{} S_{p+q}(\|X\|)$$

where $\pi : X_p \times \Delta^p \to \|X\|$ is the canonical map. To show that f is a chain map

$$f : (S_* X, D) \longrightarrow (S_*(\|X\|), d),$$

it suffices to show that

$$\begin{array}{ccc} S_q(X_p) & \xrightarrow{\quad f \quad} & S_{p+q}(\|X\|) \\ {\scriptstyle D = d + (-1)^q \delta} \Big\downarrow & & \Big\downarrow {\scriptstyle d} \\ S_{q-1}(X_p) \oplus S_q(X_{p-1}) & \xrightarrow{\quad f \quad} & S_{p+q-1}(\|X\|) \end{array}$$

commutes. Let $a \in S_q(X_p)$. Then

$$\begin{aligned} df(a) &= d\pi_* B(a \otimes \partial_p) = \pi_* B d(a \otimes \partial_p) \\ &= \pi_* B(da \otimes \partial_p + (-1)^q a \otimes d\partial_p) \end{aligned}$$

while

$$fD(a) = f(da + (-1)^q \delta a) = \pi_* B(da \otimes \partial_p) + (-1)^q \pi_* B(\delta a \otimes \partial_{p-1}).$$

Thus to show that $df = fD$ it suffices to show that

$$\pi_* B(a \otimes d\partial_p) = \pi_* B(\delta a \otimes \partial_{p-1}).$$

Now consider the diagram

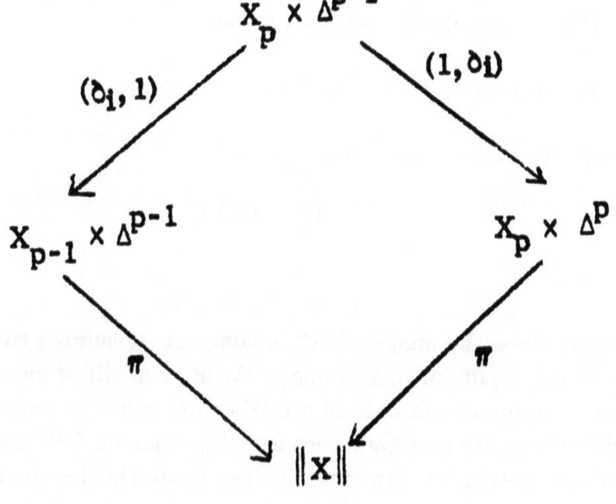

which commutes by the construction of $\|X\|$. Applying the functor S_* to the diagram and choosing $B(a \otimes \partial_{p-1}) \in S_{p+q-1}(X_p \times \Delta^{p-1})$ shows that

$$\pi_* B(\partial_i a \otimes \partial_{p-1}) = \pi_* B(a \otimes \partial_i \partial_{p-1}).$$

(This uses the naturality of B, i.e., $(\partial_i, 1)_* B = B(\partial_i, 1)_*$ and similarly for $(1, \partial_i)$.) But by definition,

$$\sum_i (-1)^i \partial_i a = \delta a \qquad \sum_i (-1)^i \partial_i \partial_{p-1} = d\partial_p$$

so

$$\pi_* B(\delta a \otimes \partial_{p-1}) = \pi_* B(a \otimes d\partial_p),$$

as was to be shown.

To prove that f is a chain equivalence, it suffices, since it is a morphism of chains of free Abelian groups, to show that f induces isomorphism in homology [Spanier, p. 192]. We prove isomorphism by filtering both chain complexes and showing that the induced map of spectral sequences is an isomorphism on the E^1 level (see [61] for more details on spectral sequences).

We filter (S_*X, D) by

$$0 = F_{-1} \subset F_0 \subset F_1 \subset \ldots, \cup_p F_p = S_*X,$$

where

$$F_p = \oplus_{i=0}^p S_* X_i.$$

In $\|X\|$, let $Y_p = \|X\|_p$ be the image of $X_p \times \Delta^p$ (under the map π). Then

$$\phi = Y_{-1} \subset Y_0 \subset Y_1 \subset \ldots, \cup_p Y_p = \|X\|$$

induces a filtration $F'_p = S_*(Y_p)$,

$$0 = F'_{-1} \subset F'_0 \subset F'_1 \subset \ldots$$

of $(S_*(\|X\|), d)$. One detail must be proved:

Lemma 4.3. $\cup_{p=0}^{\infty} F'_p = S_*(\|X\|)$.

Proof. We have to prove that any map

$$\Delta^n \longrightarrow \|X\|$$

factors

$$\Delta^n \longrightarrow Y_p \xrightarrow{\pi} \|X\|$$

for some p. Since the image of Δ^n is compact, it suffices to prove that any compact $K \subset \|X\|$ lies in Y_p for some p. We use a modified form of Whitehead's proof that a compact subspace of a CW complex lies in some p-skeleton [73]; the modifications are necessary because $\|X\|$, unlike a CW complex, need not be Hausdorff or even $T1$. (In fact, this can cause this lemma to be false if we use $|X|$ in place of $\|X\|$).

Let Z be the trivial s.s. set with Z_p =(point) for all p. Then there is a unique s.s. map $g : X \to Z$, which induces $h : \|X\| \to \|Z\|$. $\|Z\|$ contains one copy of Δ^n for each n. Now h maps $\|X\|_p$ to $\|Z\|_p = \pi(pt. \times \Delta^p)$ and furthermore, $h^{-1}(\|Z\|_p) = \|X\|_p$. Thus to be done it suffices to prove that any compact $K \subset \|Z\|$ lies in $\|Z\|_p$ for some p. Now we can use Whitehead's argument, since points are closed in $\|Z\|$. Suppose that K is not contained in any $\|Z\|_p$. Then we can pick a sequence of distinct points $z_1, z_2, \ldots$ in K such that only finitely many z_i's lie in $\|Z\|_p$ for any p. Let $W = \{z_i\}_{i=1}^{\infty}$. Now W is closed in $\|Z\|$, since $\pi^{-1}(W) \subset \coprod_p (pt. \times \Delta^p)$ is the union of finitely many points in each $(pt. \times \Delta^p)$. By the same argument, every subset of W is closed in $\|Z\|$. Hence W is discrete. Since $W \subset K$ with W closed and K compact, W is compact. Thus W is compact, discrete, and infinite, which is impossible. Q.E.D.

Returning to the proof of the theorem, we see that

$$f : (S_*X, D) \longrightarrow (S_*(\|X\|), d)$$

is filtration preserving, i.e., $f(F_p) \subset F'_p$. Hence f induces maps $f^i : E^i \to' E^i$, $i = 0, 1, \ldots, \infty$ of the respective spectral sequences. By [61], Theorem 9.1.3, if f^i is an isomorphism for some $i \geq 1$, then $f_* : H_D(S_*X) \to H_d(S_*(\|X\|))$ is an isomorphism. Thus we shall be done when we show that f^1 is an isomorphism.

On the E^0 level we have

$$E_p^0 = F_p/F_{p-1} = S_*X_p; \quad E_{p,q}^0 = S_qX_p$$
$$d_0 : E_p^0 \to E_p^0 \text{ is the map } \ d : S_qX_p \to S_{q-1}X_p.$$
$${}'E_p^0 = F'_p/F'_{p-1} = S_*Y_p/S_*Y_{p-1} = S_*(Y_p, Y_{p-1})$$

$${}'E_{p,q}^0 = S_{p+q}(Y_p, Y_{p-1}).$$
$$d'_0 : {}'E_p^0 \to {}'E_p^0 \text{ is the map}$$
$$d : S_{p+q}(Y_p, Y_{p-1}) \to S_{p+q-1}(Y_p, Y_{p-1}).$$
$$f^0 : E_{p,q}^0 \to {}'E_{p,q}^0 \text{ is the composition}$$
$$S_qX_p \xrightarrow{\otimes\partial_p} S_qX_p \otimes S_p\Delta^p \xrightarrow{B} S_{p+q}(X_p \times \Delta^p) \xrightarrow{\pi_*} S_{p+q}(Y_p, Y_{p-1}).$$

Letting $\dot{\Delta}^p$ denote the boundary of Δ^p, the functoriality of the maps shows that f^0 can also be written as the composition

$$S_qX_p \xrightarrow{\otimes\partial_p} S_q(X_p) \otimes S_p(\Delta^p, \dot{\Delta}^p) \xrightarrow{B'} S_{p+q}(X_p \times (\Delta^p, \dot{\Delta}^p))$$
$$\xrightarrow{\pi_*} S_{p+q}(Y_p, Y_{p-1}) S_{p+q}(X_p \times \Delta^p, X_p \times \dot{\Delta}^p)$$

where we now regard $\partial_p \in S_p(\Delta^p, \dot{\Delta}^p)$, and B' can be defined from B since B is natural. Furthermore, B' is a chain equivalence ([17], pp. 180-1).

The three maps $\otimes\partial_p$, B', and π_* each commute with d. Taking H_d of each term above gives the map $f^1 : E^1_{p,q} \to {}'E^1_{p,q}$; it is the composition

$$E^1_{p,q} = H_q(X_p) \xrightarrow{(\otimes\partial_p)^*} H_{p+q}(S_qX_p \otimes S_p(\Delta^p, \dot{\Delta}^p))$$
$$\xrightarrow{B'_*} H_{p+q}(X_p \times \Delta^p, X_p \times \dot{\Delta}^p) \xrightarrow{\pi_*} H_{p+q}(Y_p, Y_{p-1}) = {}'E^1_{p,q}.$$

By the Künneth theorem ([17], VI. 9. 13), together with the fact

$$\begin{cases} H_p(\Delta^p, \dot{\Delta}^p) &= Z, \quad \text{generated by } \quad \partial_p \\ H_n(\Delta^p, \dot{\Delta}^p) &= 0, \quad n \neq p \end{cases}$$

We see that $H_{p+q}(S_qX_p \otimes S_p(\Delta^p, \dot{\Delta}^p)) = H_qX_p \otimes H_p(\Delta^p, \dot{\Delta}^p)$ and that $(\otimes\partial_p)_*$ is therefore an isomorphism. B'_* is an isomorphism since B' is a chain equivalence. Thus to show that f^1 is an isomorphism, which will finish the proof, it suffices to show that

$$\pi_* : H_n(X_p \times \Delta^p, X_p \times \dot{\Delta}^p) \longrightarrow H_n(Y_p, Y_{p-1})$$

is an isomorphism for all n.

Pick a neighborhood U of $\dot{\Delta}^p$ in Δ^p and a deformation

$$h_t : U \to U; h_0 = id, \quad h_1 : U \to \dot{\Delta}^p, \quad h_t|\dot{\Delta}^p = id \quad \text{for all } t.$$

Then $X_p \times \dot{\Delta}^p$ is a strong deformation retract of $X_p \times U$, and Y_{p-1} is a s.d. retract of $\pi(X_p \times U)$. The latter follows from the fact that Y_p has the topology quotient of $\coprod_{i=0}^p X_p \times \Delta^p$ (i.e., this topology coincides with the topology of Y_p as a subspace of $\|X\|$).

Thus it suffices to show that

$$\pi_* : H_n(X_p \times \Delta^p, X_p \times U) \xrightarrow{\cong} H_n(Y_p, \pi(X_p \times U)).$$

Now excising $X_p \times \dot{\Delta}^p$ from $X_p \times U$ and Y_{p-1} from $\pi(X_p \times U)$, it suffices to show

$$\pi_* : H_n(X_p \times (\Delta^p - \dot{\Delta}^p), X_p \times (U - \dot{\Delta}^p)) \xrightarrow{\cong} H_n(Y_p - Y_{p-1}, \pi(X_p \times U) - Y_{p-1}).$$

But

$$\pi : (X_p \times (\Delta^p - \dot{\Delta}^p)) \longrightarrow (Y_p - Y_{p-1})$$

is a homeomorphism, as is

$$\pi : X_p \times (U - \dot{\Delta}^p) \to \pi(X_p \times U) - Y_{p-1} = \pi(X_p \times (U - \dot{\Delta}^p)),$$

so π_* is indeed an isomorphism. Q.E.D.

Remarks. The hypothesis of "goodness" in Theorem 4.1 is fulfilled if X is an s.s. set or s.s. CW complex, or if it is an s.s. manifold with degeneracy maps s_i which embed X_{n-1} as a submanifold of X_n. If X is *not* good, we can either use $\|X\|$ instead of $|X|$, or we can use one of several other geometric realizations, such as the *unwinding* $|X_Z|$, defined by

$$|X_Z| = \coprod_{n,(i_0,\dots,i_n)} X_n \times \Delta^n / \sim,$$

where $0 \leq i_0 < i_1 < \cdots < i_n$ are integers, and $\sim$ means all relations

$$(x; d_j t; i_0, \dots, i_n) \sim (\partial_j x; t; i_0, \dots, \widehat{i_j}, \dots, i_n),\ x \in X_n,\ t \in \Delta^{n-1}.$$

Theorem 4.4. *The map* $|X_Z| \to \|X\|$ *defined by*

$$(x; t; i_0, i_1, \dots, i_n) \longrightarrow (x; t) \quad (x \in X_n,\ t \in \Delta^n)$$

induces isomorphism in homology.

The proof, which we omit, is algebraic and uses a modification of the normalization procedure for an s.s. module [18].

The major use of the unwinding is in constructing classifying spaces for topological groups or categories, especially those with "bad" topologies (see [6], Stasheff's appendices B and C).

Example of Theorem 4.1. As in Section 2, let P be a polyhedron with ordered vertices, and $\sum(n)$ the set of its n-simplices. $\sum$ is an s.s. set, and $|\Sigma| = P$. By Theorem 4.1 there is a chain equivalence

$$f : (S_*\Sigma, D) \longrightarrow (S_*(|\Sigma|), d) = (S_*P, d).$$

Now $(S_*\Sigma, D)$ is the double complex

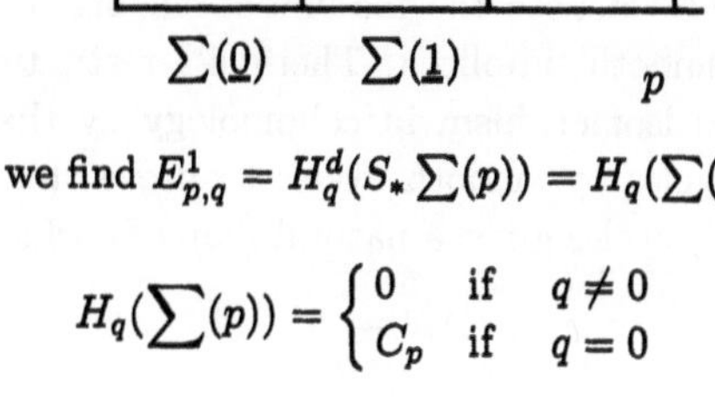

Filtering $S_*\Sigma$ by p, we find $E^1_{p,q} = H^d_q(S_* \sum(p)) = H_q(\sum(p))$. But since $\sum(p)$ is a discrete space,

$$H_q(\sum(p)) = \begin{cases} 0 & \text{if} \quad q \neq 0 \\ C_p & \text{if} \quad q = 0 \end{cases}$$

where C_p is the free Abelian group generated by the elements of $\sum(p)$. Thus $E^1 =$

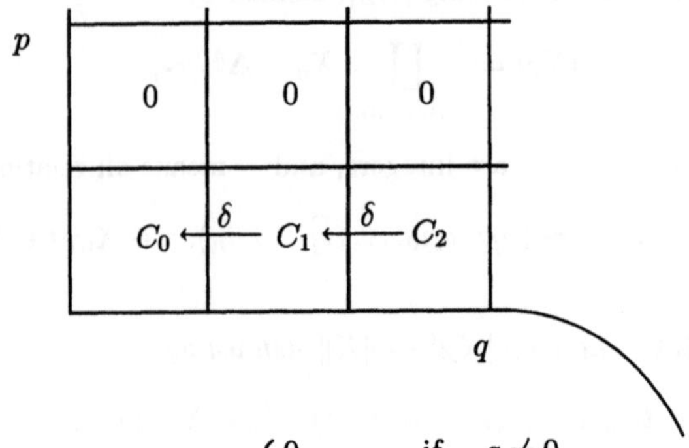

and

$$E^2_{p,q} = \begin{cases} 0 & \text{if} \quad q \neq 0 \\ H^\delta_p(C_*) & \text{if} \quad q = 0. \end{cases}$$

Since the spectral sequence collapses, we find

$$E^2_{p,0} = E^\infty_{p,0} = H_p(S_*\Sigma).$$

Thus by Theorem 4.1, $H^\delta_p(C_*) = H_p(S_*P) = H_p(P)$. Analogous results hold for cohomology or for other coefficients. Thus in this special case, Theorem 4.1 reduces to the familiar result that the (co)homology of a (geometric) simplicial complex equals the (co)homology of the free chain complex (resp. dual of the free chain complex) on its simplices.

Now let M be an s.s. manifold. Let A^q: *Manifolds*$\to$ $\mathbb{R}$-*Modules* be the functor which assigns to each manifold the vector space of its differential q-forms, and let d be the DeRham differential from A^q to A^{q+1}. Then A^qM is a co-s.s. module, and in the usual way we form the double complex $AM = \oplus A^qM(p) =$

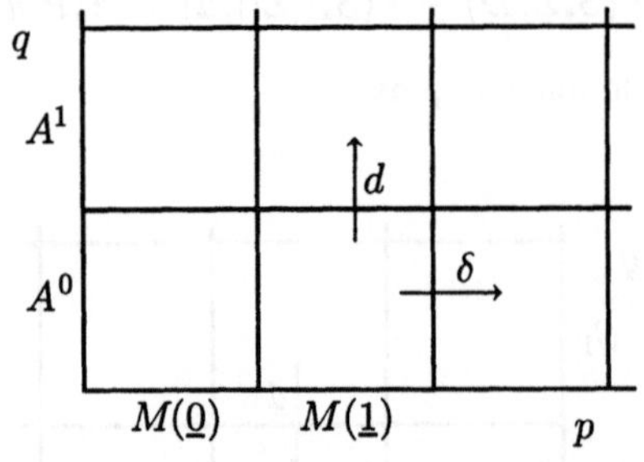

Now if N is a manifold we call a singular q-simplex $\partial \in S_q(N)$ *smooth* if $\partial : \Delta^q \to N$ can be extended to a smooth map defined on a neighborhood of Δ^q in $\mathbb{R}^q$. Let $S^*_{sm.}(N)$ denote the module of singular cochains of N which are only defined on smooth simplices. There is a restriction map $r : S^*N \to S^*_{sm.}N$ which induces isomorphism in cohomology by the DeRham theorem (one approximates cycles by smooth cycles to prove this.) Now integrating q-forms over smooth q-cycles gives a natural map of cochain complexes

$$\varphi : AN \to S^*_{sm.}N$$

of DeRham theory into smooth singular theory. By the DeRham theorem φ induces isomorphism in cohomology. Composing this with the inverse of r^* gives a natural isomorphism

$$H_d(AN) \xrightarrow{\sim} H_d(S^*N).$$

Applying these facts when $N = M(p)$, $p = 0, 1, \ldots$ and using the naturality to show that r and φ commute with δ, we see that there are maps of double complexes

$$AM \xrightarrow{\varphi} S^*_{sm.}M \xleftarrow{r} S^*M;$$

filtering each complex by p and taking H_d gives us the $E^1_{p,q}$ terms

$$H^q_d(A^*M(p)) \underset{\varphi^*}{\xrightarrow{\sim}} H^q_d(S^*_{sm.}M(p)) \underset{r^*}{\xleftarrow{\sim}} H^q_d(S^*M(p)).$$

Since φ^* and r^* induce isomorphism of E^1 terms, they induce isomorphisms

$$H_D(AM) \underset{\sim}{\xrightarrow{\varphi^*}} H_D(S^*_{sm.}M) \underset{\sim}{\xleftarrow{r^*}} H_D(S^*M).$$

Combining this with Theorem 4.1 we have proved

Theorem 4.5. *There is a natural isomorphism*

$$H_D(AM) \xrightarrow{\sim} H_D(S^*M) = H^*(|M|) \quad \text{(coefficients in } \mathbb{R}).$$

Thus we can compute the cohomology of $|M|$ using differential forms (on the $M(p)$), even though $|M|$ is *not* a manifold!

5. A semi-simplicial construction of BG for a discrete group G

Given a discrete group G, there are several constructions of a classifying space BG (which, since G is discrete, is an Eilenberg–MacLane space; see Section 1).

We start one such construction [20] by choosing a point to be the 0-skeleton of BG, and adding an oriented loop for each element $x \in G$. One then attaches 2-simplices to kill each relation of the form

$$xy = z \qquad \text{for} \quad x, y, z \in G.$$

For example, if $x, y \in G$, then one fills in the triangle

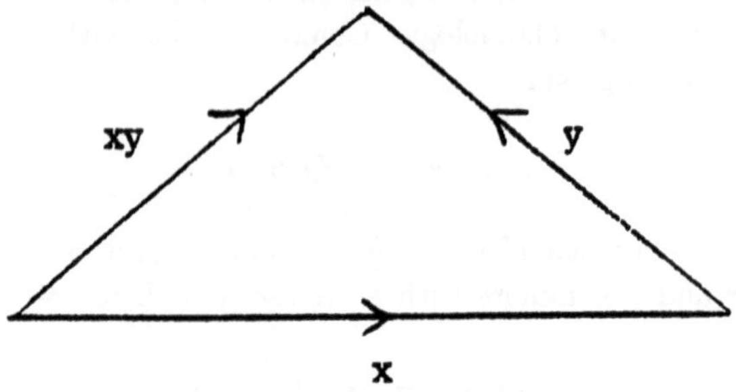

(which has all 3 vertices identified to a point) with a 2 simplex corresponding to the pair (x, y). Similarly, one kills the relation

$$x_1 x_2 \cdots x_n = y$$

by suitably attaching an n-simplex, corresponding to the n-tuple $(x_1, \ldots, x_n)$. The CW complex thus obtained is a $K(G, 1)$, hence is a BG (see Section 1).

The above construction can be described easily in semi-simplitandarcial terms. Let NG be the semi-simplicial set, that is, functor from *Ord* to *Sets*, with

$$\begin{aligned}
&NG(0) = * = \text{a point},\\
&NG(1) = G,\\
&NG(2) = G \times G,\\
&NG(p) = G^p \underset{def}{=} G \times \cdots \times G(p \text{ factors}).
\end{aligned}$$

The maps ∂_i and ϵ_i (actually $NG(\partial_i)$ and $NG(\epsilon_i)$) are defined by:

$$\begin{aligned}
&\partial_i : G^p \longrightarrow G^{p-1}, \quad i = 0, \ldots, p\\
&\partial_0(x_1, \ldots, x_p) = (x_2, \ldots, x_p)\\
&\partial_p(x_1, \ldots, x_p) = (x_1, \ldots, x_{p-1})\\
&\partial_i(x_1, \ldots, x_p) = (x_1, \ldots, x_i x_{i+1}, \ldots, x_p), \quad 1 \le i \le p-1.
\end{aligned}$$

$$\begin{aligned}
&\epsilon_i : G^{p-1} \longrightarrow G^p, \quad i = 0, \ldots, p-1\\
&\epsilon_0(x_1, \ldots, x_{p-1}) = (1, x_1, \ldots, x_{p-1})\\
&\epsilon_i(x_1, \ldots, x_{p-1}) = (x_1, \ldots, x_i, 1, x_{i+1}, \ldots, x_{p-1}), \quad 1 \le i \le p-1.
\end{aligned}$$

For example,

$$\begin{aligned}
\partial_0(x) = \partial_1(x) &= *, \quad x \in G\\
\partial_0(x, y) &= y \quad (x, y) \in G \times G\\
\partial_1(x, y) &= xy\\
\partial_2(x, y) &= x \quad .
\end{aligned}$$

We now define

$$BG = |NG| = \text{geometric realization of} \quad NG \quad [\text{cf. Section 2}].$$

That is,

$$NG = \coprod_{p=0}^{\infty} G^p \times \Delta^p / \sim,$$

where $\sim$ means to identify via face and degeneracy maps.

$$\begin{aligned} BG &= \text{point} \cup (\text{a 1-simplex for each} \quad x \in G) \\ &\cup \quad (\text{a 2-simplex for each pair} \quad (x,y) \in G \times G) \cup \cdots, \end{aligned}$$

with the 2-simplex (x, y) glued to the 1-skeleton by

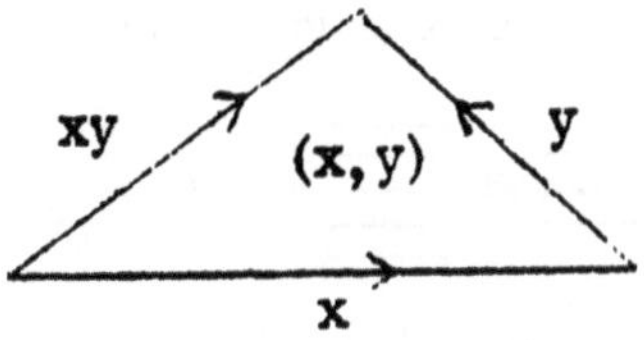

as we described above.

We can now compute $H^*(BG)$ by applying Theorem 4.1.

Theorem 5.1. $H^*(BG) = H_D(S^*NG)$*(for G a discrete group).*

Since NG is discrete, like $\sum$ in the example after Theorem 4.4, $H_D^p(S^*NG)$ reduces to $H_\delta^p(C^*)$, where

$$\begin{aligned} C^p &= \text{free module generated by } NG(p) \\ &= \text{Maps} \quad (G^p, R) \quad (\text{where } R \text{ is the coefficient group}) \\ C^* &= C^0 \xrightarrow{\delta} C^1 \xrightarrow{\delta} C^2 \xrightarrow{\delta} \cdots \end{aligned}$$

We recognize C^* as the complex used to compute $H^*_{EM}(G; R)$, the Eilenberg–MacLane cohomology of groups, hence

$$H^*_{spaces}(BG) = H^*_{EM}(G)$$

[19].

Of course, we could have avoided using Theorem 4.1 by constructing C^* directly as the cochain complex generated by the cells of BG, a CW complex. The map

$$\delta = \sum (-1)^i \partial_i,$$

an alternating sum of face maps, is just the boundary map in the cochain complex of the CW complex.

We have proven

Theorem 5.2. *If G is a discrete group, $H^*(BG; R)$ = cohomology of the complex*

$$\text{Maps}\ (*, R) \xrightarrow{\delta} \text{Maps}\ (G, R) \xrightarrow{\delta} \text{Maps}\ (G \times G, R) \longrightarrow \cdots .$$

Theorem 5.2 gives us a combinatorial way to compute $H^*(BG)$. (We can find $H_*(BG)$ similarly). Unfortunately, the combatinorics are very difficult in all but the simplest cases.

6. Construction of BG for a topological group G

The same semi-simplicial construction used to construct BG for a discrete group G carries over almost verbatim when G is a topological group (cf. supra Section 5; also [56]). As before, we construct a semi-simplicial space NG =

$$* \;\substack{\longleftarrow \\ \dashrightarrow \\ \longleftarrow}\; G \;\substack{\longleftarrow \\ \dashrightarrow \\ \longleftarrow \\ \dashrightarrow \\ \longleftarrow}\; G \times G \;\cdots$$

with the same ∂_i and ϵ_i as in Section 5, and let $BG = |NG|$ =

$$(* \times \Delta^0) \coprod (G \times \Delta^I) \coprod (G \times G \times \Delta^2) \coprod \cdots$$

modulo gluing via the maps ∂_i and ϵ_i. The only difference from the case of G discrete is that now we topologize G with its given topology, not the discrete topology, so that $G \times G \times \Delta^2$, for example, has the product topology inherited from the topology of G and the standard topology of Δ^2. Thus BG is not canonically a simplicial complex or CW complex as it is when G is discrete. (Observe, by the way, that ∂_i and ε_i just involve multiplication in G and identity maps, which are continuous.) The construction of BG as $|NG|$ is a variation of Milnor's joint construction. Other constructions of BG have been made by Dold–Lashof, Stasheff, Steenrod–Rothenber, Milgram, Segal, and Dyer. The various constructions of BG are compared in Stasheff's appendix to [6].

We prove now that $BG = |NG|$ is really a classifying space for G-bundles.

Theorem 6.1. *Let G be a Lie group. There exists a semi-simplicial space PG and a map*

$$PG \longrightarrow NG$$

such that $PG \to NG$ and the induced map

$$|PG| \longrightarrow |NG|$$

are fibrations with fiber G, and such that $|PG|$ is contractible. Hence

$$\begin{array}{ccc} |PG| & = & EG \\ \downarrow & & \\ |NG| & = & BG \end{array}$$

is a universal principal G-bundle.

Proof. We define the semi-simplicial space $PG =$

$$G \;\begin{array}{c}\longleftarrow\\ \dashrightarrow\\ \longleftarrow\end{array}\; G\times G \;\begin{array}{c}\longleftarrow\\ \dashrightarrow\\ \longleftarrow\\ \dashrightarrow\\ \longleftarrow\end{array}\; G\times G\times G\;\cdots$$

$$PG(\underline{n}) = G^{n+1},$$

where the maps

$$\partial_i : G^{p+1} \longrightarrow G^p, \quad i = 0,\ldots,p$$

are given by

$$\partial_i(x_0,\ldots,x_p) = (x_0,\ldots,\hat{x}_i,\ldots,x_p),$$

and

$$\epsilon_i : G^{p+1} \longrightarrow G^{p+2}, \quad i = 0,\ldots,p,$$

is given by

$$\epsilon_i(x_0,\ldots,x_p) = (x_0,\ldots,x_i,x_i,\ldots,x_p).$$

Now there is a map

$$\pi : PG \longrightarrow NG$$

given by

$$\begin{array}{ccccc}
PG = (G & \begin{array}{c}\longleftarrow\\ \dashrightarrow\\ \longleftarrow\end{array} & G\times G & \begin{array}{c}\longleftarrow\\ \longleftarrow\\ \longleftarrow\end{array} & G\times G\times G\;\cdots) \\
\pi\downarrow\;\;\pi\downarrow & & \pi\downarrow & & \pi\downarrow \\
NG = (* & \begin{array}{c}\longleftarrow\\ \dashrightarrow\\ \longleftarrow\end{array} & G & \begin{array}{c}\longleftarrow\\ \longleftarrow\\ \longleftarrow\end{array} & G\times G\;\;\cdots),
\end{array}$$

$$\pi(x_0,\ldots,x_p) = (x_0x_1^{-1}, x_1x_2^{-1}, x_2x_3^{-1}, \ldots, x_{p-1}x_p^{-1}).$$

To see that the above diagram commutes, hence that π is a morphism of semi-simplicial spaces, we must check that

$$\pi\partial_i = \partial_i\pi \quad \text{for all maps} \quad \partial_i$$

$$\pi\epsilon_i = \epsilon_i\pi \quad \text{for all} \quad \epsilon_i.$$

For example, for $i \neq 0, p$, we compute

$$\begin{aligned}
\pi\partial_i(x_0,\ldots,x_p) &= \pi(x_0,\ldots,\hat{x}_i,\ldots,x_p) \\
&= (x_0x_1^{-1},\ldots,x_{i-1}x_{i+1}^{-1},\ldots,x_{p-1}x_p^{-1}) \\
&= (x_0x_1^{-1},\ldots,(x_{i-1}x_i^{-1})(x_ix_{i+1}^{-1}),\ldots) \\
&= \partial_i\pi(x_0,\ldots,x_p).
\end{aligned}$$

We omit the straightforward computations that show that π commutes with the other ∂_i and ϵ_i.

G acts freely on the right on PG in a manner that preserves the fibers of π, by

$$(x_0, \ldots, x_p) \cdot x = (x_0 x, \ldots, x_p x) \quad \text{for} \quad (x_0, \ldots, x_p) \in PG(p) \quad \text{and} \quad x \in G.$$

Exercise. $\pi : PG \to NG$ is a principal G-bundle.

It can be shown [56] that $|PG|$ is contractible, and that

$$|\pi| : |PG| \longrightarrow |NG|$$

is a principal G-bundle (at least if G is a topological group with a reasonable topology, for example if G is a Lie group). By Theorem 1.1, $|\pi| : |PG| \to |NG|$ is a universal principal G-bundle.

Remark. By different methods one can show that $|NG_Z|$ classifies G-bundles when G is any topological group or category. $|NG_Z|$ denotes the unwinding defined in Section 4. See Stasheff's appendix to [6], and [56].

H^*BG : Results.

H^*BG is known for almost all Lie groups G [3]. For example, for

$$G = U(n) \quad \text{or} \quad O(n),$$

H^*BG can be computed using the Serre spectral sequence and the induction on n [3]. Without going into this and other computation methods, we give the results [cf. infra Section 9; also 53].

1. $H^*(BU(n)) = \mathbb{Z}[c_1, \ldots, c_n]$, a polynomial ring (with no relations) in $c_1, \ldots, c_n$, with $c_i \in H^{2i}(BU(n))$. c_i is called the *i-th Chern class.* By our discussion in Section 1, these must be all the characteristic classes for bundles with structure group $U(n)$.
2. $BGL(n, \mathbb{C})$ is homotopy equivalent to $BU(n)$, as can be seen in two ways:
   a) $U(n)$ has the same homotopy type as $GL(n, \mathbb{C})$, since $GL(n, \mathbb{C})$ is the product of $U(n)$ and an affine space [13].
   b) Any complex vector bundle, i.e., bundle with fiber $\mathbb{C}^n$, and structure group $GL(n, \mathbb{C})$ acting in the usual way on $\mathbb{C}^n$ can be given a Hermitian metric on the fibers. This reduces the structure group to $U(n)$. Thus $H^*BGL(n, \mathbb{C}) = H^*BU(n) = \mathbb{Z}[c_1, \ldots, c_n]$, so the Chern classes are also the characteristic classes for $GL(n, \mathbb{C})$-bundles, and in particular, for $\mathbb{C}^n$-vector bundles.
3. By similar arguments $BGL(n, \mathbb{R})$ and $BO(n)$ have the same homotopy type. With coefficients in $\mathbb{Z}_2$,

$$H^*(BGL(n, \mathbb{R}); \mathbb{Z}_2) = H^*(BO(n); \mathbb{Z}_2) = \mathbb{Z}_2[w_1, \ldots, w_n],$$

where $w_i \in H^i(BO(n); \mathbb{Z}_2)$ is the *i-th Stiefel–Whitney class*. These are the characteristic classes with coefficients in $\mathbb{Z}_2$ for unoriented $\mathbb{R}^n$-plane bundles (real vector bundles).

4. With coefficients in $\mathbb{Q}$ (similar results obtain if we use $\mathbb{R}$ instead)

$$H^*(BGL(n,\mathbb{R}); \mathbb{Q}) = H^*(BO(n); \mathbb{Q}) = \mathbb{Q}[p_1, \ldots, p_{[n/2]}],$$

where $p_i \in H^{4i}(BO(n); \mathbb{Q})$ is the *i-th Pontryagin class*, and $[n/2]$ means the greatest integer $\leq n/2$. The Pontryagin classes are the characteristic classes with coefficients in $\mathbb{Q}$ (or $\mathbb{R}$) for unoriented $\mathbb{R}^n$-plane bundles.

Note. The conceptual way to understand Pontryagin classes is that certain singular cocycles $c_1, \ldots, c_n$ can be defined on $BO(n)$ with dim $c_i = 2i$. These c_i are like Chern classes but the odd ones $c_1, c_3, c_5, \ldots$ are torsion elements, hence vanish in $H^*(BO(n); \mathbb{Q})$. The cohomology classes of the even "Chern classes" $c_2, c_4, \ldots$ are the Pontryagin classes $p_1, p_2, \ldots$.

5. If we look at *oriented* $\mathbb{R}^n$-plane bundles, the structure group is $SO(n)$. With coefficients in $\mathbb{Q}$, the characteristic classes are the elements of $H^*(BSO(n); \mathbb{Q})$. These turn out to be the Pontryagin classes $p_1, \ldots, p_{[n/2]}$ as for $O(n)$-bundles, but if n is even, we also get the *Euler class*

$$\chi \in H^n(BSO(n); \mathbb{Q})$$

which satisfies the single relation

$$\chi^2 = p_{[n/2]}.$$

$H^*(BG)$: Methods of computation

Using our semi-simplicial machinery, we find $H^*(BG)$ by applying Theorem 4.1 to the semi-simplicial space NG to get

Theorem 6.2. *Let G be a Lie group. Then there are natural isomorphisms*

$$H_D(S_*NG) \xrightarrow{\cong} H_*(BG) \quad \text{and} \quad H^*(BG) \xrightarrow{\cong} H_D(S^*NG).$$

Remark 1. This theorem is also true if G is any topological group or category, if we take $BG = |NG_Z|$. The proof uses Theorems 4.2 and 4.4.

Remark 2. Theorem 6.2 is the same as Theorem 5.1 when G is a discrete group. Of course Theorem 5.2 does not generalize to topological groups, since $H^q(G^p)$ need not be 0 when $q \neq 0$.

Graphically, $H^*(BG)$ is the cohomology of the double complex $S^*NG =$

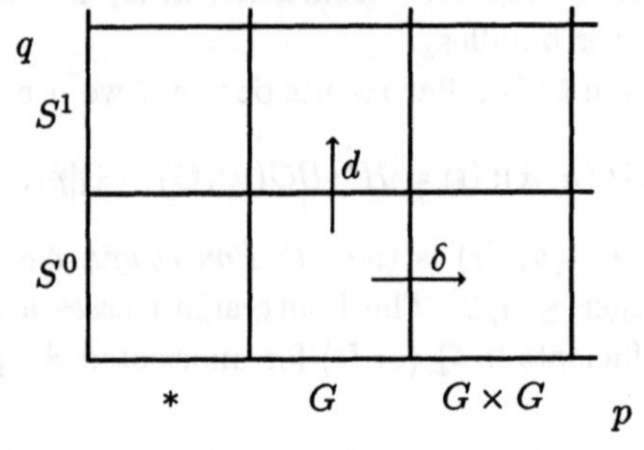

with respect to the grading $p+q$ and coboundary $D = d \pm \delta$, where $S^i =$ singular i-cochains, $\delta : S^q(G^p) \to S^q(G^{p+1})$ is the alternating sum of the face maps

$$S^q(\partial_i) : S^q(G^p) \longrightarrow S^q(G^{p+1})$$

and d is the coboundary of singular cohomology. This construction shows explicitly that $H^*(BG)$ depends on the topology of G and multiplication in G.

To compute $H_D(S^*NG)$, we can filter S^*NG either horizontally or vertically to obtain a spectral sequence converging to H^*BG.

Filtering by p, we get

$$E_1^{p,q} = H_d^q(S^*NG(p)) = H^q(G^p) \quad \text{(singular cohomology of } G^p\text{)}.$$

Filtering by q,

$$E_1^{p,q} = H_\delta^p(S^qNG),$$

a harder object to compute. (See Section 9.)

Example. Find H_*BG for $G = U(1) = S^1$. By Theorem 6.2, $H_*BG = H_D(S_*NG)$. We filter S_*NG by p to get $E_1^{p,q} = H_q(G^p)$. Now the q-th row of the double complex is

$$H_q(pt.) \longleftarrow H_q(G) \longleftarrow H_q(G^2) \cdots \qquad (*)$$

where $\delta = \sum_{i=0}^n (-1)^i \partial_i$. By Corollary 2.1.2, this complex has the same δ-homology as its normalized complex, which is obtained by dividing each $H_q(G^p)$ by the submodule of degenerate elements, i.e., by the images of $H_q(G^{p-1})$ under the degeneracy maps ϵ_i. G^p is a p-torus, and the maps ϵ_i are the inclusions $\epsilon_i : G^{p-1} \to G^p$ defined by

$$(x_1, \ldots, x_{p-1}) \longrightarrow \begin{cases} (x_1, \ldots, x_i, 1, x_{i+1}, \ldots, x_{p-1}), & i = 1, \ldots, p-1 \\ (1, x_1, \ldots, x_{p-1}), & i = 0. \end{cases}$$

Now $H_*(G^p) \cong \underbrace{H_*(G) \otimes \cdots \otimes H_*(G)}_{p\ factors}$ and $H_*(G) = H_*(S^1) = E(w) =$ exterior algebra generated by w in dimension 1. Thus

$$H_*(G^p) = E(w_1, \dots, w_p), \quad w_i \in H_1(G^p)$$

and

$$\epsilon_i : H_*(G^{p-1}) \longrightarrow H_*(G^p)$$

is the map generated by

$$\begin{aligned} \epsilon_i(w_j) &= w_j, \quad 1 \le j \le i \\ \epsilon_i(w_j) &= w_{j+1}, \quad i < j \le p-1. \end{aligned}$$

Thus every element of $H_*(G^p)$ of the form $w_{i_1} w_{i_2} \cdots w_{i_k}$ is in the image of some ϵ_i, except for the element

$$w_1 w_2 \cdots w_p \in H_p(G^p).$$

(Geometrically, all elements of $H_q(G^p)$ for $q < p$ are generated by the inclusions of lower dimensional tori via degeneracy maps).

Conclusion. After division by degeneracies, the chain complex (*) is reduced to $\mathbb{Z}$ in the q-th position. Thus

$$E_2^{p,q} = H_\delta(H_q(G^p)) = \begin{cases} \mathbb{Z}, & \text{if} \quad p = q \\ 0, & \text{otherwise.} \end{cases}$$

$E_2^{p,q} =$

$q \backslash p$	0	1	2
2	0	0	$\mathbb{Z}$
1	0	$\mathbb{Z}$	0
0	$\mathbb{Z}$	0	0

Therefore $E_2 = E_\infty = H_D(S_*NG) = H_*BG$. Therefore

$$H_q(BU(1)) = \begin{cases} \mathbb{Z} & \text{if} \quad q \quad \text{is even} \\ 0 & \text{otherwise.} \end{cases}$$

(One can show that in cohomology, $H^*(BU(1)) = \mathbb{Z}[c_1], c_1 \in H^2(BU(1))$.)

If G is a Lie group, we can apply Theorem 4.5 to compute $H^*(BG)$ by using differential forms on pt., G, $G \times G, \cdots$.

Theorem 6.3. *The DeRham map*

$$H_D(ANG) \longrightarrow H_D(S^*NG)$$

is an isomorphism, and both terms equal

$$H^*(|NG|) = H^*(BG; \mathbb{R}).$$

ANG is the double complex $\oplus A^q(G^p) =$

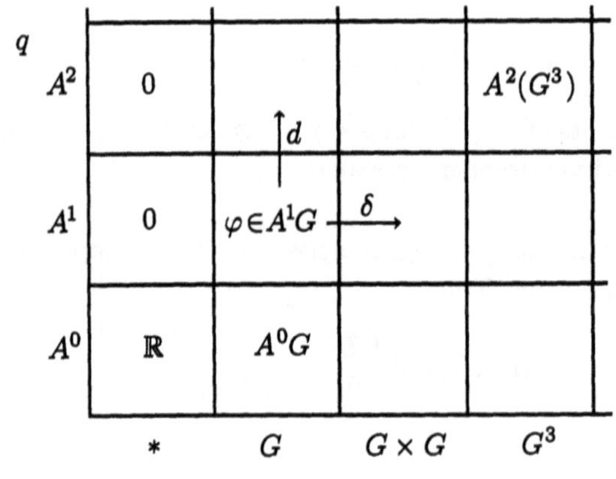

$(D = d \pm \delta$, grading double complex $= p + q$),

where $A^q(G^p) = \{$differential q-forms on $G^p\}$, d is the DeRham derivative on differential forms, and $\delta = \Sigma(-1)^i A^q(\partial_i)$ is the alternating sum of face maps on

$$A^q(\text{pt.}) \rightrightarrows A^q(G) \;\equiv\; A^q(G \times G) \quad \cdots.$$

For example, an element $\varphi \in A^1(G)$ is a D 2-cochain in ANG. In order that φ be a 2-cocycle on ANG, we must have $D\varphi = 0$, i.e., $d\varphi = 0$ and $\delta\varphi = 0$. Now $d\varphi = 0$ just means that φ is a closed 1-form on G. The map δ equals $\sum_{i=0}^{2}(-1)^i A^1(\partial_i)$, so since

$$H^*(G \times G; \mathbb{R}) = H^*(G; \mathbb{R}) \otimes H^*(G; \mathbb{R}),$$

it follows that

$$\delta\varphi = 1 \otimes \varphi - \mu^*\varphi + \varphi \otimes 1,$$

where

$$\mu : G \times G \longrightarrow G$$

is multiplication, $\mu(x, y) = xy$. Hence $\delta\varphi = 0$ if and only if $\mu^*\varphi = 1 \otimes \varphi + \varphi \otimes 1$, and the latter is just the definition that φ be a *primitive form* on G. In general, a Lie group need not have any primitive forms, so that any non-zero D-cocyles of ANG will be sums of entries in more than one box. But in special cases G does have primitive forms.

Example 1. $G = S^1 = \mathbb{R}/2\pi\,\mathbb{Z}$ with θ as the coordinate of $\mathbb{R}/2\pi\mathbb{Z}$. Then $\mu : G \times G \to G$ is given by $\mu(\theta_1, \theta_2) = \theta_1 + \theta_2$, so

$$\mu^* d\theta = d(\theta_1 + \theta_1) = d\theta_1 + d\theta_2.$$

Therefore $d\theta$ is a closed primitive form on S^1.

Example 2. $G = \mathbb{C}^*$ with coordinate z. Then $\frac{dz}{z}$ is primitive because $\mu(w, y) = wy$, so that

$$\mu^* \frac{dz}{z} = \frac{d(wy)}{wy} = \frac{wdy}{wy} + \frac{ydw}{wy} = \frac{dy}{y} + \frac{dw}{w}$$

Under the map exp: $S^1 \to C^*$ given by exp $(\theta) = e^{i\theta}$, we have $\exp^*\left(\frac{dz}{z}\right) = \frac{d(e^{i\theta})}{e^{i\theta}} = \frac{ie^{i\theta}d\theta}{e^{i\theta}} = id\theta$.

As mentioned above, in general a D-cocycle of ANG has entries in several boxes of ANG. For example, to build a D 4-cocycle on ANG, we might start by choosing $\varphi \in A^2(G \times G)$ so that $d\varphi = 0$.

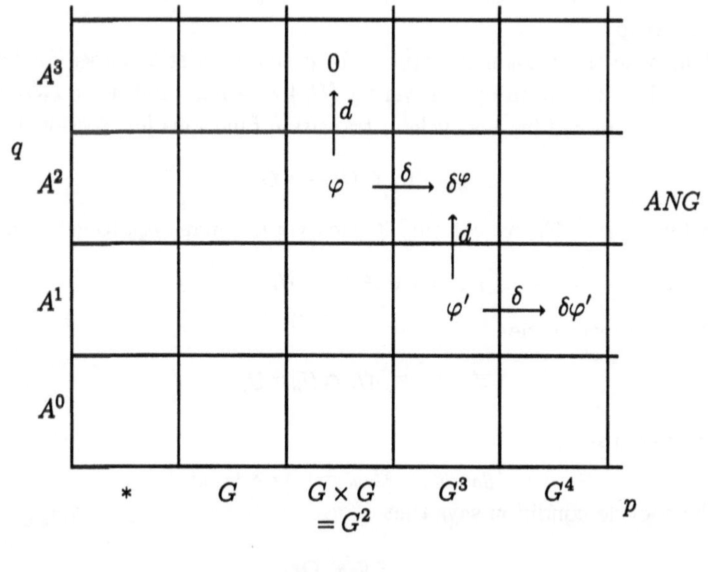

Now $d(\delta\varphi) = \delta(d\varphi) = 0$, so $\delta\varphi \in A^2(G^3)$ is a closed form. We next try to find $\varphi' \in A^1(G^3)$ such that $d\varphi' = -\delta\varphi$ (of course φ' need not exist in general).

Then

$$D(\varphi+\varphi') = (d+(-1)^q\delta)(\varphi+\varphi') = (d+\delta)\varphi + (d-\delta)\varphi'$$
$$= 0 + \delta\varphi + d\varphi' - \delta\varphi' = -\delta\varphi' \in A^1(G^4).$$

Proceeding in this way, one can build a D-cocycle $\varphi+\varphi'+\cdots$, which proceeds from φ downwards and to the right. One could also work upwards and to the left. Observe that building D-cocycles (modulo D-coboundaries) step by step in this way is equivalent to computing the spectral sequence of the double complex [cf. Section 3].

7. Construction of characteristic classes for G-bundles by semi-simplicial methods

In Section 6 we showed that H^*BG can be represented by differential forms on $NG =$

$$* \leftleftarrows G \;\substack{\longleftarrow\\ \longleftarrow\\ \longleftarrow}\; G\times G \;\cdots .$$

Given a space X and a G-bundle η on X, classified by a map

$$f_\eta : X \longrightarrow BG,$$

the characteristic classes of η are just

$$f_\eta^*(H^*BG) \subset H^*X.$$

The trouble is, how do we know what the map f_η^* is in terms of our semi-simplicial setup?

Let us take the most naive approach to characteristic classes for bundles over manifolds. Given an open covering $\{U_a\}$ of a manifold M, a G-bundle η over M is determined by a cocycle of transition functions [cf. Section 1]

$$g_{ab} : U_a \cap U_b \longrightarrow G.$$

Given a form $\varphi_1 \in AG$, we can pull it back via the maps g_{ab} to get forms

$$g_{ab}^*\varphi_1 \in A(U_a \cap U_b).$$

On each triple intersection

$$U_{abc} \underset{def.}{=} U_a \cap U_b \cap U_c$$

we have the map

$$(g_{ab}, g_{bc}) : U_{abc} \longrightarrow G \times G.$$

Since the cocycle condition says that

$$g_{ac} = g_{ab} \cdot g_{bc},$$

the map (g_{ab}, g_{bc}) gives us enough information to recover g_{ac}. Given a form

$$\varphi_2 \in A(G\times G),$$

we can pull it back via the map (g_{ab}, g_{bc}) to a form

$$(g_{ab}, g_{bc})^* \varphi_2 \in A(U_{abc}) \quad \text{for each} \quad a, b, c.$$

Thus given a general element

$$\varphi = \varphi_1 + \varphi_2 + \cdots + \varphi_k \in ANG,$$

we can pull it back via the maps g_{ab} to a form $\rho(\varphi)$ which lives in

$$(\oplus_{a,b} A(U_a \cap U_b)) \oplus (\oplus_{a,b,c} A(U_{abc})) \oplus \cdots.$$

However, this still does not live in AM.

To get forms in AM, we recall André Weil's proof of the isomorphism of Čech and DeRham cohomology [71]. (Cf. infra Section 8 for the rest of the proof. See [6] for a proof not using spectral sequences.) Given $(M, \mathcal{U})$, where M is a manifold and $\mathcal{U}$ an open covering $\{U_a\}$ of M such that the index set of the cover is ordered by $\leq$, we define the *Čech–DeRham double complex*

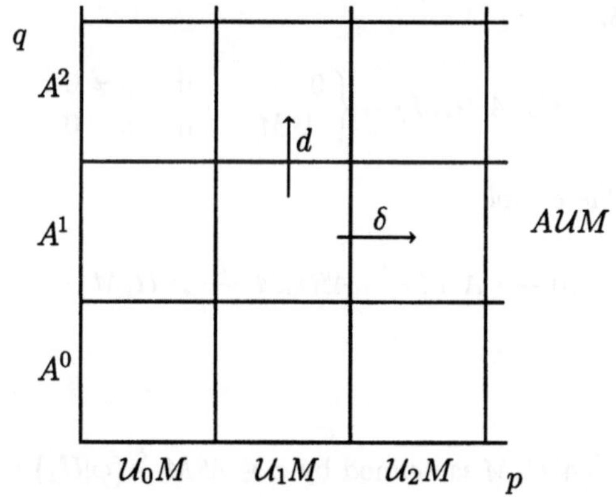

with p, q term $A^q(\mathcal{U}_p M)$.

Here $U_0 M = \coprod_a U_a$ (disjoint union)

$$\mathcal{U}_1 M = \coprod_{a \leq b} U_a \cap U_b$$

$$\mathcal{U}_2 M = \coprod_{a \leq b \leq c} U_a \cap U_b \cap U_c, \quad \textit{etc.},$$

$$d : A^q(\mathcal{U}_p M) \longrightarrow A^{q+1}(\mathcal{U}_p M)$$

is the ordinary DeRham coboundary map, and δ is a Čech coboundary map defined by

$$\delta : A^q(\mathcal{U}_{p-1} M) \longrightarrow A^q(\mathcal{U}_p M);$$

for $\varphi \in A^q(\mathcal{U}_{p-1} M)$,

$$(\delta\varphi)(U_{a_0} \cap \cdots \cap U_{a_p}) = \sum_{i=0}^{p} (-1)^i \varphi(U_{a_0} \cap \cdots \cap \widehat{U}_{a_i} \cap \cdots \cap U_{a_p}) | (U_{a_0} \cap \cdots \cap U_{a_p})$$

(Notes: | means restriction of a differential form to a subset. The formula for δ is the one used in defining cohomology of M with values in the sheaf of germs of differential q-forms on M).

Theorem 7.1. *Let $D = d \pm \delta$ in the double complex $A\mathcal{U}M$. Then $H^*_D(A\mathcal{U}M) = H^*_{DeRham}(M)$.*

Proof. Filter $A\mathcal{U}M$ by q. Then $E_1^{p,q} = H^p_\delta(A^q U_* M) = p$-th cohomology of the cochain complex

$$(0 \to A^q\mathcal{U}_*M) = (0 \to A^q\mathcal{U}_0 M \xrightarrow{\delta} A^q\mathcal{U}_1 M \xrightarrow{\delta} A^q\mathcal{U}_2 M \cdots).$$

Lemma 7.2.

$$H^p_\delta(A^q\mathcal{U}_*M) = \begin{cases} 0 & \text{if} \quad p \neq 0 \\ A^q M & \text{if} \quad p = 0. \end{cases}$$

In other words, the complex

$$0 \longrightarrow A^q M \xrightarrow{\delta} A^q\mathcal{U}_0 M \xrightarrow{\delta} A^q\mathcal{U}_1 M \cdots$$

is exact.

(The map $A^q M \xrightarrow{\delta} A^q U_0 M$ is defined by $\varphi \in A^q M \xrightarrow{\delta} \{\varphi | U_a\} \in A^q U_0 M$.)

Proof of Lemma. Define homotopy operators $h : A^q U_p M \to A^q U_{p-1} M$ as follows. Let $\{\lambda_a\}$ be a partition of unity on M subordinate to the covering $\{U_a\}$. Let $\varphi \in A^q\mathcal{U}_p M$. Then define $h\varphi \in A^q U_{p-1} M$ by

$$(h\varphi)(U_{a_0} \cap \cdots \cap U_{a_{p-1}}) = (-1)^p \sum_a \lambda_a \varphi(U_{a_0} \cap \cdots \cap U_{a_{p-1}} \cap U_a).$$

(We define $\varphi(U_{a_0} \cap \cdots \cap U_{a_{p-1}} \cap U_a)$ as $\pm\varphi(U_{b_0} \cap \cdots \cap U_{b_p})$, where $(b_0, \ldots, b_p)$ is an ordered (by $\leq$) permutation of $(a_0, \ldots, a_{p-1}, a)$, and $\pm$ is the sign of the permutation.) For $\varphi \in A^q\mathcal{U}_0 M$, define $h\varphi \in A^q M$ by

$$h\varphi = \sum_a \lambda_a \varphi(U_a).$$

Exercise. $h\delta + \delta h = id.$ Q.E.D. for lemma.

Returning to the proof of the theorem, we have shown that $E_1^{p,q} =$

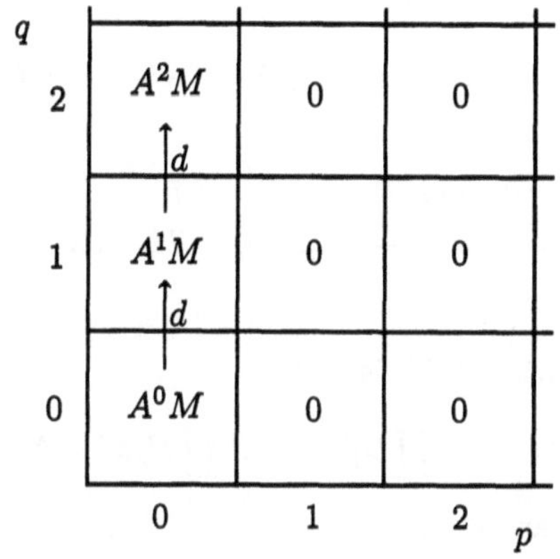

with $d_1 = d : A^qM \to A^{q+1}M$. Therefore

$$E_2^{p,q} = H_d(E_1) = H_d^q(AM) = H^q_{DeRham}(M).$$

Because E_2 is 0 outside the left column,

$$H^*_{DeRham}M = E_2 = E_\infty = H^*_D(A\mathcal{U}M).$$

Q.E.D. (Theorem 7.1)

Comment. If $\varphi \in A^q\mathcal{U}_0M$, then

$$\delta\varphi(U_a \cap U_b) = \varphi(U_b)\ |(U_a \cap U_b) - \varphi(U_a)|\ (U_a \cap U_b),$$

so $\delta\varphi = 0$ if and only if $\varphi(U_a)$ and $\varphi(U_b)$ agree on $U_a \cap U_b$, which is true if and only if φ can be patched into a q-form defined on all of M.

The appearance of the double complex $A\mathcal{U}M$ is reminiscent of our semi-simplicial machinery, and in fact, Theorem 7.1 can be stated and proved using semi-simplicial objects. We do this now. Define $\mathcal{U}M$ to be the semi-simplicial manifold

$$\mathcal{U}_0M \quad \begin{smallmatrix}\longleftarrow \\ \dashrightarrow \\ \longleftarrow\end{smallmatrix} \quad \mathcal{U}_1M \quad \begin{smallmatrix}\longleftarrow \\ \dashrightarrow \\ \longleftarrow \\ \dashrightarrow \\ \longleftarrow\end{smallmatrix} \quad \mathcal{U}_2M \quad \cdots,$$

where, as before, $\{U_a\}$ is an open cover of M with ordered indices a, and $\mathcal{U}_pM = \coprod_{a_0 \le a_1 \le \cdots \le a_p} U_{a_0} \cap \cdots \cap U_{a_p}$. (For shorthand we also write $U_{a_0 \cdots a_p} =$

$U_{a_0} \cap \cdots \cap U_{a_p}$). The maps ∂_i and ϵ_i are defined as follows:

$$\begin{aligned}
&\partial_i : \mathcal{U}_p M \longrightarrow \mathcal{U}_{p-1} M, \quad i = 0, \ldots, p \\
&\partial_i : \mathcal{U}_{a_0} \cap \cdots \cap \mathcal{U}_{a_P} \longrightarrow \mathcal{U}_{a_0} \cap \cdots \cap \widehat{\mathcal{U}}_{a_i} \cap \cdots \mathcal{U}_{a_p} \\
&\partial_i : x \longmapsto x \\
&\epsilon_i : \mathcal{U}_p M \longrightarrow \mathcal{U}_{p+1} M, \quad i = 0, \ldots, p \\
&\epsilon_i : \mathcal{U}_{a_0} \cap \cdots \cap \mathcal{U}_{a_p} \longrightarrow \mathcal{U}_{a_0} \cap \cdots \cap \mathcal{U}_{a_i} \cap \mathcal{U}_{a_i} \cap \cdots \cap \mathcal{U}_{a_p} \\
&\epsilon_i : x \longmapsto x
\end{aligned}$$

The geometric realization $|\mathcal{U}M|$ is

$$\mathcal{U}_0 M \times \Delta^0 \coprod \mathcal{U}_1 M \times \Delta^1 \coprod \mathcal{U}_2 M \times \Delta^2 \cdots$$

modulo gluing via ϵ_i and ∂_i.

Example. For $M = (0,1), \{U_a\} = \{U_a = (0, 3/4), U_b = (1/4, 1)\}$, with $a < b$, we have

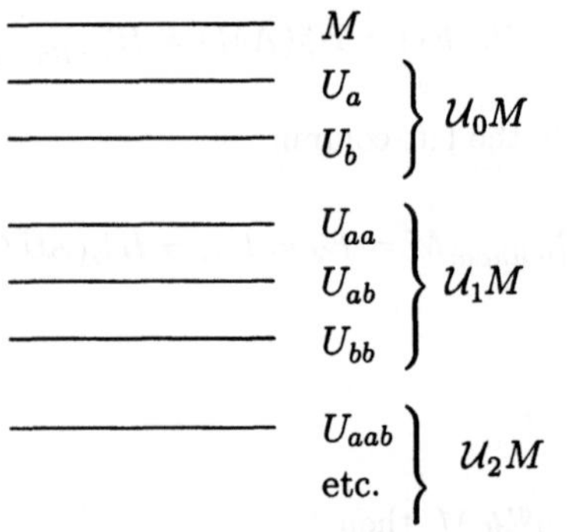

and $|\mathcal{U}M| =$ (after shrinking degenerate simplices via ϵ_i)

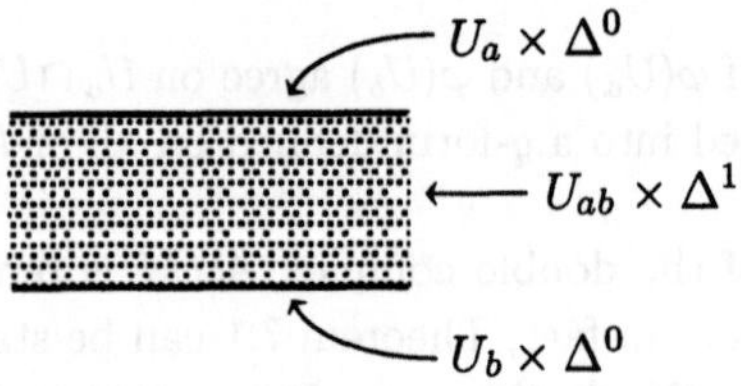

There is an obvious map

$$g : |\mathcal{U}M| \longrightarrow M$$

defined by

$$U_{a_0} \cap \cdots \cap U_{a_p} \times \Delta^p \longrightarrow U_{a_0} \cap \cdots \cap U_{a_p} \subset M.$$

We construct a map

$$f : M \longrightarrow |\mathcal{U}M|$$

by choosing a partition of unity λ_a on M, subordinate to the cover $\{U_a\}$. For $x \in M$, let $a_0 < \cdots < a_p$ be the indices for which $\lambda_a(x) > 0$. Then $t = (\lambda_{a_0}(x), \cdots, \lambda_{a_p}(x))$ is a point in Δ^p, and we define

$$f(x) = (x, t) \in U_{a_0 \cdots a_p} \times \Delta^p \to |\mathcal{U}M|.$$

Exercises

1. f and g are well-defined (i.e., are compatible with identifications via ∂_i and ϵ_i)
2. $M \xrightarrow{f} |\mathcal{U}M| \xrightarrow{g} M$ =identity
3. $|\mathcal{U}M| \to M \to |\mathcal{U}M|$ is homotopic to the identity. Hint: Construct a linear homotopy.

This proves

Theorem 7.3. *$M \xrightarrow{f} |\mathcal{U}M|$ and $|\mathcal{U}M| \xrightarrow{g} M$ are homotopy equivalences. In particular, $H^*(M) \xrightarrow{\simeq} H^*(|\mathcal{U}M|)$ is an isomorphism.*

Note. The construction of $\mathcal{U}M$ also works if M is any paracompact space rather than a manifold. Theorem 7.3 still holds in this case.

Combining Theorem 7.3 with Theorem 4.5 implies that

$$H^*(M) = H^*(|\mathcal{U}M|) = H^*_D(A(\mathcal{U}M)).$$

Using the isomorphism $H^*_{(sing.)}(M) = H^*_{DeRham}(M)$, we have reproved Theorem 7.1.

Trivial Exercise. Verify that the double complex $(A\mathcal{U}M, d, \delta)$ of Theorem 7.1 is indeed the same as $A(\mathcal{U}M)$ defined as the double complex of differential forms on the semi-simplicial manifold $\mathcal{U}M$.

Returning to characteristic classes, we see now that pulling back differential forms on NG to $A(U \cap V) \oplus A(U \cap V \cap W) \oplus \cdots$ as we did earlier corresponds to a m double complexes $\rho : ANG \to AUM$.

To treat the map ρ semi-simplicially we prove:

Theorem 7.4. *Let G be a Lie group. Let $\eta = (E \to M)$ be a G-bundle on a manifold M with cocycle $\{g_{ab}\}$ relative to the cover $\mathcal{U} = \{\mathcal{U}_a\}$ of M. Let $F : \mathcal{U}M \to NG$ be the map*

$$\begin{array}{ccccccc}
\mathcal{U}M = (\mathcal{U}_0 M & \rightleftarrows & \mathcal{U}_1 M & \overleftarrow{\underline{\overleftarrow{\longleftarrow}}} & \mathcal{U}_2 M & \cdots) \\
\downarrow F \quad \downarrow F & & \downarrow F & & \downarrow F & \\
NG = (\; * & \rightleftarrows & G & \overleftarrow{\underline{\overleftarrow{\longleftarrow}}} & G \times G & \cdots)
\end{array}$$

defined by

$$U_a \to * \quad \textit{for all} \quad a$$

$$g_{ab} : U_a \cap U_b \to G \quad \textit{for all} \quad a \le b$$

$(g_{ab}, g_{bc}) : U_a \cap U_b \cap U_c \to G \times G$, *etc. Then*

1) *F is a morphism of double complexes (i.e., it commutes with ∂_i and ϵ_i).*
2) *The induced map $|F| : M \cong |\mathcal{U}M| \to |NG| = BG$ is a classifying map for the bundle η.*
3) *$F^* : ANG \to A\mathcal{U}M$ is just the map ρ previously defined.*

Proof. Assertion 3 is obvious. We prove assertion 1 in the special case

$$\begin{array}{ccc} \mathcal{U}_2M & \begin{array}{c}\longleftarrow\\ \longleftarrow\\ \longleftarrow\\ \longleftarrow\end{array} & \mathcal{U}_3M \\ {\scriptstyle F}\downarrow & & {\scriptstyle F}\downarrow \\ G \times G & \begin{array}{c}\longleftarrow\\ \longleftarrow\\ \longleftarrow\\ \longleftarrow\end{array} & G \times G \times G \end{array}$$

(the general case is almost identical).

We restrict to $U_{abcd} \subset \mathcal{U}_3M$, and prove that $\partial_i \circ F = F \circ \partial_i, i = 0, 1, 2, 3.$

$\underline{i = 0}$

$$\begin{array}{ccc} U_{bcd} & \xleftarrow[= \text{ inclusion}]{\partial_0} & U_{abcd} \\ F = \Big\downarrow (g_{bc}, g_{cd}) & & F = \Big\downarrow (g_{ab}, g_{bc}, g_{cd}) \\ G \times G & \xleftarrow{\partial_0} & G \times G \times G \\ (y, z) & \xleftarrow{\partial_0} & (x, y, z) \end{array}$$

The diagram obviously commutes.

$\underline{i = 1}$

$$\begin{array}{ccc} U_{acd} & \xleftarrow[= \text{ inclusion}]{\partial_1} & U_{abcd} \\ F = \Big\downarrow (g_{ac}, g_{cd}) & & F = \Big\downarrow (g_{ab}, g_{bc}, g_{cd}) \\ G \times G & \xleftarrow{\partial_1} & G \times G \times G \\ (xy, z) & \xleftarrow{\partial_1} & (x, y, z) \end{array}$$

the diagram commutes because $g_{ac} = g_{ab}g_{bc}$ by the cocycle condition. The case $i = 2$ is like $i = 1$, and $i = 3$ is like $i = 0$.

Exercise. Prove that F commutes with ϵ_i, the degeneracy maps.

To prove assertion 2, we construct a semi-simplicial bundle map. We assume without loss of generality (see Section 1) that $\eta = (E \to M)$ is a *principal* G-bundle. Cover E by the opens $\pi^{-1}U_a = E|U_a$, and let $\mathcal{U}E = \mathcal{U}_0E \leftleftarrows \mathcal{U}_1E \ldots$ be the corresponding semi-simplicial manifold. The maps $E|U_a \to U_a$ induce a map $\mathcal{U}E \to \mathcal{U}M$ with fiber G. The map $|\mathcal{U}E| \to |\mathcal{U}M|$ also has fiber G. Let $\{\lambda_a\}$ be a partition of unity on M subordinate to $\{U_a\}$. Then $\{\lambda_a \circ \pi\}$ is a partition of unity on E subordinate to $\{E|U_a\}$. We use these two partitions of unity to construct maps $M \to |\mathcal{U}M|$ and $E \to |\mathcal{U}E|$ as in the proof of Theorem 7.3.

Exercise. The diagram

$$\begin{array}{ccc} E & \longrightarrow & |\mathcal{U}E| \\ \downarrow & & \downarrow \\ M & \longrightarrow & |\mathcal{U}M| \end{array}$$

commutes, and the map $E \to |\mathcal{U}E|$ commutes with the action of G.

Now recall [cf. Theorem 6.1] the semi-simplicial manifold $PG =$

$$G \;\overset{\longleftarrow}{\underset{\longleftarrow}{\dashrightarrow}}\; G \times G \;\overset{\longleftarrow}{\underset{\longleftarrow}{\longleftarrow}}\; G \times G \times G \cdots$$

defined by $\partial_i(x_0, \ldots, x_n) = (x_0, \ldots, \hat{x}_i, \ldots, x_n)$. We define a map $\mathcal{U}E \to PG$ as follows. Let $\{f_a\}$ be a set of maps which are horizontal projections for the bundle $E \xrightarrow{\pi} M$ [cf. Section 1]. Recall that this means that $f_a : (E|U_a) \to G$ (G regarded as the fiber of the bundle here), and that for all $y \in E|U_a \cap U_b$, if $\pi(y) = x$, then

$$f_b(y) = g_{ba}(x) \cdot f_a(y).$$

Now define a map $\mathcal{U}E \to PG$

$$\begin{array}{ccccc} \mathcal{U}_0E & \overset{\longleftarrow}{\underset{\longleftarrow}{\dashrightarrow}} & \mathcal{U}_1E & \overset{\longleftarrow}{\underset{\longleftarrow}{\longleftarrow}} & \mathcal{U}_2E \quad \cdots \\ \Big\downarrow & & \Big\downarrow & & \Big\downarrow \\ G & \overset{\longleftarrow}{\underset{\longleftarrow}{\dashrightarrow}} & G \times G & \overset{\longleftarrow}{\underset{\longleftarrow}{\longleftarrow}} & G \times G \times G \end{array}$$

by

$$\begin{aligned} &E|U_{a_0} \cap \cdots \cap U_{a_p} \longrightarrow G^{p+1} \quad \text{for each } a_0 \le \ldots \le a_p \\ &y \longrightarrow (f_{a_0}(y), \ldots, f_{a_p}(y)). \end{aligned}$$

Exercise. Check that $\mathcal{U}E \to PG$ commutes with ∂_i and ϵ_i.

Exercise. Check that the diagram

$$\begin{array}{ccc} \mathcal{U}E & \longrightarrow & PG \\ \downarrow & & \downarrow \\ \mathcal{U}M & \longrightarrow & NG \end{array}$$

commutes, where $PG \to NG$ is defined (as in Theorem 6.1) by

$$(x_0, x_1, \ldots, x_n) \longrightarrow (x_0 x_1^{-1}, x_1 x_2^{-1}, \ldots, x_{n-1} x_n^{-1}),$$

and other maps are as defined so far in this proof. Check also that $\mathcal{U}E \to PG$ commutes with the action of G.

This exercise proves that

$$\begin{array}{ccc} |\mathcal{U}E| & \longrightarrow & |PG| \\ \downarrow & & \downarrow \\ |\mathcal{U}M| & \longrightarrow & |NG| \end{array}$$

commutes. Composing this with the previous diagram

$$\begin{array}{ccc} E & \longrightarrow & |\mathcal{U}E| \\ \downarrow & & \downarrow \\ M & \longrightarrow & |\mathcal{U}M| \end{array}$$

and using Theorem 6.1, we get a G-bundle map

$$\begin{array}{ccccc} E & \longrightarrow & |PG| & = & EG \\ \downarrow & & \downarrow & & \downarrow \\ M & \longrightarrow & |NG| & = & BG. \end{array}$$

Hence the map $M \to |NG|$ must classify the bundle $\eta = (E \to M)$. Since the map factors $M \to |\mathcal{U}M| \to |NG| = BG$ and $M \to |\mathcal{U}M|$ is a homotopy equivalence, assertion 2 is proved. Q.E.D. (Theorem 7.4)

Corollary 7.4.1. *The map* $H^*(BG) \xrightarrow{\sim} H^*_D ANG \to H^*_D A\mathcal{U}M \xrightarrow{\sim} H^*M$ *gives the characteristic classes of the bundle* η.

Comment. Theorem 7.4 is also true if M is a paracompact space but not a manifold, provided that in assertion 3 we replace $ANG \to A\mathcal{U}M$ by $S^*NG \to S^*\mathcal{U}M$ (singular cohomology). However G must in general have a decent topology (for example $G =$ Lie group) so that $|PG| \to |NG|$ will be a (locally trivial) fibration. If G has a bad topology then $|NG_Z|$ classifies if M is any paracompact space (see Sections 4,6 and [6], Stasheff's appendix).

Example. Let $G = GL(1,\mathbb{C}) = \mathbb{C}^*$. Then as we have seen in Section 6, $\frac{dz}{z} \in A^1G$ is a primitive form on G, so in the double complex ANG (where we now mean *complex*-valued smooth forms on NG)

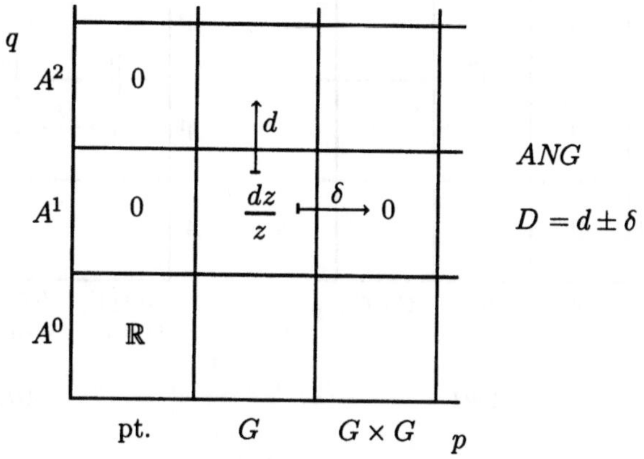

$D\left(\frac{dz}{z}\right) = 0$. Therefore dz/z is a 2-cocycle of ANG. We know from our semi-simplicial theory that $H^*_D(ANG) = H^*(BG)$, and we know from other methods [53] that $H^*BG = \mathbb{C}[c_1]$, a polynomial algebra generated by $c_1 \in H^2BG$, where c_1 is called the first Chern class. (In Section 9, we shall calculate H^*_DANG directly). Thus we can write $dz/z = xc_1$, where x is some complex constant. We shall find x by comparing the pullbacks of dz/z and c_1 on the same complex line bundle.

Now let $M = \mathbb{C}P^1$, and let $\eta = (\pi : E \to M)$ be the canonical complex line bundle over M defined as follows: Let $M = (\mathbb{C} \times \mathbb{C} - 0)/\mathbb{C}^*$

$$= \{(z_1, z_2)\}/((z_1, z_2) \sim (zz_1, zz_2))(z \in \mathbb{C}^*)$$

Let $E = ((\mathbb{C} \times \mathbb{C} - 0) \times \mathbb{C})/\mathbb{C}^* =$

$$\{(z_1, z_2; z_3)\}/((z_1, z_2; z_3) \sim (zz_1, zz_2; zz_3)).$$

Define $\pi : E \to M$ by $\pi(z_1, z_2; z_3) = (z_1, z_2)$. Cover M by U and V (ordered $U \leq V$) where

$$U = \{(1, z)\} \quad \text{and} \quad V = \{(z, 1)\}.$$

Identify U and V with $\mathbb{C}$ in the obvious way. In $E|U \cap V$, we have $(z, 1; w) = (1, z^{-1}; z^{-1}w)$ so the transition function $g_{uv} : U \cap V \to \mathbb{C}^*$ is given by $g_{uv}((z, 1)) = z^{-1}$. Coordinatizing $U \cap V$ by the map

$$U \cap V \subset V \longrightarrow \mathbb{C}, \quad (z, 1) \longmapsto z$$

we have

$$\begin{aligned} g^*_{uv}(dz/z) &= dg_{uv}/g_{uv} = d(z^{-1})/z^{-1} \\ &= -z^{-2}dz/z^{-1} = -dz/z. \end{aligned}$$

Thus the map $ANG \to \mathcal{AU}M$ defined by the cocycle of the bundle takes dz/z in A^1G to $-dz/z$ in $A^1\mathcal{U}_1M$. Let $\varphi = -dz/z \in A^1\mathcal{U}_1M$.

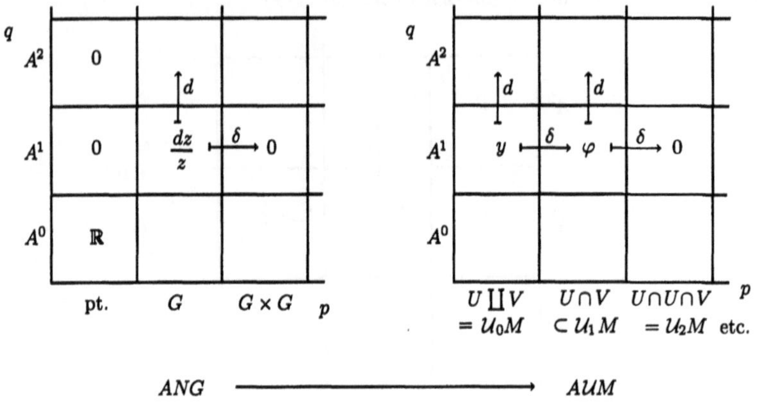

Now φ is a D 2-cocycle on $\mathcal{AU}M$, hence generates an element of $H^2(|\mathcal{U}M|) = H^2(M)$. To find out which element, we imitate the proof of Lemma 7.2 to find $y \in A^1(U \coprod V)$ such that $\delta y = \varphi$ (see diagram). As in that proof, we let $\{\lambda_u, \lambda_v\}$ be a partition of unity subordinate to $\{U, V\}$ and define $y = h(\varphi)$. We get

$$\begin{aligned} y(U) &= (-1)^1(\lambda_u)(U \cap U) + \lambda_v\varphi(U \cap V)) \\ &= -\lambda_v\varphi(U \cap V) \\ y(V) &= (-1)^1(\lambda_u\varphi(V \cap U) + \lambda_v\varphi(V \cap V)) \\ &= +\lambda_u\varphi(U \cap V) \quad (\text{since} \quad \varphi(V \cap U) = -\varphi(U \cap V)). \end{aligned}$$

$dy \in A^2(U \coprod V)$ is given by

$$\begin{aligned} dy(U) &= -(d\lambda_v)\varphi(U \cap V) - \lambda_v d\varphi(U \cap V) \\ &= -(d\lambda_v)\varphi(U \cap V) \quad (\text{since} \quad d\varphi = 0) \\ dy(V) &= (d\lambda_u)\varphi(U \cap V) \end{aligned}$$

Observe that since $\lambda_v = 1 - \lambda_u, dy(U)$ and $dy(V)$ agree on $U \cap V$. Thus dy is a form in A^2M. Now $Dy = (d + (-1)^1\delta)y = dy - \delta y = dy - \varphi$, so dy and φ are D-cohomologous in $\mathcal{AU}M$. Thus the class of φ in $H^2_D(\mathcal{AU}M)$ goes to the class of dy in H^2M under the isomorphism $H^*_D(\mathcal{AU}M) \to H^*M$.

To evaluate the class of dy in H^2M, we integrate dy over M. Observe that

$$\text{support} \quad dy \subset \quad \text{supp } d\lambda_u = \text{supp } d\lambda_v \subset U \cap V \subset V,$$

so

$$\int_M dy = \int_V dy(V) = \int_V (d\lambda_u)\varphi(U \cap V).$$

Now identifying V with $\mathbb{C}$, we have $U \cap V = \mathbb{C} - \{0\}$, and we can choose λ_u, λ_v so that supp $d\lambda_u \subset \{z \mid |z| \leq 1\} = D$. Then $\lambda_u \equiv 1$ on $(\overline{\mathbb{C} - D})$ and

$\int_V (d\lambda_u)\varphi(U\cap V) = \int_D (d\lambda_u)\varphi(U\cap V) = \int_D d(\lambda_u\varphi)$ = (by Stoke's Theorem) $\int_{\partial D} \lambda_u\varphi = \int_{|z|=1} 1\cdot(-dz/z) = -\int_{\theta=0}^{2\pi}((d(e^{i\theta}))/(e^{i\theta})) = -2\pi i.$

Now $c_1 \in H^2BG$ is defined so that its pullback to H^2M is 1 for the bundle η [53], so we have shown that

$$\frac{dz}{z} = -2\pi i c_1 \in H^2ANG = H^2BG.$$

8. More applications of semi-simplicial techniques

In Section 7 we showed that one can replace a manifold M with the semi-simplicial manifold $\mathcal{U}M$ formed from an open cover of M. We showed that $|\mathcal{U}M|$ and M have the same homotopy type, and that $H^*M = H^*_D A\mathcal{U}M$.

$\mathcal{U}M$ consists of some open sets, together with the information of how they are glued together to form M. There are also other ways to represent a space M or X semi-simplicially.

1) Making no use of the topology of X, define the semi-simplicial space $X' =$

$$X \;\substack{\longleftarrow\\ \dashrightarrow\\ \longleftarrow}\; X \;\substack{\longleftarrow\\ \dashrightarrow\\ \longleftarrow\\ \dashrightarrow\\ \longleftarrow}\; X \quad \cdots$$

where all face and degeneracy maps are identity_X. Then the degeneracy maps collapse $X\times\Delta^i$, $i>0$, onto $X\times\Delta^0$ in $|X'|$, so the geometric realization is $|X'| = X\times\Delta^0 = X$.

2) Making the most extreme use of the topology of X, define a semi-simplicial set (discrete space)

$$\Sigma X = \Sigma_0 X \;\substack{\longleftarrow\\ \dashrightarrow\\ \longleftarrow}\; \Sigma_1 X \;\substack{\longleftarrow\\ \longleftarrow\\ \longleftarrow}\; \Sigma_2 X \qquad \cdots$$

where

$$\Sigma_p = \{\text{singular } p\text{-simplices of } X\},$$

and ∂_i and ϵ_i are the usual face and degeneracy maps of singular homology. Since ΣX is discrete, $|\Sigma X|$ is a CW complex.

In computing the D-cohomology of the double complex $S^*\Sigma_* X = \oplus_{p,q} S^q\Sigma_p X$ with coefficients in the ring R, the discreteness of $\Sigma_p X$ implies that

$$\begin{aligned} E_1^{p,q} &= H_d^q(S^*\Sigma_p X) \quad (d = \text{singular coboundary map})\\ &= H^q(\Sigma_p X) \quad (\text{singular cohomology})\\ &= \begin{cases} 0 & \text{if} \quad q\neq 0\\ F(\Sigma_p X) & \text{if} \quad q = 0, \end{cases} \end{aligned}$$

where

$$\begin{aligned} F(\Sigma_p X) &= \text{the module of functions from } \quad \Sigma_p X \text{ to } R \\ &= \text{module of singular } p\text{-cochains of } X \\ &= S^p(X). \end{aligned}$$

Since the E_1 term has zeroes except when $q = 0$, we get $E_2 = E_\infty = H_D(S^*\Sigma X)$, so that $H^p_D(S^*\Sigma X) = E_2^{p,0} = H^p_\delta(E_1) = H^p_\delta(S^*X) = H^*(X)$. Applying Theorem 4.1, we see that

$$H^*(|\Sigma X|) = H^*_D(S^*\Sigma X) = H^*X,$$

so that the evaluation map on simplices

$$\text{eval} : |\Sigma X| \longrightarrow X$$

induces an isomorphism in cohomology (and similarly in homology). The map also induces isomorphism of all homotopy groups [52], [50].

3) Let us reconsider the semi-simplicial manifold $\mathcal{U}M$. If M is a differentiable n-manifold, we can choose a covering $\mathcal{U}$ of M such that U_a and every nonempty finite intersection $U_{a_0\cdots a_p}$ are diffeomorphic to $\mathbb{R}^n$; this will be taken as the definition of a *good* covering. For example, one can put a Riemannian metric on M and choose each U_a to be geodesically convex, or one can triangulate M and take the U_a to be the star neighborhoods in some subdivision. Then the semi-simplicial manifold $\mathcal{U}M$ has the form

$$\begin{array}{ccccc} \mathcal{U}_0 M & \begin{smallmatrix}\longleftarrow \\ \dashrightarrow \\ \longleftarrow\end{smallmatrix} & \mathcal{U}_1 M & \begin{smallmatrix}\longleftarrow \\ \dashrightarrow \\ \longleftarrow \\ \dashrightarrow \\ \longleftarrow\end{smallmatrix} & \mathcal{U}_2 M \quad \cdots \\ \| & & \| & & \| \\ \coprod_a \mathbb{R}^n & & \coprod_{a\le b} \mathbb{R}^n & & \coprod_{a\le b\le c} \mathbb{R}^n \\ & & \text{with } U_{ab} \neq 0 & & \text{with } U_{abc} \neq 0 \end{array}$$

where the map ∂_i is the inclusion

$$U_{a_0\cdots a_p} \longrightarrow U_{a_0\cdots \hat{a}_i \cdots a_p}.$$

The manifold $\mathcal{U}M$ is a "free resolution" of M if one regards

$$\mathcal{U}_i M = \coprod \mathbb{R}^n$$

as the analog for spaces of a free module. In fact, the cochain complex $kA^q\mathcal{U}M$ *is* a free resolution of the module A^qM by modules of the form $\oplus_a A^q\mathbb{R}^n$; the exactness of

$$0 \longrightarrow A^qM \longrightarrow A^q\mathcal{U}_0M \xrightarrow{\ \delta\ } A^q\mathcal{U}_1M \xrightarrow{\ \delta\ } \cdots$$

was proved in Lemma 7.2.

Regarding $\mathcal{AU}M$ as a double complex

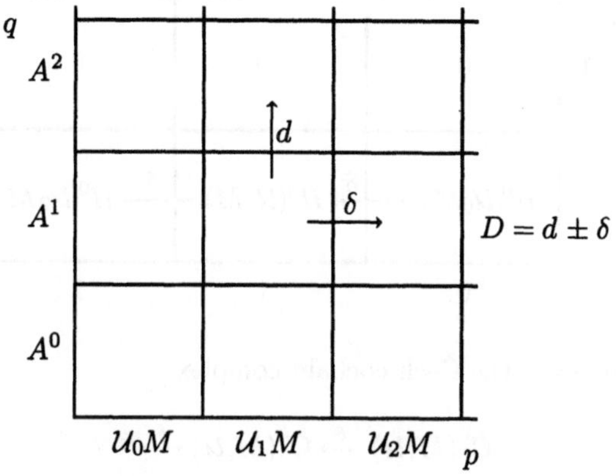

we get a simple proof, due to Weil [71], of the isomorphism between $H^*_{Cech}M$ and $H^*_{DeRham}(M)$: The rows of $\mathcal{AU}M$ are exact by Lemma 7.2 (i.e., $0 \to A^qM \to A^q\mathcal{U}_0M \xrightarrow{\delta} A^q\mathcal{U}_1M \to \cdots$ is exact) while the columns are now exact by the Poincaré Lemma, i.e.,

$$0 \longrightarrow \mathbb{R} \longrightarrow A^0\mathbb{R}^n \xrightarrow{d} A^1\mathbb{R}^n \xrightarrow{d} \cdots$$

is exact. Computing $H^*_D\mathcal{AU}M$ by spectral sequences, we can filter by q to get $E_1 = H_\delta(\mathcal{AU}M) =$

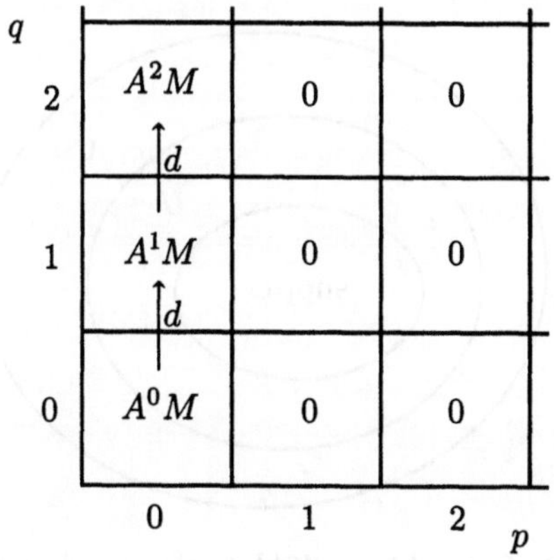

yielding $H^*_D\mathcal{AU}M = H(A^*M) = H^*_{DeRham}(M)$, or filter by p to get $E_1 =$

$H_d(\mathcal{A}\mathcal{U}M) =$

q			
1	0	0	0
0	$H^0(\mathcal{U}_0M) \xrightarrow{\delta}$	$H^0(\mathcal{U}_1M) \xrightarrow{\delta}$	$H^0(\mathcal{U}_2M)$
	0	1	2

The latter complex is the Čech cochain complex

$$\check{C}^0(M,\mathcal{U}) \xrightarrow{\delta} \check{C}^1(M,\mathcal{U}) \xrightarrow{\delta} \cdots$$

of functions which assign a constant to each nonempty intersection $U_{a_0} \cap \cdots \cap U_{a_p}$, so

$$E_2^{p,0} = H_\delta^p(\check{C}^*(M,\mathcal{U})) = \check{H}^p_{Cech}(M,\mathcal{U}) = E_\infty^{p,0} = H_D^p(\mathcal{A}\mathcal{U}M).$$

Thus $H^*_{DeRham}(M) = H^*_D(\mathcal{A}\mathcal{U}M) = \check{H}^*_{Cech}(M,\mathcal{U})$. Since this isomorphism is independent of which good cover U is used, it follows that $\check{H}^*(M,\mathcal{U}) = \check{H}^*(M)$; that is, we have also proved that one can compute Čech cohomology by using a good cover, without passing to the direct limit.

By a similar approach we prove Poincaré duality for an n-manifold M. Let $\mathcal{U}$ be a *finite* good cover of M. Let the functor A^q_c denote q-forms with compact support. This functor is *covariant* with respect to inclusions of open submanifolds, since a form φ with compact support can be extended to be 0 outside its support.

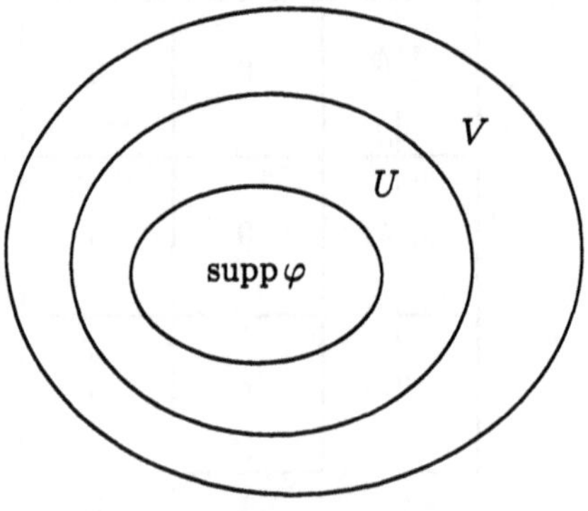

The exterior derivative $d : A^q_c \to A^{q+1}_c$ is the usual one. We define $H^q_c(M)$, the *DeRham cohomology of M with compact support*, to be $H^q_d(A^*_cM)$. Thus the double complex $A_c\mathcal{U}M$ is a chain complex with respect to δ and a cochain complex with respect to d.

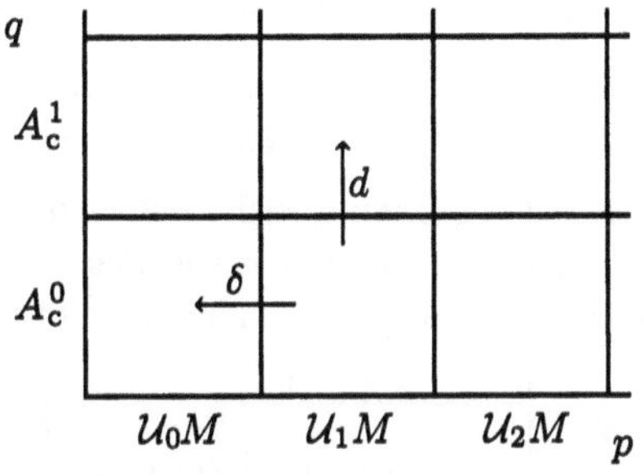

We set $D = d \pm \delta$ and grade $A_c\mathcal{U}\mathcal{U}M$ by $q-p$ to make it a D-cochain complex. Since M is an n-manifold the complex is 0 above the n-th row, so the spectral sequence for $H_D^* A_c\mathcal{U}M$ converges.

The complex

$$0 \longrightarrow (A_c^0\mathbb{R}^n \xrightarrow{d} A_c^1\mathbb{R}^n \longrightarrow \cdots \longrightarrow A_c^n\mathbb{R}^n) \xrightarrow{I} \mathbb{R} \longrightarrow 0$$

is exact, where the map I is integration of an n-form with respect to a chosen orientation of $\mathbb{R}^n$. Each column of the double complex $A_c\mathcal{U}M$ is a direct sum of copies of the cochain complez in parenthesis. Thus filtering $A_c\mathcal{U}M$ by p, $E_1^{p,q} = H_d(A_c\mathcal{U}M) = H_c(\mathcal{U}M)$ is the complex

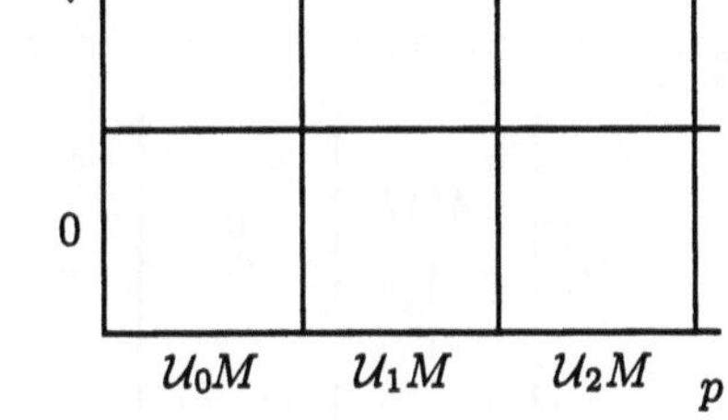

where $\oplus\mathbb{R} = H_c^n(\mathcal{U}_pM)$ means one copy of $\mathbb{R}$ for each copy of $\mathbb{R}^n$ in the space $\mathcal{U}_pM = \coprod \mathbb{R}^n$.

To define the connecting maps $\oplus\mathbb{R} \to \oplus\mathbb{R}$, we must take into account local orientations and use local coefficients, just as in Čech homology theory with local coefficients [63]. Specifically, we regard $H_c^n(U_{a_0} \cap \cdots \cap U_{a_p})$ as having coefficients in $H_c^n(U_{a_0}) \cong \mathbb{R}$. The set $\{H_c^n(U_a)\}$ is a local system on M of copies of $\mathbb{R}$ (it is essentially the orientation sheaf of M [32]), and our complex

$$\begin{array}{ccccc} H_c^n(\mathcal{U}_0 M) & \xleftarrow{\delta} & H_c^n(\mathcal{U}_1 M) & \xleftarrow{\delta} & \cdots \\ \| & & \| & & \\ \oplus\mathbb{R} & \longleftarrow & \oplus\mathbb{R} & \longleftarrow & \cdots \end{array}$$

is exactly the complex of Čech chains with coefficients in the local system. Therefore, the δ-homology of the complex is the Čech homology $\check{\mathrm{H}}_p(M;\text{ local})$ of M with coefficients in the local system. Thus $E_2^{p,q}$ of the double complex is

q		
n	$\check{H}_0(M;\text{local})$	$\check{H}_1(M;\text{local})$
$\vdots$		
0	0	0
	0	1 $\quad p$

Therefore $E_2 = E_\infty$, and

$$\check{\mathrm{H}}_p(M;\text{local}) \cong H_D^{n-p}(A_c\mathcal{U}M).$$

The rows of A_cUM,

$$0 \longleftarrow A_c^q M \longleftarrow (A_c^q\mathcal{U}_0 M \xleftarrow{\delta} A_c^q\mathcal{U}_1 M \xleftarrow{\delta} \cdots)$$

are exact by a modification of Lemma 7.2.

Thus filtering $A_c\mathcal{U}M$ by q, we get $E_1 = H_\delta(A_c\mathcal{U}M) =$

q		
2	$A_c^2 M$	0
	$\uparrow d$	
1	$A_c^1 M$	0
	$\uparrow d$	
0	$A_c^0 M$	0
	0	1 $\quad p$

and

$$E_2^{0,q} = H_d^q E_1 = H_d^q(A_c^* M) = H_c^q(M).$$

Therefore

$$H_c^q(M) = H_D^q(A_c UM) = \check{\mathrm{H}}_{n-q}(M; \mathrm{local}).$$

This is Poincaré duality.

9. Computation of HANG

Recall that ANG is the double complex

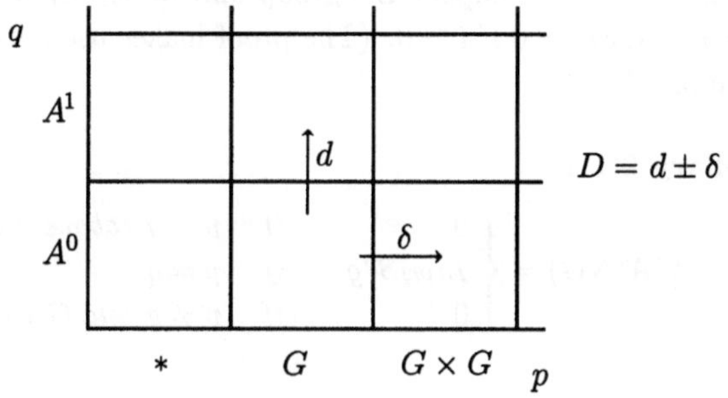

of differential forms on the semi-simplicial manifold NG (cf. Section 6). To find $H_D(ANG) = H^*(BG)$, we could filter ANG by p, getting a spectral sequence converging to $H_D(ANG)$, with $E_1 = H_d(ANG)$ and $E_1^{p,q} = H^q(G^p)$ (DeRham cohomology).

If, instead, we filter ANG by q we get another spectral sequence converging to $HANG(\underset{def.}{=} H_D(ANG))$, but with E_1 term $H_\delta(ANG)$, i.e., $E_1^{p,q} = H_\delta^p(A^q NG)$. This has been computed, and the answer has a rather nice form.

Theorem 9.1. (Bott, Hochschild).

$$H_\delta^p(A^q NG) = H_c^{p-q}(G; S^q g^*)$$

where G is any Lie group, g is its Lie algebra, g^* =$Hom_{\mathbb{R}}(g, \mathbb{R})$ is the dual of g, S^q is the q-th symmetric power (hence $S^q g^*$ is the module of degree q polynomials on g), and H_c means continuous cohomology in the sense of the cohomology of groups, in particular of Lie groups [38]. [$H_c^{p-q}(G; S^q g^*)$ is the $(p-q)$-th continuous cohomology group of G acting on the module $S^q g^*$ by the adjoint action.]

Proof. See [8].

We recall [19] that for discrete groups G, $H_{EM}^*(G; \mathbb{R})$, the Eilenberg–MacLane cohomology of the group G, is defined as the cohomology of

$$\text{Functions } (G, \mathbb{R}) \rightrightarrows \text{Functions } (G \times G, \mathbb{R}) \underset{\longrightarrow}{\rightrightarrows} \cdots$$

where the maps are the duals of those of NG. $H_{continuous}(G) = H_c(G)$ and $H_{smooth}(G)$ are defined analoagously for Lie groups G by looking only at smooth or continuous functions, respectively.

We state now two well-known facts about continuous cohomology of groups. (See [38].)

Fact 1. Let G be a Lie group and N a module on which G acts. Then $H^0_c(G;N) = \mathrm{Inv}_G N$, the G-invariants of N(i.e., $\{n \in N | gn = n \; \forall g \in G\}$). (This is the same result as in the discrete group case).

Fact 2. If G is a compact Lie group and M is any module on which G acts, then $H^i_c(G;M) = 0$ if $i > 0$. (The proof makes use left-invariant or Haar integration on G).

Corollary 9.1.1.

$$H^p_\delta(A^q NG) = \begin{cases} 0 & \text{if} \quad p < q \text{ (above diagonal)} \\ Inv_G S^p g^* & \text{if} \quad p = q \\ 0 & \text{if} \quad p \neq q \text{ and } G \text{ compact.} \end{cases}$$

Example. Let $G = GL(n, \mathbb{R})$, $g = gl(n, \mathbb{R})$. Then

$$\mathrm{Inv}_G S^* g^* = \mathbb{R}[c_1, \ldots, c_n] = \mathbb{R}[s_1, \ldots, s_n],$$

a polynomial algebra with generators c_i or s_i, where for $A \in g$, (A an $n \times n$ matrix),

$$c_i(A) = \text{coefficient of } \; t^i \; \text{ in } \det(I + tA),$$
$$s_i(A) = tr(A^i). \quad \text{(cf. appendix of [6])}$$

Thus for all G, the E_1 term of ANG, filtering by q, is

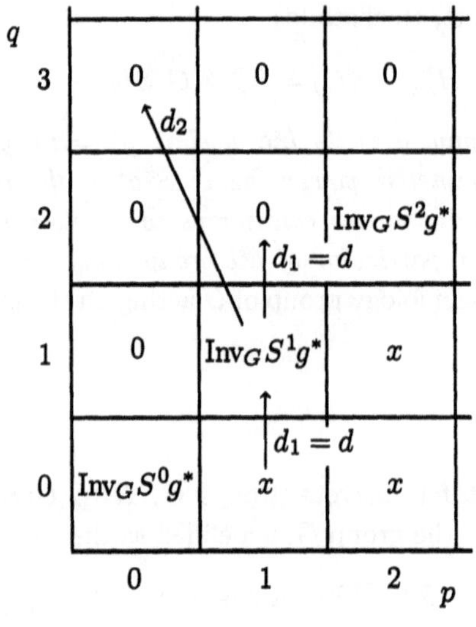

x = anything

All diagonal entries are cocycles for $d_1, d_2, d_3, \ldots$ in the spectral sequence, so

$$E_1^{p,p} \xrightarrow{onto} E_\infty^{p,p} \xrightarrow{1-1} H_D^{2p}(ANG).$$

The composition is called the *Chern–Weil homomorphism*

$$\varphi : \mathrm{Inv}_G S^p g^* \longrightarrow H^{2p}(ANG) \xrightarrow{=} H^{2p}(BG).$$

Corollary 9.1.2. *If G is compact, $\varphi : Inv_G S^p g^* \to H^{2p}(BG)$ is an isomorphism.*

Example. Let $G = O(n)$, $g = o(n) = \{n \times n$ real matrices A such that $A^t = -A\} = \{$skew-symmetric matrices$\}$. Then $tr(A^p) = 0$ for p odd, and it follows that $\mathrm{Inv}_G S^p g^* = 0$ for p odd. Thus

$$H^*(BO(n)) = \mathrm{Inv}_G S^* g^* = \mathbb{R}[c_2, c_4, \ldots, c_{[n/2]}].$$

Thus $H^*(BGL(n, \mathbb{R})) = \mathbb{R}[c_2, c_4, \ldots, c_{[n/2]}]$. $c_{2i} \in H^{4i}(BGL(n)) = i$-th Pontryagin class (cf. Section 6).

Example. Since $\mathrm{Inv}_{U(n)} u(n) = \mathbb{R}[c_1, \ldots, c_n]$, it follows that $H^*(BGL(n, \mathbb{C})) = H^*(BU(n)) = \mathbb{R}[c_1, \ldots, c_n]$, where $c_i \in H^{2i} = i$-th Chern class.

Special case. $H^*(BGL(1, \mathbb{C})) = \mathbb{R}[c_1]$, where $c_1 =$ class of $-\frac{1}{2\pi i}\frac{dz}{z}$ in $A^1\mathbb{C}^* = A^1 GL(1, \mathbb{C})$ in ANG [cf. example at end of Section 7]. Note that $\frac{dz}{z}$ is a left invariant form on $\mathbb{C}^*$, hence is an element of the dual, g^*, of the Lie algebra $gl(1, \mathbb{C}) = g$.

Exercise. Try to find universal formulas for c_1 and c_2 for $GL(2, \mathbb{C})$ bundles in terms of the transition functions g_{ab} (of the same type as $c_1 = -\frac{1}{2\pi i}\frac{dg_{ab}}{g_{ab}}$ for $GL(1, \mathbb{C})$ bundles.)

Comment. For oriented n-plane bundles with n even (structure group $G = GL^+(n, \mathbb{R})$ or $SO(n)$), there is an Euler class $\chi \in H^n(BG)$, with $\chi^2 = c_n$. For example, for $SO(2)$ bundles, $\chi \in \mathrm{Inv}^I_{SO(2)} so(2)$ is given by $\chi\begin{pmatrix} 0 & a \\ -a & 0 \end{pmatrix} = a$, where $\begin{pmatrix} 0 & a \\ -a & 0 \end{pmatrix} \in so(2)$. Observe that $\chi^2 = \det = c_2$. χ does not arise for $O(2)$ (or any $O(n)$) bundles, because conjugating $\begin{pmatrix} 0 & a \\ -a & 0 \end{pmatrix} \in o(2)$ by $\begin{pmatrix} 0 & 1 \\ 1 & 0 \end{pmatrix} \in O(2)$ yields $\begin{pmatrix} 0 & -a \\ a & 0 \end{pmatrix}$, hence takes χ to $-\chi$.

10. Discrete subgroups of Lie groups, and bundles with discrete structure group

If D is a discrete subgroup of a Lie group G, it is a difficult problem to find $H^*_{EM}(D)$, the Eilenberg–MacLane cohomology [19] of the group D. $H^*_{EM}(D)$ is isomorphic to $H^*(BD)$, where BD is, as usual, the classifying space for bundles with structure group D (see Section 5). Now the inclusion $D \to G$ induces a map $BD \to BG$, hence a map $H^*BG \to H^*BD$. Thus one can find a few elements of H^*BD, which is hard to compute, as images of elements of H^*BG, which is relatively easy to compute. For example, one can study $D = GL(n, \mathbb{Z})$ as a discrete subgroup of $G = GL(n, \mathbb{R})$, and study the images of the Pontryagin classes (H^*BG) in H^*BD. However, this procedure is not very useful because it constructs only a small fraction of the elements of H^*BD.

To get a better hold on H^*BD, one studies discrete subgroups of Lie groups in a universal way by defining for each Lie group G the discrete group G^δ (also called G_δ) which is the same group as G but with the discrete topology. The identity map $G^\delta \to G$ is continuous, and any discrete subgroup D of G maps $D \to G^\delta \to G$. This induces maps $BD \to BG^\delta$ and $H^*BG^\delta \to H^*BD$. Thus to find elements of H^*BD in a universal way, we compute H^*BG^δ and study its image in H^*BD.

What does it mean for a bundle over a space X to have structure group G^δ? It means that the transition functions $\{g_{uv} : U \cap V \to G^\delta\}$ are continuous with respect to the discrete topology on G, that is, g_{uv} is constant on each connected component of $U \cap V$. If we assume the cover is good, then g_{uv} is constant on $U \cap V$. Every G^δ bundle can be considered also as a G bundle via the map $G^\delta \to G$.

Now let

$$\begin{array}{ccc} G & \longrightarrow & P \\ & & \downarrow \pi \\ & & M \end{array}$$

be a principal G^δ bundle, i.e., a principal G-bundle whose structure group is G^δ. Let $\{U_a\}$ be a good cover of M which trivializes P. Let $\{f_a : P|U_a \to G\}$

be the horizontal projections (see Section 1).

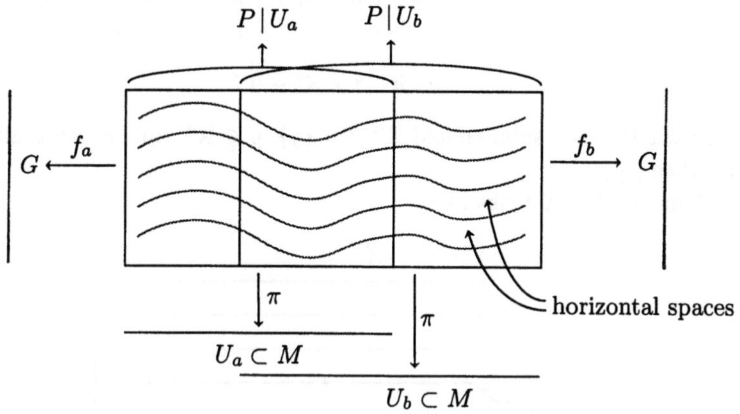

On $P|U_a$, the fibers of f_a are submanifolds diffeomorphic to U_a, and similarly on $P|U_b$. On $P|U_a \cap U_b$, $f_a = g_{ab} f_b$, where g_{ab} is a *constant* in G since the structure group is G^δ. Therefore {fibers of f_a} = {fibers of f_b} in $P|U_a \cap U_b$.

Definition. The *horizontal spaces* of P are the connected subspaces of P which are locally fibers of horizontal projections f_a. (See diagram above.) The *vertical spaces* or *fibers* of P are the ordinary fibers of the bundle $\pi : P \to M$.

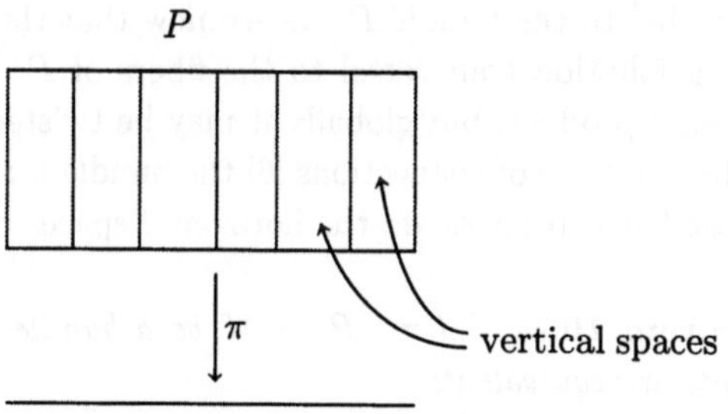

Since f_a and π are submersions and $P|U_a \underset{(\pi, f_a)}{\longrightarrow} U_a \times G$ is a diffeomorphism, the horizontal spaces and fibers are manifolds, the sum of their dimensions equals dimension (P), and they are *transversal*, i.e., at any $x \in P$,

$$TP_x = T(\text{horizontal space})_x + T(\text{fiber})_x.$$

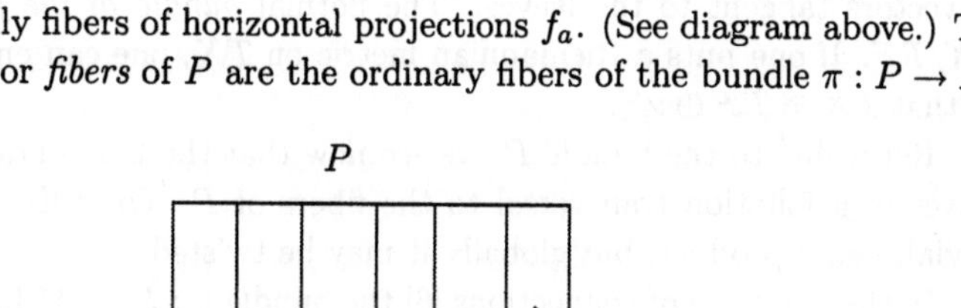

Definition. A codimension q *foliation* $\mathcal{F}$ on an n-manifold X is a decomposition of X into $(n-q)$-dimensional submanifolds, called *leaves*, such that every point of X has a neighborhood U and a diffeomorphism

$$g : \mathbb{R}^{n-q} \times \mathbb{R}^q \longrightarrow U$$

which carries each submanifold $\mathbb{R}^{n-q} \times \{y\}$, $y \in \mathbb{R}^q$, diffeomorphically into a leaf. (See [6].)

Thus locally, X looks like

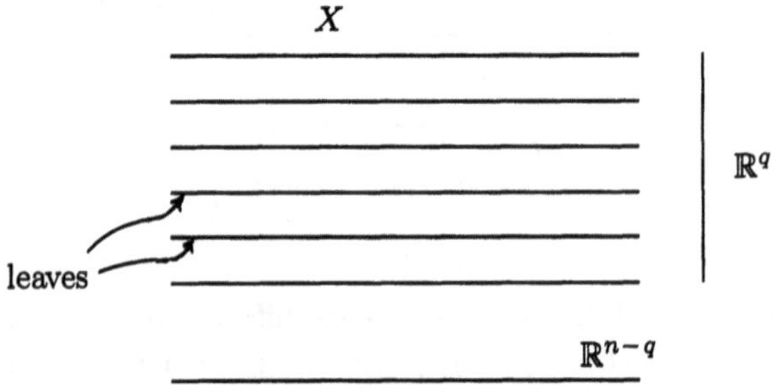

Of course, globally the foliation may be twisted.

The *tangent bundle of the foliation*, $T\mathcal{F}$, is the subbundle of TX consisting of vectors tangent to the leaves. The normal *bundle of the foliation*, $\nu\mathcal{F}$, is $TX/T\mathcal{F}$. If one puts a Riemannian metric on TX, one can embed $\nu\mathcal{F}$ in TX so that $TX \cong T\mathcal{F} \oplus \nu\mathcal{F}$.

Returning to the bundle P, we see now that the horizontal spaces are the leaves of a foliation transversal to the fibers of P. On $P|U_a$ this foliation is trivial, i.e., a product, but globally it may be twisted.

In the language of connections [6] the bundle $\pi : P \to M$ has a flat connection whose flat subspaces are the horizontal spaces or leaves. To summarize:

Theorem 10.1. *Let $\pi : P \to M$ be a bundle. Then the following three statements are equivalent:*

1. *The bundle has a discrete structure group (or its structure group can be reduced to a discrete group).*
2. *The bundle has a foliation transversal to the fibers (of π), whose leaves have the same dimension as M.*
3. *The bundle has a flat connection.*

How does one find characteristic classes of G_δ bundles? We can put a differential form ψ on G and pull it back to $P|U_a$ via the map $f_a : P|U_a \to G$. Since f_a and f_b differ by a left translation in G (by g_{ab}), $f_a^*\psi$ and $f_b^*\psi$ will agree on $P|U_a \cap U_b$ if $\psi \in \{\text{left invariant forms on } G\} \underset{def.}{=} \mathrm{Inv}_G AG$. (The action of $x \in G$ on AG is L_x^*, induced by left multiplication $L_x : G \to G$.) Thus there is a natural map $\mathrm{Inv}_G AG \to AP = \{\text{forms defined globally on } P\}$.

The module of left invariant forms $\mathrm{Inv}_G AG$ is a finite dimensional real vector space, since each such form is uniquely determined by its value on T_eG, the tangent space at the unit element $e \in G$. Writing g for the Lie algebra of G (which is defined either as T_eG or as the vector space of vector fields on G invariant under left multiplication by elements of G, with the commutator (bracket) operation $[X, Y]$), we see that $\mathrm{Inv}_G A^1 G = g^*$, the dual of g, and that $\mathrm{Inv}_G A^q G$ =left invariant q-forms $= \Lambda^q g^*$, where Λ^q denotes the q-th exterior power. That is, $\mathrm{Inv}_G A^q G$ can be regarded as the module of alternating q-linear real-valued forms on the vector space g.

We can reduce our calculations with differential forms completely to Lie algebra calculations. A standard fact from geometry is the DeRham exterior derivative d, when restricted to $\mathrm{Inv}_G A^q G = \Lambda^q g^*$, has the formula

$$d\varphi(X_1, \ldots, X_{q+1}) = \Sigma_{i<j}(-1)^{i+j}\varphi([X_i, X_j], X_1, \ldots, \widehat{X}_i, \ldots, \widehat{X}_j, \ldots, X_{q+1})$$

where $\varphi \in \Lambda^q g^*$ is a q-linear alternating form on g and $X_i \in g$. Motivated by this formula, one defines the cohomology $H^*(g; \mathbb{R})$ of *any* Lie algebra g to be the cohomology of the complex

$$\mathbb{R} = \Lambda^0 g^* \xrightarrow{d} \Lambda^1 g^* \xrightarrow{d} \Lambda^2 g^* \longrightarrow \cdots$$

where d is defined by the formula above (cf. infra, see also [14]).

We have defined a natural map $\mathrm{Inv}_G AG \to AP$. This map commutes with the DeRham d, so it induces a map $H^*(g; \mathbb{R}) = H^*(\mathrm{Inv}_G AG) \to H^*(P)$. But our goal was to construct characteristic classes for G^δ bundles, and these should lie in $H^*(M)$. How can we pull down cohomology classes from the total space P to the base space M of the bundle? We proceed in two different ways.

First, suppose the G^δ bundle $\pi : P \to M$ is trivialized as a G-bundle, that is, is G-bundle isomorphic to $M \times G$. Then π has a section

$$s : M \to M \times \{e\} \subset M \times G \cong P$$

and we can pull down forms and classes from P to M by the maps

$$s^* : AP \longrightarrow AM \quad \text{and} \quad s^* : H^*P \longrightarrow H^*M.$$

In general, both maps depend on the trivialization chosen.

Bundles with structure group G^δ which are trivialized as G-bundles have been classified by Thurston by the following scheme (see [54], p. 84, for general properties of this construction). Let G^I = path space of G = space of maps: $I = [0,1] \to G$ (with compact-open or evaluation topology). G^I maps to G by $p \to p(0)$. Let G_1 be the fiber product of G^δ and G^I:

$$\begin{array}{ccccc} G_1 & \longrightarrow & G^I & & p \\ \downarrow & & \downarrow & & \downarrow \\ G_\delta & \longrightarrow & G & & p(0) \end{array}$$

That is, $G_1 = \{(x,p) \in G_\delta \times G^I | x = p(0)\}$. $G_1 \to G_\delta$ is a homotopy equivalence. Now map $G_1 \to G$ by $(x,p) \to p(1)$. This is a Hurewicz fibration, and we call its fiber (inverse image of $e \in G$) $\overline{G}$. $\overline{G}$ is a group since we can multiply paths in G pointwise. We have $0 \to \overline{G} \to G_1 \to G \to e$, or (by a common abuse of notation) $0 \to \overline{G} \to G_\delta \to G \to e$, where we identify G_1 with G_δ since they have the same homotopy type.

Exercise. Check that a $\overline{G}$-bundle over M is given by a G-bundle $Q \to M \times I$ and by G-bundle maps

$$\begin{array}{ccccc} P & \longrightarrow & Q & \longleftarrow & M \times G \\ \downarrow & & \downarrow & & \downarrow \\ M \times \{0\} & \subset & M \times I & \longleftarrow & M \times \{1\} \end{array}$$

where P is a G^δ bundle.

Definition. A *trivialization* of a G-bundle $E \to M$ is a bundle map

$$\begin{array}{ccc} E & \longrightarrow & M \times G \\ \downarrow & & \downarrow \\ M & \underset{=}{\longrightarrow} & M \end{array}$$

(equivalently, a map $E \to G$). Two trivializations f, g are *homotopic* if there is a trivialization

$$\begin{array}{ccccc} E \times I & \xrightarrow{f_t} & M \times I \times G & \quad \text{such that} & f_0 = f \\ \downarrow & & \downarrow & & \\ M \times I & = & M \times I & \quad \text{and} & f_1 = g; \end{array}$$

an equivalent condition is that the maps $E \to G$ induced by f and g be homotopic maps.

Refer to the last exercise. By standard bundle theory [41], there exists a bundle map (not unique)

$$\begin{array}{ccc} Q & \xrightarrow{\approx} & (Q|M \times 1) \times I \\ \downarrow & \swarrow & \\ M \times I & & \end{array}$$

Since $Q|M \times 1 = M \times G$, we get a trivialization

$$\begin{array}{ccc} Q & \xrightarrow{\approx} & M \times I \times G \\ \downarrow & \swarrow & \\ M \times I & & \end{array}$$

Restricting to $P = Q|M \times \{0\}$, we get a trivialization

$$\begin{array}{ccc} P & \xrightarrow{\approx} & M \times G \\ & \searrow \quad \swarrow & \\ & M & \end{array}$$

Exercise. All trivializations of P obtained in this way are homotopic.

Thus we get a second characterization of $\overline{G}$-bundles. A $\overline{G}$-bundle over M is a G^δ bundle $P \to M$ together with a trivialization

$$\begin{array}{ccc} P & \longrightarrow & M \times G \\ & \searrow \quad \swarrow & \\ & M & \end{array}$$

of P as a G-bundle. That is, a $\overline{G}$-bundle on M is a foliation of $M \times G$ by leaves of dimension $= \dim M$, which are transversal to the fibers of $M \times G \to M$.

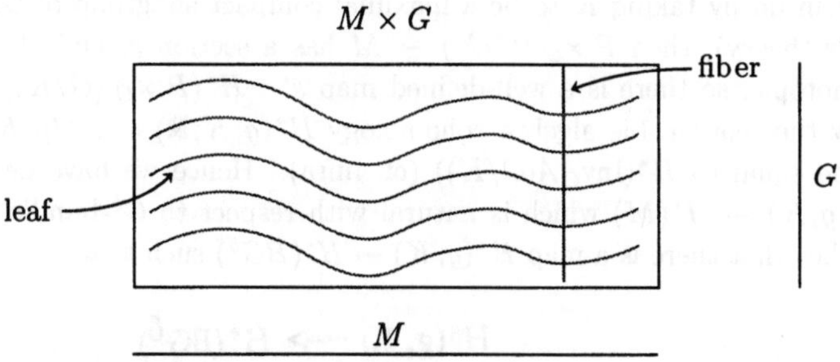

Definition. Two G-bundles $E \to M$ and $E' \to M$ are *homotopic* if there is a G-bundle $F \to M \times I$ and bundle maps

$$\begin{array}{ccc} E & \xrightarrow{\approx} & F|M \times \{0\} \\ \downarrow & & \downarrow \\ M & \longrightarrow & M \times \{0\} \end{array} \quad \text{and} \quad \begin{array}{ccc} E' & \xrightarrow{\approx} & F|M \times \{1\} \\ \downarrow & & \downarrow \\ M & \longrightarrow & M \times \{1\}. \end{array}$$

Exercise. Verify that there is indeed a canonical isomorphism of sets from {homotopy classes of $\overline{G}$ bundles for the first characterization of $\overline{G}$ bundles} to {homotopy classes of $\overline{G}$ bundles for the second characterization of $\overline{G}$ bundles}.

Using our semi-simplicial construction on the group $\overline{G}$, we construct $B\overline{G} = |N\overline{G}_Z|$ (unwinding; see Sections 4,6). $B\overline{G}$ classifies $\overline{G}$-bundles.

Our work so far has been to construct maps $H^*(g) \to H^*(M)$ which are natural with respect to maps of $\overline{G}$-bundles. The naturality implies the existence of a map $H^*(g) \to H^*(B\overline{G})$ (see [6], pp. 70–72), so that

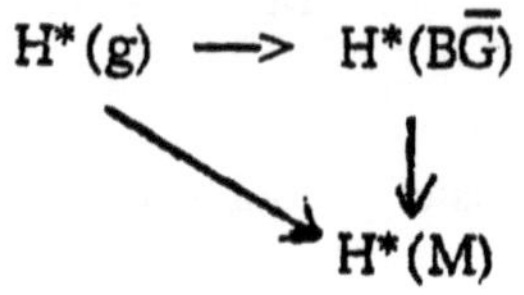

commutes, where $H^*(B\overline{G}) \to H^*(M)$ is induced by a classifying map of the bundle, $M \to B\overline{G}$.

Suppose now we try to construct characteristic classes for G^δ-bundles which are not necessarily $\overline{G}$-bundles. That is, how can we find classes in $H^*(BG^\delta)$? Suppose instead of the principal G^δ bundle $\pi : P \to M$ we consider the associated bundle $P \times_G F \to M$ with the same cocycle $\{g_{uv}\}$ but with fiber F, where F is a manifold on which G acts smoothly on the left (cf. Section 1). The same argument as before shows that $P \times_G F$ has a horizontal foliation, and that there is a natural map $\mathrm{Inv}_G AF \to A(P \times_G F)$ of G-invariant forms on F into forms on $P \times_G F$. In particular, for each subgroup $H \subset G$ one can choose $F = G/H$. In this case the module $\mathrm{Inv}_G A(G/H)$ maps into $A(P \times_G (G/H))$. Specializing further, if we choose $H = K$ so that $G/K \approx \mathbb{R}^q$ for some q (which we can do by taking K to be a maximal compact subgroup of G, using Cartan's theory), then $P \times_G (G/K) \to M$ has a section s, and all sections are homotopic, so there is a well-defined map $s^* : H^*(P \times_G (G/K)) \to H^*(M)$. Now the relative Lie algebra cohomology $H^*(g, K; \mathbb{R}) = H^*(g, K)$ is defined to be equal to $H^*(\mathrm{Inv}_G A(G/K))$ (cf. infra). Hence we have defined a map $H^*(g, K) \to H^*(M)$ which is natural with respect to G^δ-bundle maps. This implies that there is a map $H^*(g, K) \to H^*(BG^\delta)$ such that

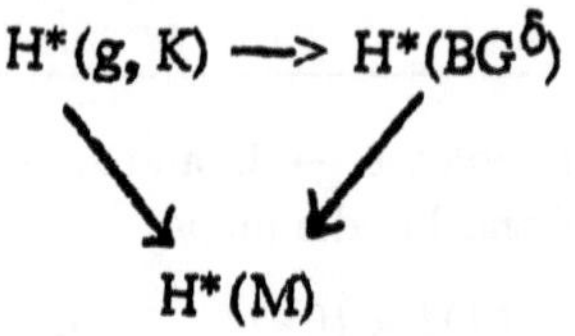

commutes for any G^δ bundle on M. In the specific case $G = GL(n, \mathbb{R})$ or $GL^+(n, \mathbb{R})$(GL^+ classifies *oriented n-plane bundles*) we have $g = gl(n, \mathbb{R})$ and $K = O(n)$ or $SO(n)$, respectively, is the maximal compact subgroup of G. We have constructed maps

$$\varphi_1 : H^*(gl(n)) \longrightarrow H^*(B\overline{G})$$
$$\varphi_2 : H^*(gl(n), K) \longrightarrow H^*(BG^\delta).$$

If $D \subset G$ is a discrete subgroup of G, for example $D = GL(n, \mathbb{Z})$, then the inclusion $D \subset G^\delta$ induces a map $H^*(g, K) \to H^*(BG^\delta) \to H^*(BD)$.

A priori, φ_1 and φ_2 need not be injective or even non-zero maps. But the deep theory of Borel and Harish-Chandra implies

Theorem 10.2. *If G is a semi-simple Lie group (e.g., $SL(n, \mathbb{R})$ with Lie algebra g and maximal compact subgroup K, then the map*

$$\varphi : H^*(g, K) \longrightarrow H^*(BG^\delta)$$

defined previously is injective.

Proof. By the theory of Borel and Harish-Chandra, there is subgroup $D \subset G$ which acts discretely on G/K by left multiplication, and such that $D\backslash G/K$ is compact. But G/K is contractible and D acts freely so

$$\begin{array}{ccc} G/K & = & ED \\ \downarrow & & \downarrow \\ D\backslash G/K & = & BD \end{array}$$

is a universal D bundle. Thus there are maps

$$H^*(g, K) \longrightarrow H^*(BG^\delta) \longrightarrow H^*(BD) = H^*(D\backslash G/K)$$

Hence it suffices to show that $f : H^*(g, K) \to H^*(D\backslash G/K)$ is injective. Well f is induced by the map $\mathrm{Inv}_G(A(G/K)) \subset \mathrm{Inv}_D(A(G/K)) = A(D\backslash G/K)$, and this map takes left-invariant volume form on G/K to a volume form on $D\backslash G/K$. Now we can assume $D\backslash G/K$ is orientable (by changing D if necessary), and it is compact, so a volume form on it represents a non-zero element of $H^*(D\backslash G/K)$ in the top dimension, $p = \dim G - \dim K$. Now p is also the top dimension of $H^*(g, K)$, and $H^p(g, K)$ and $H^p(D\backslash G/K)$ are both $\mathbb{R}$, so $H^p(g, K) \to H^p(D\backslash G/K)$ is an isomorphism. But Poincaré duality holds on $H^*(g, K)$ and on $H^*(D\backslash G/K)$, i.e., multiplication $H^i \times H^{p-i} \to H^p = \mathbb{R}$ is a nondegenerate bilinear pairing. Furthermore, $H^*(g, K) \to H^*(D\backslash G/K)$ commutes with multiplication.

Exercise. This implies that $H^*(g, K) \to H^*(D\backslash G/K)$ is injective in every dimension. Q.E.D.

Consider now the complex of singular cochains on BG^δ which are continuous in the topology of BG (since $G^\delta \to G$ can be taken to be an isomorphism of sets, BG^δ can be taken to be BG with a finer topology). We define the cohomology of this complex to be the *continuous cohomology of* BG^δ.

Theorem 10.3. (Van Est, Bott, Haefliger). *$H^*(g, k)$ is isomorphic to the continuous cohomology of BG^δ.*

Proof. [9].

11. Lie algebra cohomology

Let L be a Lie algebra. That is L is a (for now, finite dimensional real) vector space with a skew symmetric bilinear bracket product:

$$L \times L \xrightarrow{[,]} L.$$

Skew symmetric means (for $X, Y \in L$) $[X,Y] = -[Y,X]$. We also require that $[,]$ satisfy the Jacobi identity:

$$[X,[Y,Z]] = [[X,Y],Z] + [Y,[X,Z]] \quad (X,Y,Z \in L).$$

Now for $X \in L$ we have a linear map $ad_X : L \to L$ defined by

$$ad_X(Y) = [X,Y].$$

Then the Jacobi identity says that ad_X is a derivation of L : $ad_X([Y,Z]) = [ad_X(Y),Z] + [Y, ad_X(Z)]$ for $X,Y,Z \in L$.

Exercise. $ad_{[X,Y]} = ad_X ad_Y - ad_Y ad_X$ for $X, Y \in L$.

Let $A^q(L)$ be the set of q-linear alternating forms on L. (i.e., $f \in A^q(L)$ is a multilinear map $\underbrace{L \times \cdots \times L}_{q-factors} \to \mathbb{R}$ such that for $X_1, \ldots, X_q \in L$, we have $f(X_1,\ldots,X_q) = -f(X_{\partial(1)},\ldots,X_{\partial(q)})$ whenever ∂ is a permutation of q elements which interchanges two of them.)

We can define a map $d\colon A^q(L) \to A^{q+1}(L)$ by $(d\varphi)(X_1,\ldots,X_{q+1}) = \Sigma_{i<j}(-1)^{i+j}\varphi([X_i,x_j],X_1,\ldots,\widehat{X}_i,\ldots,\widehat{X}_j,\ldots,X_{q+1})$ for $\varphi \in A^q(L)$ and $X_1,\ldots,X_{q+1} \in L$. If $\varphi \in A^\circ(L) \simeq \mathbb{R}$ then $d\varphi = 0$. Then d is an antiderivation since $d^2 = 0$ and

$$d(\eta \wedge \omega) = (d\eta) \wedge \omega + (-1)^{deg\,\eta} \wedge d\omega$$

for η, $\omega \in A^*(L) = \oplus_{q=0}^{\infty} A^q(L)$. Therefore, d is determined by its value on $A^1(L)$:

$$(d\varphi)(X,Y) = -\varphi([X,Y])$$

for $\varphi \in A^1(L)$, and $X, Y \in L$.

Example. If L is the Lie algebra of left invariant vector fields of a Lie group, G, then $A^*(L)$ is identified with the left invariant differential forms on the manifold of G. Our d on $A^*(L)$ is then induced by the usual DeRham d on the differential forms.

Definition. The cohomology of a Lie algebra L, $H^*(L)$, is defined as the cohomology of the sequence:

$$0 \xrightarrow{d} A^\circ(L) \xrightarrow{d} A^1(L) \xrightarrow{d} A^2(L) \longrightarrow \cdots$$

If L is the Lie algebra of a Lie group G then there is a natural map $H^*(L) \to H^*_{DR}(G)$ = DeRham cohomology of the manifold of G. This map need not be an isomorphism.

Example. Let $G = \mathbb{R}$ under addition, with Lie algebra $g \simeq \mathbb{R}$ generated by $\partial/\partial x$. Then

$$H^q(g) = \begin{cases} \mathbb{R} & \text{for} \quad q = 0 \quad \text{(generated by 1)} \\ \mathbb{R} & \text{for} \quad q = 1 \quad \text{(generated by dx)} \\ 0 & \text{for} \quad q \geq 2. \end{cases}$$

This is not isomorphic to the DeRham cohomology, $H^*_{DR}(G) = \mathbb{R}$ only in dimension 0. However, g is also the Lie algebra of $S^1 = \mathbb{R}/\mathbb{Z}$ and $H^q_{DR}(S^1) \simeq H^q(g)$.

Theorem 11.1. *If G is compact and connected, then $H^*(g) \to H^*_{DR}(G)$ is an isomorphism.*

The idea of the proof is to get an inverse by averaging (i.e., integrating) a q-form on G to obtain a left-invariant q-form (see Chevalley–Eilenberg [14]) of the same cohomology class in $H^*_{DR}(G)$.

But we still have the problem for an arbitrary Lie algebra, L, to make (functorially) a space Y such that $H^*(L) \simeq H^*(Y)$.

Homework. Compute $H^*(gl(2,\mathbb{R}))$ or $H^*(sl(2,\mathbb{R}))$. After spending some time trying to compute directly from the definition, the reader will gain some appreciation for general techniques as computation aids. (See Hochschild–Serre spectral sequence, for instance [39].)

12. Gel'fand–Fuks

Let M be a C^∞ manifold of dimension n, and let $L = L(M)$ be the Lie algebra of all C^∞ vector fields on M with the C^∞ topology. Let us recall what the C^∞ topology means for the set of maps $\mathbb{R}^n \to \mathbb{R}$. A basic neighborhood of $f : \mathbb{R}^n \to \mathbb{R}$ is $N(f, K, r, \epsilon)$, where $K \subseteq \mathbb{R}^n$ is compact, $r \geq 0$ is an integer, and $\epsilon > 0$. Then $g \in N(f, K, r, \epsilon)$ if and only if restricted to K, the values of g and its partial derivatives $\partial^s g/\partial x^{\alpha_1} \cdots \partial x^{\alpha_s}$ of order $s \leq r$ are within ϵ of those of f. Since, locally, vector fields on M are given by n-tuples of functions $\mathbb{R}^n \to \mathbb{R}$, we obtain a topology on $L(M)$ independent of charts. We can define $A^q(L)$ to be the q-linear alternating forms on L as before and consider the cohomology of the complex $A^\circ(L) \to A^1(L) \to$.

However, this object is too large to hope to compute. So instead of all q-forms we will consider only the continuous q-forms. (For finite dimensional Lie algebras linearity of the q-forms implies continuity.)

Definition. $A_c^q(L)$ = the *continuous* q-alternating forms on $L = L(M)$. $A_c(L) = \oplus_{q=0}^{\infty} A_c^q(L)$.

Definition. The Gel'fand–Fuks cohomology of M, $H_{GF}^*(M) = H^*(A_c(L(M)))$ = cohomology of the complex:

$$0 \longrightarrow A_c^{\circ}(L) \xrightarrow{d} A_c^1(L) \xrightarrow{d} A_c^2(L) \longrightarrow \cdots$$

Think of $L(M)$ as the Lie algebra of the group of diffeomorphisms of M, Diff (M). A curve $t \to g_t$ through the identity of Diff (M) gives rise to a family of curves $t \to g_t(m)$ through each point $m \in M$. So a "tangent vector" at $id \in$ Diff (M) corresponds to a vector field on M, which is an element of $L(M)$.

Gel'fand–Fuks stated the next theorem for compact manifolds [26]. However, the only property of compact manifolds which we will require is that M possess a covering by a *finite* collection of open sets such that these open sets and all finite intersections of them are all copies of $\mathbb{R}^n$ (or empty). Such a cover will be called "*good.*" Note that if M is compact then it can be covered by a finite number of geodesically convex open sets in some Riemannian metric. But there are other obvious examples, for instance $M = \mathbb{R}^n$.

Theorem 12.1. (Gel'fand–Fuks) *If M has a good covering then $H_{GF}^r(M)$ is finite dimensional for each r.*

Theorem 12.2. (Gel'fand–Fuks) $H_{GF}^*(S^1) = E(\omega) \otimes P(y)$, *the exterior algebra generated by a class ω of dimension 3 tensored with the polynomial algebra generated by a class y of dimension 2. Further, ω and y have the following representatives. Let $X, Y, Z \in L(M)$. Then $X = f(x)\partial/\partial x$, $Y = g(x)\partial/\partial(x)$, and $Z = h(x)\partial/\partial x$, where $f, g,$ and h are smooth periodic functions $\mathbb{R} \to \mathbb{R}$ (periodic so we get a map $S^1 = \mathbb{R}/\mathbb{Z} \to \mathbb{R}$). Then:*

$$\omega(W, Y, Z) = \det \begin{vmatrix} f & f' & f'' \\ g & g' & g'' \\ h & h' & h'' \end{vmatrix}_0 .$$

Gel'fand–Fuks had said to integrate this around the circle, but we can instead evaluate at (an arbitrary point) $0 \in S^1$ and get a cohomologous representative form. See [24], [26] 9.2.

$$y(X, Y) = \int_0^{2\pi} \begin{vmatrix} f' & f'' \\ g' & g'' \end{vmatrix}_x dx$$

We can not just evaluate this one at a point since we would not get a cocycle. ω and y are clearly continuous forms since the values $\omega(X, Y, Z)$ and $y(X, Y)$ do not change much whenever $f, f', f'', g, g', g'', h, h', h''$ do not change much.

By way of contrast, an ordinary differential form on M is also a function on vector fields, but with the values in $C^\infty(M)$ = smooth functions $M \to \mathbb{R}$. Our forms, $A_c^q(L)$, have values in $\mathbb{R}$.

The plan of this proof is to compute $H^*_{GF}(\mathbb{R}^n)$ first and then apply our semi-simplicial machinery to get at $H^*_{GF}(M)$.

13. Special Case: Gel'fand–Fuks of $\mathbb{R}^1$

A vector field $X \in L(\mathbb{R}^I)$ is of the form $f(x)\partial/\partial x$, $f : \mathbb{R} \to \mathbb{R}$ smooth. Then a 1-cochain is a distribution, that is, a continuous linear functional on $C^\infty(\mathbb{R}^1) =$ smooth maps $\mathbb{R}^1 \to \mathbb{R}$.

Consider now in $A^1_c L\mathbb{R}^1$ the subcomplex $A^1_{pt} L\mathbb{R}^1$, of distributions whose support is contained at the origin 0 of $\mathbb{R}^1$. In general a q-form $\varphi \in A^q_{pt} L\mathbb{R}^1$ has support at 0 in the sense that $\varphi(X_1, \ldots, X_q) = 0$ whenever the vector fields $X_1, \ldots, X_q$ vanish in a neighborhood of 0.

Example. The Dirac δ-function: If $X = f(x)\partial/\partial x$ then $\delta(X) = f(0)$. We also have the derivatives of δ. $\delta'(X) = -f'(0)$, $\delta''(X) = f''(0)$, and $\delta^{[k]}(X) = (-1)^k f^{[k]}(0)$. The conventional $(-1)^k$ sign is used to make our notation consistent with the following:

Any smooth function $\varphi : \mathbb{R} \to \mathbb{R}$ with compact support defines a distribution (for $f : \mathbb{R}^1 \to \mathbb{R}$)

$$f \longrightarrow \varphi(f) = \int_{-\infty}^{\infty} \varphi(x)\, f(x)\, dx.$$

Since φ is smooth, φ' is defined and is the distribution:

$$f \longrightarrow \varphi'(f) = \int_{-\infty}^{\infty} \varphi'(x)\, f(x)\, dx.$$

Then by integrating by parts:

$$\varphi'(f) = \varphi(x) f(x)|_{-\infty}^{\infty} - \int_{-\infty}^{\infty} \varphi(x) f'(x)\, dx = (0 - 0) - \varphi(f').$$

This is consistent with our notation for the derivative of the δ-function: $\delta'(f) = -\delta(f')$. Now I claim that $A^1_{pt} L(\mathbb{R}^1)$ is the vector space generated by the 1-forms $\{\delta, \delta', \delta'', \ldots\}$. And $A^*_{pt} L\mathbb{R}^1$ is the exterior algebra generated by $\{\delta, \delta', \delta'', \ldots\}$. To see this we must look at what "continuous in the C^∞ topology" means. Let $\varphi \in A^1_{pt} L\mathbb{R}^1$ and $0 : \mathbb{R} \to \mathbb{R}$ be identically 0. Then by linearity $\varphi(0) = 0$. By continuity, for all $\epsilon > 0$, there exists a basic neighborhood $N(0, K, r, \eta)$ of $0 \in C^\infty(\mathbb{R})$, such that $\varphi(N(0, K, r, \eta)) \subseteq (-\epsilon, \epsilon)$. Since φ has support at the origin, we can take K to be the origin of $\mathbb{R}$. r is a positive integer and $\eta > 0$. Suppose f and its first r derivatives vanish at the origin $=K$. Then $f \in N(0, K, r, \eta)$, so that $\varphi(f) \in (-\epsilon, \epsilon)$. In fact $\varphi(f) = 0$, for otherwise, for $\lambda \in \mathbb{R}$, λf would also vanish to r-th order at the origin. So $\lambda f \in N(0, K, r, \eta)$ and $\varphi(\lambda f) = \lambda\varphi(f)$ can be $> \epsilon$ for large λ. So φ *can only depend on a finite number of derivatives* of its arguments. As a corollary, φ cannot distinguish between "flat" functions. That is, if $f : \mathbb{R} \to \mathbb{R}$, $f(0) = 0$, $f^{[k]}(0) = 0$ for all k, then for every continuous linear functional $\varphi \in A^1_{pt}L\mathbb{R}^1$, $\varphi(f(x)\partial/\partial x) = 0$.

Computation. Let $X = f(x)\partial/\partial x$, $Y = g(x)\partial/\partial x$ in $L(\mathbb{R}^1)$.

$$\begin{aligned}(\delta' \wedge \delta'')(X,Y) &= \delta'(X)\delta''(Y) - \delta''(X)\delta'(Y)\\ &= -f'(0)g''(0) - f''(0)(-g'(0))\\ &= -\det\begin{vmatrix} f' & f'' \\ g' & g'' \end{vmatrix}_0 .\end{aligned}$$

What is d on $A^*_{pt}(\mathbb{R}^1)$? Let $X, Y \in L(\mathbb{R}^1)$ as above.

$$\begin{aligned}(d\delta)(X,Y) &= -\delta([X,Y]) = -\delta([f(x)\partial/\partial x,\ g(x)\partial/\partial x])\\ &= -\delta(f(x)g'(x)(\partial/\partial x) - g(x)f'(x)(\partial/\partial x))\\ &= -f(0)g'(0) + g(0)f'(0)\\ &= \delta \wedge \delta'(X,Y).\end{aligned}$$

Thus $d\delta = \delta \wedge \delta'$. If we extend the action of $'$ to be a derivation of $A^*_{pt}L\mathbb{R}^1$ then $'$ will commute with d (Verify! Cf. the next section in which $'$ is just the Lie derivative denoted by $\theta_{\partial/\partial x}$). Then we can compute:

$$\begin{aligned}d(\delta') &= (d\delta)' = (\delta \wedge \delta')' = \delta' \wedge \delta' + \delta \wedge \delta'' = \delta \wedge \delta''\\ d(\delta'') &= (d\delta')' = (\delta \wedge \delta'')' = \delta' \wedge \delta'' + \delta \wedge \delta''' \quad \text{etc.}\end{aligned}$$

Consider the 3-form $\omega = \delta \wedge \delta' \wedge \delta''$. Then ω is a cocycle since $d\omega = -d(\delta' \wedge d\delta') = -d\delta' \wedge d\delta' + \delta' \wedge d^2\delta' = 0$. For $X = f(x)\partial/\partial x$, $Y = g(x)\partial/\partial x$, and $Z = h(x)\partial/\partial x$ we have

$$\begin{aligned}\omega(X,Y,Z) &= \delta \wedge \delta' \wedge \delta''(f(x)\partial/\partial x, g(x)\partial/\partial x, h(x)\partial/\partial x)\\ &= \det\begin{vmatrix} f & f' & f'' \\ g & g' & g'' \\ h & h' & h'' \end{vmatrix}_0\end{aligned}$$

Proposition 13.1. *$H^*(A^*_{pt}L\mathbb{R}^1)$ is generated by the trivial class 1 in dimension 0 and by ω in dimension 3.*

Proof. We will defer the proof until Section 14. See [28].

Our plan will then be:

1) to compute $H^*(A_{pt}L\mathbb{R}^n)$, and
2) to show that the inclusion $A^*_{pt}L\mathbb{R}^n \to A^*_cL\mathbb{R}^n$ induces an isomorphism in cohomology. This corresponds to the Poincaré Lemma in DeRham theory: we will study the contraction of $\mathbb{R}^n$ given by $f_t : \mathbb{R}^n \to \mathbb{R}^n$, $x \to tx$, for $t \in \mathbb{R}$, $x \in \mathbb{R}^n$.

14. Techniques and basic formulae

Let g be a Lie algebra, $A^q(g) = q$-linear alternating forms on g. Recall that, if g is the Lie algebra of a Lie group G, then $A^q(g) \simeq \mathrm{Inv}_G A^q(G) =$ left invariant q-forms on the underlying manifold of G. For $g = L(M) = C^\infty$ vector fields on a manifold M, we can only take continuous forms.

Definition. For each $X \in g$ we define the *interior product*:

$$\iota_X : A^q(g) \to A^{q-1}(g) \quad \text{by}$$
$$(\iota_X\varphi)(X_1, \ldots, X_{q-1}) = \varphi(X, X_1, \ldots, X_{q-1})$$

for $\varphi \in A^q(g); X_1, \ldots, X_{q-1} \in g$. We have $\iota_X : A^\circ(g) \to 0$ and $\iota_X : A^1(g) \to \mathbb{R} = A^\circ(g)$ is a duality.

Exercise. ι_X is an anti-derivation. That is, $\iota_X^2 = 0$ and

$$\iota_X(\omega \wedge \eta) = (\iota_X\omega) \wedge \eta + (-1)^{deg\,\omega}\omega \wedge (\iota_X\eta).$$

Definition. For $X \in g$, we define the *Lie derivative* $\theta_X : A^q(g) \to A^q(g)$ by the relation

$$\theta_X = \iota_X d + d\iota_X.$$

Remark. θ_X is chain homotopic to 0 via ι_X so that θ_X is 0 in $H^*(g)$.

Exercise. Prove the following useful formulae: (for $X, Y \in g$)

$$d\theta_X = \theta_X d$$
$$\theta_X \quad \text{and} \quad \iota_X \quad \text{are linear in} \quad X$$
$$\theta_{[X,Y]} = \theta_X\theta_Y - \theta_Y\theta_X \equiv [\theta_X, \theta_Y]$$
$$\iota_{[X,Y]} = \theta_X\iota_Y - \iota_Y\theta_X \equiv [\theta_X, \iota_Y]$$
$$(\theta_X, \varphi)(Y) = -\varphi([X,Y]), \quad \text{for} \quad \varphi \in A^1(g)$$
$$\theta_X \quad \text{is a derivation of} \quad A^*(g).$$

Remark. Suppose θ_X splits up $A^*(g)$ into eigenspaces:

$$A^*(g) = \underset{\lambda}{\oplus} A_\lambda$$

where θ_X acts on A_λ by scalar multiplication by $\lambda \in \mathbb{R}$. Then $d(A_\lambda) \subseteq A_\lambda$ and $H^*(A(g)) = \oplus_\lambda H^*(A_\lambda)$. Furthermore, if $\lambda \neq 0$, $H^*(A_\lambda) = 0$ because $id = (1/\lambda)\theta_X = (1/\lambda)(\iota_X d + d\iota_X)$ is homotopic to 0 in cohomology. Therefore, $H^*(A_\circ) \simeq H^*(A(g))$, where the isomorphism is induced by the inclusion of the zero eigenspace $A_\circ$ into $A(g)$.

Using this principle, we can compute $H^*(A_{pt}(L\mathbb{R}^1))$, where $A_{pt}(L\mathbb{R}^1) =$ continuous forms on $L\mathbb{R}^1$ with support at the origin. Let $R = x(\partial/\partial x)$ be the radial vector field on $\mathbb{R}^1$. Then for $X = f(x)\partial/\partial x$ we have:

$$\begin{aligned}(\theta_R\delta)(X) = -\delta([R, X]) &= -\delta(xf'(x)(\partial/\partial x) - f(x)(\partial/\partial x) \\ &= f(0) = \delta(X).\end{aligned}$$

Therefore, $\theta_R\delta = \delta$.

Since δ' is just $\theta_{\partial/\partial x}(\delta)$, we have:

$$\begin{aligned}\theta_R\delta' = \theta_R\theta_{\partial/\partial x}\delta &= \theta_{\partial/\partial x}(\theta_R\delta) + \theta_{[x(\partial/\partial x),\partial/\partial x]}(\delta) \\ &= \theta_{\partial/\partial x}(\delta) + \theta_{-\partial/\partial x}(\delta) = 0.\end{aligned}$$

Inductively, we obtain:

$$\theta_R\delta^{[k]} = (1-k)\delta^{[k]},$$

which gives an eigenspace decomposition of $A^1_{pt}(L\mathbb{R}^1)$. The eigenvalue $(1-k)$ is called the *weight* of $\delta^{[k]}$. Since θ_R is a derivation, the weight of a product is the sum of the weights: $A_\lambda \cdot A_\mu \subseteq A_{\lambda+\mu}$. Here are some examples:

weight

	1	0	-1	-2	-3	-4	...
1-forms	$\{\delta$	δ'	δ''	δ'''	δ''''	...	
2-forms	$\{\delta\wedge\delta'$	$\delta\wedge\delta''$	$\delta\wedge\delta'''$	$\delta\wedge\delta''''$	...		
			$\delta'\wedge\delta'''$	$\delta\wedge\delta''''$	...		
3-forms	$\{$		...				
		$\delta\wedge\delta'\wedge\delta''$	...				

We note that the 0-eigenspace of $A^*_{pt}L\mathbb{R}^1$ is generated by $\{1, \delta', \delta\wedge\delta'', \delta\wedge\delta'\wedge\delta''\}$ as a vector space. Since d acts by $1 \to 0, \delta' \to \delta\wedge\delta''$, $\delta\wedge\delta'' \to 0$, $\delta\wedge\delta'\wedge\delta'' \to 0$, the cohomology, $H^*(A_\circ)$, of the 0-eigenspace (and, therefore, $= H^*(A_{pt}L\mathbb{R}'))$ is generated by 1 in dimension 0 and $\omega = \delta\wedge\delta'\wedge\delta''$ in dimension 3. This proves

15. Distributions again

Let us return to the larger problem of $A^*_c L\mathbb{R}^1$, the continuous forms (with aribitrary support) on $L\mathbb{R}^1$.

Exercise. $\varphi \in A^q_c L\mathbb{R}^1$ is continuous $\Rightarrow \varphi$ has compact support. In fact the values of φ only depend on a finite number of derivatives of its arguments on a compact set K of $\mathbb{R}^1$ (The *support* of φ is the intersection of all closed sets K such that for all $g : \mathbb{R} \to \mathbb{R}$ smooth, vanishing on a neighborhood of K we have $\varphi(g) = 0$. Use the K in the definition of the C^∞-topology.)

Theorem 15.1. (Schwartz [55]. Cf. Gel'fand [23].) *Every continuous linear functional* $\varphi : C^\infty(\mathbb{R}) \to \mathbb{R}$ *(where* $C^\infty(\mathbb{R})$ *= smooth maps* $\mathbb{R} \to \mathbb{R}$ *in the* C^∞ *topology) is a finite linear combination of terms of the following form:*

$$f \longrightarrow \int_{-\infty}^{\infty} \mu(x) f^{[k]}(x)\,dx$$

where $f^{[k]}(x)$ *is the k-th derivative of* f *and* $\mu(x)$ *is a continuous function* $\mathbb{R} \to \mathbb{R}$ *with compact support. In Schwartz's language, every distribution with compact support is the sum of derivatives of continuous distributions.*

Example. What about the Dirac δ? How can δ be put in the above form? Let $\mu : \mathbb{R} \to \mathbb{R}$ be a C^∞ except at the origin such that

$$\mu(x) = \begin{cases} 0 & \text{for} \quad x \le 0 \quad \text{or} \quad x \ge 3/4 \\ x & \text{for} \quad 0 \le x \le 1/4. \end{cases}$$

Then μ looks like:

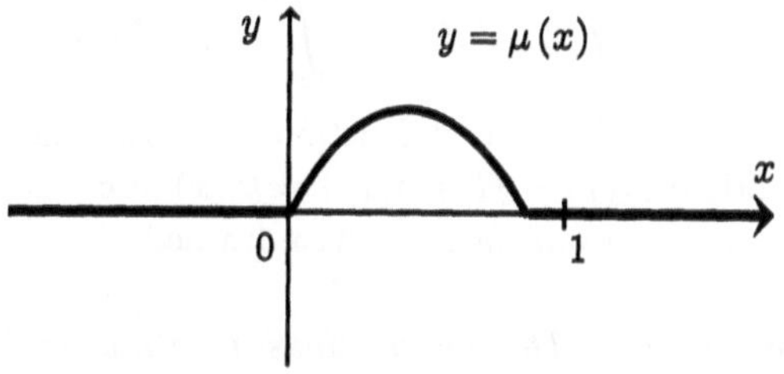

And μ' looks like:

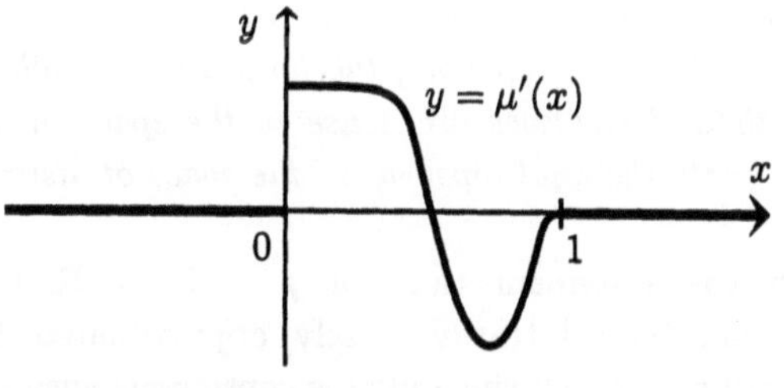

μ' is discontinuous at 0. And intuitively μ'' looks like the sum of a smooth function, ν, and a Dirac-δ spike at 0.

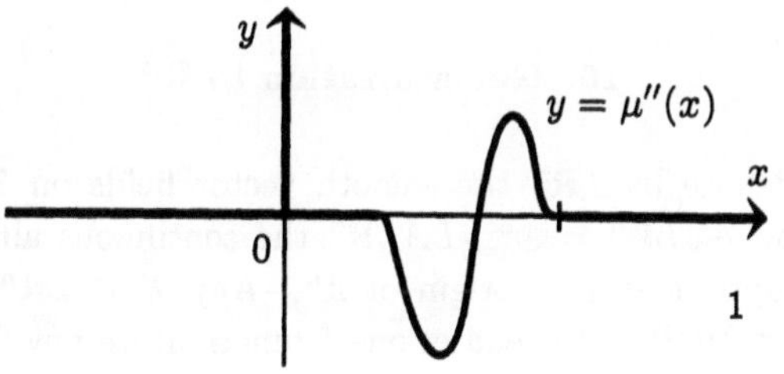

Although μ'' is not rigorously defined as a function, it is well-defined as a distribution.

$$\begin{aligned}
\mu''(f) = +\mu(f'') &= \int_{-\infty}^{\infty} \mu(x)f''(x)dx = \int_0^1 \mu(x)f''(x)dx \\
&= \mu(x)f'(x)|_0^1 - \left\{\int_0^1 \mu'(x)f'(x)dx\right\} \\
&= (0-0) - \left\{\mu'(x)\ f(x)|_0^1 - \int_0^1 \mu''(x)f(x)dx\right\} \\
&= -0 + f(0) + \int_{-\infty}^{\infty} \nu(x)f(x)dx
\end{aligned}$$

so that as distributions it makes sense to say $\mu'' = \nu + \delta$. We have shown:

$$\delta(f) = \int_{-\infty}^{\infty} \mu(x)f''(x)dx - \int_{-\infty}^{\infty} \nu(x)f(x)dx.$$

We can write this as

$$\delta = \int_{-\infty}^{\infty} \mu(x)\delta_x''dx - \int_{-\infty}^{\infty} \nu(x)\delta_x dx,$$

where δ_x and δ_x'' are just the Dirac-δ function and its second derivative, but at x instead of the origin $\delta_x(f) = f(x)$, $\delta_x''(f) = f''(x)$, etc.

We repeat some important facts we have learned:

Proposition 15.2. *The distributions (continuous linear functionals $C^\infty(\mathbb{R}) \to \mathbb{R}$) which have support at the origin are all given by finite linear combinations of the δ-function and its derivatives.*

Proposition 15.3. *Furthermore, the finite linear combinations of the δ_x for $x \in \mathbb{R}$ and their derivatives are dense in the space of distributions with compact support (with the dual topology on the space of distributions).*

This is just the statement that for $f : \mathbb{R} \to \mathbb{R}$, $\int \mu(x)\delta_x^{[k]}(f)dx = \int \mu(x)f^{[k]}(x)dx$ can be arbitrarily closely approximated by a finite sum $\sum_{i=1}^{N} \mu(x_i)f^{[k]}(x_i)(x_i - x_{i-1})$, since $\mu(x)$ is continuous with compact support. The dual topology is the one which makes all the evaluations: {distributions with compact support} $\to \mathbb{R}$ by $\varphi \to \varphi(f)$ continuous for each $f \in C^\infty(\mathbb{R} \to \mathbb{R})$.

16. Generalization to $\mathbb{R}^n$

Again we denote by $L\mathbb{R}^n$ the smooth vector fields on $\mathbb{R}^n$ with the C^∞ topology, and by $A_{pt}L\mathbb{R}^n = \oplus_{q=0}^{\infty} LA_{pt}^q\mathbb{R}^n$ the continuous alternating q-forms on $L\mathbb{R}^n$ with support at pt= origin of $\mathbb{R}^n$. Any $X \in L\mathbb{R}^n$ looks like $X = \sum_{i=1}^{n} f_i(x_i, \dots, x_n)\partial/\partial x_i$. Instead of one δ function, we now have n of them:

Definition. $\delta^i(X) = f_i(0,\cdots,0)$, for X as above, and for $i = 1,\ldots,n$. Instead of ordinary derivatives we now have n partial derivatives:

Definition.

$$\delta^i_j = \theta_{\partial/\partial x_j}(\delta^i)$$
$$\delta^i_{jk} = \theta_{\partial/\partial x_k}(\theta_{\partial/\partial x_j}\delta^i) = \theta_{\partial/\partial x_k}\delta^i_j$$
$$\delta^i_{jkl} = \theta_{\partial/\partial x_\ell}(\delta^i_{jk})$$

and so forth inductively, for $i, j, k, l, = 1,\ldots,n$.

Exercise. For $X \in L\mathbb{R}^1$ as above, $\delta^i_j(X) = -\partial f_i/\partial x_j$, $\delta^i_{jk}(X) = \partial^2 f_i/\partial x_j \partial x_k$, etc. Thus, $\delta^i_{j_1\cdots j_r}$ is symmetric in its lower indices.

In fact, $A_{pt}L\mathbb{R}^n$ is the exterior algebra generated by the 1-forms $\{\delta^i, \delta^i_j, \delta^i_{jk}, \cdots\}$ in $A^1_{pt}L\mathbb{R}^n$. The derivative d of this complex is computed as follows. Let $X = \sum_{\alpha=1}^n (f_\alpha)\partial/\partial x_\alpha$ and $Y = \sum_{\beta=1}^n (g_\beta)\partial/\partial x_\beta$ be in $L\mathbb{R}^n$. Then

$$\begin{aligned}
(d\delta^i)(X,Y) &= -\delta^i([X,Y]) \\
&= -\delta^i\left(\sum_{\alpha,\beta} f_\alpha(\partial g_\beta/\partial x_\alpha)\partial/\partial x_\beta - g_\beta(\partial f_\alpha/\partial x_\beta)\partial/\partial x_\alpha\right) \\
&= -\left(\sum_\alpha f_\alpha(\partial g_i/\partial x_\alpha) - \sum_\beta g_\beta(\partial f_i/\partial x_\beta)\right) \\
&= \sum_j (\delta^j \wedge \partial^i_j)(X,Y).
\end{aligned}$$

Therefore, $d\delta^i = \sum_j(\delta^j \wedge \delta^i_j)$. We will often omit the summation sign when summing over repeated indices.

$$(d\delta^i_k) = d\theta_{\partial/\partial x_k}\delta^i = \theta_{\partial/\partial x_k}(d\delta^i) = \theta_{\partial/\partial x_k}(\delta^j \wedge \delta^i_j) = \delta^j_k \wedge \delta^i_j + \delta^j \wedge \delta^i_{jk},$$

since $\theta_{\partial/\partial x_k}$ is a derivation. d is defined inductively on $A^1_{pt}L\mathbb{R}^n$, and extends to all of $A^*_{pt}L\mathbb{R}^n = \Lambda^* A^1_{pt}L\mathbb{R}^n$ since d is an anti derivation. For the geometers, notice that $d\delta^i_k - \delta^i_k \wedge \delta^i_j$ looks like a curvature formula.

Question. Is $H^*(A_{pt}L\mathbb{R}^n)$ finite dimensional? Consider the radial vector field $R = \sum_i x_i(\partial/\partial x_i)$. Then just as in the $n = 1$ case we have $\theta_R\delta^i = \delta^i, \theta_R\delta_j = 0, \theta_R\delta^i_{jk} = -\delta^i_{jk}, \theta_R\delta^i_{j_1\cdots j_r} = (1-r)\delta^i_{j_1\cdots j_r}$. Then on $A^*_{pt}L\mathbb{R}^n$, θ_R acts with eigenvalues $n, n-1,\ldots,0,-1,\ldots$. For example, $\delta^1 \wedge \delta^2 \wedge \cdots \wedge \delta^n$ has weight n. Each eigenspace A_λ of $A_{pt}L\mathbb{R}^n$ is finite dimensional and stable under d. $H^*(A_{pt}L\mathbb{R}^n) = H^*(\oplus_{\lambda=-\infty}^n A_\lambda) = \oplus_\lambda H^*(A_\lambda) = H^*(A_\circ)$ is finite dimensional since θ_R is multiplication by λ on A_λ and $\theta_R = \iota_R d + d\iota_R$ is homotopic to 0. For $\lambda \neq 0$, therefore, $H^*(A_\lambda) = 0$. Therefore,

Proposition 16.1. *$H^*(A_{pt}L\mathbb{R}^n)$ is finite dimensional.*

17. Digression: Tautological forms

Consider the frame bundle $F(M) \xrightarrow{\pi} M$ with fiber $GL(n,\mathbb{R})$ and structure group also $GL(n,\mathbb{R})$. This is the principal bundle associated to the tangent bundle of a manifold M. A point of $F(M)$ looks like $(x; v_1, \ldots, v_n)$ where $x \in M$ and $v_1, \ldots, v_n$ form a basis for the tangent space $T_x M$ at x. π is projection to $x \in M$.

There are some canonical 1-forms $\omega^1, \ldots, \omega^n$ on $F(M)$ defined as follows: Let X be a tangent vector at $(x; v_1, \ldots, v_n) \in F(M)$. Then $\pi_* X$ is a tangent vector at $x \in M$ and can be written in the basis $v_1, \ldots, v_n$. Then $\omega^i(X)$ is defined by:

$$\pi_*(X) = \omega^1(X)v^1 + \cdots + \omega^n(X)v^n.$$

Second approach. Consider the charts $f : \mathbb{R}^n \to M$. For $k = 0, 1, 2, \cdots$ we can identify two such maps f and g whenever

$$f(0) = g(0), \frac{\partial f}{\partial x_i}\Big|_0 = \frac{\partial g}{\partial x_i}\Big|_0, \ldots, \frac{\partial^k f}{\partial x_{i_1} \cdots \partial x_{i_k}}\Big|_0 = \frac{\partial^k g}{\partial x_{i_1} \cdots \partial x_{i_k}}\Big|_0$$

where the requirements are well-defined via any chart on M. The equivalence class of f is called the *k-jet* of f, $j^k f$. We denote by $F_k(M)$ the collection of k-jects of charts $f : \mathbb{R}^n \to M$.

The set, $F_0(M)$, of 0-jets is just M via the target $= f(0)$. The set, $F_1(M)$, of 1-jets is just the frame bundle $F(M)$. We get a tower of fiber bundles:

$$\begin{array}{c} \vdots \\ \downarrow \\ F_3(M) \\ \downarrow \\ F_2(M) \\ \downarrow \\ F_1(M) = F(M) \\ \downarrow \pi \\ M = F_0(M) \end{array}$$

On $F_2(M)$ we have the tautologous forms ω^i which pull back from $F_1(M)$. But we also have others: $\omega^i_j = j$-th partial of ω^i. This construction may be a little clearer in coordinates. Locally a map $g : \mathbb{R}^n \to M$ looks like $g : \mathbb{R}^n \to \mathbb{R}^n$ by

$$(x_1, \ldots, x_n) \to (g_1(x_1, \ldots, x_n), \ldots, g_n(x_1, \ldots, x_n))$$

Then

$$j^0 g = (g_1(0), \ldots, g_n(0)) = \text{target}$$

$$j^1 g = \left(g_i(0); \frac{\partial g_i}{\partial x_j}\Big|_0 \right)_{i,j}$$

$$j^2 g = \left(g_i(0); \frac{\partial g_i}{\partial x_j}\Big|_0; \frac{\partial^2 gi}{\partial x_j \partial x_k}\Big|_0 \right)_{i,j,k} \quad \text{etc.}$$

Let $f : \mathbb{R}^n \to M$ be a chart. Let X be a tangent vector to $j^1 f \in F_1(M)$. X is represented by a curve $t \to j^1 f_t$ with $f_0 = f$, and f_t a family of charts. Differentiating formally,

$$\left.\frac{\partial f_t}{\partial t}\right|_{t=0} = \frac{\partial x_i}{\partial t} \cdot \frac{\partial f_0}{\partial x_i}.$$

Then $\omega^i(X) \equiv \partial x^i/\partial t = i$-th component of $\pi_* X$ (which is represented by the curve $t \to f_t(0)$) when we are at the frame $(f_1(0), \ldots, f_n(0), \partial f(0)/\partial x_1, \ldots, \partial f(0)/\partial x_n)$.

On $F_2(M)$, f_t has well-defined derivatives of order 2, so that $\partial x_i/\partial t$ has 1st order partials. We define ω^i_j on $F_2(M)$:

$$\omega^i_j(X) = -\frac{\partial}{\partial x_j}\left(\frac{\partial x^i}{\partial t}\right).$$

On $F_3(M)$, $\omega^i_{jk}(X) = \partial^2(\partial x^i/\partial t)/(\partial x_j \partial x_k)$ is defined, and so forth. The reader may have noticed that we have differentiated $\partial f_t/\partial t = \sum(\partial x_i/\partial t)(\partial f/\partial x_i)$ formally, and $(\partial x_i/\partial t)$ does not really make sense until we interpret x as a function of t. However, $f_t(x_0) = f(x(t, x_0))$ will hold if we set $x(t, x_0) = f^{-1} \circ f_t(x_0)$, a curve through each $x_0 \in \mathbb{R}^n$ in a neighborhood of the origin. $x(0, x_0) = x_0$ since $f_0 = f$. Then, the expressions $\partial x_i/\partial t$, $(\partial/\partial(x_0)_j)(\partial x_i/\partial t)$, etc. make sense. We have a commutative diagram:

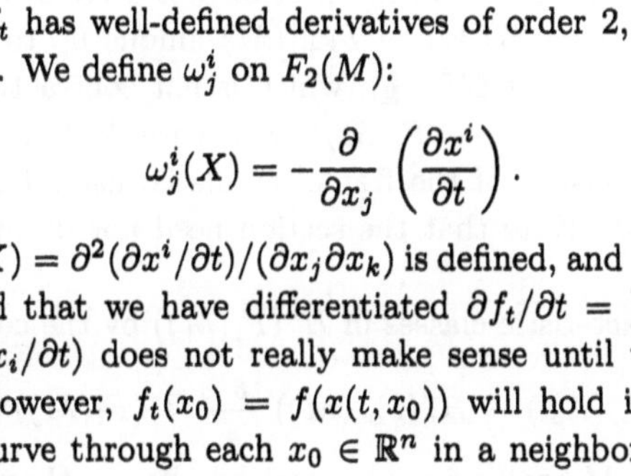

where φ is given by $\varphi_t = f^{-1} \circ f_t : \mathbb{R}^n \to \mathbb{R}^n$. $t \to \varphi_t$ is a curve through the identity of Diff $(\mathbb{R}^n)$ whose derivative (at $t = 0$) is a vector field

$$Y = (\partial x_1/\partial t)_{t=0}\, \partial/\partial x_1 + \cdots + (\partial x_n/\partial t)_{t=0}\, \partial/\partial x_n$$

on $\mathbb{R}^n$ written in our coordinates $x_1, \ldots, x_n$ of $\mathbb{R}^n$; $x(t, x_0) = \varphi_t(x_0) = f^{-1} \circ f_t(x_0)$.

If we scrap the abusive (but common) notations for partial derivatives, we can define our tautological forms more rigorously. For our tangent vector X on $F_k(M)$ represented by $t \to j^k f_t$, $\varphi_t = f^{-1} \circ f_t$ is defined up to its k-jet at the origin. After differentiating, Y is defined only up to its $k-1$ jet at $0 \in \mathbb{R}^n$. Thus, $\delta^i(Y), \delta^i_j(Y), \ldots, \delta_{j_1 \cdots j_{k-1}}(Y)$ are defined (where $\delta^i, \delta^i_j, \ldots \in A^1_{pt} L\mathbb{R}^n$ as before). For example, $\delta^i(Y) = \delta^i(\sum_j (\partial x_j/\partial t)\partial/\partial x_j) = \partial x_i/\partial t$.

Definition. $\omega^i(X) \equiv \delta^i(Y), \omega^i_j(X) \equiv \delta^i_j(Y), \ldots$, so that $\omega^i, \omega^i_j, \omega^i_{j_1, \cdots, j_k - 1}$ are well-defined one-forms on $F_k(M)$.

Thus we have a natural map $\delta^i \to \omega^i, \delta^i_j \to \omega^i_j, \delta^i_{jk} \to \omega^i_{jk}, \ldots$ of $A^*_{pt}L\mathbb{R}^n$-tautological forms on $F_\infty(M)$, where $F_\infty(M) = \varprojlim_k F_k(M)$, and $A^*F_\infty(M) = \varinjlim_k A^*F_k(M)$ (here A^* means "differential forms on").

18. Sections of the frame tower and characteristic classes

The fibers of $F_{k+1}(M) \to F_k(M)$ are all affine spaces for $k \geq 1$ (since invertibility is only a condition on the 1-jet or first derivatives). So a partition of unity gives us sections $F_k(M) \to F_{k+1}(M)$ unique up to homotopy. Since the fiber of $F_1(M) \to M$ is $GL_{(n,\mathbb{R})}$, which is not contractible, we do not in general get a section $\partial : M \to F_1(M)$. When we do have one, M is called parallelizable, for a section of the frame bundle is just a trivialization of the tangent bundle of M. Note that the section need not be unique, even up to homotopy.

We obtain characteristic classes in $H^*(F_1(M))$ by the composition:

$$H^*(A_{pt}) \to H^*(F_\infty(M)) \xrightarrow{\partial^*} H^*(F_1(M))$$

However, only when M has a trivial tangent bundle can these classes be pulled down to $H^*(M)$ where we want them. Unless we modify our approach we will not get very interesting classes on M.

Remark. $H^*(A^*_{pt}L\mathbb{R}^n)$ can be described slightly differently. Let $\mathfrak{a}_n$ be the Lie algebra of formal-power-series vector fields on $\mathbb{R}^n$. Any $X \in \mathfrak{a}_n$ is of the form $X = \sum f_i \frac{\partial}{\partial x_i}$ where $f_i \in \mathbb{R}[[x_1, \ldots, x_n]]$ is a formal power series in $x_1, \ldots, x_n$. (No convergence is required.) Under $[\,,\,]$ $\mathfrak{a}_n$ is a topological Lie algebra. The topology is defined by stipulating, for instance, that $x_1^{200}\frac{\partial}{\partial x_1}$ be small. More formally, $\mathbb{R}[[x_1, \ldots, x_n]]$ has a topology relative to the ideal $m = (x_1, \ldots, x_n)$. The basic neighborhoods of 0 are the powers of m.

Exercise. The algebra of continuous forms, $A^*_c(\mathfrak{a}_n)$, on $\mathfrak{a}_n$ is just our friend $A^*_{pt}L\mathbb{R}^n$.

An $\mathbb{R}$-vector space basis for $\mathfrak{a}_n$ is $\{\partial/\partial x_i, (x_i)\partial/\partial x_j, (x_i, x_j)\partial/\partial x_k, \ldots\}_{i,j,\ldots}$ Note that $gl(n, \mathbb{R})$ sits in $\mathfrak{a}_n$ generated by the $\{(x_i)\partial/\partial x_j\}_{i,j}$ part.

Exercise. Verify that the map of a matrix $(\alpha_i^j) \in gl_{n,\mathbb{R}}$ to $\sum \alpha_i^j(x_i\partial/\partial x_j) \in \mathfrak{a}_n$ is a Lie algebra isomorphism onto the $\{(x_i)\partial/\partial x_j\}_{i,j}$ subspace. Furthermore, the skew symmetric matrices $so(n, \mathbb{R}) \subseteq gl(n, \mathbb{R})$ are mapped onto the subspace generated by $\{(x_i)\partial/\partial x_j - (x_j)\partial/\partial x_i\}_{i,j}$.

We can, therefore, speak of the relative cohomology $H^*(\mathfrak{a}_n, so(n))$ as the cohomology of $A^0_c(\mathfrak{a}_n, so_n) \xrightarrow{d} A^1_i(\mathfrak{a}_n, so_n) \xrightarrow{d} \cdots$ where

$$A^q_c(\mathfrak{a}_n, so(n)) = \{\varphi \in A^q_c(\mathfrak{a}_n) | \iota_x\varphi = 0 = \theta_X\varphi \quad \text{for all} \quad X \in so_n\}.$$

Then we have a natural map:

$$H^*(\mathfrak{a}_n, so(n)) \longrightarrow H^*(F_\infty(M)/SO(n))$$

where $F_\infty(M)/SO(n)$ is obtained from $F_\infty(M)$ by identifying points which are on the same orbit of the right action of $SO(n)$. (If $f : \mathbb{R}^n \to M$ represents a point in $F_\infty(M)$ and if $\alpha \in SO(n)$ then $f \circ \alpha : \mathbb{R}^n \to M$ represents another point of $F_\infty(M)$.) Note that the forms on $F_\infty(M)/SO(n)$ correspond to the differential forms on $F_\infty(M)$ which are zero when applied to vectors tangent to the fiber $\simeq SO(n)$ of $F_\infty(M) \to SO(n)$, and are also invariant under the right action of $SO(n)$. Since $SO(n)$ is connected, invariance (being fixed) by $SO(n)$ is equivalent to invariance (being killed) by $so(n)$. At least when M is oriented, $F_\infty(M)/S(n) \to M$ has a section, and up to homotopy, there are exactly two sections, one for each orientation. This is because we can reduce the structure group of $F_1(M) \to M$ to be $GL(n,\mathbb{R})^+$ = matrices with positive determinant. Then the fiber of $F_1(M) \to M$ is $GL(n,\mathbb{R})^+/SO(n)$ is contractible because $SO(n)$ is a maximal compact subgroup of $GL(n,\mathbb{R})^+$. We get characteristic classes on M by the composition $H^*(\mathfrak{a}_n, so_n) \to H^*(F_\infty(M)/SO(n)) \xrightarrow{\partial^*} H^*(M)$.

Remark. If M is not orientable, then we must replace $SO(n)$ with $O(n)$ in the above discussion. $F_1(M)/O(n) \to M$ always has a section since fiber is $GL(n,\mathbb{R})/O(n)$. This is contractible since $O(n)$ is a maximal compact subgroup of $GL(n,\mathbb{R})$. $H^*(\mathfrak{a}_n, O(n))$ is defined as the cohomology of

$$A_c^0(\mathfrak{a}_n, O(n)) \xrightarrow{d} A_c^1(\mathfrak{a}_n, O(n)) \xrightarrow{d} \cdots .$$

Think of $\mathfrak{a}_n$ as the Lie algebra of the pseudogroup G of infinite jets of maps $\mathbb{R}^n \to \mathbb{R}^n$. Then $A_c^q(\mathfrak{a}_n)$ corresponds to left invariant (continuous) differential q-forms on G. The subalgebra, $so(n)$ of $\mathfrak{a}_n$ corresponds to the subgroup $SO(n)$ or $O(n)$ of G. (See Haefliger [36] for details.) Then $A_c^q(\mathfrak{a}_n, O(n)) =$ {left invariant q-forms on $G/O(n)$} $\simeq$ left invariant q-forms on G which are zero in the $O(n)$ directions and invariant under the right actions of $O(n)$ on G. We define $A_c^q(\mathfrak{a}_n, SO(n))$ similarly. Since $SO(n)$ is connected, $A_c^q(\mathfrak{a}_n, SO(n)) = A_c^q(\mathfrak{a}_n, so(n))$ as defined earlier. In fact, we will see that $H^*(\mathfrak{a}_n, so_n)$ contains the Pontryagin classes up to dimension n and the Euler class (if n is even). But there are also more classes in $H^*(\mathfrak{a}_n, so_n)$ above dimension $n = \dim M$, which must, therefore, go to 0 in $H^*(M)$. However, these will show up non-trivially for foliations. The same holds for $H^*(\mathfrak{a}_n, O_{(n)})$ except that there is, of course, no Euler class. These characteristic classes come to $H^*(M)$ entirely by the way M is glued together by the charts $\mathbb{R}^n \to M$. We will return to this point later when we consider foliations of M.

19. Return to $H^*(A_{pt}L\mathbb{R}^n)$

Recall that $A_{pt}L\mathbb{R}^n = \oplus_{q=0}^{\infty} A_{pt}^q L\mathbb{R}^n$ = continuous alternating q-forms with support at the origin of $\mathbb{R}^n$, on the vector fields on $\mathbb{R}^n$. This is the same as the continuous q-forms on the Lie algebra $\mathfrak{a}_n$ of formal vector fields on $\mathbb{R}^n$. So we identify $A_{pt}(L\mathbb{R}^n)$ with $A_c^*(\mathfrak{a}_n)$ and $H^*(A_{pt}L\mathbb{R}^n)$ with $H^*(\mathfrak{a}_n)$. Let $V = \mathbb{R}^n$. A basis for the dual, V^*, is the coordinate functions $\{x_1, \ldots, x_n\}$. Identifying V with its tangent space at the origin, $\{\partial/\partial x_1, \ldots, \partial/\partial x_n\}$ is a basis for V. Then $\mathfrak{a}_n$ is the direct product

$$\mathfrak{a}_n = \prod_{p=-1}^{\infty} S^{p+1}(V^*) \otimes V,$$

where $S^{p+1}(V^*)$ = polynomials of degree $p+1$ in $x_1, \ldots, x_n$, and the direct product (instead of sum) reflects the formal power series in $\mathfrak{a}_n$. $\mathfrak{a}_n$ is the infinite jets (inverse limit of k-jets) at the origin of vector fields on $\mathbb{R}^n$. Note that $gl(n,\mathbb{R})$ is a subalgebra of $\mathfrak{a}_n$ (as the $p = 0$ factor, $V^* \otimes V$). Thus, we regard $\mathfrak{a}_n$ as the $gl(n,\mathbb{R})$-module.

Let $R = \sum (x_i)\partial/\partial x_i$, the radial vector field, which generates the center, Z, of $gl(n,\mathbb{R})$. $\mathfrak{a}_n$ breaks up under the action of R into (finite dimensional) eigenspaces $S^{p+1}(V^*) \otimes V$ with eigenvalue, or *weight*, p. For example, for $\partial/\partial x_1 \in S^{-1+1}(V^*) \otimes V = V, [R, \partial/\partial x_1] = [\sum(x_i)\partial/\partial x_i, \partial/\partial x_1] = -\partial/\partial x_1$ with eigenvalue $p = -1$. Since $R \in Z$, each eigenspace is preserved by $gl(n,\mathbb{R})$. Since $gl(n,\mathbb{R}) = sl(n,\mathbb{R}) \oplus Z$, where $sl(n,\mathbb{R})$ is semi-simple, we see that $\mathfrak{a}_n$ is completely reducible as $gl(n,\mathbb{R})$-module. Therefore, $gl(n,\mathbb{R})$ is called a *reductive subalgebra* of $\mathfrak{a}_n$. (See Appendix. See also [39].) We use the Hochschild–Serre spectral sequence relative to a reductive subalgebra. The spectral sequence converges to $H^*(\mathfrak{a}_n)$.

Consider the 1-forms δ_j^i the 2-forms Ω_j^i defined by

$$\Omega_j^i = d\delta_j^i - \sum_k \delta_j^k \wedge \delta_k^i = \delta^k \wedge \delta_{jk}^i.$$

(Recall that $\delta^i, \delta_j^i, \delta_{jk}^i, \ldots \in A^1(\mathfrak{a}_n)$ are dual to $\partial/\partial x_i, -(x_j)\partial/\partial x_i,$ $(x_j x_k)\partial/\partial_{x_1,\ldots} \in \mathfrak{a}_n$.) Let $\underline{W} \subseteq A_{pt}^* L\mathbb{R}^n$ be the subalgebra they generate. Then $d(\underline{W}) \subseteq \underline{W}$ and the inclusion $\underline{W} \to A_{pt}^* L\mathbb{R}^n$ induces the isomorphism in cohomology:

Theorem 19.1.

$$H^*(A_{pt}L\mathbb{R}^n) \simeq H^*(W).$$

Proof. Denote the continuous $\mathbb{R}$-valued q-forms on $\mathfrak{a}_n$ by $C^q(\mathfrak{a}_n;\mathbb{R}) = C^q(\mathfrak{a}_n)$. $C^*(\mathfrak{a}_n;\mathbb{R}) = C^*(\mathfrak{a}_n) = \oplus C^q(\mathfrak{a}_n)$. We filter $C^*(\mathfrak{a}_n)$ by the subspaces:

$$A^p = \oplus_{q=-\infty}^{\infty} A^{p,q}$$

where

$$A^{p,q} = \begin{cases} 0 \quad if \quad q<0 \\ C^{p+q}(\mathfrak{a}_n) \text{ if} p \leq 0 \\ \{\omega \in C^{p+q}(\mathfrak{a}_n) | \omega(X_1, \ldots, X_{p+q}) = 0 \\ \text{whenever at least } \quad q+1 \text{ of the} \\ X_i \quad \text{are in} \quad gl(n, \mathbb{R}) \end{cases}$$

The corresponding spectral sequence begins with

$$E_0^{p,q} = C^q(gl(n, \mathbb{R}); A^{p,0})$$
$$E_1^{p,q} = H^q(gl(n, \mathbb{R}); A^{p,0})$$

where $A^{p,0}$ is the coefficient module. (See appendix for definition of cohomology of a Lie algebra with coefficients in a module.) Since $A^{p,0}$ are the p-cochains which vanish when any one of their arguments is in the subalgebra $gl(n, \mathbb{R})$, $A^{p,0} = C^p(\mathfrak{a}_n, gl(n, \mathbb{R}); \mathbb{R}) \equiv$ cochains "relative to $gl(n, \mathbb{R})$."

For convenience, we define:

$$g^{(p)} = S^{p+1}V^* \otimes V \subseteq \mathfrak{a}_n$$
$$g_{(p)} = g^{(p)*} = S^{p+1}V \otimes V^* \subseteq C^1(\mathfrak{a}_n)$$

for $p = -1, 0, 1, \ldots$. The radial vector field $R \in gl(n, \mathbb{R})$ acts on $g^{(p)}$ and $g_{(p)}$ with eigenvalue p and $-p$ respectively. $g^{(0)} = gl(n, \mathbb{R})$.

$$C^1(\mathfrak{a}_n) = \oplus_{p=-1}^{\infty} g_{(p)}$$
$$C^p(\mathfrak{a}_n) = \Lambda^p C^1(\mathfrak{a}_n) \quad \text{is a direct sum of subspaces}$$
$$= \oplus_{p_{-1}+p_0+\cdots=p} \Lambda^{p-1} g_{(-1)} \otimes \Lambda^{p_0} g_{(0)} \otimes \Lambda^{p_1} g_{(1)} \otimes \cdots.$$

The *weight* (eigenvalue for R) of each of these subspaces is $p_{-1} - p_1 - 2p_2 - 3p_3 - \cdots$. $A^{p,0}$ is the subspace of $C^p(\mathfrak{a}_n)$ for which $p_0 = 0$, i.e., no factor of $g_{(0)} = gl(n, \mathbb{R})^*$. Since the coefficient module, $A^{p,0}$, of $E_1^{p,q} = H^q(gl(n, \mathbb{R}); A^{p,0})$ breaks up into eigenspaces for R, we can replace the coefficient module in the E_1 term by the submodule of $gl(n, \mathbb{R})$-invariants. (See Appendix. Also [39].)

$$E_1^{p,q} \simeq H^q(gl(n, \mathbb{R}), \ C^p(\mathfrak{a}_n, gl(n, \mathbb{R})))$$
$$= H^q(gl(n, \mathbb{R}), \ C^p(\mathfrak{a}_n, gl(n, \mathbb{R}))^{gl(n, \mathbb{R})}).$$

$C^p(\mathfrak{a}_n, gl(n, \mathbb{R}))^{gl(n, \mathbb{R})} \equiv \{\omega \in C^p(\mathfrak{a}_n, gl(n, \mathbb{R})) | X \cdot \omega = 0 \text{ for all } x + gl(n, \mathbb{R})\}$ is computed from scratch in Godbillon [30]. However, we will apply a theorem on the invariant theory of $gl(n, \mathbb{R})$. (See Atiyah et al. [2]; Weyl [72]). What are the invariants of the $gl(n, \mathbb{R})$-module,

$$\underbrace{V \otimes \cdots \otimes V}_{r \ times} \otimes \underbrace{V^* \otimes \cdots \otimes V^*}_{s \ times}?$$

(See Appendix for definition of the $gl(n, \mathbb{R})$ action on V^* and tensor products.) There are none (except 0) if $r \neq s$ (apply R = identity matrix). So suppose

$r = s$. Let $\{v_i\}$ be a basis for V and $\{v^i\}$, the dual basis for V^*. Then the invariants are spanned as a vector space by

$$\sum_{i_1,\cdots,i_r}^{n} v_{i_1} \otimes \cdots \otimes v_{i_r} \otimes v^{i_{\partial(1)}} \otimes \cdots \otimes v^{i_{\partial(1)}} \otimes \cdots \otimes v^{i_{\partial(r)}},$$

for each permutation, ∂, of $1, \ldots, r$.

Exercise. Show that the elements $\sum_{i,j} v_i \otimes v_j \otimes v^i \otimes v^j$ and $\sum_{i,j} v_i \otimes v_j \otimes v^j \otimes v^i$ of $V \otimes V \otimes V^* \otimes V^*$ are invariant (i.e., killed) under the action of $gl(n, \mathbb{R})$.

Recall that

$$A^{p,0} = c^p(\mathfrak{a}_n, gl(n,\mathbb{R})) = \oplus_{p_{-1}+p_1+p_2+\cdots=p} \Lambda^{p-1} V^* \otimes \Lambda^{p_1}(S^2 V \otimes V^*) \otimes \Lambda^{p_2}(S^3 V \otimes V^*) \otimes \cdots$$

where no factor appears from $\Lambda^* S^1 V \otimes V^* = C^*(gl(n,\mathbb{R})) \subseteq C^*(\mathfrak{a}_n)$. Each summand of $A^{p,0}$ is a sub $gl(n,\mathbb{R})$-module of

$$\underbrace{V \otimes \cdots \otimes V}_{(2p_1 + 3p_2 + \cdots \text{ factors})} \otimes \underbrace{V^* \otimes \cdots \otimes V^*}_{(p_{-1} + p_1 + p_2 + \cdots \text{ factors})}.$$

(Verify that the action via θ of $gl(n,\mathbb{R}) \subseteq \mathfrak{a}_n$ on $A^{P,0} \subseteq C^p(\mathfrak{a}_n)$ is the same as the usual action of $gl(n,\mathbb{R})$ on $V \otimes \cdots \otimes V \otimes V^* \otimes \cdots \otimes V^*$.) In particular, for there to be any invariants under the action of $gl(n,\mathbb{R})$ we must have

$$2p_1 + 3p_2 + \cdots = p_{-1} + p_1 + p_2 + \cdots.$$

This is just a restatement of the requirement that the weight $= p_{-1} - p_1 - 2p_2 - \cdots$ be zero. We get only invariants of the following form (where $\{\delta^i\}$ is a basis of V^*, $\{\delta^i_{jk}\}$ is a basis of $S^2 V \otimes V^* \subseteq V \otimes V \otimes V^*$, etc. and all indices must be in pairs, one upper and one lower for summation):

$$\delta^a \wedge \delta^i_{ai} = \Omega^i_i = \text{trace}\,(\Omega)$$
$$\delta^a \wedge \delta^i_{ai} \wedge \delta^b \wedge \delta^j_{bj} = \Omega^i_i \Omega^j_j = (tr(\Omega))^2$$
$$\delta^a \wedge \delta^i_{aj} \wedge \delta^b \wedge \delta^j_{bi} = \Omega^i_j \Omega^j_i = tr(\Omega^2)$$

etc. (Summation over repeated indices)

We get only the (invariant) polynomials in the $\Omega^i_j = \delta^a \wedge \delta^i_{aj}$. In fact, we recognize the traces of the powers of the matrix Ω, whose entries are Ω^i_j. To prove that these are the *only* invariants, we play off the symmetry in the lower indices of $\delta^i_{jk}, \delta^i_{jkl}, \ldots$ against the anti-symmetry of the wedge product, $\wedge$. For example interchanging the dummy variables a and b in the following possible invariant:

$$\begin{aligned} \delta^a \wedge \delta^b \wedge \delta^i_{ab} \wedge \delta^j_{ij} &= (\delta^b \wedge \delta^a) \wedge \delta^i_{ba} \wedge \delta^j_{ij} \\ &= -(\delta^a \wedge \delta^b) \wedge \delta^i_{ba} \wedge \delta^j_{ij} \\ &= -\delta^a \wedge \delta^b \wedge \delta^i_{ab} \wedge \delta^j_{ij} \end{aligned}$$

So both sides are 0. Similarly, (plus the pigeon-hole principle) we find there can be no factors of δ^i_{jkl} , etc. We are only left with polynomials in the Ω^i_j. We speak of polynomials since $\{\Omega^i_j\}$ have degree 2, and, therefore, commute.

$\underline{W}$ is generated by the δ^i_j's in $C^*(gl(n,\mathbb{R}))$ and the Ω^i_j's in $C^*(\mathfrak{a}_n, gl(n,\mathbb{R}))$. The filtration on $C^*(\mathfrak{a}_n)$ induces one on W with E_0 term:

$$E_0^{p,q} = C^q(gl(n,\mathbb{R}), P^{p/2}(\Omega^i_j)),$$

where $P^{p/2}(\Omega^i_j) = 0$ if p is odd. If p is even, $P^{p/2}(\Omega^i_j)$ is the homogeneous polynomials in the Ω^i_j's of degree $p/2$ as a polynomial (degree p as a form in $C^*(\mathfrak{a}_n)$). $P^{p/2}(\Omega^i_j) \subseteq C^p(\mathfrak{a}_n, gl(n,\mathbb{R}))$. The E_1 term is

$$\begin{aligned} E_1^{p,q} &= H^q(gl(n,\mathbb{R}),\ P^{p/2}(\Omega^i_j)) \\ &= H^q(gl(n,\mathbb{R}),\ P^{p/2}(\Omega^i_j)^{gl(n,\mathbb{R})}). \end{aligned}$$

Since $C^p(\mathfrak{a}_{n,gl(n,\mathbb{R})})^{gl(n,\mathbb{R})} = P^{p/2}(\Omega^i_j)^{gl(n,\mathbb{R})}$, the inclusion $W \to C^*(\mathfrak{a})$ induces isomorphism on the E_1 terms (commuting with the succeeding differentials since the filtrations are the same) and, therefore, on the E_∞ terms. Therefore:

$$H^*(W) \simeq H^*(\mathfrak{a}_n).$$

Q.E.D.

Corollary 19.1.1. *$H^q(\mathfrak{a}_n)$ is finite dimensional and is 0 for $q > 2n + n^2$.*

Proof. There are only n^2 of the δ^i_j and at most n of the $\Omega^i_j = \delta^a \wedge \delta^i_{aj}$ can be multiplied together without getting 0. Each Ω^i_j has dimension 2.

Corollary 19.1.2. *Every class in $H^*(\mathfrak{a}_n)$ contains a representative cocycle which depends only on the 2-jets of its arguments. That is, we never need use representative forms like $\delta^i_{jkl}, \delta^i_{jklm}, \cdots$ in cohomology.*

Exercise. Let δ and Ω be the matrices whose (i,j)-th entries are the 1-forms δ^i_j and the 2-forms Ω^i_j, respectively. Then $d\delta = \Omega - \delta \wedge \delta$, and $d\Omega = \sum_{i,j} X^i_j(x^j)\partial/\partial x^i \in gl(n,\mathbb{R}) \subseteq \mathfrak{a}_n$, let X be the corresponding matrix, (X^i_j), of scalars. Then $\iota_{\tilde{X}}\delta = -X$, $\iota_{\tilde{X}}\Omega = 0$, $\theta_{\tilde{X}}\delta = -[X,\delta]$, and $\theta_{\tilde{X}}\Omega = -[X,\Omega]$.

Let W be the free (anti-) commutative algebra generated by the symbols $\{\delta^i_j\}$, $\{\Omega^i_j\}$ of dimensions 1 and 2 respectively, $i,j = 1,2,\ldots,n$. Define $d : W \to W$ by

$$d\delta^i_j = \Omega^i_j - \delta^i_k \wedge \delta^k_j.$$

Then in order that $d^2 = 0$ we must define:

$$\begin{aligned} d\Omega^i_j &= d(\delta^i_k \wedge \delta^k_j) = (d\delta^i_k) \wedge \delta^k_j - \delta^i_k \wedge (d\delta^k_j) \\ &= (\Omega^i_k - \delta^i_1 \wedge \delta^1_k) \wedge \delta^k_j - \delta^i_k \wedge (\Omega^k_j - \delta^k \wedge \delta^1_j) \\ &= \Omega^i_k \wedge \delta^k_j - \delta^i_k \wedge \Omega^k_j. \end{aligned}$$

W is called the *Weil algebra* of $gl_{(n,\mathbb{R})}$. In general, a Lie algebra g has Weil algebra:

$$W(g) = \Lambda^* g^* \otimes S^* g^*$$

where $\Lambda^* g^*$ (resp. $S^* g^*$) is the alternating (resp. symmetric) tensor algebra of the dual g^* to g. The degree of an element in $\Lambda^p g^* \otimes S^q g^*$ is taken to be $p+2q$. In our case, $\{\Omega^i_j\}$ generate $S^1 g^*$ (degree $=2 \Rightarrow$ they commute), and $\{\delta^i_j\}$ generate $\Lambda^1 g^*$ (degree $=1 \Rightarrow$ they anti-commute).

Proposition 19.2.

$$H^*(W) = \begin{cases} \mathbb{R} & \textit{in} \quad \dim = 0 \\ & \textit{is acyclic.} \\ 0 & \textit{in} \quad \dim \geq 1. \end{cases}$$

Proof. Consider the antiderivation k of degree -1 defined by $k(\delta^i_j) = 0$, $k(\Omega^i_j) = \delta^i_j$. Then

$$\begin{aligned} (dk + kd)(\delta^i_j) &= k(\Omega^i_j - \delta^i_k \wedge \delta^k_j) = \delta^i_j \\ (dk + kd)(\Omega^i_j) &= d(\delta^i_j) + k(\Omega^i_k \wedge \delta^k_j - \delta^i_k \wedge \Omega^k_j) \\ &= (\Omega^i_j - \delta^i_k \wedge \delta^k_j) + (\delta^i_k \wedge \delta^k_j - (-\delta^i_k \wedge \delta^k_j)) \\ &= \Omega^i_j + \delta^i_k \wedge \delta^k_j. \end{aligned}$$

We say for now that an element $\varphi \in W$ has weight q if $\varphi \in \oplus_{r=0}^q \Lambda^*(gl_n)^* \otimes S^r(gl_n^*)$. That is, φ is a polynomial in the $\{\Omega^i_j\}$ (with coefficients in the exterior algebra generated by the $\{\delta^i_j\}$) of degree $\leq q$. The weight of the element $0 \in W$ is considered to be -1.

Now suppose φ is homogeneous of degree $m > 0$ (recall that δ^i_j has degree 1, and Ω^i_j has degree 2). Then $(kd + dk)\varphi$ is still homogeneous of degree m. $(kd + dk)$ is a derivation of degree 0. Check that if φ has weight $q \geq 0$ then:

$$\varphi_1 = \varphi - \frac{1}{m-q}(\delta k + k\delta)\varphi \quad \text{has weight} \quad \leq q - 1.$$

If φ were a cocycle then so is φ_1, and φ and φ_1 differ by a coboundary. We continue inductively

$$\varphi_2 = \varphi_1 - \frac{1}{m-(q-1)}(\delta k - k\delta)\varphi_1 \quad \text{has weight} \quad \leq q - 2.$$

Finally, φ_q has weight 0 and $\varphi_{q+1} = 0$. If φ is a cocycle, then φ differs from $\varphi_{q+1} = 0$ by a coboundary. (Cf. H. Cartan [12] p. 58 for the generalization to $W(g)$ for any Lie algebra g.)

We have a homomorphism (commuting with d) $W \to W$ (onto). What is the kernel? Since $\Omega^i_j \to d\delta^i_j - \delta^k_j \wedge \delta^i_k = \delta^k \wedge \delta^i_{jk}$, if we multiply more than n of the $\{\Omega^i_j\}$ together we get 0. This is the only added relation. The kernel of $W \to \underline{W}$ is the ideal in W generated by all polynomials in the $\{\Omega^i_j\}$ of degree

$> 2n$, where we count each Ω^i_j as degree 2. Therefore, we truncate W to get $\underline{W}$ in this way.

$$\begin{aligned}\underline{W} &= W/(\text{polynomials of degree} > 2n \quad \text{in}\, \{\Omega^i_j\}) \\ &= \oplus^n_{q=0}(\Lambda^* gl(n,\mathbb{R})^* \otimes s^q gl(n,\mathbb{R})^*).\end{aligned}$$

In the spectral sequence converging to $H^*(\underline{W})$ we had $E_1^{p,q} = 0$ if p is odd or if $p > 2n$. Otherwise:

$$\begin{aligned}E_1^{2r,q} &\simeq H^q(gl(n,\mathbb{R}), (S^r gl(n,\mathbb{R})^*)^{gl(n,\mathbb{R})}) \\ &\simeq H^q(gl(n,\mathbb{R})) \otimes (S^r gl(n,\mathbb{R})^*)^{gl(n,\mathbb{R})}.\end{aligned}$$

Since $E_1^{p,q} = 0$ if p is odd, all the d_1 maps are 0 and $E_2^{p,q} = E_1^{p,q}$. Now the invariant polynomials on $gl(n,\mathbb{R})$,

$$(S^* gl(n,\mathbb{R})^*)^{gl(n,\mathbb{R})} = P(c_1,\ldots,c_n)$$

is a polynomial algebra generated by $c_1, c_2, \ldots, c_n$ where $c_i \in S^i gl(n,\mathbb{R})^*$ is $\pm$ the coefficient of λ^{n-i} in the characteristic polynomial $\det|\lambda I - \Omega|$ of the $n \times n$ matrix $\Omega = (\Omega^k_j)$ of 2-forms. (See Bott [6].) For instance $c_1(\Omega) = \Omega^i_j = tr(\Omega)$, $c_n(\Omega) = \pm\det(\Omega)$. After normalizing c_i by some constant factor, it will be the usual i-th Chern class. This will become clearer when we interpret (δ^k_j) as a connection matrix and (Ω^k_j) as the corresponding curvature matrix. (See Bott [6] for definition of Chern classes via a curvature matrix.) The $E_2 = E_1$ term is, thus:

$$\begin{aligned}E_2 &= H^*(gl(n,\mathbb{R})) \otimes \underline{P}(c_1,\ldots,c_n) \\ E_2 &= E(h_1,h_2,h_3,\ldots,h_n) \otimes \underline{P}(c_1,\ldots,c_n)\end{aligned}$$

where $E(h_1,h_2,\ldots,h_n)$ is the exterior algebra in the h_i of degree $2i-1$, and $P(c_1,\ldots,c_n)$ is the polynomial algebra in the c_i, degree $> 2n$ (counting each c_i as degree $2i$).

Homework. Use the Hochschild–Serre spectral sequence [39] relative to the reductive subalgebra $gl(n-1,\mathbb{R})$ of $gl(n,\mathbb{R})$ to show that $H^*(gl_{n,\mathbb{R}}) = E(h_1,\ldots,h_n)$. Use induction on n and the theorem on invariants of $V \otimes \cdots \otimes V \otimes V^* \otimes \cdots \otimes V^*$.

20. The unitary trick to compute $H^*(gl(n,\mathbb{R}))$.

We can also use the unitary trick by extending the base field to $\mathbb{C}$. ([23] p. 337). Then $gl(n,\mathbb{C}) = gl(n,\mathbb{R}) \otimes \mathbb{C}$ has cohomology $H^*(gl(n,\mathbb{C})) = H^*(gl(n,\mathbb{R})) \otimes \mathbb{C}$. But $gl(n,\mathbb{C}) \simeq u(n) \otimes_{\mathbb{R}} \mathbb{C}$, where $u(n) = \{X \in gl(n,\mathbb{C}) | X^t = -\overline{X}\}$ is the Lie algebra of the unitary group $U(n)$. Since $U(n)$ is compact and connected, we have

$$H^*(u(n)) = H^*(U(n))$$

where $H^*(U(n))$ is the real cohomology of the topological space $U(n)$, which we now compute. $U(n)$ maps onto S^{2n-1} = unit ball in $\mathbb{C}^n$ with fiber $U(n-1)$:

$$\begin{array}{ccc} U(n-1) & \longrightarrow & U(n) \\ & & \downarrow \\ & & S^{2n-1} \end{array}$$

Now the Serre spectral sequence for a fibration [58] (converging to $H^*(U(n))$) has E_2 term:

$$\begin{aligned} E_2^{p,q} &= H^p(S^{2n-1}, H^q(U(n-1))) \\ &= H^p(S^{2n-1}) \otimes H^q(U(n-1)) \end{aligned}$$

where, for $n \geq 2$, we have $\pi_1(S^{2n-1}) = 0$, so we do not need local coefficients. By an induction hypothesis $H^q(U(n-1)) = E(h_1, \ldots, h_{n-1})$. We sketch for $n = 3$. $H^*(U(2)) = E(h_1, h_2)$ and $H^*(S^{2n-1}) = E(h_3)$.

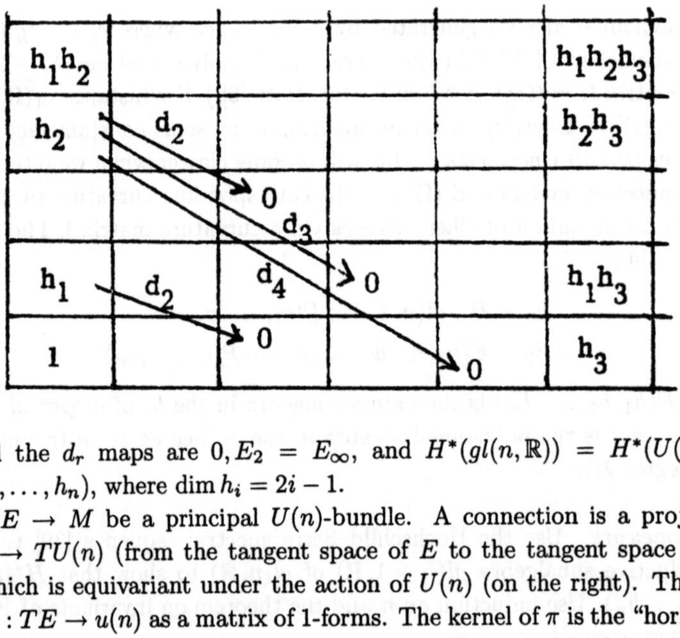

Since all the d_r maps are $0, E_2 = E_\infty$, and $H^*(gl(n,\mathbb{R})) = H^*(U(n)) = E(h_1, h_2, \ldots, h_n)$, where $\dim h_i = 2i - 1$.

Let $E \to M$ be a principal $U(n)$-bundle. A connection is a projection $\pi : TE \to TU(n)$ (from the tangent space of E to the tangent space of the fiber) which is equivariant under the action of $U(n)$ (on the right). Thus, we regard $\pi : TE \to u(n)$ as a matrix of 1-forms. The kernel of π is the "horizontal space." The corresponding curvature is a matrix of 2-forms $\Omega = d\pi - \pi \wedge \pi$ (Cf. Bott [6]), which takes pairs of vectors in $\Lambda^2 TE$ into $u(n)$. π dualizes to a map

$$\Lambda^1(u(n))^* = u(n)^* \to A^1(E) = \text{1-forms on } E.$$

Also the curvature dualizes to a map

$$S^1(u(n))^* = u(n)^* \to A^2(E).$$

We extend these to a map

$$W(u(n)) = \Lambda^* u(n)^* \otimes S^* u(n)^* \to A^*(E),$$

which will commute with d. Now, we wish to find a space Y_n with the same cohomology as $H^*(\mathfrak{a}_n) = H^*(\underline{W}(u(n))$. Since $W(u(n))$ is acyclic, it provides a good model for cochains on the total space of the universal principal $U(n)$-bundle, $EU(n) \to BU(n)$, where $E(U(n))$ is contractible. (Note that $BU(n) = BGL(n, \mathbb{C})$ since any $GL(n, \mathbb{C})$-bundle can be reduced to a $U(n)$-bundle.) Now for $E \to M$ with M an n-manifold, polynomials in the components of the matrix Ω of degree $> n$ vanish (see Section 21; cf. Bott [6] section 6) so

$$\underline{W}(u(n)) \to A^*(E),$$

where $\underline{W}(u(n)) = W(u(n))/(S^{n+1}(u(n))^*)$ is the truncated Weil algebra of $u(n)$. Since this map is functorial, there exists a map

$$\underline{W}(u(n)) \to A^*(Y_n),$$

where $Y_n \to BU(n)^{2n}$ is the restriction of $EU(n)$ to the $2n$-skeleton, $BU(n)^{2n}$, of $BU(n)$. $Y_n \to BU(n)^{2n}$ classifies $U(n)$-principal bundles over n-manifolds.

$$\begin{array}{ccc} Y_n & \longrightarrow & EU(n) \\ \downarrow & & \downarrow \\ BU(n)^{2n} & \longrightarrow & BU(n) \simeq BGL(n, \mathbb{C}) \end{array}$$

In fact, we have the following

Theorem 20.1. (Gel'fand–Fuks) $H^*(\mathfrak{a}_n) = H^*(\underline{W}(u(n))) = H^*(Y_n)$ *has the same cohomology as the complex:*

$$E(h_1, \ldots, h_n) \otimes P(c_1, \ldots, c_n),$$

where $E(h_1, \ldots, h_n)$ *is the exterior algebra generated by* $\{h_i\}$ *of degree* $2i - 1$, $P(c_1, \ldots, c_n)$ *is the polynomial algebra with generators* $\{c_i\}$ *of dimension* $2i$ *divided by the relations* $c_1^{a_1} \cdots c_n^{a_n} = 0$ *whenever* $\sum_{k=1}^{n}(ka_k) > n$, *and the differential is given by* $d(h_i) = c_i$.

We recognize $E(h_1, \ldots, h_n) \otimes \underline{P}(c_p, \ldots, c_n)$ as the E_2 term of the spectral sequence converging to $H^*(\mathfrak{a}_n)$. The d_r maps are 0 for r odd, and $d_{2i}(h_i) = c_i$. (Gel'fand–Fuks [28] p. 337–340. Cf. H. Cartan [12] p. 27, Godbillon [30] p. 5, 10, 14). Since $H^*(W(u(n)))$ is acyclic, we can choose $h_i \in W(u(n))$ such that $dh_i = c_i$ and the same will hold for their images under $W(u(n)) \to \underline{W}(u(n))$.

Example. $n = 1$. $H^*(\mathfrak{a}_1) = H^*(E(h_1) \otimes P(c_1)/c_1^2 = 0)$ with $\dim h_1 = 1$, $\dim c_1 = 2$, $d(h_1) = c_1$, $d(c_1) = 0$, $d(h_1 c_1) = c_1^2 = 0$.

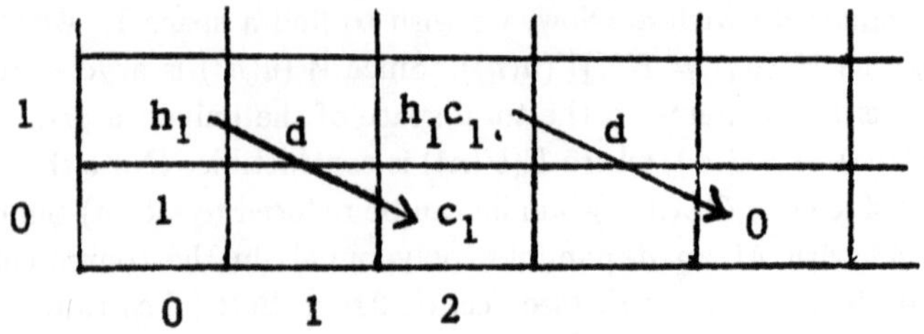

There is cohomology generated by 1 in dim= 0, and by $\omega = h_1 c_1$ in dim= 3.

Example. $n = 1$, $U(1) = S^1$. $BU(1) = \mathbb{C}P^\infty$, the infinite Grassmannian has 2-skeleton: $BU(1)^2 = \mathbb{C}P^1 = S^2$. Then the pullback $Y_1 = S^3$ and the map $S^3 \to S^2$ is the Hopf fibration:

$$\begin{array}{ccc} Y_1 = S^3 & \longrightarrow & EU(1) \\ \downarrow & & \downarrow \\ BU(1)^2 = S^2 & \longrightarrow & BU(1) = \mathbb{C}P^\infty \end{array}$$

$$H^*(Y_1) = H^*(S^3) = \left\{ \begin{array}{cl} \mathbb{R} & \text{in dim } 0, 3 \\ 0 & \text{otherwise} \end{array} \right\} = H^*(\mathfrak{a}_1).$$

Example. $n = 2$. $H^*(\mathfrak{a}_2) = H^*(E(h_1, h_2) \otimes P(c_1, c_2)/(c_1^3, c_1c_2, c_2^2))$.

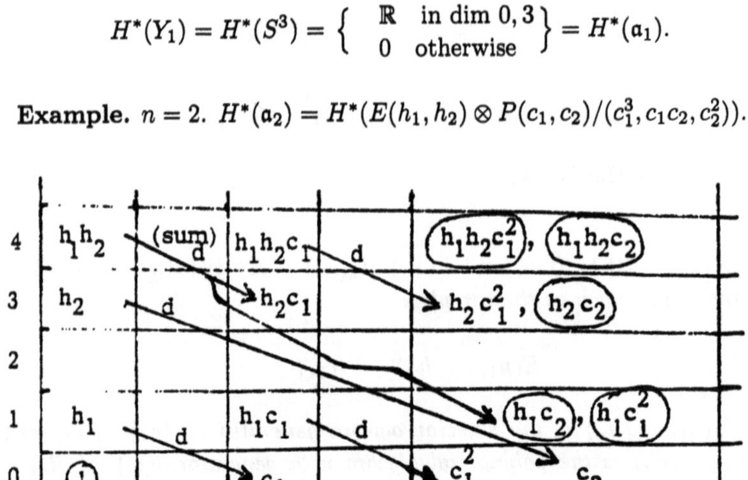

The generators of cohomology are circled. There are 1 in dim= 0, 2 in dim= 5, 1 in dim= 7, and 2 in dim= 8. Vey computed an explicit basis for $H^*(\mathfrak{a}_n)$ ([30] p. 13, [37] p. 383; cf. [28] p. 340.)

Proposition 20.2. *The* $\{h_{i_1} \cdots h_{i_r} c_{j_1} \cdots c_{j_s}\}$ *such that*

$$1 \le i_1 < i_2 < \cdots < i_r \le n \quad \textit{for} \quad r = 1, \ldots, n,$$
$$1 \le j_1 \le j_2 \le \cdots \le j_s \le n \quad \textit{for} \quad s = 1, \ldots, n,$$
$$j_1 + \cdots + j_s \le n < i_1 + j_1 + \cdots + j_s, \textit{ and } i_1 < j_1$$

form a basis of $H^*(\mathfrak{a}_n)$.

Then the following two propositions are immediate:

Proposition 20.3. (Gel'fand–Fuks) *Except for the trivial class in dimension 0,* $H^*(\mathfrak{a}_n)$ *starts in dimension* $2n+1$ *and goes no higher than* n^2+2n.

Proposition 20.4. (Gel'fand–Fuks) *Multiplication is trivial in* $H^*(\mathfrak{a}_n)$.

Proof. Note that for $h_{i_1}\cdots h_{i_r}c_{j_1}\cdots c_{j_s}$ in the Vey basis, $i_1 < j_1$ implies $j_1+\cdots+j_s > (n/2)$. Thus, all products are truncated.

21. Connection with connections

Recall our tower of frame bundles $\cdots \to F_2(M) \to F_1(M) \to M$ over a manifold M. Choosing a connection on $\pi : F_1(M) \to M$ is the same as choosing a section $\partial : F_1(M) \to F_2(M)$ which is equivariant under the action of $gl(n,\mathbb{R})$. The forms ω^i_j on $F_2(M)$ pull back to a connection form, the matrix, $(\partial^*\omega^i_j)$, on $F_1(M)$. A tangent vector X on $F_1(M)$ is mapped to the matrix $(\partial^*\omega^i_j(X)) \in gl(n,\mathbb{R}) \simeq$ tangent space of the fiber $\simeq GL(n,\mathbb{R})$. This projection $T(F_1(M)) \to T$ (fiber) is a connection on the bundle $F_1(M) \to M$. (Verify!) Then the invariant polynomials in the curvature forms,

$$\Omega^i_j = d\omega^i_j - \omega^k_j \wedge \omega^i_k = \omega^k \wedge \omega^i_{kj}$$

(where we have pulled ω^k_j and ω^i_{kj} down to $F_1(M)$ by suitable sections), will be basic forms (i.e., come from the base, M, via π^*) and give us characteristic classes in $H^*(M)$.

22. Reduction of $H^*_{GF}(\mathbb{R}^n)$ to $H^*(W)$

The object of this section is to show that the inclusion

$$A^*_{pt}L\mathbb{R}^n \longrightarrow A^*_cL\mathbb{R}^n$$

induces isomorphism in cohomology. (See Bott [7].) To simplify the notation we first take $n=1$. For $x \in \mathbb{R}^1$, we denote by δ_x the evaluation at $x : \delta_x(f) = f(x)$, for $f : \mathbb{R} \to \mathbb{R}$. $\delta'_x, \delta''_x, \ldots, \delta^{[k]}_x, \ldots$ will be the evaluation of the 1st, 2nd, $\ldots$, kth, $\ldots$ derivatives at $x \in \mathbb{R}^1$. $\delta'_x(f) = f'(x)$, etc. (We will forget about the $\pm$ sign in this section for convenience.) In this context, Schwartz's theorem says that any continuous linear functional on $C^\infty(\mathbb{R}^1)$ can be written as a finite sum of ones of the form:

$$\varphi = \int_{-\infty}^{\infty} \mu(x)\, \delta^{[k]}_x dx$$

where $\mu(x)$ is continuous $\mathbb{R} \to \mathbb{R}$ with compact support. This means for $f : \mathbb{R} \to \mathbb{R}$ smooth, $\varphi(f) = \int \mu(x)\delta_x^{[k]} f = \int \mu(x) f^{[k]}(x)dx$. Then since μ is continuous with compact support,

$$d\varphi = \int_{-\infty}^{\infty} \mu(x)\,[d(\delta_x^{[k]})]dx$$

where we already have seen how to compute $d(\delta_x^{[k]})$ in the $A^*_{pt}L\mathbb{R}^1$ case.

What are the 2-forms? For $x, y \in \mathbb{R}^1$ we can of course have $\delta_x^{[k_1]} \wedge \delta_y^{[k_2]}$ as a 2-form. We can also take finite linear combinations of these with x, y, k_1, and k_2 varying. Limits of these are integrals:

$$\varphi = \int_{-\infty}^{\infty}\int_{-\infty}^{\infty} \mu(x,y)\,\delta_x^{[k_1]} \wedge \delta_y^{[k_2]} dxdy.$$

At first $\mu(x, y)$ may have some Dirac-δ spikes, but as in the 1-form case, we find that the most general continuous 2-form on $L\mathbb{R}^1$ can be written as a finite sum of φ's as above with $\mu(x,y)$ *continuous*. This is the Schwartz kernel theorem. To generalize to the $\mathbb{R}^n$ case we replace $\delta_x^{[k]}$ with some $(\partial^k/\partial x_{i_1} \cdots \partial x_{i_k})\delta_x^j$ where j indicates which component of $\mathbb{R}^n$ and $x = (x_1, \ldots, x_n) \in \mathbb{R}^n$.

The general continuous q-form on $L\mathbb{R}^1$ will be finite sums of forms of the form:

$$\int \mu(x_1, \ldots, x_q)\,\delta_{x_1}^{[k_1]} \wedge \cdots \wedge \delta_{x_q}^{[k_q]} dx_1 \cdots dx_q.$$

When evaluating such an expression $\delta_{x_1}^{[k_1]} \wedge \cdots \wedge \delta_{x_q}^{[k_q]}$ on q vector fields $f(x)\frac{\partial}{\partial x}, \ldots, f_q(x)\frac{\partial}{\partial x}$ we must not forget the alternation.

Consider the function (for $t \in \mathbb{R}$) $f_t : \mathbb{R}^1 \to \mathbb{R}^1$ by $x \to tx$. By the chain rule, we compute f_t^* for $t \neq 0$. I claim:

$$f_t^*\delta_x = t\delta_{(x/t)} \quad \text{for} \quad t \neq 0$$

This is because $f_t^* : \frac{\partial}{\partial x}|_x \longrightarrow t\frac{\delta}{\delta x}|_{tx}$. As $t \to \infty$, the support (i.e., x/t) goes to 0, but the form blows up because of the factor of t in front. We also have by the chain rule:

$$f_t^*\delta'_x = t(\delta'_{x/t})(1/t) = \delta'_{x/t}.$$

Namely, $(f_t^*\delta'_x)(f(y)\partial/\partial y) = \delta'_x(tf(y/t)\partial/\partial y) = tf'(x/t)(1/t)$. We also have

$$f_t^*\delta''_x = (1/t)\delta''_{x/t},$$

and in general:

$$f_t^*\delta_x^{[k]} = t^{(1-k)}\delta_{x/t}^{[k]}.$$

As $t \to \infty$ there is no blowing up in $f_t^*\delta_x^{[k]}$ for $k \geq 1$. Only $f_t^*\delta_x$ blows up for $k = 0$. We are reminded of our earlier decomposition into eigenspaces of θ_R where $R = (x)\partial/\partial x$, the infinitesimal generator of f_t. Consider the q-form $\varphi^q = \int \mu(x_1, \ldots, x_q)\delta_{x_1}^{[k_1]} \wedge \cdots \wedge \delta_{x_q}^{[k_q]} dx_1 \cdots dx_q$. A priori $f_t^*\varphi^q$ can grow no faster than t^q as $t \to \infty$. But we can say more. It never grows faster than t^1.

Lemma 22.1. *Let $f_t^* : A^q(L\mathbb{R}^1) \to A^q(L\mathbb{R}^1)$ as above. Then $(1/t)f_t^*$ will have a limit as $t \to \infty$. Furthermore, this limit is in $A_{pt}^* L\mathbb{R}^1$ and will be an eigenvector for $\theta_{(x)\partial/\partial x}$ with eigenvalue (or weight) 1.*

Recall that $A_{pt}^*(L\mathbb{R}^1)$ breaks up a direct sum of eigenspaces $A_{pt}(\lambda)$ for $\lambda = 1, 0, -1, -2, \ldots$ of the operator $\theta_{(x)\partial/\partial x}$. To repeat the lemma, $\lim_{t\to\infty}(1/t)f_t^*$ exists and takes values in $A_{pt}(1)$.

Exercise. $\lim_{t\to\infty}(1/t)f_t^*$ is identity on $A_{pt}(1)$, and 0 on $A_{pt}(k)$ for $k \leq 0$. Suppose we have shown that the limit exists and is finite. Since the supports shrink to zero, the limit has support at the origin. Further, $\gamma = \lim_{t\to\infty}(1/t)f_t^*$ is a projection, $\gamma \circ \gamma = \gamma$, since for $s, t \in \mathbb{R}$, $f_s^* \circ f_t^* = f_{st}^*$, and (assuming that the limits as $t \to \infty$ exist) the composition of limits = limit of compositions.

We obtain a projection which commutes with d:

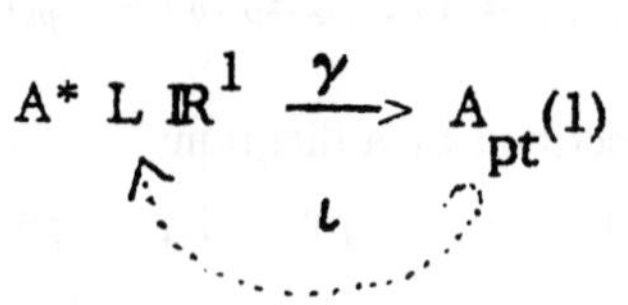

which is split by the inclusion ι. $\gamma \circ \iota =$ identity. Since $H^*(A_{pt}(1)) = 0$ (Recall that $\theta_{(x)\partial/\partial x} = \iota_{(x)\partial/\partial x} d + d\iota_{(x)\partial/\partial x}$ is homotopic to 0 but is 1 on $A_{pt}(1)$.) If u is a cocycle in $A^*L\mathbb{R}^1$ then γu is a coboundary, dv, for $v \in A_{pt}(1)$. Also $u - \iota dv \in \ker \gamma$.

Idea of proof of Lemma. Since $\delta'_x, \delta''_x, \ldots$ give us no problem, the worst case is illustrated by the form: $\varphi = \int \mu(x,y)\delta_x \wedge \delta_y dxdy$. Since μ is continuous with compact support, $f_t^*\varphi = t^2 \int \mu(x,y)\delta_{x/t} \wedge \delta_{y/t} dxdy$. Now as $t \to \infty$, we have the t^2 factor blowing up to order 2. However, we also have $\delta_{x/t} \wedge \delta_{y/t} \to \delta_0 \wedge \delta_0 = 0$ by the alternation. Explicitly, let $x = f(x)\partial/\partial x$, $y = g(y)\partial/\partial y$ in $L\mathbb{R}^1$. Then $f(x) = a_0 + a_1 x + x^2(h_1(x))$ and $g(y) = b_0 + b_1 y + y^2 h_2(y)$

$$
\begin{aligned}
&(\delta_{x/t} \wedge \delta_{y/t}(X,Y) = f(x/t)g(y/t) - f(y/t)g(x/t) \\
&= (a_0 + a_1(x/t) + (x/t)^2 h_1(x/t))(b_0 + b_1(y/t) + (y/t)^2 h_2(y/t)) \\
&- (a_0 + a_1(y/t) + (y^2/t)h_1(y/t))(b_0 + b_1(x/t) + (x/t)^2 h_2(x/t)) = \frac{1}{t}\{\cdots\}
\end{aligned}
$$

Because of this $1/t$ factor $f_t^*\varphi$ can blow up only like $t^2 \cdot (1/t) = t^1$.

Remark. In the $\mathbb{R}^n$ case the worst that $f_t^*\varphi$ can blow up is to order t^n as illustrated by the term $\delta_{x_1}^1 \wedge \cdots \wedge \delta_{x_n}^n$ where superscript means component in $\mathbb{R}^n$, not derivative. In this case the image of γ is the n-eigenspace of θ_R where $R = x_1\partial/\partial x_1 + \cdots + x_n\partial/\partial x_n$ is the infinitesimal generator of f_t : $(x_1, \ldots, x_n) \to (tx_1, \ldots, tx_n)$.

Granting this lemma, one extends it as follows:

Lemma 22.2. *Let $F^n \supseteq F^{n-1} \supseteq \cdots \supseteq F^0 \supseteq F^{-1} \supseteq \cdots$ be a filtration of $A_c^* L\mathbb{R}^1$ defined by:*

$$F^k = \{\omega \in A_c^* L\mathbb{R}^1 | \gamma_k \omega = \lim_{t\to\infty} \frac{1}{t^k}\omega \quad \textit{is well defined}\}.$$

Then one has a split exact sequence:

$$0 \longrightarrow F^{k-1} \longrightarrow F^k \xrightarrow{\gamma_k} A_{pt}(k) \longrightarrow 0.$$

Remark. By 22.1, we have $A_c^* L\mathbb{R}^1 = F^1$. In the $\mathbb{R}^n$ case, we would have $A_c^* L\mathbb{R}^n = F^n$.

Proposition 22.3. *If $A_{pt}^* L\mathbb{R}^1$ is graded by weight then the associated graded group of $\{F^k\}_{k=-\infty}^1$ is $\oplus_{k=-\infty}^1 A_{pt}(k) = A_{pt}^* L\mathbb{R}^1$.*

The situation is described by a diagram:

$$\begin{array}{ccccccc}
\cdots & F^{-1} & \hookrightarrow & F^0 & \hookrightarrow & F^1 & = A_c^* L\mathbb{R}^1 \\
& \iota\uparrow\downarrow\gamma_{-1} & & \iota\uparrow\downarrow\gamma_0 & & \iota\uparrow\downarrow\gamma_1 & \\
\cdots & A_{pt}(-1) & & A_{pt}(0) & & A_{pt}(1) &
\end{array}$$

where $\gamma_1 = \lim_{t\to\infty} (1/t) f_t^*$ defined on all of $A_c^* L\mathbb{R}^1 = F^1$, $\gamma_k = \lim_{t\to\infty} t^{-k} f_t^*$ defined on $F^k = \ker(\gamma_{k+1})$ for $k = 0, -1, \ldots$, and ι is inclusion. We have already shown that any cocycle, u, in F^1 is cohomologous to a cocycle, $u_2 = u - i\gamma_1 u$, in F^0, since $H^*(A_{pt}(1)) = 0 \Rightarrow \gamma_1 u$ is a coboundary. Repeating the process, $u_3 = u_2 - \iota\gamma_0 u_2$ is a cocycle in F^{-1}. After we show that $H^*(F^{-1}) = 0$, u_3 will be a coboundary in F^{-1}. Then the map $H^*(AL\mathbb{R}^1) \to H^*(A_{pt}(0))$ induced by $u \to \gamma_0(u - \iota\gamma_1 u)$ is the isomorphism inverse to ι^*. $H^*(A_{pt}(0))$ has already been computed in 20.1 and 20.2.

To complete the proof, we need a homotopy formula to show $H^*(F^{-1}) = 0$. Again let $R = (x)\partial/\partial x$ be the infinitesimal generator of f_t.

Lemma 22.4. $\frac{d}{dt} f_t^* = (1/t)\{\iota_R d + d\iota_R\} f_t^*$.

Proof.

$$t\frac{d}{dt} f_t^* = \lim_{\Delta t \to 0} t \frac{f_{t+(t\Delta t)}^* - f_t^*}{(t\Delta t)} = \lim_{\Delta t\to 0} \left(\frac{f_{1+\Delta t}^* - 1}{\Delta t} \right) f_t^*.$$

Now

$$\lim_{\Delta t\to 0} \frac{f_{t+\Delta t}^* - 1}{\Delta t} = \frac{d}{dt}\Big|_{t=1} f_t^* \quad \text{and } \theta_R = \iota_R d + d\iota_R.$$

So it suffices to show $d/dt|_{t=1}(f_t^*) = \theta_R$. But this is just the statement that θ_R is the Lie derivative in the R direction. For example, let us practice what these symbols mean on a vector field $f(x)\partial/\partial x$ and 1-form $\delta_a^{[k]}$ = evaluation of k-th derivative at $x = a$ in $\mathbb{R}$.

$$\begin{aligned}\frac{d}{dt}\Big|_{t=1}[(f_t^*\delta_a^{[k]})\left(f(x)\frac{\partial}{\partial x}\right) &= \frac{d}{dt}\Big|_{t=1}(t^{1-k}f^{[k]}(a/t))\\ &= [(1-k)t^{-k}f^{[k]}(a/t) + t^{1-k}f^{k+1}_{(a/t)}(-at^{-2})]_{t=1}\\ &= (1-k)f^{[k]}(a) - af^{[k+1]}(a)\end{aligned}$$

$(1)_k$

On the other hand,

$$\begin{aligned}(\theta_{x(\partial/\partial x)}\delta_a^{[k]})(f(x)\frac{\partial}{\partial x}) &= -\delta_a^k[(x(\partial/\partial x), f(x)\partial/\partial x])\\ &= -\delta_a^k(xf'(x)\frac{\partial}{\partial x} - f(x)\frac{\partial}{\partial x})\\ &= \frac{\partial^k}{\partial x^k}\Big|_a(f(x) = xf'(x))\end{aligned}$$

$(2)_k$

For $k = 0$, $(2)_0 = f(a) - af'(a) = (1)_0$. The inductive step is given by $(2)_{k+1} = d/dx|_a\{(2)_k\} = d/dx|_a\{(1-k)f^{[k]}(x) - xf^{[k+1]}(x)\}$ (by inductive hypothesis) $= (1-k)f^{[k+1]}(a) - (f^{[k+1]}(a) + af^{[k+2]}(a)) = (1)_{k+1}$. Therefore, $(1)_k = (2)_k$ for all $k \geq 0$. We integrate from 1 to T to obtain:

$$\begin{aligned}\int_1^T \frac{d}{dt}f_t^* &= \int_1^T \{\iota_R d + d\iota_R\}f_t^*(dt/t)\\ f_T^* - 1 &= \int_1^T \iota_R f_t^* d + d\iota_R f_t^*(dt/t)\\ &= K_T d + dK_T\\ \text{where}\quad K_T(\omega) &= \int_1^T \iota_R f_t^*(\omega)(dt/t)\end{aligned}$$

In general, $K_\infty = \lim_{T\to\infty} K_T$ need not exist, but if $\omega \in F^{-1}$, $K_\infty(\omega)$ will exist and also be in F^{-1}. Since $\lim_{T\to\infty} f_T^* = 0$ on F^{-1}, we will have shown $H^*(F^{-1}) = 0$. For $\omega \in F^{-1}$:

$$K_T(\omega) = \int_1^T \iota_R t f_t^*(\omega)(dt/t^2).$$

Changing variables, let $\lambda = (1/\lambda)$, $d\lambda = -(dt/t^2)$.

$$K_T(\omega) = \int_{1/T}^1 \iota_R((1/\lambda)f_{1/\lambda}^*(\omega))d\lambda$$

which converges to

$$K_\infty(\omega) = \int_0^1 \iota_R((1/\lambda)f_{1/\lambda}^*(\omega))d\lambda$$

Since $\lim_{\lambda\to\infty}(1/\lambda)f^*_{1/\lambda}(\omega)$ is defined for $\omega \in F^{-1}$.

But we still have to show that $K_\infty(F^{-1}) \subset F^{-1}$. First note that ι_R commutes with f^*_t, and $f^*_s f^*_t = f^*_{st}$.

$$\begin{aligned} f^*_t \circ K_\infty &= \int_0^1 f^*_t \iota_R (1/\lambda) f^*_{1/\lambda} d\lambda \\ &= \int_0^1 (1/\lambda)\iota_R f^*_{(t/\lambda)} d\lambda \\ &= \int_0^1 (t/\lambda)\iota_R f^*_{(t/\lambda)} (d\lambda/t) \\ &= \int_0^{1/t} (1/\nu)\iota_R f^*_{(1/\nu)} d\nu \to 0 \quad \text{as } t\to\infty \quad \text{on } F^{-1} \end{aligned}$$

where $\nu = \lambda/t$, $d\nu = d\lambda/t$.

23. Cohomology of subalgebras of $L\mathbb{R}^n$

Generalizing the argument for $n = 1$, we have

Proposition 23.1. *The inclusion*

$$A^*_{pt}L\mathbb{R}^n \hookrightarrow A^*_c L\mathbb{R}^n$$

induces isomorphism in cohomology:

$$H^*_{GF}(\mathbb{R}^n) = H^*(\mathfrak{a}_n) = H^*(Y_n).$$

Y_n is the restriction of the total space of universal principal $U(n)$-bundle to the $2n$-skeleton of $BU(n)$, as in Section 20.

Corollary 23.1.1. $H^*_{GF}(\mathbb{R}^n)$ *is finite dimensional.*

We can refine the proof to certain subalgebras of $L\mathbb{R}^n$ which correspond to certain subgroups of Diff $\mathbb{R}^n$ = diffeomorphisms $\mathbb{R}^n \to \mathbb{R}^n$. Recall that $L\mathbb{R}^n$ is the analogue of the Lie algebra for the infinite dimensional group Diff $\mathbb{R}^n$. However, there are some unsolved cases.

Example. The Hamiltonian case. Let $x_1, \ldots, x_n, y_1, \ldots, y_n$ be coordinates on $\mathbb{R}^{2n}$. Let $\omega = \sum_{i=1}^n dx_i \wedge dy_i$. The diffeomorphisms $\mathbb{R}^{2n} \to \mathbb{R}^{2n}$ which preserve ω form a subgroup of Diff $\mathbb{R}^{2n}$ whose Lie algebra is $L_H\mathbb{R}^{2n} = \{X \in L\mathbb{R}^n | \theta_X\omega = 0\}$. Since θ_X is the Lie derivative, this says ω is constant along a flow which is tangent to the vector field X. We can ask what the inclusion

$$A^*_{pt}L_H\mathbb{R}^n \to A^*_c L_H\mathbb{R}^n$$

induces in cohomology. But here the argument involving the contraction $f_t : \mathbb{R}^{2n} \to \mathbb{R}^{2n}$ breaks down because f_t does not preserve ω. By way of contrast, for the subalgebra $\{X \in L\mathbb{R}^n | \theta_X\omega = \lambda\omega \text{ for some } \lambda \in \mathbb{R}\}$ of $L\mathbb{R}^n$, the contraction argument remains valid. Note that $H^*(A_{pt}L_H\mathbb{R}^n)$ is just $H_c^*(\mathfrak{a}_{2n;H})$ where $\mathfrak{a}_{2n;H}$ is the subalgebra of $\mathfrak{a}_{2n}$ containing those $X = \sum_i(f_i\partial/\partial x_i + g_i\partial/\partial y_i)$ such that $\theta_X\omega = 0$. f_i and g_i are formal power series in $x_1, \ldots, x_n, y_1, \ldots, y_n$.

The first question we would like to know about $H^q(A_cL_H\mathbb{R}^n)$ is whether or not it is finite dimensional. Gel'fand used a computer to find some very exotic classes in there. The requirement that $\theta_X\omega = 0$ imposes certain symmetry conditions on the $\delta^i_{j_1\cdots j_r}$ of $A_cL_H\mathbb{R}^n$. (See [29].) Another problem is to take $\omega =$ volume form on $\mathbb{R}^n$ and consider the volume-preserving diffeomorphisms. This corresponds to the Lie algebra of vector fields X on $\mathbb{R}^n$ such that $\theta_X\omega = 0$. It is hard to adapt the Gel'fand–Fuks original proof to other groups with the contraction property. But we will use the abstract nonsense approach to vector fields on manifolds.

24. Gel'fand–Fuks on manifolds. Semi-simplicial methods

Let M be a manifold, $U(M) = \{U_a\}_a$ be a good covering of M. Recall that this means that $U_a, U_a \cap U_b, U_a \cap U_b \cap U_c, \ldots$ are all copies of $\mathbb{R}^n$ (or else empty). We also assume $a, b, \ldots$ taken over some finite, linearly ordered index set. Then we have our semi-simplicial manifold:

$$U_0(M) = \coprod_a U_a \quad \text{where } \coprod \text{ means disjoint union.}$$

$$U_1(M) = \coprod_{a \le b} (U_a \cap U_b)$$

$$U_2(M) = \coprod_{a \le b \le c} (U_a \cap U_b \cap U_c), \text{ etc.}$$

The maps $\partial_0, \ldots, \partial_k : U_k(M) \to U_{k-1}(M)$ are defined by the inclusions $\partial_i : U_{a_0} \cap \cdots \cap U_{a_k} \to U_{a_0} \cap \cdots \cap \widehat{U}_{a_i} \cap \cdots \cap U_{a_k}$ where $\widehat{\ }$ means omit as usual. We also have the degeneracy maps $\epsilon_0, \ldots, \epsilon_k : U_k(M) \to U_{k+1}(M)$ given by the identity maps $\epsilon_i : U_{a_0} \cap \cdots \cap U_{a_k} \to U_{a_0} \cap \cdots \cap U_{a_i} \cap \cdots \cap U_{a_k}$.

Example. $M = (0, 1)$, the open interval, can be covered by $U_a = (0, 3/4)$ and $U_b = (1/4, 1)$. We have

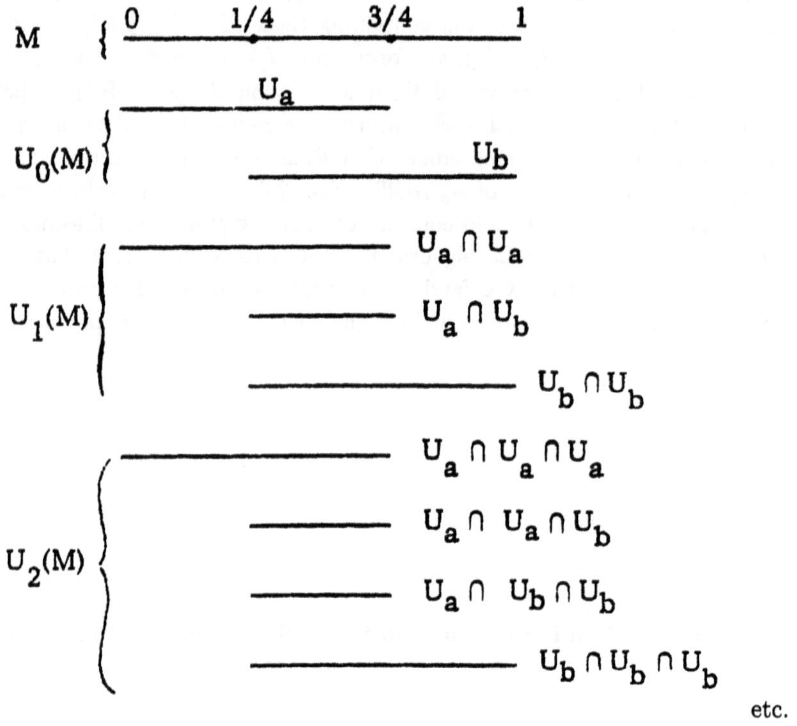

Notice that for $U_2(M)$ and higher we get more and more copies of $\mathbb{R}^n$, but these will all be degenerate. Because of the repeated subscripts, we will not see anything new besides $(0,3/4), (1/4,3/4)$, and $(1/4,1)$. Now to the semi-simplicial manifold,

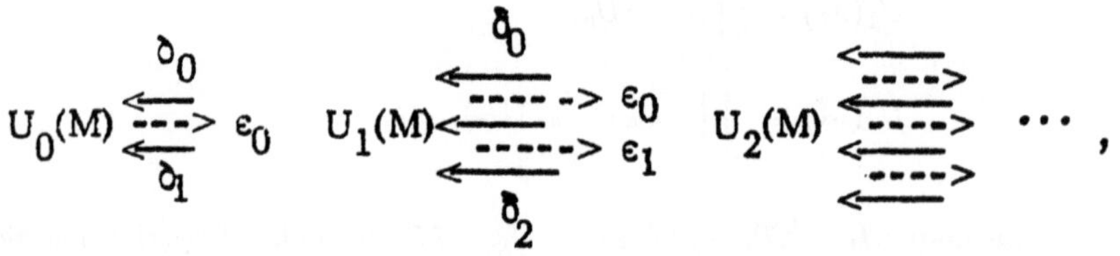

we apply the functor $L =$ "vector fields on." L is a contravariant functor with respect to the maps ∂_i because vector fields pull back, or restrict, to open subsets. Contrast this situation with the fact that vectors at a point (but not vector fields) are pushed foward by differentiable maps. We obtain a co-semi-simplicial Lie algebra:

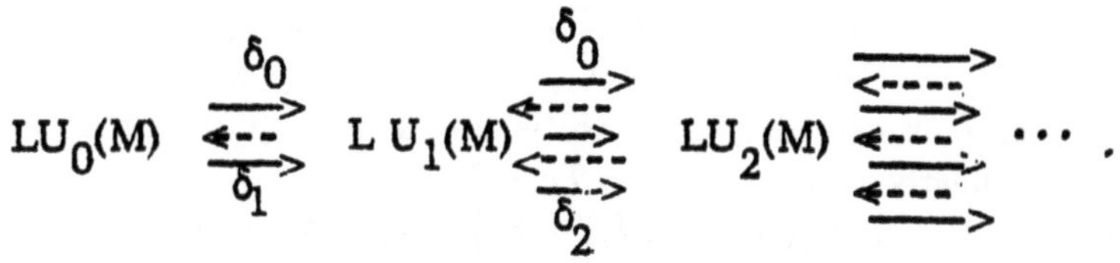

We now apply the contravariant functor $A^* = \oplus_p A^p$ where A^p = "continuous cochains on" to obtain the semi-simplicial module:

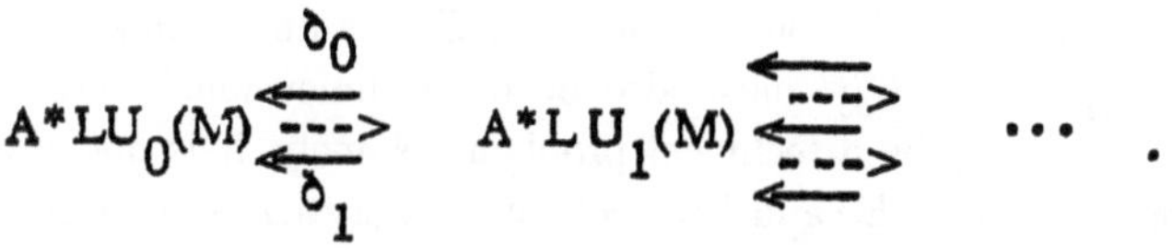

Remark. Although we are doing cohomology, the map $\delta = \sum_i (-1)^i \partial_i$ will act like homology, $\delta : A^p LU_q(M) \to A^p LU_{q-1}(M)$. We will also have the d of the Gel'fand–Fuks theory, $d : A^p LU_q(M) \to A^{p+1} LU_q(M)$. On the double complex $\{A^p LU_q(M)\}_{p,q}$ we have the total derivative $D = d \pm \delta$ as usual. We have two spectral sequences convergeing to H_D^* (double complex), one by the filtering by p and the other by filtering with respect to q.

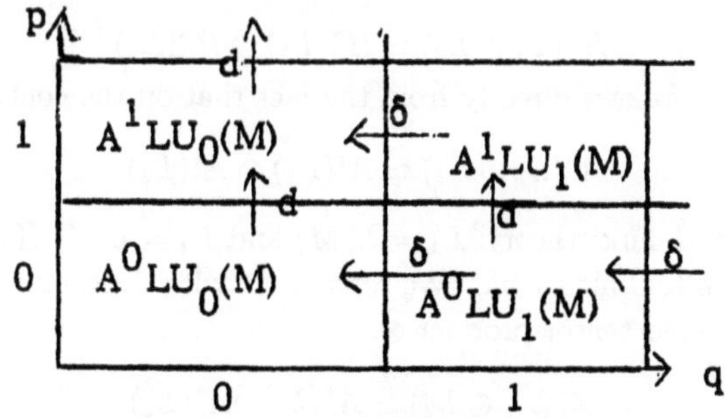

Because d points up, but δ to the left, the total D-cocycles of degree n lie on a line $p - q = n$ instead of the usual situation $p + q$ = total degree. Filtering with respect to p we find that:

$$E_1^{p,q} = H_q^\delta(A^p LU_*(M)) = \begin{cases} A^p LM & \text{if } q = 0 \\ 0 & \text{if } q > 0. \end{cases}$$

To see this let $\{\rho_a\}$ be a partition of unity subordinate to $\{U_a\}$. Then the following sequence is exact:

$$0 \longleftarrow A^p L(M) \overset{\iota}{\longleftarrow} A^p LU_0(M) \overset{\delta}{\longleftarrow} A^p LU_1(M) \overset{\delta}{\longleftarrow} \cdots$$

where ι is induced by the inclusions $U_a \hookrightarrow M$. A contracting homotopy T is contructed using $\{\rho_a\}$. For $\varphi \in A^pLM$ and $X_1, \ldots, X_p \in LU_0(M)$, define $T : A^pLM \to A^pLU_0(M)$ by $(T\varphi)(X_1, \ldots, X_p) = \sum_a \varphi(\rho_a X_1|_{U_a}, \ldots, \rho_a^{X_p}|U_a)$ where $\rho_a X_i|_{U_a}$ is extended to M to be 0 outside U_a. $T : A^pLU_k(M) \to A^pLU_{k+1}$ similarly, and we get $Id = T\delta + \delta T$.

Then $E_2^{p,q} = H_d^p H_q^\delta(A^*LU_*(M)) = H_{GF}^p(M)$ in the column $q = 0$, and 0 for $q > 0$. Then $E_\infty^{p,0} = E_2^{p,0} = H_{GF}^p(M)$ and $E_\infty^{P,q} = 0$ for $q > 0$. What is the spectral sequence going the other way (i.e., filtering with respect to q)?

$$'E_1^{p,q} = H_d^p(A^*LU_q(M)) = H_{GF}^p(U_qM)$$
$$'E_2^{p,q} = H_q^\delta H_d^p(A^*LU_*(M)) = H_q^\delta H_{GF}^p(U_*M).$$

In Section 25 we will show that $\oplus_{p-q=n}{'E_2^{p,q}}$ is finite dimensional for each n. Then $H_{GF}^n(M) = E_\infty^{n,0}$ must also be finite dimensional. (At this point, we are still bothered by a technical problem. Namely, it is not obvious that the spectral sequences, E_r and $'E_r$, both converge to $H_D(ALU(M))$ because the usual convergence criteria for spectral sequences do not hold in this case. In particular, any D-cocycle in the double complex $ALU(M)$ which has infinitely many non-zero entries on the line $p - q = n$ will escape detection by $'E_1 = H_d(ALU(M))$. (We will ignore this problem here and continue.)

25. The Künneth formula

If L_1 and L_2 are two finite dimensional Lie algebras then we have a Künneth Formula:

$$H^*(L_1 \oplus L_2) \simeq H^*(L_1) \otimes H^*(L_2)$$

and this relation follows directly from the fact that on the cochain level:

$$A^*(L_1 \oplus L_2) \simeq A^*(L_1) \otimes A^*(L_2).$$

Now the Gel'fand–Fuks theory, $L_1 = L(M)$ and $L_2 = L(N)$ (for manifolds M and N) are infinite dimensional. We replace A^* with continuous forms A_c^* and $\otimes$ with a completed tensor product $\widehat{\otimes}$.

$$A_c^*(L_1 \oplus L_2) \simeq A_c^*(L_1) \widehat{\otimes} A_c^*(L_2)$$

and $L_1 + L_2 = L(M \coprod N) =$ vector fields on the disjoint union of M and N. It is difficult to prove directly the Künneth relation. However, our earlier techniques can be extended to prove the Künneth formula for the special case of a disjoint union of $\mathbb{R}^n$'s. We have

$$H^*(\oplus_{j=1}^k L\mathbb{R}^n) = \otimes_{j=1}^k H^*(L\mathbb{R}^n),$$

so that

$$H_{GF}^*(\coprod_{j=1}^k \mathbb{R}^n) = \otimes_{j=1}^k H^*(\mathfrak{a}_n).$$

The idea of the proof is to apply the contraction f_t to all k copies of $\mathbb{R}^n$ at once. A point x in some copy of $\mathbb{R}^n$ is sent to tx in the same copy of $\mathbb{R}^n$. Our forms may then blow up to order kn instead of n, but we use the same argument as before to get down to weight 0.

Let $G = \widetilde{H^*_{GF}}(\mathbb{R}^n) = H^*_{GF}(\mathbb{R}^n)$ for $* > 0$ and 0 in dimension $* = 0$ be reduced Gel'fand–Fuks cohomology. Then

$$\widetilde{H^*_{GF}}(\mathbb{R}^n \coprod \mathbb{R}^n) = G \oplus G \oplus (G \otimes G)_{(alternating\ part)} \oplus \cdots$$

In order that we interpret the δ maps of our complex, we must take this description functorial. Let Y be a disjoint union of $\mathbb{R}^n$'s. Consider the functor

$$Y \to H_0(Y, G) \oplus H_0^\alpha(Y \times Y, G \otimes G) \oplus H_0^\alpha(Y \times Y \times Y, G \otimes G \otimes G),$$

where α means alternating under a permutation. Now since multiplication of Gel'fand–Fuks classes in any one component $\mathbb{R}^n$ of Y is trivial, we must replace the above description by

$$Y \to H_0(Y, G) \oplus H_0^\alpha(Y \times Y, \Delta; G \otimes G) \oplus \cdots$$

where Δ is the diagonal of $Y \times Y$. $H_0(Y, G)$ represents those classes of $\widetilde{H^*_{GF}}(Y)$ which can be represented by a GF class on some one component $\mathbb{R}^n$ of Y (times 1's on the other components). $H_0^\alpha(Y \times Y, \Delta; G \otimes G)$ represents the classes of $\widetilde{H^*_{GF}}(Y)$ which are an (alternating) product of two GF classes on some two components $\mathbb{R}^n$'s of Y (times 1's on the other components). Next come products of 3, etc. In general, for $Y = \mathbb{R}^n \coprod \cdots \coprod \mathbb{R}^n$,

$$\begin{aligned} H^*_{GF}(Y) = H_0(Y, G) + H_0^\alpha(Y^2, Y_1^2; G \otimes G) + H_0^\alpha(Y^3, Y_2^3; G \otimes G \otimes G) \\ \cdots + H_0^\alpha(Y^k, Y_{k-1}^k, \underbrace{G \otimes \cdots \otimes G}_{k-times}) + \cdots, \end{aligned}$$

where $Y^k = Y \times \cdots \times Y$ (k times), and $Y_{k-1}^k = \{(y_1, \ldots, y_k) \in Y^k | y_i = y_j$ for some $i \neq j\}$ generalizes $\Delta \subseteq Y^2$. This generalized diagonal expresses the fact that multiplying G-F classes on the same component $\mathbb{R}^n$ gives 0 since multiplication in $H^*_{GF}(\mathbb{R}^n)$ is trivial. This is the Künneth formula made functorial with respect to maps.

Recall our second spectral sequence which converges to $H^*_{GF}(M)$ and begins with:

$$'E_1^{p,q} = H^p_{GF}(U_q(M))$$
$$'E_2^{p,q} = H_q^\delta H^p_{GF}(U_*(M))$$

But now we use our functorial description of $H^*_{GF}(U_q(M))$:

$$\begin{array}{lll} 'E_1^{*,q} \simeq H_0(U_q(M); G) & \oplus\, H_0^\alpha(U_q(M) \times U_q(M), \Delta; G \otimes G) & \oplus \cdots \\ \downarrow \delta & \downarrow \delta & \downarrow \delta \\ 'E_1^{*,q-1} \simeq H_0(U_{q-1}(M); G) & \oplus\, H_0^\alpha(U_{q-1}(M) \times U_{q-1}(M), \Delta; G \otimes G) & \oplus \cdots \end{array}$$

taking the δ-homology (Čech) we have

$$'E_2^{*,q} = H_q^\delta \widetilde{H_{GF}^*}(U_*M) = Hq(M;G) \oplus H_q^\alpha(M \times M, \Delta; G \otimes G) \otimes \cdots$$

since $U(M)$ is a good covering of M in the Čech theory. This is the original spectral sequence found in Gel'fand–Fuks. ([26] and [27].)

$$'E_2^{*,q} = \oplus_k H_q^\alpha(M^k, M_{k-1}^k; \otimes^k G).$$

These $'E_2$ terms appear only in certain rectangles (different rectangles for each summand k) shown on the following diagram for $k = 1, 2$:

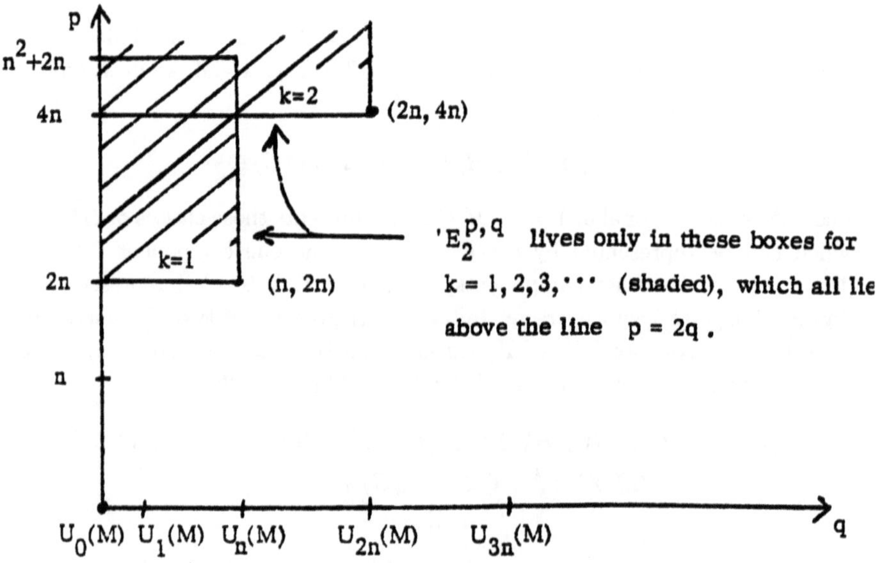

Recall that $G^p \equiv \widetilde{H_{GF}^p}(\mathbb{R}^n) = 0$ whenever $p \leq 2n$ (or whenever $p > n^2 + 2n$). Therefore, $\otimes^k G = 0$ in dimensions $p \leq 2kn$. This gives the lower edge of our boxes. But we also have (for $k = 1$), $H^q(M, G) = 0$ in dimensions $q > n$ since M is n-dimensional. In general, $H^q(M^k, M_{k-1}^k; \otimes^k G) = 0$ whenever $q > kn = \dim M$. This gives the right edge of our boxes. The boxes all live above the line $p = 2q$, so any line $p - q =$ some constant r can only hit finitely many boxes. Since each box is finite dimensional, so is $\oplus'_{p-q=r} E_2^{p,q}$. Therefore, by the argument at the end of Section 24 (cf. [26] 8.7) we have:

Theorem 25.1. (Gel'fand–Fuks) $H_{GF}^r(M)$ *is finite dimensional for each* r.

26. The conjecture

The problem is to represent the Gel'fand–Fuks cohomology $H^*_{GF}(M)$ of a manifold, M, as the usual cohomology of some topological space, Z. That is, is there a Z such that $H^*_{GF}(M) \simeq H^*(Z, \mathbb{R})$? For $M = \mathbb{R}^n$, we can set $Z = Y_n$, which was constructed for Theorem 20.1. Indeed, $H^*_{GF}(\mathbb{R}^n) = H^*(\mathfrak{a}_n) = H^*(Y_n)$.

Secondly, we are led to a Künneth formula $H^*_{GF}(M \coprod N) = H^*_{GF}(M) \otimes H^*_{GF}(N)$ for disjoint unions which we have proved for $M = N = \mathbb{R}^n$. This sort of behavior appears in the cohomology of *function spaces.* That is $Z =$ Maps $\{M, Y\}$, for some fixed Y. For then Maps $\{M \coprod N, Y\}$ Maps $\{M, Y\} \times$ Maps $\{N, Y\}$ for disjoint unions, which suggests a Künneth formula. Futhermore, for $M = \mathbb{R}^n$, H^*(Maps $\{\mathbb{R}^n, Y\}) = H^*(Y) = H^*_{GF}(\mathbb{R}^n)$ will hold if $Y = Y_n$. But for general M, we will have to take account of the gluing of the charts on M by semi simplicial methods.

Consider the following situation: a fiber bundle with fiber Y:

$$\begin{array}{ccc} Y & \longrightarrow & E \\ & & \downarrow \pi \\ & & M \end{array}$$

Instead of the function space Maps $\{M, Y\}$ we will consider the sections $\Gamma(M)$ of the bundle. Now replace M by $U(M)$ for a good cover. Restricting the bundle to each of the opens $=\mathbb{R}^n$, $E|_{U_a} = U_a \times Y$ is trivial, since $U_a \simeq \mathbb{R}^n$ is contractible (same for $U_a \cap U_b$ etc.) Then $\Gamma(U_a)$ =Maps $\{U_a, Y\} \sim Y$. Replacing M by $U(M)$, we get a co-semi-simplicial space

$$\Gamma(U_0(M)) \rightrightarrows \Gamma(U_1(M)) \underset{\longrightarrow}{\rightrightarrows} \cdots$$

whose geometric realization has the homotopy type of $\Gamma(M)$. Now we apply the functor $S^* =$ singular cochains to get a semi-simplicial module for each positive integer p.

$$S^p\Gamma_0 M \overset{\delta_0}{\underset{\delta_1}{\leftleftarrows}} S^p\Gamma U_1 M \overset{\delta_0}{\underset{\delta_2}{\leftleftarrows}} \cdots$$

The resulting double complex is half homology and half cohomology since $d : S^p\Gamma U_q M \to S^{p+1}\Gamma U_q M$ is the singular coboundary and $\delta : S^p\Gamma U_q M \to S^p\Gamma U_{q-1} M$ is the alternating sum of the δ_i maps. The total derivative $D = d \pm \delta$ has total degree n cochains along the $p - q = n$ lines:

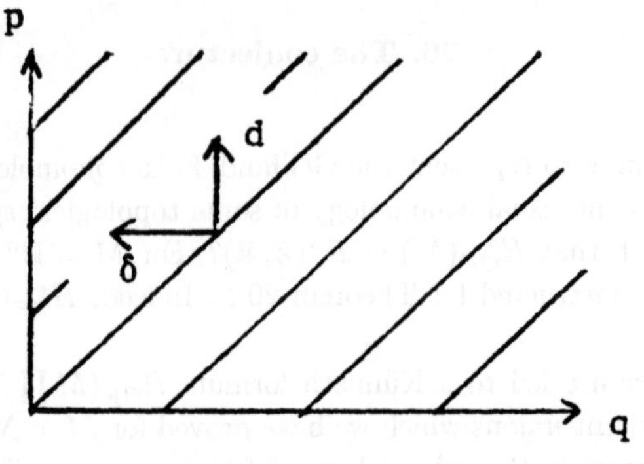

Therefore, we need certain dimension restrictions to get convergence of the spectral sequence. The spectral sequence of Anderson [1] does exactly this for us. $H^*_D(S^*\Gamma UM) = H^*(\Gamma M)$ under the condition that the fiber Y be acyclic through dim M. $\Gamma(U(M))$ has the homotopy type of copies of the fiber Y, since the base $U_a, U_a \cap U_b$, etc. are all contractible. Therefore, the E_2 term is the same as the ${}'E_2^{p,q} = H_q^\delta H_{GF}^p(U(M))$ term of Sections 24 and 25, if we take fiber, Y, as a space with the same cohomology as $H^*(\mathfrak{a}_n) = H^*_{GF}(\mathbb{R}^n)$. We let $Y = Y_n$. How can we make a bundle over M with fiber Y_n? Let $F(M) \to M$ be the frame bundle of M (= the principal bundle associated to the tangent bundle of M). Since $GL(n, \mathbb{R}) \subseteq GL(n, \mathbb{C})$ acts on Y_n we can form Y_nM, the associated bundle with fiber Y_n by taking

$$\begin{array}{ccc} Y_n & \longrightarrow & Y_nM \;= F(M)X_{GL(n,\mathbb{R})}Y_n \\ & & \downarrow \\ & & M \end{array}$$

We are now in a position to state the

Conjecture 26.1. $H^*_{GF}(M) = H^*(\Gamma(Y_nM))$.

Exercise. Why is the conjecture true for $M = \mathbb{R}^n$?

Example. $M = S^1$. $Y_1 = S^3$. $F(M) = S^1 \times GL(1)$ is trivial. Therefore, $Y_1S^1 = S^1 \times S^3$. Hence, the sections, $\Gamma(Y_1S^1) = \text{Maps}\,(S^1, S^3)$. This function space has the cohomology of $H^*(S^3) \otimes H^*(\Omega S^3) = E(\omega) \otimes P(y) =$ Exterior algebra generated by ω in dim $= 3$, tensored with the polynomial algebra generated by y in dim $=2$. This is because we have a Serre fibration

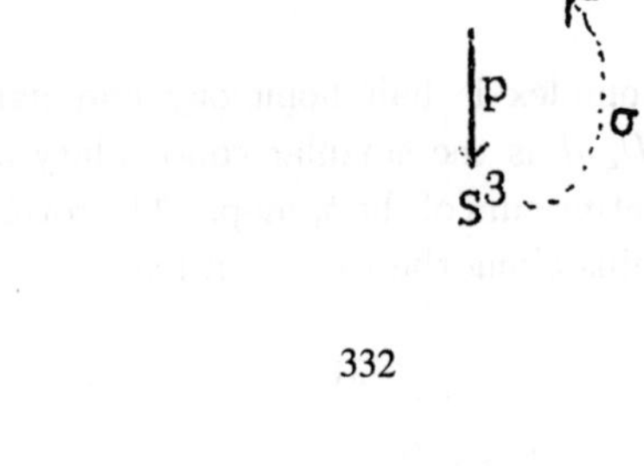

where p is evaluation at a base point of S^1. The fiber is ΩS^3 = loop space of S^3. The section ∂ takes a point $x \in S^3$ to the constant map $S^1 \to \{x\}$. The Serre spectral sequence ([58]) has E_2 term $H^*(S^3) \otimes H^*(\Omega S^3) = E(\omega) \otimes P(y)$ (S^3 is simply connected so we do not need local coefficients). ω persits to E_∞ because of ∂. $H^*(\Omega S^3)$ is calculated similarly as in [58].

For $H^*_{GF}(S^n)$, $n \geq 2$ this computation gives a different answer than that announced by Gel'fand–Fuks. However, in this case, the conjecture can be verified by another method due to Haefliger to compute $H^*_{GF}(S^n)$.

27. Foliations

In the concluding sections, we wish to interpret $H^*(\mathfrak{a}_n)$ as characteristic classes for foliations. Let $E \to M$ be a sub bundle of the tangent bundle $TM \to M$ of an n-dimensional manifold M. E is integrable if and only if $\Gamma(E)$ = {sections of $E \to M$} is closed under bracket production [,]. The Frobenius Theorem gives a local description:

Theorem 27.1. (Frobenius) *Locally, every integrable $E \subset TM$ is given as the kernel of df where f is a submersion.*

A submersion is a map $f : M^n \to N^q$ between C^1 manifolds such that for all $x \in M$, $(df) : T_xM \to T_{f(x)}N$ is onto (in particular $q \leq n$). The implicit function theorem tells us that locally (i.e., on some small chart), any submersion looks like a projection $\mathbb{R}^n \to \mathbb{R}^q$. Now a projection $\mathbb{R}^n \to \mathbb{R}^q$ foliates $\mathbb{R}^n$, where the leaves are $\mathbb{R}^{n-q}$'s, the inverse images of points in $\mathbb{R}^q$. Therefore, any integrable E is tangent to a foliation, defining a system of local projections as follows: choose a covering $\{U_a\}$ of M such that

1) $f_a : U_a \to \mathbb{R}^q$ is a submersion (q = codimension of $E \subset TM$)
2) $E|_{U_a} = \ker\,(df_a)$

Then on $U_a \cap U_b$ we have two maps into $\mathbb{R}^q$, f_a and f_b. By the implicit function theorem, f_b is a function of f_a (and vice versa) on some smaller neighborhood. So for any $x \in U_a \cap U_b$, there will be a smaller neighborhood U of x such that:

$$f_b|_U = \gamma^U_{ba} \circ f_a|_U$$

where γ^u_{ba} is a local diffeomorphism of $\mathbb{R}^q$, $\gamma^U_{ba} : f_a(u) \to f_b(U)$. This is easier to state in terms of the germs. For all $x \in U_a \cap U_b$, there exists a germ of a diffeomorphism γ^x_{ba} from $f_a(x)$ to $f_b(x)$ such that

$$f^x_b = \gamma^x_{ba} \circ f^x_a$$

where f^x_a and f^x_b are the germs of f_a and f_b at x. Again by the implicit function theorem, we must have a cocycle condition:

$$\gamma^x_{ab} = \gamma^x_{ac} \circ \gamma^x_{cb} \quad \text{for} \quad x \in U_a \cap U_b \cap U_c.$$

This is just like bundle theory except that the structure (pseudo-) group (where γ^x_{ab} takes its values) is Γ_q, the germs of diffeomorphisms $\mathbb{R}^q \to \mathbb{R}^q$. If Γ_q is given the sheaf topology, the map $x \to \gamma^x_{ab}$ is continuous. We can imitate the classifying space construction for groups ([34] and [35]). For foliations, the differential, $d(\gamma^x_{ab})$ of the germ $\gamma^x_{ab} : \mathbb{R}^n \to \mathbb{R}^n$ is the $GL(q, \mathbb{R})$-valued cocycle of the normal bundle, TM/E, of the foliation.

Remark. This procedure extends the definition of a manifold so that the f_U's are submersions, not necessarily diffeomorphisms (charts). In this context, a codimension $q = n$ foliation of an n-dimensional manifold (where the "leaves" are just points) is the old definition of manifold, and the normal bundle to this trivial foliation is just the tangent bundle.

Now recall our procedure for characteristic classes for q-manifolds N^q via tautologous forms on the frame bundles. The mysterious classes were all in dimension $> q$, and, therefore, trivial in $H^*(N^q)$. However, these classes may be non-trivial on M^n with a codimension q foliation, $q < n$. A codim q foliation is like a q-manifold structure on M^n and we get a natural map:

$$H^*(\mathfrak{a}_q, so_q) \longrightarrow H^*(M)$$

Through dimension $2q$, $H^*(\mathfrak{a}_q, so_q)$ gives rise to the usual (Pontryagin, Euler) classes of the normal bundle TM/E of the foliation. (Via a Riemannian metric on M, the tangent bundle splits $TM = E \oplus TM/E$, and the normal bundle can be realized as vectors perpendicular to the foliation.) Technically, we must assume that the normal bundle be oriented, for then we can choose our γ^x_{ab} to be orientation preserving on $\mathbb{R}^q$. Otherwise we must replace SO_q above by O_q and we do not get an Euler class. Above dimension $2q$, $H^*(\mathfrak{a}_q, so_q)$ gives rise to exotic classes in $H^*(M)$.

For example, $H^*(\mathfrak{a}_1) = \mathbb{R}$ in dimensions 0 and 3. Since $SO_{(1)}$ is trivial class 1 in dim $=0$ and ω in dim $=3$. This is the Godbillon–Vey class for codimension 1 foliations (cf. [31]). An example of a foliated 3-manifold for which ω is non zero was found by Roussarie (see [6]). A large problem in the theory is to find examples of foliated manifolds with non-trivial exotic classes. Thurston has found many interesting examples of these (See [66], [68].)

Remark. In our earlier construction of canonical 1-forms on the higher frame bundles $\cdots \to F_2(M) \to F_1(M) \to M$, we consider the charts of diffeomorphisms $\mathbb{R}^n \to M$ (dim $M = n$). But we could equally well have taken our maps in the other direction, from open sets of M to $\mathbb{R}^n$.

For a fixed codim q foliation of M, consider the space G of germs of the local projections $f_a : u_a \to \mathbb{R}^q$ (which are constant along the leaves). We might as well assume that the target of each germ is $0 \in \mathbb{R}^q$ or compose with a translation to force this.

We can now take equivalence classes (jets) of the f_a's.

$F_0(M) = $ 0-jets $=$ M via the source map

$F_1(M) = $ 1-jets $=$ frame bundle of normal bundle of the foliation

$F_k(M) = $ k-jets

$F_\infty(M) = \varprojlim F_R(M)$ via the projections $F_k(M) \to F_{k-1}(M)$.

In our earlier case, $q = n = \dim M$, $F_1(M) = $ frame bundle of M is the frame bundle of normal bundle of the foliation by points. Then as before, we have tautologous forms ω^i on our new $F_1(M), \omega^i_j$ on $F_2(M)$, ω^i_{jk} on $F_3(M)$, and so forth, $i, j, k, \ldots = 1, 2, \ldots, q$. And we get a natural map $A^*_c(\mathfrak{a}_q) \to F_\infty(M)$ given by $\delta^i \to \omega^i$, $\delta^i_j \to \omega^i_j$ etc. (Cf. [36].) For a tangent vector X on $F_R(M)$ represented by $f_t : U_t \to \mathbb{R}^q$, $U_t \subset M$. We have the following picture:

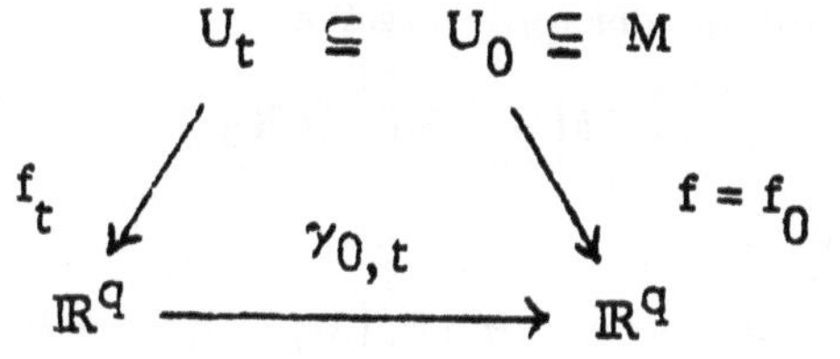

Let X be a tangent vector at the point $j^k_x f_0 \in F_R(M)$, where $x \in M$ is the source of $f_0 : U_0 \to \mathbb{R}^q$, $f_0(x) = 0$. X is represented by a curve $t \to j^k_{x(t)} f_t$ (where $f_t : U_t \to \mathbb{R}^q$; at $t = 0$ we get f_0; $f_t(x(t)) = 0$; and t is small enough so that $x(t) \in U_0$, $x(0) = x$). Let $\gamma_{0,t}$ be the diffeomorphism of $\mathbb{R}^q$ such that $f_0 = \gamma_{0,t} \circ f_t$ holds in a neighborhood of $x(t)$. Then $t \to \gamma_{0,t}(p)$ defines a family of C^1 curves through points p near $0 \in \mathbb{R}^q$. (For example, $\gamma_{0,t}(0) = f_0(x(t))$ is the curve through the origin.) Let $Y = d\gamma_{0,t}/dt$ be the vector field tangent to them. Since f_0, f_t, and $\gamma_{0,t}$ are defined only up to their k-jets at $0 \in \mathbb{R}^q$, Y is defined only up to order $k - 1$. Therefore, $\delta^i(Y), \delta^i_{j_1}(Y), \ldots, \delta^i_{j_1 \cdots j_{k-1}}(Y)$ are defined and we define these to be the values of $\omega^i(X), \omega^i_j(X), \ldots, \omega^i_{j_1 \cdots j_{k-1}}(X)$ on $F_k(M)$. Thus ω^i is defined on $F_1(M)$, ω^i and ω^i_j are defined on $F_2(M)$, etc. This is the map $A^*_c(\mathfrak{a}_n) \to$ tautological differential forms on $F_\infty(M)$.

$$
\begin{array}{c}
F_\infty(M) \\
\downarrow \\
\vdots \\
\downarrow \\
F_3(M) \sim \omega^i_{jk} \\
\downarrow \\
F_2(M) \sim \omega^i_j \\
\downarrow \\
F_1(M) \sim \omega^i \\
\downarrow \\
M
\end{array}
$$

The induced map in cohomology is $H^*(A_c(\mathfrak{a}_q)) \to H^*(F_\infty(M)) \to H^*(F_1(M))$ where the second arrow is via the sections $F_1(M) \to F_{i+1}(M)$ $i = 1, 2, \cdots$ which are defined uniquely up to homotopy. (The fibers of $F_{i+1}(M) \to F_i(M)$ are all affine spaces and thus contractible.) Now if the foliation of M has a trivialized normal bundle, then we have a section $M \to F_1(M)$ which pulls classes all the way down to $H^*(M)$. We get a map

$$H^*(A_c(\mathfrak{a}_q)) \to H^*(M).$$

Otherwise, we must divide out by the (maximal compact) subgroup $O(q)$ of $J \equiv$ infinite jets at the origin of diffeomorphisms $\mathbb{R}^q \to \mathbb{R}^q$ taking 0 to 0. If the normal bundle to the foliation is oriented, we can get away with $SO(q)$, since the structure group, J, of $F_\infty(M)$ can be reduced to the orientation preserving jets.

This defines (up to homotopy) a section:

$$M \longrightarrow F_1(M)/SO(q).$$

The natural map

$$H^*(\mathfrak{a}_q, so(q)) \to H^*(F_\infty(M)/SO(q)) \to H^*(M)$$

gives characteristic classes on M induced directly by the foliation. In the unoriented case, we get a map $H^*(\mathfrak{a}_q, O_q)) \to H^*(M)$. On $F_2(M)$ we have the formula (as in $A_c^*(\mathfrak{a}_q)$)

$$d\omega^i = -\sum \omega^i_k \wedge \omega^k$$

On $F_1(M)$ we do not have canonical ω^i_k forms, but the Frobenius theorem tells us that for an (integrable) foliation, there exists forms θ^i_k such that

$$d\omega^i = -\sum \theta^i_k \wedge \omega^k$$

The θ^i_k can be given as the pullbacks of the ω^i_k via a section $F_1(M) \to F_2(M)$. This section is equivariant with the action of $GL(n, \mathbb{R})$ if and only if the (θ^i_j) matrices transform by the adjoint action of $GL(n, \mathbb{R})$, that is the matrix (θ^i_j) is a connection on $F_1(M)$. Then the curvature is a matrix defined by

$$\Omega^i_j = d\theta^i_j - \theta^k_j \wedge \theta^i_k = \omega^k \wedge \theta^i_{jk}$$

where θ^i_{jk} is the pullback of ω^i_{jk} from $F_3(M)$ via a section. The invariant polynomials on the curvature matrix (Ω^i_j) are generated by $c_1, \ldots, c_q$ which are (up to a scalar multiple) the coefficients of the characteristic polynomial of (Ω^i_j). The image of c_i in $H^{2i}(M)$ is the i-th Chern class of the normal bundle to the foliation. (See [6] for the definition of Chern classes via a curvature matrix.)

We obtain a refinement of the vanishing theorem (cf. [5]) that the ring of Pontryagin classes of the normal bundle to codimension-q foliation vanish in dim $> 2q$ in $H^*(M)$. Let

$$\begin{aligned} W_q &= E(h_1 h_2, \ldots, h_q) \otimes p(c_1, c_2, \ldots, c_q) \\ &= \text{exterior alegbra in } \{h_i\} \text{ tensor the polynomial algebra in } \{c_i\} \end{aligned}$$

We set $\dim h_i = 2i - 1$, $\dim c_i = 2i$. $d : W_q \to W_q$ is defined by $d(h_i) = c_i$, $d(c_i) = 0$. Recall that $H^*(W_q) = 0$. Let

$$WO_q = E(h_1, h_3, \ldots, h_{2k+1}) \otimes P(c_1, c_2, \ldots, c_q).$$

be the same except that we throw away all the even h_{2i}. $2k + 1 \leq q$ but $2k + 3 > q$. Let

$$W_q = W_q/\text{terms of dim} > 2q \text{ in the } c_i\text{'s}.$$
$$WO_q = WO_q/\text{terms of dim} > 2q \text{ in the } c_i\text{'s}.$$

We have seen (Theorem 20.1) that

$$H^*(\mathfrak{a}_q) = H^*(W_q)'$$

Also, it turns out (see [[10]) that

$$H^*(\mathfrak{a}_q, O(q) = H^*(WO_q)$$

$$H^*(\mathfrak{a}_q, so_q) = \begin{cases} H^*(WO_q), & q \text{ odd} \\ H^*(WO_q)[\chi]/(\chi^2 - c_q), & q \text{ even.} \end{cases}$$

In $H^*(\mathfrak{a}_q, so_q)$ we find the Pontryagin classes $c_2, c_4, \ldots$ in dimensions 4,8,..., and, if q is even, the Euler class, χ, of the normal bundle. The Pontryagin classes all vanish in dimension $> 2q$ by the truncation. In $H^*(\mathfrak{a}_q)$ all of the Chern classes $c_1, \ldots, c_q$ (not just the odd ones) are coboundaries. This is reasonable since $H^*(\mathfrak{a}_q)$ gives characteristic classes in $H^*(M)$ when the normal bundle is trivial. Exotic classes (e.g., $h_1 c_q$) occur from dimensions $2q + 1$ through $q^2 + 2q$.

28. Haefliger's classifying space for foliations

We begin with the notion of a topological category. (Cf. [6], [56], [34], and [35].) This means that the objects, θ, and the morphisms, m, are both toopological spaces and that the maps

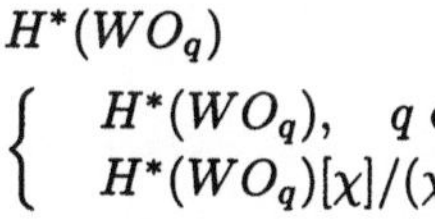

and the composition of (composable) morphisms are all continuous. The dotted arrow takes an object, A, to the idenity morphism $A \to A$.

Example. Let G be a topological group. Let θ = points of G, and let m = left translations. For any two objects $g, h \in G$ there is exactly one morphism, $L_{hg} - 1$, between them. Therefore, $m \simeq G \times G$. The three maps

$$\begin{array}{rcl} G \leftleftarrows G \times G & & \text{given by:} \\ g \longleftarrow (g,h) & & \text{source} \\ x \longrightarrow (x,x) & & \text{identity} \\ h \longleftarrow (g,h) & & \text{target} \end{array}$$

are all continuous.

Another Example. Let $\theta =^*$ a single point, and let $m = G$ with composition as in G. We get three fairly trivial maps:

$$* \leftleftarrows\!\!\dashrightarrow G \qquad \text{where} \quad * \dashrightarrow \mathrm{id} \in G.$$

We now wish to define a topological category, Γ_q. The objects and morphisms are given by

$$\begin{aligned} \theta(\Gamma_q) &= \text{points of } \mathbb{R}^q \\ m(\Gamma_q) &= \text{germs of diffeomorphisms of } \mathbb{R}^q. \end{aligned}$$

$m(\Gamma_q)$ has two possible topologies of interest. In the soft (C^∞) topology, two germs are close if their source and target, as well as derivatives up to some order, are close. But we will take here the sheaf topology: if $f : U \to V$ is a diffeomorphism with $x \in U$, let f^x denote the germ at x. A neighborhood of f^x is $\{f^y\}_{y \in U}$. This is a very non-Hausdorff topology and the induced topology on the germs with source x is discrete. The maps

$$\mathbb{R}^q = \theta \underset{\text{target}}{\overset{\text{source}}{\leftleftarrows}} m$$

are local homeomorphisms and give m a (non-Hausdorff) smooth n-manifold structure. Recall how we constructed the classifying space BG from the semi-simplicial space NG

$$* \leftleftarrows G \substack{\longleftarrow \\ \longleftarrow \\ \longleftarrow} G \times G \cdots$$

Now consider the semi-simplicial space $N\Gamma_q$

$$m_0\Gamma_q \underset{\text{target}}{\overset{\text{source}}{\leftleftarrows}} m_1(\Gamma_q) \substack{\longleftarrow \\ \longleftarrow \\ \longleftarrow} m_2(\Gamma_q) \substack{\longleftarrow \\ \longleftarrow \\ \longleftarrow \\ \longleftarrow} \cdots$$

where

$m_0\Gamma_q = \mathbb{R}^q = \text{objects} \quad \theta(\Gamma_q)$

$m_1(\Gamma_q) = m(\Gamma_q) = \text{germs of diffeomorphisms of} \quad \mathbb{R}^q$

$m_2(\Gamma_q) =$ pairs of composable germs (i.e, target one = source of other)

$m_3(\Gamma_q) =$ composable triples, etc.

The $k+1$ maps $m_k(\Gamma_q) \to m_{k-1}(\Gamma_q)$ are given by the same formulas as for NG.

$$\partial_i(g_1, \ldots, g_k) = \begin{cases} (g_2, \ldots, g_k) & i = 0 \\ (g_1, \ldots, g_i \circ g_{i+1}, \ldots, g_k) & 0 < i < k \\ (g_1, \ldots, g_{k-1}) & i = k. \end{cases}$$

The ordinary geometric realization is $|N\Gamma_q|$, but to get a classifying space we must use the unwound space $B\Gamma_q = |(N\Gamma_q)_z|$ (see Sections 4 and 6 above, and [6], Stasheff's appendix). Γ_q with the sheaf topology is the "structure group" for foliations in the sense that the map

$$U_a \cap U_b \longrightarrow m(\Gamma_q)$$

given by

$$x \longrightarrow \gamma^x_{ab}$$

is continuous in the sheaf topology. A foliation on M induces a homotopy class of maps

$$M \longrightarrow B\Gamma_q,$$

so that the characteristic classes for foliations are to be found in $H^*(B\Gamma_q)$. Thus we have a map $H^*(\mathfrak{a}_q, O(q)) \to H^*(B\Gamma_q)$. However, we must identify foliations by a suitable notion of homotopy to make a homotopy class of maps $M \to B\Gamma_q$ represent a unique (homotopy class of) foliation. (See [6], [35].)

Actually, for any topological space, X, a map $X \to B\Gamma_q$ makes sense, but what would a foliation on X be? We see that $B\Gamma_q$ actually classifies Haefliger structures (defined below), of which foliations are a special case. If X is paracompact, we have ([35] p. 141):

Theorem 28.1. (Haefliger). *Homotopy classes of codim q Haefliger structures on X are in one-to-one correspondence with homotopy classes of continuous maps $X \to B\Gamma_q$.*

Definition. A codimension q Haefliger structure on a space X is defined by an open cover $\{U_a\}$ of X and continuous maps

$$f_a : U_a \longrightarrow \mathbb{R}^q$$

not necessarily submersions, and a continuous map

$$\begin{aligned} U_a \cap U_b &\longrightarrow m(\Gamma_q) \\ x &\longrightarrow \gamma_{ab}^x \end{aligned}$$

such that

(1) $f_a^x = \gamma_{ab}^x f_b^x$
(2) $\gamma_{ab}^x = \gamma_{ac}^x \gamma_{cb}^x$ for $x \in U_a \cap U_b \cap U_c$

For a foliation, (2) follows from (1) since the f_a are submersions in this special case. Even if all the f_a are constant maps to $0 \in \mathbb{R}^q$, we get a non-trivial Haefliger structure if we take the γ_{ab}^x to be non-trivial. The normal bundle of a Haefliger structure is the vector bundle whose $GL(q, \mathbb{R})$-valued transition functions, g_{ab}, are given by $d(\gamma_{ab}^x)$, the differential of γ_{ab}^x as a map $\mathbb{R}^q \to \mathbb{R}^q$.

The advantage of passing to Haefliger structures is that any continuous map, $N \to M$, between manifolds pulls back Haefliger structures on M to ones on N. For foliations, this only works if $N \to M$ is transverse to the leaves on M.

We will not in general have smooth leaves from a Haefliger structure. Locally the level sets (inverse images of points of $\mathbb{R}^q$) are defined, but these can have singularities. Nevertheless, these can be smoothed out in certain cases. By the Phillips–Gromov theorem, a Haefliger structure on an open (no compact component) manifold M whose normal bundle is a subbundle of the tangent bundle of M is homotopic to a smooth foliation ([35]). Thurston has a similar result for compact manifolds, when codim. of the foliation ≥ 2. [67]

Remark. The derivative of a germ $\nu : \Gamma_q \to GL(q, \mathbb{R})$ induces maps

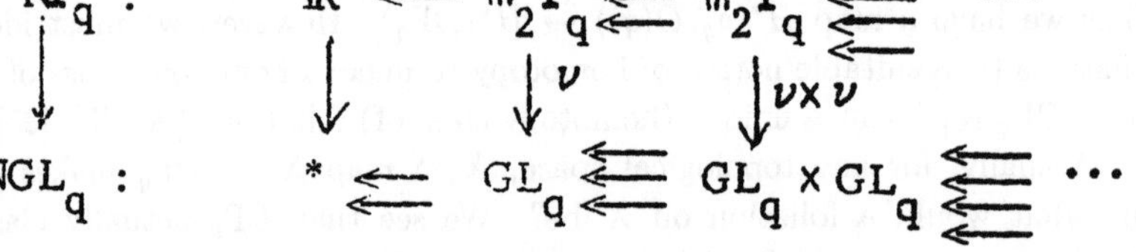

and, therefore, induces a map

$$B\Gamma_q \xrightarrow{\nu} BGL_q$$

so that if $f : M \to B\Gamma_q$ classifies a foliation on M, then $\nu \circ f$ classifies the normal bundle. Then the vanishing theorem ([5]) says

Theorem 28.2. (Bott) *$\nu^* : H^*(BGL_q) \to H^*(B\Gamma_q)$ is 0 in dimensions $> 2q$.*

$H^*(BGL_q)$ and $H^*(B\Gamma_q)$ are equal to the total cohomology of the double complexes $S^p(GL_q \times \cdots \times GL_q)$ (k factors) and $S^p(m_k(\Gamma_q))$, where S^p = singular p-cochains. (See Theorems 4.1, 4.2, 4.4.) Now I claim

$$\nu^* : H^p(GL_q \times \cdots \times GL_q) \to H^p(m_k(\Gamma_q))$$

is 0 in dimension $p > q$. This is because ν^* factors through a map:

$$\begin{array}{ccc} H^p(GL_q \times \cdots \times GL_q) & \xrightarrow{\nu^*} & H^p(m_k(\Gamma_q)) \\ \simeq \downarrow & & \uparrow \\ H^p_{DR}(GL_q \times \cdots \times GL_q) & \longrightarrow & H^p_{DR}(m_k(\Gamma_q)) \end{array}$$

where H^p_{DR} = DeRham cohomology and the bottom arrow is zero for $p > q = \dim m_k(\Gamma_q)$ (sheaf topology!).

Why do we then have a vanishing theorem for dim $> 2q$? Recall that a 3-form on $m_2(\Gamma_q)$ counts as a form of total dimension 3+2 =5. For definiteness, we illustrate with q=4.

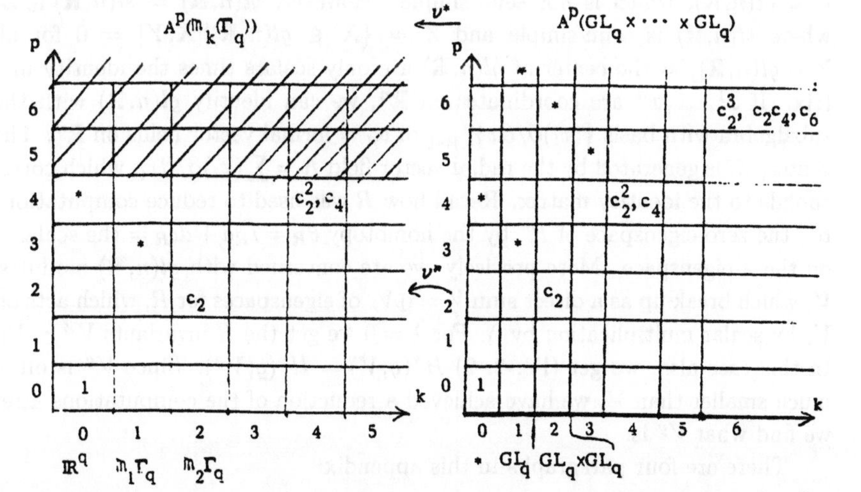

In the double complex, $A^p(m_k(\Gamma_q)) = 0$ for $p > q$ (in the shaded region). In $A^p(GL_q \times \cdots \times GL_q)$, the Pontryagin classes appear on the diagonal as E_∞ terms which are extended to total $D = d \pm \delta$ cocycles by adding forms, *, up and to the left only. (See Section 3.) For $q = 4$ as shown, the classes c_2^3, c_2c_4, c_6

together with the forms to their upper left go to 0 (the shaded region). The highest degrees shown on the left which are non-zero are c_2^2 and c_4 in the (4,4) box with total degree 4+4 =8 =2q.

Appendix

Introduction

We briefly review some of the standard facts about Lie algebras. (See [59], [22].) We define the cohomology of a Lie algebra, g, with coefficients in a g-module, V, denoted by $H^*(g, V)$ or $H^*(g, \rho)$ where $\rho : g \to \text{End } V$ is the corresponding representation. If k is the base field (characteristic 0, but we only need $k = \mathbb{R}$ or $k = \mathbb{C}$) and g acts trivially on k, then $H^*(g, k)$ is just $H^*(g)$ as defined earlier. The basic results are (II.2) the following are equivalent:

(i) g has no Abelian ideals. (Abelian means $[X, Y] = 0$ for all X, Y.)
(ii) Every g-module, V, is a direct sum of simple submodules $V = \oplus V_i$, where each V_i has non-trivial submodules.

Such g are called *semi-simple* Lie algebras. Let $V^g = \{v \in V | \rho_X(v) = 0$ for all $X \in g\}$ be the *invariants* of V. Then we also show (IV. 2. 1) if g is semi-simple then $H^*(g, V) \simeq H^*(g, V^g) =$ as many copies of $H^*(g)$ as the dimension of V^g. (See [14] for details.)

In the application to Gel'fand–Fuks cohomology, we will want to use $g = gl(n, \mathbb{R})$, which is not semi-simple. However, $gl(n, \mathbb{R}) = sl(n, \mathbb{R}) \oplus Z$, where $sl(n, \mathbb{R})$ is semi-simple and $Z = \{X \in gl(n, \mathbb{R}) | [X, Y] = 0$ for all $Y \in gl(n, \mathbb{R})\} =$ the center of $gl(n, \mathbb{R})$ are only scalars times the identity matrix. If $x^1, \ldots, x^n$ are coordinates on $\mathbb{R}^n$, we can identify $gl(n, \mathbb{R})$ with the subalgebra with basis $\{(x_i)\partial/\partial x\}_{i,j=1}^n$ of $\mathfrak{a}_n =$ formal vector fields on $\mathbb{R}^n$. The center, Z, is generated by the radial vector field $R = \sum(x_i)\partial/\partial x_i$, which corresponds to the identity matrix. Recall how R was used to reduce computations to "the zero-eigenspace of R" by the homotopy $\theta_R = \iota_R d + d\iota_R =$ the scalar λ on the λ-eigenspace. More precisely, we are concerned with $gl(n, \mathbb{R})$-modules, V, which break up as a direct sum $V = \oplus V_\lambda$ of eigenspaces for R, which acts on V_λ by scalar multiplication by λ. For $\lambda = 0$ we get the R-invariants $V^Z = V_0$. In this case also, we get (IV. 2. 2) $H^*(g, V) = H^*(g, V^g)$. Since V^g is often much smaller than V, we have achieved a reduction of the computations after we find what V^g is.

There are four paragraphs in this appendix:

I. Bilinear form of a representation
II. No Abelian ideals in $g \Leftrightarrow$ all g-modules are semi-simple
III. Special case: $sl(n, k)$
IV. Cohomology with coefficients in a module

I. In this section, we show that the bilinear form, B_ρ, of a faithful representation of a semi-simple Lie algebra is non-degenerate.

Definitions. g will denote a finite dimensional Lie algebra over a field k of characteristic 0. Define $C_g^1 = D^1 g = g$, and $C^r(g) = [g, C^{r-1}(g)]$, $D^r(g) = [D^{r-1}g, D^{r-1}g]$ for $r > 1$ an integer. Then g is *nilpotent* (resp. *solvable*) if $C^n(g) = 0$ (resp. $D^n(g) = 0$) for some integer n. g is *semi-simple (s.s)* if g has no Abelian ideals (other than 0). V will denote a finite dimensional k-vector space. A *representation* of g is a Lie algebra homomorphism $\rho : g \to \mathrm{End}_k(V)$. ρ equips V with a *g-module* structure. ρ is *simple* if V contains no non-trivial sub g-modules. ρ is *semi-simple* if V is a direct sum of simple sub g-modules. By finite dimensionality, this is equivalent to saying that for any sub g-module W of V, $V \simeq W \oplus V/W$. $\rho : g \to \mathrm{End}_k(V)$ is *faithful* if it is one-to-one. If V is a g-module (via ρ) we often write $X \cdot v$ instead of $\rho_X(v)$ for $X \in g, v \in V$. A g-module structure is defined on V^* = dual of a g-module V, given by $(X \cdot f)(v) = -f(X \cdot v)$, for $X \in g$, $f \in V^*$, $v \in V$. If V and W are g-modules, then $V \oplus W$ and $V \otimes W$ are g-modules: For $X \in g$, $v \in V$, $w \in W$, we define $X \cdot (v \oplus w) = (X \cdot v) \oplus (X \cdot w)$ and $X \cdot (v \otimes w) = (X \cdot v) \otimes w + v \otimes (X \cdot w)$. (These are just the rules for the derivative of sums and products when g is the Lie algebra of a Lie group acting on V and W.) In particular if V is a g-module, then so is the tensor algebra $\oplus(V \otimes \cdots \otimes V \otimes V^* \otimes \cdots \otimes V^*)$. The alternating and symmetric summands, $\Lambda^* V$ and $S^* V$ (and also $\Lambda^* V^*$ and $S^* V^*$) are, thus, g-modules.

Theorem I.1. *If g is semi-simple and ρ is faithful, then the quadratic form $B_\rho(X, Y) = Tr_V(\rho_X \rho_Y)$, for X and $Y \in g$, is non-degenerate.*

Proof. Let $b = \{X \in g | Tr(\rho_x \rho_Y) = 0 \text{ for all } Y \in g\}$. To show $b = 0$. b is an ideal since $B_\rho([X, Y], Z) = B_\rho(X, [Y, Z])$ for all $X, Y, Z \in g$. We will show that b is solvable. But then if $b \neq 0$, the last non-zero derived, $D^{n-1}(b)$ will be an Abelian ideal of g, contradicting semi-simplicity. We will show that b is solvable by showing that $[b, b]$ is nilpotent. We use only the fact that $B_\rho(X, Y) = 0$ for all X in b and for all $Y \in [b, b]$. (Cf. Cartan's criterion in any book on Lie algebras, e.g., [ ].)

To show that $[b, b]$ is nilpotent, it suffices to show that any $y \in [b, b]$ is nilpotent (via ρ) as an endomorphism of V. For then left and right multiplication, L_y and R_y, are both nilpotent on $\mathrm{End}(V)$, and, therefore, so is $ad_y = L_y - R_y$ (since $L_y R_y = R_y L_y$, $(L_y - R_y)^{m+n+1} = 0$ if $L_y^m = 0 = R_y^n$). Then by Engel's theorem (proven after the present proof), $[b, b]$ is nilpotent because it has a flag: $0 = b_0 \subset b_1 \subset b_2 \subset \cdots \subset b_r = [b, b]$ with $ad_X(b_i) \subseteq b_{i-1}$ for $i = 1, \ldots, r$, and for all $X \in [b, b]$.

Note that this theorem is linear in the sense that the hypothesis and the conclusion each remain true or remain false when the base field is extended or restricted. Thus, we can assume k is algebraically closed. In fact, by choosing bases for g and V we can restrict k to a subfield k' so that (1) g and V are still defined over k' as a Lie algebra and module, and (2) k' is finitely generated over $\mathbb{Q} \subseteq k$. By embedding $k' \to \mathbb{C}$ and extending, we can even reduce to the case $k = \mathbb{C}$.

Lemma. *Let $y \in End_k(V)$. Then there exists a unique diagonalizable s and nilpotent n in $End_k(V)$ such that $y = s + n$ and $sn = ns$. Moreover, s and n are polynomials $S(y)$ and $N(y)$ of y with $N(0) = 0 = S(0)$.*

Proof. The characteristic polynomial of y in the indeterminant T is det $(T - y) = \Pi(T - \lambda_i)^{m_i}$, where λ_i are the eigenvalues of y with multiplicity m_i. For each i, let V_i = ker $(y - \lambda_i)^{m_i} \subseteq V$ be the eigenspace of λ_i, $V = \oplus V_i$. Define $s|_{V_i} = \lambda_i$ and $n = y - s$ is $y - \lambda_i$ on V_i. Since $(y - \lambda_i)^{m_i} = 0$ on V_i, n is nilpotent. Also, $ns = sn$. If $y = s + n$ is any other such decomposition, then it is easy to check that s must be as above. Finally, let $S(T)$ be any polynomial satisfying $S(T) = \lambda_i \bmod (T - \lambda_i)^{m_i}$ and $S(0) = 0$. Let $N(T) = T - S(T)$. Then $N(0) = 0 = S(0)$ and $s = S(y)$ since on $V_i, (y - \lambda_i)^{m_i} = 0$ so that $S(y) = \lambda_i$ on V_i. Also $N(y) = y - S(y) = y - s = n$. Q. E. D. for Lemma.

Since ρ is faithful, we will occasionally forget to write it and identify elements of the Lie algebra with the corresponding endomorphisms of V. Our $Y \in [b, b]$ has a decomposition $Y = s + n$ as in the lemma where s (resp. n) is a diagonalizable (resp. nilpotent) endomorphism of V. We must show $s = 0$. I claim it is enough to show $tr(Y \circ \varphi(s)) = 0$ for all $\varphi \in \mathrm{Hom}_{\mathbb{Q}}(k, \mathbb{Q})$ where $\varphi(s)$ is defined since s is diagonalizeable: If Y and s have eigenvalue λ_i with multiplicity m_i on $V_i \subseteq V$, then $\varphi(s)$ is defined to be $\varphi(\lambda_i)$ on V_i. Also $Y \circ \varphi(s)$ has eigenvalue $\lambda_i \varphi(\lambda_i)$ on V_i. Therefore:

$$0 = tr(Y \circ \varphi(s)) = \sum m_i \lambda_i \varphi(\lambda_i).$$

Applying φ again,

$$0 = \varphi(0) = \sum m_i \, \varphi(\lambda_i)\varphi(\lambda_i)$$

Since $\varphi(\lambda_i) \in \mathbb{Q}$ where φ is identity. But then $\varphi(\lambda_i) = 0$ for all i. Since $\varphi(\lambda_i) \in \mathbb{Q}$ was arbitrary, $\lambda_i = 0$ for all i. Thus $s = 0$, and $Y = n$ is nilpotent.

The trouble is that $\varphi(s) \in \mathrm{End}_k(V)$ may not be in $b \subseteq \mathrm{End}_k(V)$ (or even in g), so we cannot apply the definition of b directly. Since $Y \in [b, b]$, $Y = \sum_r [X_r, Z_r]$ for $X_r, Z_r \in b$.

$$tr(Y \circ \varphi(s)) = \sum_r tr([X_r, Z_r] \circ \varphi(s)) = \sum tr(X_r \circ [Z_r, \varphi(s)]).$$

Since $X_r \in b$ it suffices to show $[Z_r, \varphi(s)] \in b$ to get 0. In fact it is in $[b, b]$. Let ad denote the adjoint representation of $\mathrm{End}_k(V)$ on itself. Since ad is linear, $ad_Y = ad_s + ad_n$ (for $Y = n + s$). This is the canonical decomposition (in the lemma) of ad_Y because (1) $ad_n = L_n - R_n$ is nilpotent, (2) $ad_S = \lambda_i - \lambda_j$ on the eigenspace $V_i \otimes V_j^* \subseteq V \otimes V^* \simeq \mathrm{End}(V)$, and (3) $[ad_S, ad_n] = ad_{[s,n]} = ad_0 = 0$.

Thus $ad_s = P(ad_Y)$ where $P(T)$ is a polynomial with $P(0) = 0$. On $V_i \otimes V_j^*$ we have:

$$\begin{aligned} ad_{\varphi(s)} &= \varphi(\lambda_i) - \varphi(\lambda_j) = \varphi(\lambda_i - \lambda_j) \quad \text{(by linearity of } \varphi) \\ &= \varphi(ad_s) \quad \text{on} \quad V_i \otimes V_j^* \end{aligned}$$

Thus, $ad_{\varphi(s)} = \varphi(ad_s) = Q(ad_s)$ where Q is a polynomial satisfying $Q(\lambda_i - \lambda_j) = \varphi(\lambda_i - \lambda_j)$. Then:

$$\begin{aligned} [\varphi(s), Z_r] &= ad_{\varphi(s)}(Z_r) = \varphi(ad_s)(Z_r) \\ &= Q(P(ad_Y))(Z_r) \in [b, b] \end{aligned}$$

Since Q and P are polynomials, and $Y, Z_r \in b$. Q. E. D.

Definition. For $\rho = ad =$ the adjoint representation of g on itself, $B_{ad}(X,Y) = tr(ad_X ad_Y)$ is called the *Killing form* of g.

Corollary I.1.1. *g is semi-simple if and only if the Killing form is non-degenerate.*

Proof. The kernel of ad is just the center of g, which is 0 if g is semi-simple (no Abelian ideals except 0). Apply I.1. Conversely, suppose $X \neq 0$ is in an Abelian ideal $\mathfrak{a} \subseteq g$. Let $Y \in g$. Then $ad_X \circ ad_Y$ takes g into $\mathfrak{a}$, and $\mathfrak{a}$ into 0. Thus, $(ad_X \circ ad_Y) = 0$. Since $Y \in g$ was arbitrary, the Killing form is degenerate: $B_{ad}(X,Y) = 0$ for all $y \in g$.

We have omitted the proof of the fact that g is nilpotent if and only if ad_X is nilpotent for each $X \in g$. This is a corollary of:

Theorem I.2. (Engel): *Let $\rho : g \to End(V)$ be a representation such that $\rho(X)$ is nilpotent for each $X \in g$. Then there exists a flag: $0 \subseteq V_0 \subseteq V_1 \subseteq \cdots \subseteq V_r = V$ such that $\rho(g) : V_i \to V_{i-1}$.*

Proof. The converse is trivial. The problem is to get one flag that works for all $X \in g$. By induction it suffices to show if $V \neq 0$ then there exists $v \in V, v \neq 0$ such that $\rho(g)(v) = 0$. Since the theorem only involves the image $\rho(g)$ of g, we assume $g \subseteq$ End (V). For all $X \in g$, ad_X is nilpotent since $ad_X = L_X - R_X$ where left and right multiplication by X on End(V), L_X and R_X, are both nilpotent and commute.

Let $h \subset g$ be a maximal subalgebra, $h \neq g$. Then I claim h is an ideal and codim $(h) = 1$. We can assume the theorem for h, since it has smaller dimension than g, by an induction hypothesis. Since h acts on g/h by nilpotents, there exists an $X \in g - h$ such that for all $Y \in h, [Y, X] \in h$. Since $h + (X)$ is a larger subalgebra than $h, h + (X) = g$. h is an ideal of codim. 1 since $[h, X] \subseteq h$.

Finally, let $W = V^h = \{v \in V | X \cdot v = 0 \text{ for all } X \in h\}$. Then $g(W) \subset W$ since h is an ideal. $W \neq 0$ by induction since dim $h <$ dim$g$. For our $X \in$

$g - h$, X is nilpotent so it kills something $\neq 0$ in W. That something is then killed by $g = h + (X)$. Q. E. D.

II. In this section, we show that if g is semi-simple, then every finite dimensional g-module is semi-simple, and conversely. First we prove the result for g as a module over itself via the adjoint action.

Theorem II. 1. *Let $\mathfrak{a}$ be an ideal of the semi-simple Lie algebra g. Let $\mathfrak{a}^\perp$ be the orthogonal complement of $\mathfrak{a}$ with respect to the non-degenerate bilinear form $B_{ad}(XY) = tr(ad_X ad_Y)$. That is, $\mathfrak{a}^\perp = \{x \in g | B_{ad}(X,Y) = 0$ for all $Y \in \mathfrak{a}\}$. Then $\mathfrak{a}^\perp$ is an ideal of g and $g = \mathfrak{a} \oplus \mathfrak{a}^\perp$.*

Proof. Let $X \in \mathfrak{a}^\perp$, $Y \in g$, $Z \in \mathfrak{a}$. Then $B([X,Y],Z) = B(X,[Y,Z]) = 0$ since $[Y,Z] \in \mathfrak{a}$. Since Z is arbitrary in $\mathfrak{a}$, $[X,Y] \in \mathfrak{a}^\perp$, so $\mathfrak{a}^\perp$ is an ideal. Extending a basis for $\mathfrak{a}$ to a basis for g and writing B_{ad} as a symmetric matrix (non-singular), $\mathfrak{a}^\perp$ is the image of the vector space complement to $\mathfrak{a}$ by B_{ad}^{-1}. So $\dim \mathfrak{a} + \dim \mathfrak{a}^\perp = \dim g$. So it suffices to show $\mathfrak{a} \cap \mathfrak{a}^\perp = 0$. But ad is faithful, and $tr(ad_X ad_Y) = 0$ for all $X, Y \in \mathfrak{a} \cap \mathfrak{a}^\perp$. But this implies $\mathfrak{a} \cap \mathfrak{a}^\perp$ is solvable by the same proof that b was solvable in the proof of I.1, since in that proof all we used was the fact that for all $X, Y \in b$, $tr(\rho_X \rho_Y) = 0$, and we now take $\rho = ad$. (Cf. Cartan's Criterion.)

Corollary II. 1.1. *If g is semi-simple then $g = [g,g]$.*

Proof. g contains an ideal isomorphic to $[g,g]^\perp \simeq g/[g,g]$, which is Abelian.

Theorem II. 2. *Let g be a finite dimensional semi-simple Lie algebra. Then every finite dimensional representation is semi-simple.*

Proof. By finite dimensionality it suffices to show that every exact sequence of g modules and g-module homomorphisms splits:

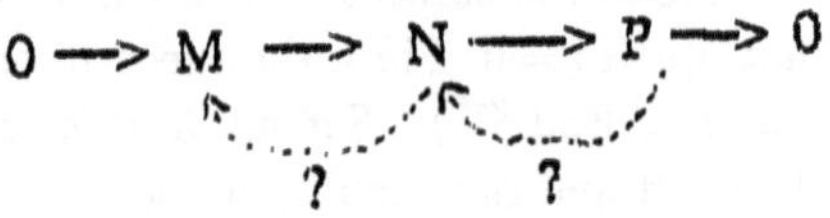

1) Reduction to the case $P = k$, the base field with the trivial g-module structure. Let $W = \{f \in \mathrm{Hom}_k(N,M) | f|_M$ is the scalar multiplication$\}$ and $V = \{f \in \mathrm{Hom}_k(N,M) | F|_M = 0\}$. For $f \in \mathrm{Hom}_k(N,M)$, $X \in g, X \circ f \in \mathrm{Hom}_k(N,M)$ is defined by $(X \cdot f)(n) = X(f(n)) - f(Xn)$ for $n \in N$. Then $X(W) \subset W$ (in fact $X(W) \subset V$) and $X(V) \subset V$. And we have an exact sequence of g-modules and g-homomorphisms.

$$0 \longrightarrow V \longrightarrow W \longrightarrow k \longrightarrow 0.$$

Assuming such sequences split, let f be the image of $1 \in k$ by a splitting (g-homomorphism) map. Since g acts trivially on k, g acts trivially on f. For all $X \in g$, $n \in N$, $0 = (X \cdot f)(n) = X(f(n))$ implies $f \in \mathrm{Hom}_g(N, M)$ is the required splitting map since $f|_M = 1$. It remains to show $0 \to V \to W \to k \to 0$ splits.

2) Reduce to the case V is simple. If $V' \subset V$ is a ($\neq 0$) sub g-module, then

$$0 \longrightarrow V/V' \longrightarrow W/V' \longrightarrow k \longrightarrow 0 \qquad (\psi : k \to W/V')$$

splits by an inductive hypothesis on the dimension of W (one supposes the theorem for modules of smaller dimension). Let $V^0 \subseteq W$ be the inverse image of $\psi(k) \subseteq W/V'$. Then $V' \hookrightarrow V^0$. Also $0 \to V' \to V^0 \to k \to 0$ splits by the same inductive hypothesis on dimension. This gives the required map $k \to W$.

3) Reduce to the case ρ faithful on V. Let $h = \{X \in g | X \cdot (V) = 0\}$ be the kernel of $\rho|_V$. Then h is semi-simple, for an Abelian ideal of h is one of $g = h \oplus h^\perp$. Now for $X \in h$, $X \cdot (W) \subset V$ because

$$\begin{array}{ccccc} & & W & \longrightarrow & k \\ & & \downarrow \rho_X & & \downarrow 0 \\ V & \hookrightarrow & W & \longrightarrow & k \end{array}$$

commutes, i.e., $X \cdot W \to 0$ in k so it comes from V. Also $h(V) = 0$. But h is semi-simple so $h = [h, h]$ takes W to 0 and we can replace g by g/h throughout.

Now let B_ρ be the (non-degenerate) form for $\rho : g \hookrightarrow \mathrm{End}(V)$ (we assume ρ to be faithful). Let $\{Y_i\}$ and $\{Z_i\}$ be dual bases of g so that $B_\rho(Y_i, Z_j) \equiv Tr(\rho_{Y_i}\rho_{Z_j}) = \delta_{ij}$, the Kronecker delta. The *Universal enveloping* alegbra, U_g, of g is the (associative) tensor algebra $k \oplus g \oplus (g \otimes g) \oplus \cdots$ divided by the relations generated by $X \otimes Y - Y \otimes X - [X, Y]$, for $X, Y \in g$. If V is a g-module (via ρ), then V is also a $U(g)$-module: For $X = X_1 \otimes X_2 \otimes \cdots \otimes X_r \in U(g)$, $v \in V$, $\rho_X(v) = \rho_{X_1} \circ \rho_{X_2} \circ \cdots \circ \rho_{X_r}(v)$. Conversely, any module of the associative algebra $U(g)$, is also a module of the Lie algebra g because of the relations in $U(g)$. In particular $U(g)$, itself is a g-module. The invariants, $U(g)^g$ is just the center of $U(g)$. Define the *Casimir* element $K = K_\rho = \sum_i Y_i \otimes Z_i \in U(g)$. Note that for all $X \in g, X = \sum_i B_\rho(Z_i, X)Y_i = \sum_i B_\rho(Y_i, X)Z_i$. *$K$ is well-defined,* for if $\{R_i\}$ and $\{S_i\}$ are another pair of dual bases then $\sum_j R_j \otimes S_j = \sum_{i,j} B(Z_i, R_j)Y_i \otimes S_j = \sum_{i,j} Y_i \otimes B(R_j, Z_i)S_j = \sum_i Y_i \otimes Z_i$. *$K$ is in the center* *of* $U(g)$ because for all $X \in g$,

$$X \otimes K - K \otimes X = \sum_i (X \otimes Y_i \otimes Z_i - Y_i \otimes X \otimes Z_i + Y_i \otimes X \otimes Z_i$$
$$- Y_i \otimes Z_i \otimes X) = \sum_i ([X, Y_i] \otimes Z_i + Y_i \otimes [X, Z_i])$$
$$= \sum_i (B_\rho([X, Y_i], Z_i) Y_i \otimes Z_i + Y_i \otimes B_\rho(Y_i, [X, Z_i]) Z_i) = 0$$

since $B_\rho([Y, X], Z) = B_\rho(Y, [X, Z])$ for all $X, Y, Z \in g$.

Consequently, the *Casimir* operator, $\rho(K) = \sum_i \rho_{Y_i} \circ \rho_{Z_i}$ commutes with the action of g on W. Now $\rho(K)|_V$ is either 0 or an isomorphism of V since ker $(\rho(k)|_V)$ is a sub g-module of V and V is simple. But $\rho(K)|_V \neq 0$ since $Tr_V(\rho(K)) = Tr(\sum_i \rho_{Y_i} \circ \rho_{Z_i}) = \sum_i B_\rho(Y_i, Z_i) = \dim g \neq 0$. Therefore, $\rho(K)|_V$ is an isomorphism of V. Since $\rho(K)(W)$ goes to 0 in k (where g acts trivially), $\rho(K)(W) \subseteq V$. Then $\rho(K)|_V)^{-1} \circ \rho(K)$ is the required splitting map:

$$0 \to V \to W \to k \to 0$$

Q. E. D.

Converse II. 2.1. *Suppose every representation of* g *is semi-simple. Then* g *is semi-simple.*

Proof. g is a g-module under the adjoint representation. If g has any Abelian ideals then g is a direct sum $g = g_1 \oplus \cdots \oplus g_r$ of simple submodules (ideals), where, say g_1 is Abelian, therefore, 1-dimensional. $g_1 \simeq k$, the base field. Define the representation $\rho : g \to \mathrm{End}_k(k^2)$ by:

$$a \xrightarrow{\rho} \begin{pmatrix} 1 & 0 \\ a & 1 \end{pmatrix}$$

for $a \in g_1 \simeq k$, and $\rho = 0$ on $g_2, \ldots, g_r$. Then ρ is not a semi-simple representation, since $\begin{pmatrix} 0 \\ 1 \end{pmatrix}$ generates a submodule, V, of k^2 without a complementary submodule, W, such that $k^2 = V \oplus W$ as g-modules.

III. We compute the Killing form of $sl(n, k)$ and show that $sl(n, k)$ is semi-simple.

Exercise. Let M, N be finite dimensional k-vector spaces. Let $x \in \mathrm{End}(M), Y \in \mathrm{End}(N)$. Define $X \otimes Y \in \mathrm{End}(M \otimes N)$ by $(X \otimes Y)(m \otimes n) = X(m) \otimes Y(n)$ for $m \in M$, $n \in N$. Then $Tr_{M \otimes N}(X \otimes Y) = Tr_M(X) Tr_N(Y)$.

Let $V = k^n$. Any $X \in \mathrm{End}(V)$ acts on V by the canonical representation, ρ. X acts on the dual V^* by $(\rho^*_X(b^*))(v) = -b^*(X \cdot v)$ for $b^* \in V^*$, $v \in V$.

So we write $\rho^*_X b^* = -b^*X$. X acts on $V \otimes V^*$ =End(V) as a derivation: $X \cdot (a \otimes b^*) = Xa \otimes b^* - a \otimes b^*X = [X, a \otimes b^*]$, where (by choosing a basis for V) we write $a \in V$ as a column, $b^* \in V^*$ as a row, and X as a matrix. This is just the usual adjoint representation, ad, if we identify $a \otimes b^* \in V \otimes V^*$ with the $n \times n$ matrix ab^* ∈End(V).

Computation. Suppose $X, Y \in sl(n,k) \subseteq \text{End}(V)$ have trace =0. Then

$$\begin{aligned} Tr_{V\otimes V^*}(ad_X ad_Y) &= Tr_{V\otimes V^*}(\rho_X \otimes 1 + 1 \otimes \rho^*_X)(\rho_Y \otimes 1 + 1 \otimes \rho^*_Y) \\ &= Tr_{V\otimes V^*}(\rho_X\rho_Y \otimes 1 + \rho_X \otimes \rho^*_Y + \rho_Y \otimes \rho^*_X \\ &\quad + 1 \otimes \rho^*_X\rho^*_Y) = Tr_V(\rho_X\rho_Y)Tr_{V^*}(1) + Tr_V(\rho_X)Tr_{V^*}(\rho^*_Y) \\ &\quad + Tr_V(\rho_Y)Tr_{V^*}(\rho_X) + Tr_V(1)Tr_{V^*}(\rho^*_X\rho^*_Y) \\ &= Tr(XY)\cdot n + 0 + 0 + nTr(Y^tX^t) \\ &= 2n\,Tr(XY) = 2n\,B_\rho(X,Y) \end{aligned}$$

On the other hand, $V \otimes V^* \simeq sl(n,k) \oplus Z$ where Z = center of $V \otimes V^*$ =End(V) is generated by the identity matrix. For all X ∈End (V), $ad_X(Z) = 0$.

$$Tr_{V\otimes V^*}(ad_X ad_Y) = Tr_{sl(n,k)}(ad_X ad_Y) + Tr_Z(ad_X ad_Y) = B_{ad}(X,Y) + 0$$

Therefore, we have

Proposition III.1. *The Killing form,* $B_{ad}(X,Y) = 2nB_\rho(X,Y) = 2nTr(XY)$ *is non-degenerate, and* $sl(n,k)$ *is semi-simple. The corresponding Casimir elements also differ by a scalar,* $K_\rho = 2n \cdot K_{ad}$.

IV. In this section, we define Lie algebra cohomology with coefficients in a module and describe conditions under which we can replace the coefficient module by the submodule of invariants.

Notations and basic lemmas. Let g be a finite dimensional Lie algebra and V a finite dimensional vector space over a field k of characteristic 0. Let $\rho : g \to \text{End}(V)$ be a Lie homomorphism, giving V a g-module structure. $\Lambda^* g^* = \oplus_{n=0}^{\infty}(\Lambda^n g^*)$ is the exterior algebra of the dual, g^*, of g. The adjoint representation of g on itself extends to a representations of g on $\Lambda^* g$ (see Section I). We also denote the dual representation on $\Lambda^* g^*$ by $ad : g \to$End$(\Lambda^* g^*)$. In Section 14 we called this action θ, but calling it "ad" instead will remind us where it comes from and will distinguish it from the other Greek letter, ρ, used in this section. Explicitly, for $\omega \in \Lambda^q g^*$ and $X_1, \ldots, X_q \in g$, we have

$$\begin{aligned} (ad_X\omega)(X_1,\ldots,X_q) = -\{&\omega([X,X_1],X_2,X_3,\ldots,X_q) \\ &+ \omega(X_1,[X,X_2],X_3,\ldots,X_q) + \ldots\}. \end{aligned}$$

For each $X \in g$, ad_X is a derivation of degree 0 on $\Lambda^* g^*$. $ad_X : \Lambda^q g^* \to \Lambda^q g^*$, and $ad_X(\omega \wedge \eta) = (ad_X\omega)\wedge\eta + \omega \wedge ad_X\eta$. For each $X \in g$, $\iota_X : \Lambda^q g^* \to \Lambda^{q-1} g^*$ is defined by

$$(\iota_X\omega)(X_1,\ldots,X_{q-1}) = \omega(X, X_1, \ldots, X_{q-1})$$

for $\omega \in \Lambda^q g^*$, $X_1, \ldots, X_{q-1} \in g$. Then ι_X is seen to be an anti-derivation on $\Lambda^* g^*$ of degree -1 (i.e., lowering degree by 1), and $ad_X \iota_Y = \iota_Y ad_X + \iota_{[X,Y]}$. Define $d : \Lambda^q g^* \to \Lambda^{q+1} g^*$ as the anti-derivation of $\Lambda^* g^*$ such that $d|_{\Lambda^0 g^*} \equiv 0$ and $d|_{\Lambda^1 g^*}$ is defined by $(d\omega)(X,Y) = -\omega([X,Y])$ for $X, Y \in g$, $\omega \in \Lambda^1 g^*$. In general, for $\omega \in \Lambda^q g^*$, $X_1, \ldots, X_{q+1} \in g$,

$$d\omega(X_1, \ldots, X_{q+1}) = \sum_{i<j} (-1)^{i+j} ([X_i, X_j], X_1, \ldots, \widehat{X}_i, \ldots, \widehat{X}_j, \ldots, X_{q+1}).$$

Since d and ι_X (for $X \in g$) are anti-derivations of degrees +1 and -1, respectively, $d\iota_X + \iota_X d$ is a derivation of degree 0 of $\Lambda^* g^*$. Since $ad_X = d\iota_X + \iota_X d$ on $\Lambda^0 g^*$ and $\Lambda^1 g^*$, this formula is true on $\Lambda^* g^*$. For $\tilde{x} \in \Lambda^1 g^*$, define $e(\tilde{x})$ on $\Lambda^* g^*$ to be exterior multiplication (on the left) by $\tilde{x}$. That is $e(\tilde{x})(\omega) = \tilde{x} \wedge \omega$, $(\omega \in \Lambda^* g^*)$.

Lemma IV. 1. *For $Z \in g$,*

$$ad_Z = -\sum_{j=1}^{n} e(\tilde{x}_j)\iota_{[Z,X_j]}$$

where $X_1, \ldots, X_n$ is any basis of g and $\tilde{x}_1, \ldots, \tilde{x}_n$ the dual basis for $\Lambda^1 g^*$.

Proof. Apply both sides to $f \in \Lambda^q g^*$ and check on q linearly independent elements of g. We can choose these from our basis $X_1, \ldots, X_n$, and by renumbering, we can suppose we have chosen the first q elements.

$$\begin{aligned} -\sum_{j=1}^{n} & e(\tilde{x}_j)(\iota_{[Z,X_j]} f)(X_1, \ldots, X_q) \\ &= -\sum_{j=1}^{q} (\iota_{[Z,X_j]} f)(X_1, \ldots, \widehat{X}_j, \ldots, X_n)(-1)^{j-1} \\ &= -\sum_{j=1}^{q} f([Z, X_j], X_1, \ldots, \widehat{X}_j, \ldots, X_n)(-1)^{j-1} \\ &= -\sum_{j=1}^{q} f(X_1, \ldots, [Z, X_j], \ldots, X_n)(-1)^{j-1}\,(-1)^{j-1} \\ &= (ad_Z f)(X_1, \ldots, X_q) \end{aligned}$$

Definitions. g acts on $\Lambda^* g^* \otimes V$ by $\mathcal{L}_X \equiv ad_X \otimes 1 + 1 \otimes \rho_X$, for $X \in g$. (Cf. definition of tensor product of g-modules in I.) Explictly, for $\omega \otimes v \in \Lambda^* g^* \otimes V$, $\mathcal{L}_X(\omega \otimes v) = ad_X \omega \otimes v + \omega \otimes \rho_X v$.

Define $\delta_V : \Lambda^q g^* \otimes V \to \Lambda^{q+1} g^* \otimes V$ to be $\sum_{j=1}^{n} e(\tilde{x}_j) \otimes \rho_{X_j}$, where $X_1, \ldots, X_n$ is a basis for g, and $\tilde{x}_1, \ldots, \tilde{x}_n \in g^*$ is the dual basis. δ_V is independent of this choice of basis.

Define $\delta = d \otimes 1 + \delta_V : \Lambda^q g^* \otimes V \to \Lambda^{q+1} g^* \otimes V$. Then δ satisfies

$$\mathcal{L}_X = \delta i_X + \iota_X \delta,$$

for all $X \in g$, where we have written ι_X for $\iota_X \otimes 1$ on $\Lambda^* g^* \otimes V$.

Define $H^*(g, \rho) = H^*(g, V)$ to be the cohomology of the complex:

$$0 \longrightarrow \Lambda^0 g^* \otimes V \xrightarrow{\delta} \Lambda^1 g^* \otimes V \xrightarrow{\delta} \Lambda^2 g^* \otimes V \longrightarrow \cdots$$

For example, check that $H^0(g, \rho) \simeq V^g \equiv \{v \in V | \rho_X v = 0 \text{ for all } X \in g\}$.

Theorem IV. 2. *Let g be a semi-simple finite dimensional Lie algebra, and let V be a finite dimensional g-module such that $V^g = 0$. Then $H^q(g, V) = 0$ for all q.*

Proof. (See [14].) Since g is semi-simple, $V = \oplus V_i$, a sum of simple g-modules, and $\rho|_{V_i} \neq 0$. Since $H^*(g, \oplus V_i) \simeq \oplus_i H^*(g, V_i)$, it suffices to take V to be simple. Let $h \subseteq g$ be the complementary ideal to ker ρ, $g = h \oplus \ker\rho$. $h \neq 0$ since $\rho \neq 0$. h is semi-simple since any Abelian ideal of h is also one of g. Since $\rho|_h$ is faithful, I. 1 tells us that there exist dual bases $\{Y_i\}$ and $\{Z_i\}$ of h such that $B_\rho(Y_i, Z_j) = \delta_{ij}$, the Kronecker delta. In the proof of II. 2 we defined the Casimir element $K = K_\rho = \sum_{i=1}^{\dim h} (Y_i \otimes Z_i) \in U(h) \subseteq U(g)$ and showed that K is well-defined and is in the center of $U(h)$. Since K is then in the center of $U(g)$, the Casimir operator $\rho(K) = \sum \rho_{Y_i} \circ \rho Z_i$ commutes with ρ_X for all $X \in g$. Since V is simple, ker $\rho(K)$ is a submodule of V. Therefore, $\rho(K)$ is either 0 or an isomorphism of V. But $\rho(K) \neq 0$, since its trace is $\sum_i tr(\rho_{Y_i} \circ \rho_{Z_i}) = \dim h \neq 0$. Thus $\rho(K)$ is an isomorphism of V.

Let $\Gamma = 1 \otimes \rho(K)$ on $\Lambda^* g^* \otimes V$. Γ commutes with $ad_X \otimes 1$ and $1 \otimes \rho_X$ for all $X \in g$. Thus, Γ commutes with $\mathcal{L}_X = ad_X \otimes 1 + 1 \otimes \rho_X$. Also, Γ commutes with $d \otimes 1$ and $\delta_v = \sum_j e(\tilde{x}_j) \otimes \rho_{X_j}$ (for $\{X_j\}$ a basis of g and $\{\tilde{x}_j\}$ the dual basis). Therefore, Γ commutes with $\delta = d \otimes 1 + \delta_v$. $\delta\Gamma = \Gamma\delta$ implies $\Gamma^{-1}\delta = \delta\Gamma^{-1}$ implies Γ^* and Γ^{-1*} are defined on $H^*(g, \rho)$, and $\Gamma^* \circ \Gamma^{-1*} = (\Gamma \circ \Gamma^{-1})^* = id^*$. But we will show that Γ is homotopic to zero: $\Gamma^* = 0^*$. Hence, $H^*(g, \rho) = 0$.

Let $T = \sum_{i=1}^{\dim h} \iota Z_i \otimes \rho_{Y_i}$. I claim that $\Gamma = \delta T + T\delta$.

$$\begin{aligned} \delta T + T\delta &= \delta(\sum_i \iota_{z_i} \otimes \rho_{Y_i}) + (\sum_i \iota_{Z_i} \otimes \rho_{Y_i})\delta \\ &= \delta(\sum \iota_{Z_i} \otimes \rho_{Y_i}) + (1 \otimes \rho Y_i)(-\delta(\iota_{z_i} \otimes 1) + \mathcal{L} Z_i), \end{aligned}$$

$$\begin{aligned} \text{since } \mathcal{L}_{Z_i} = \delta\, \iota\, Z_i + \iota_{Z_i}\delta &= \sum (d \otimes 1 + \delta_v)(\iota_{Z_i} \otimes \rho_{Y_i}) \\ &- \sum (1 \otimes \rho_{Y_i})(d \otimes 1 + \delta_v)(\iota_{Z_i} \otimes 1) + \sum (1 \otimes \rho_{Y_i})(\mathcal{L} Z_i) \end{aligned}$$

Since $d \otimes 1$ and $1 \otimes \rho_{Y_i}$ commute, we get some cancellation, and, since $\mathcal{L}_{Z_i} = 1 \otimes \rho_{Z_i} + ad_{Z_i} \otimes 1$, we obtain:

$$\begin{aligned}\delta T + T\delta = \delta_V(\sum \iota_{Z_i} \otimes \rho_{Y_i}) - \sum(1 \otimes \rho_{Y_i})\delta_V(\iota_{Z_i} \otimes 1) \\ + \sum(1 \otimes \rho_{Y_i})(1 \otimes \rho_{Z_i} + ad_{Z_i} \otimes 1) = \delta_V(\sum \iota_{Z_i} \otimes \rho_{Y_i}) \\ - \sum(1 \otimes \rho_{Y_i})\delta_V(\iota_{Z_i} \otimes 1) + \Gamma + \sum(ad_{Z_i} \otimes \rho_{Y_i}).\end{aligned}$$

Now express δ_V in terms of a basis $\{X_i\}$ of g with dual basis $\{X_i\}$. Also, express ad_{Z_i} in terms of the same basis via Lemma IV. 1.

$$\begin{aligned}\delta T + T\delta - \Gamma &= \sum_{i,j}(e(\tilde{x}_j)\iota_{Z_i} \otimes \rho_{X_j}\rho_{Y_i} - \sum_{i,j}(e(\tilde{x}_j)\iota_{Z_i} \otimes \rho_{Y_i}\rho_{X_j}) \\ &- \sum_{i,j}(e(\tilde{x}_j)\iota_{[Z_i,X_j]} \otimes \rho_{Y_i}) \\ &= \sum_{i,j} e(\tilde{x}_j) \otimes 1\{\iota_{Z_i} \otimes (\rho_{X_j}\rho_{Y_i} - \rho_{Y_i}\rho_{X_j}) - \iota_{[Z_i,X_j]} \otimes \rho_{Y_i}\} \\ &= \sum_{i,j} e(\tilde{x}_j) \otimes 1\,\{\iota_{Z_i} \otimes \rho_{[X_j,Y_i]} - \iota_{[Z_i,X_j]} \otimes \rho_{Y_i}\} = 0\end{aligned}$$

because

$$\begin{aligned}\sum_i \iota_{[Z_i,X_j]} \otimes \rho_{Y_i} &= \sum_{i,k} \iota_{B([Z_i,X_j],Y_k)Z_k} \otimes \rho_{Y_i} \\ &= \sum_{i,k} \iota Z_k \otimes \rho_{B([Z_i,X_j],Y_k)Y_i} \\ &= \sum_{i,k} \iota_{Z_k} \otimes \rho_{B(Z_i,[X_j,Y_k])Y_i} \\ &= \sum_k \iota_{Z_k} \otimes \rho_{[X_j,Y_k]} \\ &= \sum_i \iota_{Z_i} \otimes \rho_{[X_j,Y_i]}.\end{aligned}$$

Q. E. D.

Corollary IV. 2. 1. *If g is semi-simple, and V is a finite dimensional g-module, then $H^*(g,V) = H^*(g,V^g)$, where $V^g \subseteq V$ are the g-invariants, $V^g = \{v \in V | X \cdot v = 0$ for all $X \in g\}$.*

Proof. $V = V^g \oplus W$, $W^g = 0$, $H^*(g,W) = 0$, and $H^*(g,V) \simeq H^*(g,V^g) \oplus H^*(g,W)$.

Remark. Let $g = gl(n,k) = sl(n,k) \oplus Z$, where $sl(n,k)$ is semi-simple and Z is the center, generated by the identity matrix R. We wish to extend the above corollary to $g = gl(n,k)$. However, we need the additional condition that the g-module V break up as a sum $V = \oplus_\lambda V_\lambda$ of eigenspaces for $R \in Z \subseteq g$ with $\rho_{R|V_\lambda}$ = multiplication by $\lambda \in k$. We also assume V_0 (but not necessarily V)

is finite dimensional. $\mathcal{L}_R = ad_R \otimes 1 + 1 \otimes \rho_R = 0 + 1 \otimes \rho_R$ is multiplication by λ on $\Lambda^* g^* \otimes V_\lambda$. But $\mathcal{L}_R = i_R\delta + \delta i_R$ is homotopic to zero. Therefore, $H^*(g,V) = \oplus_\lambda H^*(g, V_\lambda) = H^*(g, V_0)$. Since $sl(n,k)$ is semi-simple, V_0 breaks up as a sum of $sl(n,k)$-modules (in fact as $gl(n,k)$-modules since $\rho_R = 0$ on V_0), $V_0 = V_0^{sl(n,k)} \oplus W = V^{gl(n,k)} \oplus W$, where $W^{sl(n,k)} = W^{gl(n,k)} = 0$. Since $R \in \ker(\rho|_W)$, a complementary ideal h can be chosen $\subseteq sl(n,k)$. Just as in IV. 2, we get a Casimir operator ($\rho|_W$ restricted to h is faithful) which is an isomorphism of W, commutes with δ and is homotopic to the identity. $H^*(g,W) = 0$. Finally, we have in this case,

Corollary IV. 2.2. $H^*(g,V) = H^*(g,V^g)$.

For more details, see the section on reductive algebras and the spectral sequence relative to a reductive subalgebra in [39].

References

[1] Anderson, D. W., *A generalization of the Eilenberg–Moore spectral sequences.* Bull. AMS, **78**, 784–786, 1972.

[2] Atiyah, M., Bott, R. and Patodi, V., *On the heat equation and the index theorem.* Invent. Math., **19**, 279–330, 1973.

[3] Borel, A. and Hirzebruch, F., *Characteristic classes and homogeneous spaces, II.* AM. Journal of Mathematics, **81**, 315–382, 1959.

[4] Bott, R., *The Lefschetz formula and exotic characteristic classes.* Proc. Diff. Geom. Conference, Rome 1971.

[5] —, *On a topological obstruction to integrability.* Proc. International Congress Math.; (Nice 1970) Gauthier-Villars, **1**, 27–36, Paris 1971.

[6] —, *Lectures on characteristic classes and foliations (Notes by Lawrence Conlon).* Lect. Notes in Math.; Springer–Verlag, (279), 1–94, New York 1972.

[7] —, *On the Gel'fand Fuks cohomology.* Proc. Diff. Geom. Conference, Stanford 1973.

[8] —, *On the Chern–Weil homomorphism and the continuous cohomology of Lie groups.* Advances in Math, **11**, 289–303, 1973.

[9] —, *Some remarks on continuous cohomology.* Proc. of International Conference on Topology and Related Topics, Univ. of Tokyo, Japan 1973.

[10] Bott, R. and Haefliger, A., *On characteristic classes of* Γ*-foliations.* Bull. AMS, **78**, 1039–1044, 1972.

[11] Brown, E. H., *Abstract homotopy theory.* Trans. AMS, **119**, 79–85, 1965.

[12] Cartan, H., *Notions d'algèbre différentielle. La transgression dans une groupe de Lie.* Colloque de Topologie, Bruxelles, 15–27, 57–71, 1950.

[13] Chevalley, C., "Theory of Lie Groups," Princeton Univ. Press, Princeton 1946.

[14] Chevalley, C. and Eilenberg, S., *Cohomology theory of Lie groups and Lie algebras.* Trans. AMS, **63**, 85–124, 1948.

[15] Chern, S. and Simons, J., *Some cohomology classes in principal fiber bundles and their applications to Riemannian geometry.* Proc. Nat. Acad. Sci., U. S. A., **68**, 791–794, 1971.

[16] DeRham, G., *Sur la théorie des intersections et les integrales multiples.* Comment. Math. Helv., **4**, 151–157, 1932.

[17] Dold, A., "Lectures on Algebraic Topology," Springer, New York 1972.

[18] Dold, A. and Puppe, D., *Homologie Nicht-Additiver Funktoren.* Anwendungen. Ann. Inst. Fourier, Grenoble, **11**, 201–312, 1961.

[19] Eilenberg, S., *Homologie des groupes.* Séminarie Henri Cartan, Ecole Normal Superieure, Paris 1950/1951.

[20] Eilenberg, S. and MacLane, S., *Relation between homology and homotopy groups of spheres.* Ann. of Math., **46**(2), 480–509, 1945.

[21] Eilenberg, S. and Zilber, J., *Semi-simplicial complexes and singular homology.* Ann. of Math., **51**, 499–513, 1950.

[22] Fogarty, J., "Invariant Theory." Benjamin, New York 1969.

[23] Gel'fand, I., "Generalized Functions." Academic Press, New York 1964.

[24] Gel'fand, I. and Fuks, D., *Cohomologies of the Lie algebra of vector fields in a circle.* Funct. Anal., **2**, 342–3, 1968.

[25] —, *Cohomologies of the Lie algebra of vector fields on a manifold.* Funct. Anal., **3**, 155, 1969.

[26] —, *The cohomology of the Lie algebra of tangent vector fields on a smooth manifold,* I. J. Funct. Anal., **3**, 194–210, 1969.

[27] —, *The cohomology of the Lie algebra of tangent vector fields on a smooth manifold,* II. J. Funct. Anal., **4**, 110–116, 1970.

[28] —, *The cohomology of the Lie algebra of formal vector fields.* Izv. Ann. SSSR, **34**, 327–342, 1970.

[29] Gel'fand, I., Fuks D. and Kalinin, D., *Cohomology of the Lie algebra of formal Hamiltonian vector fields.* Funct. Anal., **6**, 25–29, 1972.

[30] Godbillon, C., *Cohomologie d'algèbres de Lie de champs de vecteurs formels.* Sém. Bourbaki, 421–01 to 421–19, 1972.

[31] Godbillon, C. and Vey, J., *Un invariant des feuilletages de codimension un.* C. R. Acad. Sci., **273**, 92, 1971.

[32] Greenberg, M. J., "Lectures on Algebraic Topology." Benjamin, New York 1967.

[33] Guillemin, V., *Cohomology of vector fields on a manifold.* Advances in Math., **10**, 192–220, 1973.

[34] Haefliger, A., *Feuilletages sur les variétés ouvertes.* Topology, **9**, 183–194, 1970.

[35] —, *Homotopy and integrability.* Lecture Notes in Math; Springer-Verlag (197), 133-163, New York 1971.

[36] —, *Sur les classes caractéristiques des feuilletages.* Sém. Bourbaki, 412–01 to 412–21, 1972.

[37] Heitsch, J., *Deformations of secondary characteristic classes.* Topology, **12**, 381–388, 1973.

[38] Hochschild, G. and Mostow, G. D., *Cohomology of Lie groups.* III. J. of Math., **6**, 367–401, 1962.

[39] Hochschild, G., and Serre, J.-P., *Cohomology of Lie algebras.* Ann. of Math., **57**, 591–603, 1953.

[40] Hu, S.-T., "Elements of General Topology." Holden–Day, San Francisco, 1964.

[41] Husemoller, D., "Fiber Bundles." McGraw–Hill, New York 1966.

[42] Kamber, F. and Tondeur, P., *Cohomologie des algebres de Weil relatives tronquées. Algèbres de Weil semi-simpliciales. Homomorphisme caracteristique d'un fibré principal feuilleté.* Comptes Rendu Acad. Sci. Paris, t. 276, 1973.

[43] —, *Characteristic invariants of foliated bundles.* Manuscripta Math., **11**, 51–89, 1974.

[44] Kan, D., *Abstract homotopy II.* Proc. Nat. Acad. Sci., U. S. A., **42**, 255–258, 1956.

[45] Kelley, J., "General Topology." Van Nostrand, New York 1955.

[46] Koszul, J. -L., *Homologie et cohomologie des algèbres de Lie.* Bulletin de Soc. Math. de France, **78**, 5–127, 1950.

[47] Lamotke, K., "Semisimpliziale algebraische Topologie." Springer, Berlin 1968.

[48] Lawson, H. B., Jr., *Foliations.* Bull. Ams.,**80**, 369–418, 1974.

[49] Mather, J., *Loops and foliations.* Proc. of Int. Conf. on Topology and Related Topics, Univ. of Tokyo, Japan 1973.

[50] May, J. P., "Simplicial Objects in Algebraic Topology." Van Nostrand, Princeton 1967.

[51] Milnor, J., *Construction of universal bundles, I, II,* Ann. Math., **63**, 272–284 and 430–436, 1956.

[52] —, *The realization of a semi-simplicial complex.* Ann. Math., **65**, 357–362, 1957.

[53] —, "Notes on Characteristic Classes." Princeton Univ. Press, Princeton 1957.

[54] Mosher, R. and Tangora, M., "Cohomology Operations." Harper and Row, 1968.

[55] Schwartz, L., "Théorie des Distributions." Hermann, Paris, 1950.

[56] Segal, G., *Classifying spaces and spectral sequences.* Inst. des Hautes Etudes Scientifiques, Publ. Math., **34**, 105–112.

[57] —, "Categories and cohomology theories." Topology, **13**, 293–312, 1974.

[58] Serre, J.-P., *Homologie singulière des espaces fibrés.* Ann. of Math., **54**, 425–505, 1951.

[59] —, "Lie Algebras and Lie Groups." Benjamin, New York 1965.

[60] Shulman, H., "On characteristic classes." Thesis, Univ. of Cal., Berkeley 1972.

[61] Spanier, E., "Algebraic Topology." McGraw Hill, New York 1966.

[62] Stasheff, J., *Homotopy associativity of H-spaces II.* Trans. AMS, **108**, 293–312, 1963.

[63] Steenrod, N., *Homology with local coefficients.* Annals. of Math., **44**, 610–627, 1943.

[64] —, "The topology of fibre bundles." Princeton Univ. Press, Princeton 1951.

[65] tom Dieck, T., "On the homotopy type of classifying spaces." Manu. Math., **11**, 41–49, 1974.

[66] Thurston, W., *Noncobordant foliations of* S^3. Bull. Am. Math. Soc., **78**, 511–514.

[67] —, *The theory of foliations of codimension greater than one.* Proc. of Stanford U. Conf. on Diff. Geom., 1973.

[68] —, *Foliations and groups of diffeomorphisms.* Bull. AMS, **80**, 304–312, 1974.

[69] Trauber, P., "The continuous cohomology of the Lie algebra of vector fields on a manifold." Ph.D. thesis, Princeton Univ., 1973.

[70] Van Est, W., *Une application d'une méthode de Cartan–Leray.* Indag. Math., **17**, 542–544, 1955.

[71] Weil, A., *Sur les théorèmes de DeRham.* Comment. Math. Helv., **26**, 119–145, 1952.

[72] Weyl, H., "The Classical Groups." Princeton Univ. Press, Princeton 1946.

[73] Whitehead, J. H. C., *Combinatorial Homotopy, I.* Bull. AMS., **55**, 213–245, 1949.

CENTRO INTERNAZIONALE MATEMATICO ESTIVO
(C.I.M.E.)

DIFFERENTIAL OPERATORS ON MANIFOLDS

Some Aspects of Invariant Theory in Differential Geometry

R. BOTT

C. I. M. E **Lectures delivered at Varenna, August 1975.**

Table of Contents

Some Aspects of Invariant Theory in Differential Geometry

Raoul Bott

1. Introduction.

A central theme in the development of topology and geometry has been the relation between Riemannian curvature and the topological shape of a manifold. Roughly speaking, certain universal polynomials in the curvature tensor R of a Riemannian structure g give rise to differential forms $P_i(g)$ of dimension $4k$ on M -- the Pontriagin forms of g -- whose cohomology classes in the deRham theory of M, are independent of g, and so can be denoted by $P_i(M)$. These classes are therefore invariants of the differentiable manifolds and play a central role in all aspects of global differential topology, and analysis.

My first aim in these lectures will be to present, in a self-contained manner, a relatively new way of getting at these classes, which as noted only recently, say 1972, by several of us (Bernstein, Rosenfeld, Haefliger, Malgrange) essentially simultaneously, as a consequence of the work of Gelfand-Fuks and Godbillon-Vey. Of course once observed it was also clear to all geometers concerned with G - structures and would have been no surprise to E. Cartan.

The essential point is the following one. Traditionally the theory of characteristic classes is developed for Lie groups and bundles in general, and then specialization to the orthogonal sub-bundle $O(M) \subset F(M)$ of the frame

* I wish to thank the NSF for partial support under Grant MPS 74-11896 of the work reported here.

bundle $F(M)$ of a manifold yields the characteristic classes of M. On the other hand, a smooth manifold naturally carries more structure than its tangent frame bundle, for instance for each $k \geq 1$, M has a natural k-th order frame bundle $J_k(M)$ associated to it, and our observation is simply that already in $J_2(M)$ one finds a natural curvature class $R(M)$ of M, out of which the $P_i(M)$ can be constructed without choice. This $R(M)$ is furthermore the universal curvature of a Riemannian structure on M, in the sense that under the natural map

$$O(M) \xrightarrow{\text{Levicivita}} J_k(M)$$

furnished by the Levicivita connection of a Riemannian structure g, $R(M)$ goes over into $R(g)$.

I myself learned about $R(M)$ many years ago in a lecture by Kobayashi, [45], however, its significance became clear to me only in the framework of Gelfand-Fuks, whose theory surveys all the possible invariants of M which might be culled from the higher order contact bundles of M. Of course their conclusion is that although "interesting" classes can be found in $J_k M$, their dimension is too high to yield new invariants of manifolds.

On the other hand, as we will see in the last section, these invariants can also be applied to foliations of codimension n and there indeed they can be nontrivial.

I will try to keep these lectures as self-contained as possible, but do expect some experience with manifold theory in my audience. Basically, I have in mind an audience, say of analysts who feel ill at ease in global differential geometry and especially Lie groups etc.. I will therefore start with deRham theory and the basic integrability criterion in Section 2 . Thereafter in Section 3 I will construct and discuss the structure equation for some of the classical groups in considerable detail. From there we move on to the group Diff(M) in Section 4 and in Sections 5, 6, 7, I will outline the computation of the characteristic classes cohomology from this point of view of jet-spaces. These are essentially the computations initiated by Gelfand-Fuks. Section 8 deals with foliations and vanishing phenomena in general.

By and large these lectures then, are a pedestrian account of a point of view developed jointly with Andre Haefliger over the past few years. A superb but rather condensed version can be found in Haefliger's Bourbaki Seminar Lecture [36]. For a short survey of this material see our note [10]. For an account of some of this material, but slanted more to topology than geometry see [0].

Finally, I have appended the bibliography from [0], to related material, which attests to the interest in this general subject in recent years.

August 1975

Harvard University
Cambridge, Mass.

2. Some basic theorems in differential geometry.

I will assume the basic facts of deRham theory. Thus every C^∞-manifold comes equipped with a complex of differential operators

$$(2.1)\qquad \mathfrak{F} \xrightarrow{d} \Gamma\Lambda^0 T^* \xrightarrow{d} \Gamma\Lambda^1 T^* \cdots \qquad \Gamma\Lambda^n T^* \longrightarrow 0$$

where $\Lambda^i T^*$ denotes the i-th exterior power of the cotangent bundle of M, and Γ denotes the space of C^∞-sections of a vector bundle. Furthermore $d^2 = 0$ and

$$(2.2)\qquad H^q_{deRham}(M) = \text{Ker } d/\text{Im } d \quad \text{in} \quad \Gamma\Lambda^q T^* .$$

We also write $\Omega^q(M)$ for $\Gamma\Lambda^q T^*(M)$ and refer to this space as the space of q-forms on M. If $U \subset M$ in an open subset of M covered by coordinates $x^1, \ldots, x^n$, then $\omega \in \Omega^q(U)$ by

$$(2.3)\qquad \omega = \sum \omega_{i_1 \ldots i_q} dx^{i_1} \wedge \cdots \wedge dx^{i_q}$$

with $\omega_{i_1 \ldots i_q}$ a C^∞- function on U, alternating in its q-tupled indices, and the differential operator d has the local representation:

$$(2.4)\qquad d\omega = \sum \left(\frac{\partial}{\partial x^\lambda} \omega_{i_1 \ldots i_q} \right) dx^\lambda \wedge dx^{i_1} \wedge \cdots \wedge dx^{i_q} .$$

The formula $d^2 = 0$ then follows from the symmetry of the second partials, and the naturality of (2.4) follows easily from the chain rule.

In short one may think of $\omega \in \Omega^q M$, as a collection of q-forms defined on each open set of U of an atlas for M, and agreeing on the overlaps of two open sets of the atlas.

Given a map $f: M \longrightarrow N$, the deRham groups behave contravariantly, and

$$f^*: H^* \Omega N \longrightarrow H^* \Omega M ,$$

is invariant under a smooth homotopy of f. In particular as $\mathbb{R}^n$ is contractible

(2.5) $$H^*(\mathbb{R}^n) = \mathbb{R} \text{ in } \dim 0 .$$

This is of course the well known Poincaré lemma. Finally recall its infinitesimal version. Namely, if X is a smooth vector field on M, then the Lie derivative $\mathcal{L}(X)$ in the direction X, acts on ΩM by

(2.6) $$\mathcal{L}(X) = \iota(X)\, d + d\, \iota(X)$$

where $\iota(X)$ is the antiderivation which on 1-forms is given by the duality

$$\iota(X)\, \omega = \omega(X) .$$

To a topologist $\iota(X)$ thus plays the role of a homotopy operator for the operator $\mathcal{L}(X)$ showing that $\mathcal{L}(X)$ acts trivially on $H^* \Omega M$.

Next let me recall some of the most basic existance theorems which are the standard tools of the geometer:

1) Implicit function theorem:

If $f_1, \cdots, f_q$ are smooth functions with $df_1 \wedge \cdots \wedge df_q \neq 0$ at $p \in M$, then they can be completed to a coordinate system near p by the addition of $n-q$ function $f_{q+1}, \cdots, f_n$.

2) The fundamental existence theorem of ordinary differential equations:

If X is a vector field on M with $X_p \neq 0$, then there exists a coordinate system $x^1, \cdots, x^n$ near p so that near p, $X \equiv \frac{\partial}{\partial x_1}$.

Recall that to create this coordinate system one may proceed as follows. Choose some coordinates $y_1, y_2, \cdots, y_n$ centered at p, and by a possible renumbering arrange it so that

$$dy_1(X_p) \neq 0 .$$

The plane $P : y_1 = 0$ then is transversal to X in the vicinity of p, and we can define the desired coordinates $x_i, \cdots, x_n$ by :

$x_1(q)$ = time needed to flow along X from q to P. Next, letting $\overline{q}$ be the intersection of this flow line with P, set $x_i(q) = y_i(q)$ for $i > 1$.

3) A fundamental extension of this theorem due to Deana-Clebsch and known as the Frobenius theorem is the following:

Assume that $X_1, \cdots, X_m$ are vector fields on M, which span a q-dimensional space $\{X_1, \cdots, X_m\}$ at each point q near $p \in M$. Assume further that near p

$$[X_i, X_j] \in \{X_1, \cdots, X_m\} .$$

Then there exist coordinates

$$x_1, x_2, \cdots, x_m ; x_{m+1}, \cdots, x_n$$

centered at p, such that near p

$$\{X_1, \cdots, X_m\} = \frac{\partial}{\partial x_1}, \cdots, \frac{\partial}{\partial x_m} .$$

There are many equivalent formulations of this result all of them useful in various contexts. Let me state some of them:

VARIANT 1. Let $V \subset TM$ be a sub-bundle of TM such that if $X, Y \in \Gamma(V)$ then $[X, Y] \subset \Gamma(V)$. Then, locally

$$V = \text{Ker } df$$

where f is a submersion onto a manifold of dimension T_pM/V_pM.

VARIANT 2. Suppose $\theta^1, \cdots, \theta^r$ are independent 1-forms on M such that $d\theta^i$ is contained in the <u>ideal</u> $(\theta^1, \cdots, \theta^r)$ spanned by the θ^i's in $\Omega(M)$. Then there exist coordinates $x_1, \cdots, x_r$ near any $p \in M$ such that sufficiently close to p the $dx^1, \cdots, dx^r$ span the same space as the $\theta^1, \cdots, \theta^r$.

VARIANT 3. Given $\theta^1, \cdots, \theta^q$ with $\omega = \theta^1 \wedge \cdots \wedge \theta^q|_p \neq 0$, and

$$d\omega \wedge \theta^q = 0 \quad \text{near } p.$$

<u>Then the conclusions of Variant 2 hold.</u>

This theorem is proved by induction and the following type of argument: Suppose X and Y are two vectorfields satisfying our hypotheses . Then we may, by Theorem I, assume that in a suitable coordinate system centered at p

$$X = \frac{\partial}{\partial x_1} ,$$

and after subtracting an appropriate multiple of X from Y that Y is of the form

$$Y = \sum_{i>1} a_i \frac{\partial}{\partial x^i}$$

in the same region.

It follows that

$$[X, Y] = \sum_{i>1} a_{i,1} \frac{\partial}{\partial x^i}$$

and therefore does not involve $\frac{\partial}{\partial x_1}$. Being in the span of X, Y it follows that

$$[X,Y] = \lambda Y , \qquad \text{near } p$$

where λ is a smooth function. Hence for any smooth function φ with

$$[X, e^{\varphi}Y] = \left(\frac{\partial\varphi'}{\partial X_1} + \lambda\right) e^{\varphi} Y ,$$

and we can clearly choose a φ with $\frac{\partial\varphi}{\partial x_1} = \lambda$ by simply integrating in the x_1 direction. In short we can span the same space as the original X and Y did, with a new X and Y ($= e^{\varphi}Y$) such that

$$[X,Y] = 0 . \tag{2.7}$$

simply restricts to forms from $\mathfrak{g}$ to $\mathfrak{g}'$. For instance, in our case this procedure leads, to the following results:

EXAMPLE 1. The Special Linear Group $SL(n, \mathbb{R})$.

This is the subgroup of $A \in GL(n, \mathbb{R})$ with $\det A = 1$. Because $A \to \det A$ is a homomorphism of $GL(n, \mathbb{R})$ onto $GL(1, \mathbb{R})$ the induced map on forms sends the left invariant form θ_1^1 of $GL(1, \mathbb{R})$ into a left invariant form θ in $GL(n, \mathbb{R})$ which will vanish on $SL(n, \mathbb{R})$. To find this θ we need to understand the map det only near the identity; and there:

$$d \det|_e = dx_i^i ,$$

as follows directly from Kramer's rule.

Hence $\theta = \Sigma_i \theta_i^i$ and the structure equations of $SL(n, \mathbb{R})$ are

$$(3.15) \qquad d\theta_j^i + \sum \theta_\alpha^i \wedge \theta_j^\alpha = 0 , \quad \text{with} \quad \sum_i \theta_{ii} = 0 .$$

EXAMPLE 2. The Orthogonal Group $O(n)$.

This is the subgroup of matrices A with

$$A^t \cdot A = 1 .$$

In our coordinates this implies the relation $\Sigma_\alpha x_i^\alpha x_j^\alpha = \delta_j^i$, or in matrix notation:

$$(3.14) \qquad x^t \cdot x = 1 .$$

Thus on $O(n)$ the matrix of 1-forms $\theta = y\,dx$ is given by $x^t \cdot dx$.

We now plan to change the coordinates $x_2, \cdots, x_n$ — leaving x_1 alone — so that $Y = \frac{\partial}{\partial x_2}$. For this purpose we again let P be a neighborhood of p on the hypersurface $x^1 = 0$. The field Y is tangent to P at p because $Y \cdot x_1 = 0$ near p. Hence we can choose new coordinates $x_2, \cdots, x_n$ on P near p so that

$$Y|P = \frac{\partial}{\partial x_2} .$$

It follows that near p

$$Y = \sum a_i(x_1, \cdots, x_n) \frac{\partial}{\partial x^i}$$

with

$$a_i(0, x_2, \cdots, x_n) = \begin{cases} 0 & i > 2 \\ 1 & i = 2 \end{cases}$$

and, in view of (2.7)

$$\frac{\partial}{\partial x_1} a_i(x_1, x_2, \cdots, x_n) = 0 .$$

Hence in this new coordinate system $Y = \frac{\partial}{\partial x^2}$ as was to be shown.

Q.E.D.

3. The full linear group.

In this Section I would like to construct the left invariant forms on the full linear group $GL(n, \mathbb{R})$ from scratch because they play such a fundamental role in all phases of modern geometry. The underlying space of $GL(n, \mathbb{R})$ is of course the set of real $n \times n$ matrices with determinant $\neq 0$. Thus $GL(n, \mathbb{R})$ is naturally an open subset of the Euclidean space $\mathbb{R}^{n^2}$, and as such has a natural structure of a real analytic manifold. Under matrix multiplication, which is a clearly real analytic law $GL(n, \mathbb{R})$ is furthermore a group and in fact therefore a Lie group, i.e., a group object in the category of differentiable or real analytic manifolds. Now for any such object G, the deRham complex ΩG, contains a natural finite dimensional subcomplex formed by the left invariant forms $\mathrm{Inv}_G \cdot \Omega(G)$ on G, i.e., those forms $\omega \in \Omega G$ for which

$$(3.1) \qquad \ell_g^* \omega = \omega$$

for every $g \in G$, ℓg denoting the map

$$\ell_g : G \longrightarrow G,$$

given by

$$\ell_g(x) = \{gx\}.$$

Because G acts transitively on itself, it is clear that an invariant form ω is determined by its value at one point and therefore for instance by its value at the identity $e \in G$. Hence if we denote the tangent space to G at e by $\mathfrak{g}$, then it is clear that restriction of ω^q to $\wedge^q \mathfrak{g}$ imbeds

$$\mathrm{Inv}_G\, \Omega^q G \text{ into } \wedge^q \mathfrak{g}^*,$$

and it is not hard to show that this map is onto, so that as a vector space over $\mathbb{R}$

(3.2) $$\mathrm{Inv}_G\, \Omega G \simeq \wedge \mathfrak{g}^* .$$

Thus the particular nature of G is not apparent at this stage. On the other hand because each ℓg commutes with d, $\mathrm{Inv}_G\, \Omega G$ is stable under the deRham d, and so furnishes us with a local invariant of G. In particular this local invariant can be put in concrete form as follows. Let π^α be a base for $\mathfrak{g}^*$. Then we must have

(3.3) $$d\pi^\alpha = \sum_{\beta\gamma} c^\alpha_{\beta\gamma}\, \pi^\beta \wedge \pi^\gamma$$

where the $c^\alpha_{\beta\gamma} \in \mathbb{R}$ are skew symmetric in β, γ, and these numbers in a sense capture the local nature of G. Traditionally one of course favors the dual point of view. Namely, one extends the elements of $\mathfrak{g} = TG$ to left-invariant vector fields. These are furthermore seen to be closed under the bracket so that $\mathfrak{g}$ becomes a finite dimensional Lie algebra, "the Lie algebra of G", and in terms of a dual base $X_\alpha \in \mathfrak{g}$ to our π^α the structure constants of G now appear in the form

(3.4) $$[X_\alpha, X_\beta] = \sum_\gamma c^\gamma_{\alpha\beta} X_\gamma .$$

Thus (3.3) and (3.4) are duals of each other, and in this context the dual of the Jacobi identity

(3.5) $$[Z,[X,Y]] = [[Z,X],Y] + [X,[Z,Y]]$$

is seen to be

(3.6) $$d^2 = 0 \qquad \text{in } \wedge \mathfrak{g}^* .$$

Of course from the topologists point of view the complex $(\wedge \mathfrak{g}^*, d)$ which we will denote by $\Omega(\mathfrak{g})$, has the advantage that we are naturally lead to consider its cohomology $H^*(\Omega\mathfrak{g})$, and to the question of how $H^*(\Omega\mathfrak{g})$ is related to $H^*(G)$. But before we turn to these matters let us construct the left invariant forms and the structure constants for $GL(n, \mathbb{R})$ — the arche-type of all Lie groups.

For this purpose it is convenient to think of the (i,j)-th entry of a matrix A as a smooth function $x^i_j(A)$, $ij = 1, \cdots, n$. These functions are then global coordinates on $G = GL(n, \mathbb{R})$ and they satisfy the rule

(3.7) $$x^i_j(A \cdot B) = \sum_\alpha x^i_\alpha(A) \cdot x^\alpha_j(B) .$$

Hence under ℓ_A these functions transform by:

(3.8) $$\ell^*_A x^i_j = \sum_\alpha x^i_\alpha(A) \cdot x^\alpha_j .$$

Their differentials therefore transform similarly

(3.9) $$\ell^*_A dx^i = \sum_\alpha x^i_\alpha(A) dx^\alpha$$

and in particular are not left invariant. But it is not difficult to correct for this here. Indeed, let y^i_j be the function on G given by $y^i_j(A) = x^i_j(A^{-1})$, so that the y's and x's satisfy the identity

(3.10) $$\sum_{\alpha} y^i_{\alpha} \cdot x^{\alpha}_j \equiv \delta^i_j$$

where δ^i_j is the usual Kronnecker delta

$$\delta^i_j = \begin{Bmatrix} 0 & i \neq j \\ 1 & i = j \end{Bmatrix} .$$

It follows that

$$y^i_j(AB) = \sum_{\alpha} y^i_{\alpha}(B)\, y^{\alpha}(A)$$

so that

$$\ell^*_A\, y^i_j = \sum_{\alpha} y^{\alpha}_j(A) \cdot y^{\alpha}_i$$

Hence if we set

(3.11) $$\theta^i_j = \sum_k y^i_k \cdot dx^k_j$$

then

$$\ell^*_A\, \theta^i_j = \sum_{\alpha} \ell^*_A y^i_k \cdot \ell^*_A dx^k_j$$

$$= \sum y^{\alpha}_k(A) \cdot y^i_{\alpha} x^k_{\beta}(A)\, dx^{\beta}_k$$

$$= \theta^i_j .$$

Thus the θ^i_j are n^2 left invariant 1-forms on G , which agree with dx^i_j at $1 \in G$, and so form a basis for $\mathfrak{g}^*$.

To differentiate our forms, let me rewrite this development in matrix shorthand . Thus x stands for the matrix of function $\|x_{ij}\|$ etc.

(and is therefore ultimately just the identity map !) Then we have

$$xy \equiv 1 \,,$$

$$\theta \equiv y \cdot dx \,,$$

whence

$$d\theta = dy \cdot dx$$

but

$$dx \cdot y + x\,dy = 0$$

which gives

$$y\,dx\,y + dy = 0$$

or

$$\begin{aligned} d\theta &= -\, y dx \wedge y\, dx \\ &= -\, \theta \wedge \theta \,. \end{aligned}$$

In terms of our θ^i_j this amounts to

$$\boxed{d\theta^i_j + \sum \theta^i_k \wedge \theta^k_j = 0 \,,} \tag{3.12}$$

and these are the "structure equations" or "Maurer-Cartan equations" for $GL(n, \mathbb{R})$.

The corresponding structure <u>constants</u> are given by

$$c^{\,i_j}_{\substack{\alpha\lambda \\ \beta\mu}} = -\frac{1}{2}\{\delta^\lambda_\beta - \delta^\alpha_\mu\} \,, \tag{3.13}$$

but by and large (3.12) is of course a better way of recording these than (3.13).

Once the structure equations of a Lie group G are written down it is an easy matter to deduce them for a subgroup $G' \subset G$. One

Differentiating therefore yields

(3.14) $$\theta^t + \theta = 0 \qquad \text{on } O(n) .$$

Hence the structure equations for $O(n)$ are given by

(3.15) $$d\theta^i_j + \sum \theta^i_\alpha \wedge \theta^\alpha_j = 0 , \quad \text{with} \quad \theta^i_j + \theta^j_i = 0 .$$

EXAMPLE 3. <u>The Symplectic Subgroups</u> $Sp(J; \mathbb{R})$.

Let $n = 2m$ and let J be a nondegenerate skew form on $\mathbb{R}^{2m}$. The subgroup preserving this form, i.e., the matrices A with

(3.16) $$A^t J A = J$$

then define the symplectic subgroup $Sp(J; \mathbb{R})$ of $GL(n; \mathbb{R})$. Proceeding as before one obtains the relations

(3.17) $$x^t J x = J \qquad \text{on } Sp(J)$$

whence $\theta = y\,dx = J^{-1} x^t J\,dx$ on $Sp(J)$. This time differentiating (3.17) yields the "symmetry":

(3.18) $$J\theta + \theta^t J^t = 0 ,$$

characteristic of $Sp(J; \mathbb{R})$.

If we choose J to be given by the matrix

$$J = \begin{pmatrix} 0 & 1 \\ -1 & 0 \end{pmatrix}$$

each entry indicating an $m \times m$ block, then $J^t = -1J$ so that (3.18) takes the form:

$$J\theta - \theta^t J = 0 . \tag{3.19}$$

Pictorially this implies that the matrix θ has the symmetries:

$$\theta = \begin{pmatrix} \theta_1^1 & \theta_2^1 \\ \theta_1^2 & \theta_1^{1t} \end{pmatrix} \quad \text{with} \quad \begin{aligned} \theta_2^1 &= -\theta_2^{1t} \\ \theta_1^2 &= -\theta_1^{2t} \end{aligned} \tag{3.20}$$

and modulo these, the structure equations are given by the old formulae (3.12).

EXAMPLE 4. <u>The Conformal Group</u> $CO(n) \subset GL(n, \mathbb{R})$,

The conformal group $CO(n) \subset GL(n, \mathbb{R})$ is given by the subgroup generated by $O(n)$ and the scalar multiples of 1 in $GL(n, \mathbb{R})$. Thus $CO(n) = O(n) \times \mathbb{R}^*$, and the defining equations are:

$$x^t \cdot x = 1/n \quad \mathrm{Trace}(x^t x) \cdot 1 . \tag{3.21}$$

Proceeding as before, this leads to the relation

$$\theta^t + \theta = 2\,(\mathrm{Trace}\ \theta)\, 1 \quad ,$$

so that here a basis is given by

$$\theta_j^i, \quad i \neq j, \quad \text{and} \quad \theta = \frac{1}{n}\left(\sum_\alpha \theta_\alpha\right) .$$

EXAMPLE 5. <u>The Subgroup Preserving a Subspace</u>; $GL(n, p; \mathbb{R})$.

As a final instance consider the subgroup $GL(n, p; \mathbb{R})$ preserving $\mathbb{R}^p \subset \mathbb{R}^n$ under the standard action of $GL(n, \mathbb{R})$ on $\mathbb{R}^n$. If $e_1, \cdots, e_n$

is the standard basis then this action is given by:

$$Ae_i = \sum_\alpha x_i^\alpha(A) \cdot e_\alpha$$

so that the defining relations of $GL(n, p; \mathbb{R})$ are simply given by

$$x_i^\alpha = 0 \qquad \text{for} \quad \begin{matrix} \alpha > p \\ i \leq n \end{matrix} .$$

Hence the matrix x has the picture

$$(3.22) \qquad \left(\begin{array}{c|c} * & p \quad * \\ \hline & n-p \\ 0 & * \end{array} \right) \qquad \text{on} \quad GL(n, p; \mathbb{R})$$

and θ has the corresponding zeroes i.e., $\theta_j^\alpha \equiv 0$ on $GL(n, p)$ for $\alpha > p$, $i \leq n$.

These examples should suffice for us and I will therefore continue towards our main goal, by constructing the corresponding structure equations for the Lie groupoids which underlie C^∞-geometry in the next sections.

Let me close this section however, by briefly discussing the cohomology of these complexes of left invariant forms, and how this concept might be of use in geometry.

First a remark on $H^*(\Omega \mathfrak{g})$. In general, computing this space from the structure constants of $\mathfrak{g}$ is a difficult task and usually has to be attacked indirectly. For the "reductive Lie-algebras $\mathfrak{g}$" this problem can however be reduced to a problem in invariant theory. The argument takes the following form :

First observe that right translation r_g by an element $g \in G$ commutes with all left translations. Hence r_g induces a representation on $\Omega(\mathfrak{g})$ which preserves d which is the dual of the usual adjoint representation of G on $\mathfrak{g}$. The infinitesimal right translations are given by the Lie-derivatives $\mathcal{L}(v)$, of left-invariant vector fields v in $\mathfrak{g}$. We may therefore consider the sub-complexes

$$\mathrm{Inv}^{\mathfrak{g}}\, \Omega(\mathfrak{g}) \supset \mathrm{Inv}^{G}\, \Omega(\mathfrak{g})$$

given by the infinitesimally and globally invariant elements under the adjoint action of G on $\Omega(\mathfrak{g})$, and of course these coincide if G is connected. In any case determining these spaces is clearly a question in the realm of invariant theory.

On the other hand, I claim that the d these spaces of left and right invariant forms inherit from $\Omega(\mathfrak{g})$ is trivial, so that all homology questions disappear on them.

To see this, recall that for any $X \in \mathfrak{g}$, (considered now as a left-invariant vector field on G !)

$$\mathcal{L}(X) = \iota(X)d + d\,\iota(X) ,$$

on all of $\Omega(G)$. On $\Omega^1(\mathfrak{g}) \subset \Omega(G)$, $\iota(X)$ is constant, so that there

(3.22) $$\mathcal{L}(X)\,\theta = \iota(X)\,d\theta ,$$

which enables one to express d on $\Omega^1(\mathfrak{g})$ in terms of the $\mathcal{L}(X^{\alpha})$ where

X^{α} ranges over a base for $\mathfrak{g}$. Indeed if X_{α} and π^{α} are dual bases we see from (3.22) that

$$d\theta = \frac{1}{2}\sum_{\beta} \pi^{\beta} \mathcal{L}(X_{\beta})\,\theta \tag{3.23}$$

for any $\theta \in \Omega^{*}(\mathfrak{g})$.

Hence invariance of θ, which means $\mathcal{L}(X)\,\theta = 0$ for all X, certainly implies $d\theta = 0$.

Now then the reductive action of $\mathfrak{g}$ on $\Omega\mathfrak{g}$ implies that $\Omega\mathfrak{g}$ breaks up into a direct sum of subcomplexes

$$\Omega\mathfrak{g} = \bigoplus_{\lambda} (\Omega\,\mathfrak{g})_{\lambda} \tag{3.23}$$

where λ ranges over the irreducible representations of $\mathfrak{g}$, and $\mathfrak{g}$ acts on $(\Omega\,\mathfrak{g})_{\lambda}$ only via λ. It follows that

$$H(\Omega\,\mathfrak{g}) = \bigoplus_{\lambda} H_{\lambda}(\Omega\,\mathfrak{g})$$

has a corresponding decomposition. On the other hand $\mathcal{L}(X)$ acts trivially on $H(\Omega\mathfrak{g})$ in view of the homotopy formula (2.6). Hence $\Omega\mathfrak{g}$ has the homotopy type of $\Omega\mathfrak{g}_{\lambda}$ with λ - the trivial $\mathfrak{g}$ - module, i.e., with $\mathrm{Inv}^{\mathfrak{g}}\,\Omega(\mathfrak{g})$.

In short then, one has :

PROPOSITION 3.1. For a reductive Lie algebra :

$$H^{*}(\Omega\mathfrak{g}) \cong \mathrm{Inv}^{\mathfrak{g}}\,\Omega(\mathfrak{g})\,. \tag{3.24}$$

Of course this still leaves a lot of work to be done, and in fact although (3.24) goes back to E. Cartan, the actual computations of the invariants in $\Omega(\mathfrak{g})$ when G is one of the classical simple Lie groups was only carried out by R. Brauer in the 1930's . We will return to this problem later; here let me rather turn to the geometric analogues of (3.24) and the unitary trick.

Here the problem is to compare the cohomology of $\Omega\mathfrak{g}$ with that of ΩG , and the result analogous to (3.24) is given by:

PROPOSITION 3.2. If G is a compact connected Lie group then

$$H^*(\Omega\mathfrak{g}) \cong H^*(\Omega G) \quad . \tag{3.25}$$

The argument for (3.25) is to decompose ΩG qua G module under left translation - into a direct sum of subcomplexes

$$\Omega G = \bigoplus_{\lambda} \Omega_{\lambda} G$$

λ ranging this time over the irreducible representations of G . As G is compact this is possible and the projection on $\Omega_{\lambda} G$ being given by an integral operator, and in particular $\Omega\mathfrak{g}$ is precisely $\Omega_{\lambda} G$ with λ the trivial representation and the projection of ΩG to $\Omega\mathfrak{g}$ is simply given by averaging a form θ over G :

$$\theta \longrightarrow \int_G \ell_g^*(\theta)\, dg \quad .$$

This result then brings the Lie algebra cohomology into line with the topological cohomology for compact groups. For non-compact ones, however, there is in general very little relation between $H^*(\Omega\mathfrak{g})$ and

$H^*(\Omega$. Here it is only for the semi-simple groups that one can still get at $H(\Omega g)$ via (3.24) and a principle H. Weyl calls the unitary trick.

First of all, given a Lie algebra g over $\mathbb{R}$, we may construct a new Lie algebra over the complex numbers $\mathbb{C}$ simply by extending the base field and we denote this new Lie algebra by $g \underset{\mathbb{R}}{\otimes} \mathbb{C}$ or simply $g^{\mathbb{C}}$. Thus if the X_α generate g and g has structure constants $c^\gamma_{\alpha\beta}$:

$$[X_\alpha, X_\beta] = \sum_\gamma c^\gamma_{\alpha\beta} X_\gamma , \tag{3.26}$$

then $g^{\mathbb{C}}$ is also generated by X_α over $\mathbb{C}$ and (3.26) still holds, the [,] now however being linear over $\mathbb{C}$. The complex $\Omega(g^{\mathbb{C}}) = \Omega(g) \underset{\mathbb{R}}{\otimes} \mathbb{C}$ is naturally a complex of $\mathbb{C}$ - modules and $H^*(\Omega g^{\mathbb{C}})$ also. Therefore it follows from

$$\Omega(g^{\mathbb{C}}) = \Omega(g) \otimes \mathbb{C} ,$$

that

$$H^*(\Omega g^{\mathbb{C}}) = H^*(\Omega g) \underset{\mathbb{R}}{\otimes} \mathbb{C} . \tag{3.27}$$

In short, complexification commutes with homology and hence two Lie algebras with isomorphic complexifications have isomorphic cohomology. Now the semi-simple Lie algebras g have the marvelous property that their complexification $g^{\mathbb{C}}$ are simultaneously the complexification of a Lie algebra $k(g)$ of a certain compact connected Lie group $K(g)$, so that for these Lie algebras one has :

$$H^*(\Omega g) = H^* \{K(g)\} . \tag{3.28}$$

This trick also works for $g\ell$ (IR) because

(3. 29) $$g\ell(n;\mathbb{R}) \otimes \mathbb{C} \cong g\ell(n;\mathbb{C}) \cong u(n) \otimes_{\mathbb{R}} \mathbb{C}$$

where $u(n)$ denotes the Lie algebra of the unitary group.

Indeed, using the same type of analysis as in our subgroups of $GL(n;\mathbb{R})$, we identify $U(n) \in GL(n;\mathbb{C})$ with the complex matrices A subject to

$$\bar{A}^t \cdot A = 1 ,$$

where the bar denotes conjugation, and hence the Lie algebra of $U(n)$ with the skew Hermitian matrices satisfying:

$$\bar{A}^t + A = 0 .$$

It follows that iA, is Hermitian for any $A \in u(n)$ and hence, as every matrix is the sum of a Hermitian and a skew Hermetian matrix, that $u(n) \otimes \mathbb{C} \cong g\ell(n, \mathbb{C})$. The upshot is therefore that

(3.30) $$H^*\{\Omega\, g\ell(n;\mathbb{R})\} = H^*\{u(n)\} ,$$

and this is how, by and large, topologists prefer to compute the left hand side. In any case let me conclude this section by recording the answer here for future reference:

THEOREM: $$H^*\{\Omega\, g\ell(n, \mathbb{R})\} = E(h_1, \cdots, h_n)$$

where E denotes the exterior algebra generated by the elements

$h_i \in H^{2i-1}$.

Representative forms for the h_i can be given to be

$$h_i = \text{Trace}\,\{\underbrace{\theta \wedge \theta \wedge \cdots \wedge \theta}_{2i-1 \text{ factors}}\}$$

and $\theta = \|\theta^i_j\|$ is the matrix of our left invariant forms.

4. The Group Diff M .

We turn now to the analogues of these classical constructions for certain function spaces and in particular, the group Diff (M) of diffeomorphisms of a C^∞-manifold.

Here Gelfand-Fuks proposed the complex $\Omega_c LM$, of continuous forms on the Lie algebra LM of all C^∞-vector fields as the proper analogue of Ωg in the finite dimensional case.

Precisely then, let L(M) be given the C^∞ topology, so that $X_n \to Y$, if and only if on every compact $K \subset M$, the components of X_n relative to a finite set of coordinate systems covering K , as well as all their derivatives, converge to the components and corresponding derivatives of Y . With this understood $\Omega_c^q LM$ is defined as the alternating continuous q-forms on LM , and as $X, Y \to [X,Y]$ is clearly a continuous map $LM \times LM \to LM$, the old formula for d on Ωg preserves this complex. Recall that explicitly:

$$(4.1) \qquad d\varphi(X_0, X_1, \cdots, X_{q+1}) = \sum (-1)^{i+1} \varphi([X_i, X_j], X_1, \ldots, \hat{X}_i, \cdots, \hat{X}_j, \cdots, X_{q+1}) .$$

To check that this complex which we also denote by A(M) is the proper analogue of Ωg , note that Ωg can also be characterized by the following universal property:

For any map (C^∞ of course)

$$f : W \to G ,$$

there is an induced map of complexes

$$\Omega g \xrightarrow{f^{\#}} \Omega W$$

which is:

a) Natural ; that is given.

$$W' \xrightarrow{g} W \xrightarrow{f^{\#}} G,$$

then

(4.2) $$(f \circ g)^{\#} = g^{*} \circ f^{\#}$$

where g^* denotes the deRham homomorphism induced by g,

b) Left invariant: that is for any diagram

$$W \xrightarrow{f} G \xrightarrow{\ell_g} G$$

where ℓ_g denotes left translation by g, one has

(4.3) $$(\ell_g \circ f)^{\#} = f^{\#},$$

c) Universal in this context. That is, given a function Φ which to each map $f : W \to G$, assigns a form $\Phi(f) \in \Omega(W)$ subject to a) and b):

$$g^{*}\Phi(f) = \Phi(gf), \quad \text{and} \quad \Phi(f \circ \ell_g) = \Phi(f),$$

then there exists a unique $\varphi \in \Omega g$ with

(4.4) $$\Phi(f) = f^*\varphi \ .$$

Indeed to construct φ, just consider the identity map 1

$$G \longrightarrow G$$

and set $\Phi(1) = \varphi$.

Now these criteria also make perfect sense in the infinite dimensional case, once we decide <u>when</u> a map

$$W \xrightarrow{\ f\ } \text{Diff}(M)$$

is to be considered as <u>smooth</u>, and here the natural criterion is clearly that the adjoint map f^t,

$$f^t : W \times M \longrightarrow M$$

given by

$$f^t(w, m) = f(w) \cdot (m)$$

be smooth. Note by the way that this criterion works equally well for the function spaces of all C^∞ maps from M to N, as well as for many of its natural subspaces.

To return to our main task we have to construct a map

(4.5) $$f^{\#} : A(M) \longrightarrow \Omega(W)$$

subject to (4.2) and (4.3) for any map

$$f : W \longrightarrow \text{Diff}(M) \ .$$

The construction of $f^{\#}$ proceeds as follows: Consider the map

$$f^t : W \times M \longrightarrow M$$

and let us think of $W \times M$ as a fibering over W, with fiber M, and let

$$\begin{array}{c} W \times M \\ \big\downarrow \pi \\ W \end{array}$$

denote the natural projection. The product structure furthermore gives us a natural decomposition of the tangent space to $W \times M$ into a vertical and horizontal space

$$T_{(w,m)}(W \times M) \cong T_w \oplus T_m .$$

Now consider the map f^t, and let H_f be the sub-bundle of the tangent space to $W \times M$ which is in the kernel of df^t:

(4.6) $$H_f = \operatorname{Ker} df^t .$$

Restricted to a fiber of π, f^t is a diffeomorphism. Hence the vertical space intersects H_f only in 0. Furthermore df^t is an isomorphism of $T_m M$ at (w, m) onto $T_n M, n = f(w) \cdot m$. In short, H_f is indeed a sub-bundle of dimension equal to $\dim W$, and transversal to the vertical tangent space in $W \times M$. Pictorially the situation is described by:

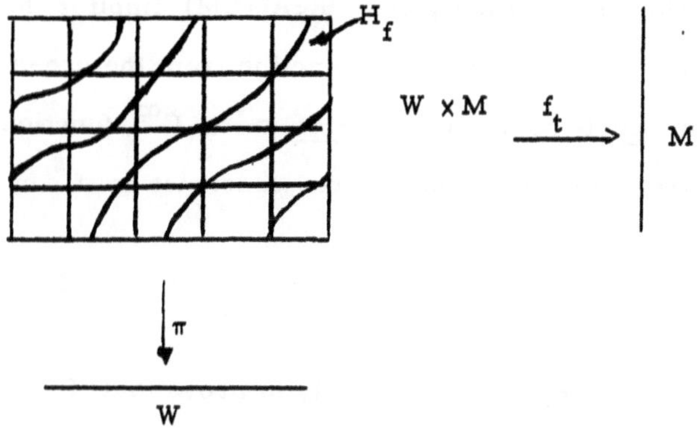

Figure 4.1

the slanted lines indicating the sub-bundle H_f.

In this situation then note that every vector field X on W has two natural liftings to $W \times M$, namely the lift HX given by the horizontal lift relative to the product structure, and the lift $H_f X$ relative to the horizontal complement H_f. Thus $H_f X$ is uniquely characterized by the fact that it maps to X under π, and is contained in H_f. Now then define

(4.7) $$V_f X = H_f X - HX$$

as the <u>vertical component of</u> $H_f X$ in our product decomposition.

Note that V_f is a pointwise operation sending a tangent vector X_w to a vector field $V_f X_w$ on $W \times M$, tangent to the fiber $w \times M$. Thus V_f is a construction from smooth vector fields on X to smooth vector fields tangent to the fibers on $W \times M$.

Now the continuity of a $\theta \in A^q(LM)$ implies that we may interpret such a θ as a q-multilinear function from the cross-sections of the tangent bundle along the fibers of π to C^∞ functions on W, and so interpreted we define the desired construction $f^\#$ by:

(4.8) $$f^\#\theta(X_1, \cdots, X_q) = \theta(V_f X_1, \cdots, V_f X_q) .$$

Therefore it remains only to prove that $f^\#$ commutes with d, is natural and invariant.

Let me explain the "commuting with d" on 1-forms $\theta \in A^1(M)$. The higher ones proceed analogously.

We have, by definition,

(4.10) $$\begin{aligned} f^\# d\theta(X, Y) &= d\theta(V_f X, V_f Y) \\ &= -\theta([V_f X, V_f Y]) \end{aligned}$$

while, again by definition,

(4.11) $$d(f^\#\theta)(X, Y) = X\,\theta(V_f Y) - Y\,\theta(V_f X) - \theta(V_f[X, Y]) .$$

To bring these into accord we therefore need to recall first of all that H_f is an integrable sub-bundle of $W \times M$.

Hence

$$[H_f X, H_f Y] = H_f[X, Y]$$

and similarly of course,

$$[HX, HY] = H[X, Y] .$$

It now follows from $H_f X = V_f X + HX$ that

$$H_f[X, Y] = [V_f X, V_f Y] + [HX, V_f Y] + [HY, V_f X] + H[X, Y] ,$$

whence,

$$(4.12) \qquad V_f[X, Y] = [V_f X, V_f Y] + [HX, V_f Y] - [H(Y), V_f X] .$$

Finally, it follows again from the continuity of θ that

$$(4.13) \qquad X\theta(U) = \theta\{\mathcal{L}(HX) \cdot U\}$$

for any section U of the bundle along the fibers of π. Hence using (4.12) and (4.13), (4.10) takes the form

$$\begin{aligned} d(f^{\#}\theta)(X, Y) &= \theta\{[HX, V_f Y] - [HY, V_f X] - [HX, V_f Y] \\ &\quad + [HY, V_f X] - [V_f X, V_f Y]\} \\ &= f^{\#} d\theta(X, Y) . \end{aligned}$$

Q.E.D.

The proof of naturality is quite straightforward, and of invariance also. Indeed, if $\varphi \in \mathrm{Diff}(M)$ then

$$(\varphi \circ f)^t = \varphi \circ f^t .$$

Hence $\mathrm{Ker}\, d(\varphi \circ f)^t = \mathrm{Ker}\, df^t$, so that $H_{\varphi \circ f} = H_f$, and we are done.

There are two immediate extensions of this construction of which we will need only the first one. The second one was pointed out and used to great advantage by G. Segal.

The first extension deals with the space

$$\mathrm{Im}(U, M)$$

of imbeddings of one n-dimensional manifold U into another n-dimensional M.

The construction of H_f, V_f and therefore $f^{\#}$ clearly goes through, so that for any smooth map

$$f: W \longrightarrow \mathrm{Im}(U, M)$$

there is a natural, left invariant homomorphism

$$f^{\#} A(U) \longrightarrow \Omega(W) \quad . \tag{4.14}$$

Actually, this is the G. Segal remark, our $f^{\#}$ is only the first component of a more general

$$f' : A(U) \longrightarrow \Omega(W; A(M))$$

defined as follows:

Given vector fields $X_1, \cdots, X_p$ on W, and vector field $Y_1, \cdots, Y_q$ on M, as well as a $p+q$ form $\theta \in A^{p+q}(U)$, define

$$f^{(p,q)}\,\theta(X_1, \cdots, X_p; Y_1, \cdots, Y_q) = \theta\,\{V_f X_1, \ldots, V_f X_p, \tilde{Y}_1, \cdots, \tilde{Y}_q\}$$

where the $\tilde{Y}_1, \cdots, \tilde{Y}_q$ are the unique vertical field mapped to $Y_1, \cdots, Y_q$, via df.

The sum of these $f^{p,q}$ now defines f^{-1}, which is again seen to be natural, left invariant and to commute with d.

We sum this whole development up in the following proposition:

PROPOSITION 4.1. The Gelfand-Fuks Complex $A(U) = \overset{\bullet}{\Omega} LU$ of an n-manifold U, has the property that to any smooth map

$$f : W \longrightarrow Im(U, M)$$

there is associated an induced homomorphism

$$f^{\#} : A(U) \longrightarrow \Omega(W) \tag{4.15}$$

which is natural and left-invariant in the sense of 4.2 and 4.3 respectively.

The Universality condition (4.3) is harder to treat in this context and I will not pursue it further here. Rather, let us get to know the complex A(M) by explicitly writing down the structure equations of the simplest of its subcomplexes.

The complex in question is the subcomplex $A_p(U)$ furnished by these forms in A(U) whose support is at a fixed point $p \in U$.

For this purpose choose a fixed coordinate system $x_1, \cdots, x_n$, centered at p, so that every $X \in L(U)$ is locally given by

$$X = \sum_i a^i \frac{\partial}{\partial x^i} \tag{4.16}$$

where the a^i are C^∞ functions near p.

From the most elementary properties of distributions it then follows that all the 1-forms of $A_p(U)$ are generated by the finite partial derivatives of the δ - function applied to the components of X . Hence $A_p(U)$ is really independent of U , and therefore a purely local and universal complex which we will often simply denote by A_p :

$$A_p = A_p(U) \qquad \text{for any } U \ni p .$$

Gelfand and Fuks take the position that A_p is again the continuous analogue of $\Omega\mathfrak{g}$. Indeed if $\mathfrak{a}_n$ denotes the Lie algebra of formal vector-fields

$$X = \sum a_i \frac{\partial}{\partial x_i}$$

where the a_i are formal power series in the x's , and if we topologise $\mathfrak{a}_n$ in the usual power series topology, - x^α is small if $|\alpha|$ is large - then it is obvious that

$$(4.17) \qquad A_p = \Omega_c(\mathfrak{a}_n) ,$$

is the complex of continuous alternating forms on $\mathfrak{a}_n$.

Well, in any case, to get on with an explicit computation of $H^*A_p = H^*_c(\mathfrak{a}_n)$ we proceed as follows:

In terms of our local coordinates x^i at p , define 1-forms

$$\theta^i_\alpha \in A^1_p = \Omega^1_c(\mathfrak{a}_n) , \quad i = 1, \cdots, n ,$$

$$\alpha = (\alpha_1, \alpha_2, \cdots, \alpha_k)$$

by:

$$\theta^i_\alpha(X) = (-1)^{|\alpha|} \frac{\partial^{|\alpha|}}{\partial x^\alpha} a^i \Big|_p . \tag{4.18}$$

Here the α's are k-tuples of non negative integers and $|\alpha| = \Sigma \alpha_i$.

Our earlier statement then translates to the assertion that

<u>The</u> θ^i_α <u>span</u> A^1_p <u>and hence generate all of</u> A_p .

Note that of course the θ's are symmetric in their lower indexes so that $\theta^i_{12} \equiv \theta^i_{21}$, etc .

The sign has been chosen so that the lower indices correspond to Lie - derivatives of θ^i in the direction $\frac{\partial}{\partial x^{\alpha_1}}$, $\frac{\partial}{\partial x^{\alpha_k}}$. That is, we have

$$\mathcal{L}\left(\frac{\partial}{\partial x^j}\right) \theta^i = \theta^i_j$$

$$\mathcal{L}\left(\frac{\partial}{\partial x^j}\right) \cdot \mathcal{L}\left(\frac{\partial}{\partial x^k}\right) \theta^i = \theta^i_{jk} , \quad \text{etc.}$$

where $\mathcal{L}$ is still given by (2.6) and so <u>commutes</u> with d .

Now because the $\left(\frac{\partial}{\partial x^i}\right)$ commute with one another, the $\mathcal{L}\left(\frac{\partial}{\partial x^i}\right)$ also commute.

Hence the map

$$\frac{\partial}{\partial x^{\alpha_1}} \otimes \cdots \otimes \frac{\partial}{\partial x^{\alpha_n}} \longrightarrow \mathcal{L}\left(\frac{\partial}{\partial x^{\alpha}}\right) \cdots \mathcal{L}\left(\frac{\partial}{\partial x^{\alpha_n}}\right)$$

factors through the symmetric, or polynomial, ring $\mathbb{R}\left[\frac{\partial}{\partial x^1},\dots,\frac{\partial}{\partial x^n}\right]$ generated by the $\frac{\partial}{\partial x^i}$. over $\mathbb{R}$, and it follows that under $\mathcal{L}$, the complex A^1_p is naturally an $\mathbb{R}\left[\frac{\partial}{\partial x^i}\right]$ module.

From this point of view (4.17) can therefore be stated as the

PROPOSITION 4.2. <u>Under $\mathcal{L}$ the 1-forms A^1_p constitute a free module with n generators $\theta^1,\dots,\theta^n$ over the polynomial ring</u> $\mathbb{R}\left[\frac{\partial}{\partial x^i}\right]$:

$$A^1_p \simeq \mathbb{R}\left[\frac{\partial}{\partial x_1},\dots,\frac{\partial}{\partial x_n}\right](\theta^1,\dots,\theta^2)$$

It follows now that structure equations of A_p are completely determined once they are written down for θ^i , and hence by the following,

PROPOSITION 4.3.

$$d\theta^i + \theta^i_j \wedge \theta^j = 0\ .^{\dagger} \tag{4.19}$$

Proof. By definition

$$d\theta^i(X,Y) = -\theta^i([X,Y])$$

On the other hand if $X = a^\alpha \frac{\partial}{\partial x^\alpha}$, $Y = b^\beta \frac{\partial}{\partial x^\beta}$, then by explicit computation,

$$[X,Y] = \left\{a^\alpha b,^\beta_\alpha - b^\alpha a,^\beta_\alpha\right\}\frac{\partial}{\partial x^\beta} \tag{4.20}$$

where $,_\alpha$ denotes $\frac{\partial}{\partial x^\alpha}$ as usual. Hence

$$\theta^i[X,Y] = a^\alpha b,^i_\alpha - b^\alpha a,^i_\alpha$$

---

† I will be using the usual summation convention for repeated indices from now.

while

$$\theta^i_j \wedge \theta^j (X,Y) = -a,^i_j b^j + a^j b,^i_j \quad . \qquad \text{Q. E. D.}$$

To see how the formulae (4.18), (4.19) determines all other structure equations, one need only apply $\mathcal{L}\left(\frac{\partial}{\partial x^k}\right)$ etc., to both sides of (4.19) and recall that $\mathcal{L}$ acts by derivatives. Explicitly one obtains:

$$\text{(4.21)} \qquad d\,\theta^i_k + \theta^i_{jk} \wedge \theta^j + \theta^i_j \wedge \theta^j_k = 0$$

$$\text{(4.22)} \qquad d\,\theta^i_{k\ell} + \theta^i_{jk\ell} \wedge \theta^j + \theta^i_{jk} \wedge \theta^j_\ell + \theta^i_{j\ell} \wedge \theta^j_k + \theta^i_j \wedge \theta^j_{k\ell} = 0 \; .$$

in the first two cases.

Clearly now (4.20) bears a considerable similarity to the structure equations of $GL(n, \mathbb{R})$ and to bring this out further let us define the 2-forms $R^i_j \in A^2_p$ by

$$\text{(4.23)} \qquad R^i_j \equiv d\,\theta^i_j + \theta^i_k \wedge \theta^k_j \quad .$$

Then (4.21) takes the form

$$\text{(4.24)} \qquad R^i_j \equiv -\,\theta^i_{jk} \wedge \theta^k$$

while the structure equations of $GL(n, \mathbb{R})$ of course were $R^i_j \equiv 0$. In fact these forms behave just as the curvature tensor of a "torsion free connection" does, and as we will see later the R^i_j do play a role of the universal or formal curvature forms of a manifold. Here let us just as an exercise in the calculus of $A_p(U)$ derive the classical curvature identifies :

PROPOSITION 4.3. The forms $R^i_j \in A^2_p$ have the following properties :

(4.25) $$R^i_j \wedge \theta^j = 0 \quad , \quad dR^i_j = R^i_k \wedge \theta^k_j - \theta^i_k \wedge R^k_j$$

which correspond to the two Bianchi Identities in Torsion free Geometry.

Proof: By (4.24)

$$R^i_j \wedge \theta^j = - \theta^i_{jk} \wedge \theta^k \wedge \theta^j \equiv 0$$

because θ^i_{jk} is symmetric in its lower indices.

To obtain the second identity consider the equation (4.22) and multiply it with θ^ℓ summing over the repeated index ℓ . One obtains

(4.26) $$d\theta^i_{k\ell} \wedge \theta^\ell + \theta^i_{jk} \wedge \theta^j_\ell \wedge \theta^\ell + \theta^i_{j\ell} \wedge \theta^j_k \wedge \theta^\ell + \theta^i_j \wedge \theta^j_{k\ell} \wedge \theta^\ell = 0 \ .$$

where the second form of (4.22) has cancelled out by the symmetry of $\theta_{jk\ell}$ in j, k, and ℓ . But (4.26) simply restates (4.25) in the equivalent form

$$- dR^i_k + \theta^i_j \wedge R^i_k - R^i_j \wedge \theta^j_k = 0$$

On the cohomological side (4.24) and (4.25) implies that the subalgebra

$$\underline{W} = \theta^i_j , R^i_j$$

generated by the indicated forms in A_p is, first of all, finite dimensional and secondly closed under d. Indeed by (4.24) any monomial in the R's of length $> n$ vanishes, while (4.25) together with (4.23) expresses d of the θ's and

R's in terms of the θ's and R's . Actually this subcomplex already carries all the cohomology of A_p .

THEOREM : (Gelfand-Fuks) The inclusion

$$\underline{W} \subset A_p(U)$$

induces an isomorphism in cohomology.

I will sketch a proof of this fact in Section 7 dealing with the fundamental theorem on invariant theory and some of its applications. The proof is very similar to the one given in [2] of the Gilkey theorem.

Here I will only carry the computation of $H^*(\underline{W})$ a bit further along.

For this purpose let us define the ith Chern polynomial as a function on $GL(n, \mathbb{R})$ by the formula:

$$\Sigma t^i c_i(A) = \det(1 + tA) \quad . \tag{4.27}$$

It is then clear that each c_i is a polynomial in the entries A^i_j of A , for example

$$\begin{cases} c_1(A) = A^i_i \\ c_2(A) = (- A^i_j A^i_j + (A^i_i)^2)/2 \\ c_n(A) = \det A \end{cases} \tag{4.28}$$

Now as the R^i_j are two-forms and so commute we can evaluate these polynomials c_i on the R^i_j's to obtain forms

(4.29) $$c_i(R) \in A^{2i}_p \quad .$$

We call these the Chern classes in A_p and in terms of them the Pontrajgin forms $p_i(R) \in A^{4i}_p$ are defined simply by

(4.30) $$p_i(R) = c_{2i}(R) ,$$

of course in the literature one often encounters these forms with factors of $(-1)^i \left(\frac{\sqrt{-1}}{2\pi}\right)^i$, but this is not to the point here, where they are, for the moment, only elements of $\underline{W} \subset A^{4i}_p$. In any case in terms of these $c_i(R)$ one can reduce $\underline{W}$ further to a complex WU_n defined as follows.

Consider the polynomial ring $\mathbb{R}[c_1, \cdots, c_n]$ with weight of $c_i = 2i$, and let $\underline{\mathbb{R}}[c_1, \cdots, c_n]$ be the quotient of $\mathbb{R}[c_1, \cdots, c_n]$ by the ideal of elements with weight $> 2n$. Here of course the weight w of a monomial $c_1^{\alpha_1}, \cdots, c_n^{\alpha_n}$ in $\alpha_1 \cdot 1 + \alpha_2 \cdot 2 +, \cdots, + \alpha_n \cdot n$.

With this understood one sets

$$WU_n = \mathbb{R}[c_1, \cdots, c_n] \otimes E(h_1, \cdots, h_n)$$

where $E(h_1, \cdots, h_n)$ denotes an exterior algebra with generators h_i in dimensions $2i - 1$. Finally a differential operator d is introduced in

WU_n by setting

(4.30)
$$d\, c_i \otimes 1 = 0$$
$$d\, h_i \otimes 1 = 1 \otimes c_i \ .$$

The cohomology of A_p is now made rather explicit by:

PROPOSITION 4.4 <u>The map</u> $1 \otimes c_i \to c_i(R)$ <u>has an extension to a map of complexes</u>

$$WU_n \longrightarrow \underline{W} \subset A_p$$

<u>and any such extension induces an isomorphism</u>

(4.32)
$$H^*(WU_n) \xrightarrow{\cong} H^*(A_p) \simeq H^*(\mathfrak{a}_n)$$

As an example, consider the case of $n = 1$. Then $WU_1 \subset \underline{W}$ can be taken to be

$$c_1 = R_1^1 \ , \quad h_1 = \theta_1^1 \ ,$$

for we clearly have $d\, h_1 = c_1$. In this case $H(WU_1)$ is generated by 1, and $c_1 h_1$ in dim 3. Thus $U = \mathbb{R}^1$

(4.33)
$$H^* A_p(\mathbb{R}^1) = \begin{cases} \mathbb{R} \text{ in dim } 0 \\ \mathbb{R} \text{ in dim } 3 \end{cases} .$$

The computations of course become more complicated in higher dimensional cases but were carried out by Vey, [30] who produced an explicit base for $H^* A_p(\mathbb{R}^n)$ in terms of the monomials in the h's and c's of WU_n.

So much then for our brief excursion into the A_p . We have learned first of all, that this "subcomplex of minimal support" has structure equations which are easy to write down, and which are very analogous to the equations one meets in differential geometry. Secondly we have seen that its cohomology is finite dimensional and computable.

Now in the computation of $H^*A(M)$, the cohomology of $A_p(M)$ plays a fundamental role, which I will have no time to explore here. Let me just give those of you familiar with some topology the barest outlines of the final outcome.

Already Gelfand-Fuks observed that every manifold M carries a bundle EM , which is associated to its principal frame-bundle and whose fibers E_pM , have cohomology isomorphic to that of $A_p(M)$:

$$H^*(E_pM) \simeq H^*A_p(M)$$

Thus E is in a sense a geometric realization of the family of complexes $A_p(M)$ $p \in M$, and in terms of E , the final result on H^*AM , is given by the following theorem which was conjectured independently two years ago by Fuks and myself and for which two independent proofs have recently been given by Haefliger and Segal.

THEOREM (Haefliger, Segal). Let ΓE denote the space of continuous sections of E . Then

$$H^*(AM) \simeq H^*(\Gamma E) \quad . \tag{4.34}$$

This beautiful result is, of course, nevertheless a negative one, as it reduces the mysterious $H^*(AM)$ to the complicated but still rather well understood $H^*(\Gamma E)$.

Observe that (4. 34) implies, in particular, that

$$H^*(A\mathbb{R}^n) = H^*(\Gamma E) = H^*(E_p) = H^*(A_p\mathbb{R}^n)$$

the second equality being true because E is contractible to E_p over p . Finally let us check (4. 34) for the circle, where Gelfand and Fuks made their own computations. In this case E_p is the three sphere S^3 , which conforms with (4. 33). Further as the tangent bundle of S^3 is trivial, we have

$$E = S^3 \times S^1$$

whence $\Gamma E = \mathrm{Map}\,(S^1, S^3)$, the space of maps from S^1 to S^3 . Now this function space has well known homology, to wit:

(4. 35) $$H^*\mathrm{Map}(S^1, S^3) = \mathbb{R}\,[\mu] \otimes E(\omega) ,$$

is a polynomial ring with generator μ in dimension 2, tensored with an exterior algebra with generator ω in dimension 3.

Gelfand and Fuks arrived at the same result, giving the following lovely explicit representatives for μ and ω .

Let us take S^1 in the form $\mathbb{R}^1/\mathbb{Z}$, so that vector fields $f\frac{\partial}{\partial x}$

can be identified with functions f of period 1. With this understood

(4.36)
$$\mu(f, g) = \int_0^1 \det\begin{vmatrix} f' & g' \\ f'' & g'' \end{vmatrix} dx$$

$$\omega(f, g, h) = \det\begin{vmatrix} f & g & h \\ f' & g' & h' \\ f'' & g'' & h'' \end{vmatrix}_0$$

where the 0 in the second formula denotes evaluation at 0. Of course, any other point would do just as well, and in fact one could also integrate this expression for ω. On the other hand μ can never be represented by a formula that does not involve the behavior of f and g on all of S^1.

5. The Jet-spaces $J_k M$.

In this section we will first of all heuristically identify A_p with the complex of left invariant forms on the space of jets $J M$ under the action of the group Diff (M) . Thereafter I will briefly review connections from this point of view and so bring an explanation of why the structure equations of A_p bore such a resemblance to those of a torsion free connection.

From our present point of view the Jet-spaces over M should be thought of as the simplest finite dimensional manifolds on which Diff (M) acts transitively and compatibly with its action on M . Thus if we think of M as the homogenous space

$$M = \mathrm{Diff}(M) / \mathrm{Diff}(M ; p) , \tag{5.1}$$

of Diff (M) modulo the subgroup Diff (M ; p) keeping a point $p \in M$ fixed, then $J_k M$ should be thought of as the homogenous space with stability group $\mathrm{Diff}(M ; p^{(k+1)})$, whose elements f keep p fixed to the (k+1) order:

$$J_k M = \mathrm{Diff}(M) / \mathrm{Diff}(M ; p^{k+1}) \tag{5.2}$$

In this way one obtains, heuristically at least, a tower of manifolds "over M " :

$$M \xleftarrow{\pi} J_1 M \xleftarrow{\pi} J_2 M \xleftarrow{\pi} J_3 M \xleftarrow{\pi} \tag{5.3}$$

each fibered over its predecessor with fiber of type

$$\mathrm{Diff}\,(M\,;\,p^{r})\,/\,\mathrm{Diff}\,(M\,;\,p^{r+1}) \quad .$$

Let me continue heuristically for the moment, to show that the left invariant forms on these $J_k M$ are indeed related to the complex $A_p(M)$ of the last section. Consider then, first of all, the finite dimensional analogue of our problem. We are given a Lie group G, a closed subgroup H and we wish to compute the left invariant forms on G/H. In analogy with our previous notation one writes

$$(5.4) \qquad \Omega(\mathfrak{g}\,;\,H) = \mathrm{Inv}_G\,\Omega(G/H)$$

for this complex, $\mathfrak{g}$ and $\mathfrak{h}$ denoting the Lie algebras of G and H respectively.

Now just as before the transitivity of the left action of G on G/H implies that a form $\omega \in \Omega(\mathfrak{g}\,;\,H)$ is determined by its value at the coset $H \in G/H$, and there

$$(5.5) \qquad T_H \cdot G/H = \mathfrak{g}/\mathfrak{h}$$

Hence we get an imbedding of

$$\Omega(\mathfrak{g}\,;\,H) \subset \Lambda(\mathfrak{g}/\mathfrak{h})^{*} \quad .$$

On the other hand, unlike before this map is not onto, because the left action of H on $T_H\,G/H$ is in general nontrivial and so one can at best expect that

$$\Omega(\mathfrak{g}\,;H) = \mathrm{Inv}_H \wedge(\mathfrak{g}/\mathfrak{h})^* \quad .$$

Finally note that the action of $h \in H$ on $T_H(G/H)$ is compatible with the action of right translation, $r(h^{-1})$, on the sequence:

(5.6)
$$0 \longrightarrow \mathfrak{h} \longrightarrow \mathfrak{g} \longrightarrow (\mathfrak{g}/\mathfrak{h}) \longrightarrow 0 \quad ,$$

because the maps

$$G \longrightarrow G$$

given by $\ell(h)$ and $\ell(h) \cdot r(h^{-1})$ have equal projection on G/H. Putting this all together yields the following recipe for the left-invariant forms on G/H:

PROPOSITION 5.1. The left-invariant forms on G/H, can be identified with the complex of H - basic forms on $\mathfrak{g}$, that is the complex;

(5.7)
$$\Omega(\mathfrak{g}\,;H) = \mathrm{Inv}_H \wedge(\mathfrak{g}/\mathfrak{h})^*$$

of invariants under the adjoint action: $h \longrightarrow \ell(h) \cdot r(h^{-1})$.

Note finally that when H is connected

(5.8)
$$\mathrm{Inv}_H \wedge(\mathfrak{g}/\mathfrak{h})^* = \mathrm{Inv}_{\mathfrak{h}} \wedge(\mathfrak{g}/\mathfrak{h})^*$$

where $\mathfrak{h}$ acts on $\wedge(\mathfrak{g}/\mathfrak{h})^*$ via $\mathcal{L}(h)$.

So much then for the finite dimensional case. We follow it now by analogy in our situation.

If the Lie algebra of Diff M is LM then the Lie algebra of Diff $(M ; p^k)$ is the subalgebra $L(M ; p^k)$ of vector fields X on M which vanish at p to the k-th order. The analogue of (5.6) is therefore the exact sequence

$$0 \longrightarrow L(M ; p^{k+1}) \longrightarrow L(M) \longrightarrow L(M)/L(M ; p^{k+1}) \longrightarrow 0 \tag{5.9}$$

Observe that in terms of our forms θ^i_α in $A_p(M)$, the algebra $L(M ; p^k)$ is precisely the subspace annihilated by the 1-forms

$$\theta^i, \theta_{\alpha_1}, \cdots, \theta_{\alpha_1} \tag{5.10}$$

Hence the finite dimensional space $A_p(M ; k)$ generated by these corresponds to $\wedge^*(\mathfrak{g}/\mathfrak{h})$ in this situation. To get at the invariant forms on $J_k M$ it remains therefore only to look at the invariant forms in $A_p(M ; k)$ under $L(M ; p^k)$ acting on $A_p(M ; k)$. Now under the Lie bracket

$$\left[L(M ; p^k) , L(M ; p^{k'})\right] \subset L(M ; p^{k+k'-1}) , \tag{5.11}$$

as is clear enough because one differentiation is involved. For example;

$$\left[x^k \frac{\partial}{\partial x} , x^\ell \frac{\partial}{\partial x}\right] = x^{k+\ell-1} \cdot (\ell - k) \frac{\partial}{\partial x} . \tag{5.12}$$

As a consequence we see that $L(M ; p^{k+2})$ acts trivially on $A_p(M ; k)$ while $L(M ; p^{k+1})$ act <u>nontrivially</u> on $A_p(M ; k)$. In any case however $L(M ; p^{k+1})$ acts trivially on $A_p(M ; k-1)$ so that according to this argument we obtain the result:

(5.13) <u>The left invariant forms on</u> $J_k M$ <u>contain the subcomplex</u> $A_p(M\,;\,k-1)$.

Even more explicitly, we expect the following situation:

$\Omega J_k M$ contains a finite dimensional subalgebra of left invariant forms under the left action of Diff (M) , which contains, as a subalgebra, the complex spanned by the 1-forms, θ^i_α , <u>with at most</u> $k-1$ <u>lower indices in</u> $A_p(M)$.

To sum up, on $J_1 M$ we expect the invariant 1-forms θ^i , on $J_2 M$ the θ^i_j appears, etc. On $J_1 M$ the $d\,\theta^i$ are of course also invariant however <u>cannot be expressed</u> in terms of invariant 1-forms - i.e., $d\,\theta^i$ is not decomposable in the invariant forms on $J_1(M)$. On the other hand in $J_2(M)$ we of course have our old structure equations:

$$d\,\theta^i + \theta^i_j \wedge \theta^j = 0 \quad ,$$

and this state of affairs continues for each k : If θ^i_α is defined on $J_k(M)$ $d\,\theta^i_\alpha$ is decomposable in the invariant forms on $J_{k+1}(M)$.

The remarks so far have been heuristic because the definition

$$J_k M = \text{Diff } M \,/\, \text{Diff }(M\,;\,p^{k+1})$$

brings us the finite dimensional object $J_k M$ as a quotient of two infinite dimensional ones. The more honest way is therefore to construct $J_k M$

explicitly as a finite dimensional manifold over M, which behaves naturally under diffeomorphisms. Precisely, we need a construction J_k which to each M associates a manifold J_kM together with a map

$$\pi : J_kM \longrightarrow M$$

such that:

(5.14) <u>Every open imbedding</u>

$$\varphi : M \longrightarrow N$$

<u>induces a map</u>

$$J_k(\varphi) : J_k(M) \longrightarrow J_k(N) ,$$

commuting with π. Further,

(5.15) <u>given maps,</u>

$$M \xrightarrow{\varphi} N \xrightarrow{\psi} P$$

<u>these induced maps satisfy the naturality condition:</u>

$$J_k(\varphi \circ \psi) = J_k(\varphi) \circ J_k(\psi) .$$

(5.16) <u>Finally, if</u> $U \supset 0$ <u>is a neighborhood of</u> 0 <u>in</u> $\mathbb{R}^n$ <u>then there is a natural map</u>

$$j_k : \mathrm{Im}(U ; M) \longrightarrow J_k(M)$$

<u>whose fibers are precisely the imbeddings having the same k - jet at</u> 0 .

Recall here that two maps $f, g : U \longrightarrow V$ have the same k - jet at $p \in U$ if and only if they and their derivatives up to order k agree at p.

Now because of the assumed functoriality of J_k and the fact that every manifold is patched together from open sets in $\mathbb{R}^n$ it is sufficient to construct J_k explicitly on $\mathbb{R}^n$ and then to extend it to any manifold M, by using the J_k's of the patching maps of M.

To construct $J_k(\mathbb{R}^n)$ one proceeds as follows. Suppose that $x^1, \cdots, x^n$, is a system of linear coordinates on $\mathbb{R}^n$. Then $J_k(\mathbb{R}^n)$ will be defined as a subset of

$$\mathbb{R}^n \times \mathbb{R}^{n^2} \times \mathbb{R}^{n\binom{n}{2}} \times \cdots \times \mathbb{R}^{n\binom{n}{k}},$$

(coordinatised by the x^i_α with "léngth" $\ell(\alpha)$ of the multiindex $\alpha \leq k$), given by the inequality:

$$\det \|x^i_j\| \neq 0 . \tag{5.17}$$

Topologically $J_k(\mathbb{R}^n)$ is then simply $GL(n)$ crossed with a large Euclidean space. On the other hand the action $J_k(\varphi)$ of $\varphi \in \mathrm{Diff}(\mathbb{R}^n)$ - and this is of course the heart of the matter - is more complicated to describe. However in principle this action is completely described by the chain rule.

Precisely, recall that this famous formula:

$$\frac{\partial u^i}{\partial t^j} = \frac{\partial u^i}{\partial x^k} \cdot \frac{\partial x^k}{\partial t^j} \tag{5.18}$$

is valid when the x depend smoothly on the t's and the u's smoothly on the x's . Iterating one obtains universal polynomials which express $\frac{\partial^{|\alpha|} u^i}{\partial t^\alpha}$ in terms of $\frac{\partial u^i}{\partial x^\beta}$ and $\frac{\partial x^i}{\partial t^\lambda}$ with $|\beta|$ and $|\lambda| \le |\alpha|$:

$$\frac{\partial^{|\alpha|} u^i}{\partial t^\alpha} = \Phi_\alpha\left(\frac{\partial^{|\beta|} u^i}{\partial x^\beta} \; ; \; \frac{\partial^{|\lambda|} x^i}{\partial t^\lambda}\right) \tag{5.19}$$

and these universal polynomials determine the action of $J_k(\varphi)$, according to the following principle.

Let $\varphi : \mathbb{R}^n \longrightarrow \mathbb{R}^n$ be a diffeomorphism, and let $\varphi^* x^i$ be denoted by u^i . With this understood the

$$\frac{\partial^{|\beta|} u^i}{\partial x^\beta}$$

become well defined functions of x , and so intrepreted one now sets

$$J_k(\varphi)^* x^i_\alpha = \Phi_\alpha\left(\frac{\partial^{|\beta|} u^i}{\partial x^\beta} \; ; \; x^i_\lambda\right) \tag{5.20}$$

Thus, for example, when $k = 1$:

$$\begin{cases} J_1(\varphi)^* x^i = u^i & ; \; u^i = \varphi^* x^2 \\ J_1(\varphi)^* x^i_j = \frac{\partial u^i}{\partial x^k} \cdot x^k_j & . \end{cases} \tag{5.21}$$

When $k = 2$, one obtains in addition to formulas identical to (5.21) the formula

$$J_2(\varphi)^* x^i_{jk} = \frac{\partial u^i}{\partial x^m} x^m_{jk} + \frac{\partial^2 u^i}{\partial x^m \partial x^n} \cdot x^m_k x^n_j \quad . \tag{5.22}$$

Concerning (5. 21) one now has the following straightforward

LEMMA 5.1 . The correspondence

$$\varphi \longrightarrow J_k(\varphi)$$

defines a homomorphism from Diff $\mathbb{R}^n$ to Diff $\{J_k(\mathbb{R}^n)\}$. Furthermore these representations of Diff $(\mathbb{R}^n)$ are compatible with natural projections

$$\longleftarrow J_k(\mathbb{R}^n) \xleftarrow{\pi} J_{k+1}(\mathbb{R}^n) \longleftarrow \cdots$$

under which x^i_α in $J_k(\mathbb{R}^n)$ maps to x^i_α in $J_{k+1}(\mathbb{R}^n)$.

To prove that the $J_k(\varphi)$'s are diffeomorphism one simply computes its Jacobian, which in view of the triangular nature of $J_k(\varphi)$ is simply a high power of the Jacobian of φ .

The representation, or naturality, property of φ then follows from the chain rule. And, of course, with this lemma one now extends J_k functorially to manifolds without trouble.

Let me close this section by writing down explicit formulas for our invariant forms θ^i and θ^i_j on $J_2(\mathbb{R}^n)$ in terms of the x^i_α . The procedure is quite analogous to our construction of the θ^i_j on $GL(n, \mathbb{R})$ in terms of the x^i_j 's . Of course elimination is more complicated if one proceeds quite straightforwardly as I plan to do.

PROPOSITION 5.1. Let x^i_α $\ell(\alpha) \leq 2$, denote the coordinates of

$J_2(\mathbb{R}^n)$ and let y^i_j be functions defined on $J_2(\mathbb{R}^n)$ by the formula

$$y^i_j x^j_k = \delta^i_k .$$

Then in terms of these the forms

$$\theta^i = y^i_k dx^k$$

(5.23)

$$\theta^i_j = y^i_k \left(dx^k_j - x^k_{jm} \theta^m\right)$$

are invariant under $J_2(\varphi)^*$, $\varphi \in \mathrm{Diff}(\mathbb{R}^n)$ and correspond to the analogously named forms in A_p. In particular they satisfy the equation

(5.24) $$d\theta^i + \theta^i_j \wedge \theta^j = 0 .$$

The formula (5.24) of course follows directly from (5.23) and so establishes (5.24) directly within our "honest", constrictive approach.

Proof: If we write u^i_k for $\frac{\partial u^i}{\partial x^k}$ etc., and abbreviate $J_2(\varphi^*)$ to φ^*, the transformation laws of (5.21) and (5.22) are:

$$\varphi^* x^i \equiv u^i \; ; \; \varphi^* x^i_j = u^i_k x^k_j , \quad \text{and}$$

$$\varphi^* x^i_{jk} = u^i_m x^m_{jk} + u^i_{mn} x^m_k x^n_j .$$

Now from $y^i_j x^j_k = \delta^i_k$ we conclude that

(5.25) $$(\varphi^* y^i_j)\, u^j_\alpha \cdot x^\alpha_k = \delta^i_k$$

Hence

$$\begin{aligned}\varphi^*(y^i_k \cdot dx^k) &= (\varphi^* y^i_k) \cdot d\{\varphi^* x^k\}\\ &= (\varphi^* y^i_k)\, u^k_\alpha \cdot dx^\alpha\\ &= y^i_\alpha\, dx^\alpha \qquad \text{by (5.25)}\end{aligned}$$

Next consider $y^i_k\, dx^k_j$. We have

$$\begin{aligned}\varphi^*(y^i_k\, dx^k_j) &= (\varphi^* y^i_k) \cdot d(\varphi^* x^k_j)\\ &= \varphi^*(y^i_k)\, d(u^k_\alpha x^\alpha_j)\\ &= \varphi^*(y^i_k) \left\{u^k_\alpha \cdot dx^\alpha_j + u^k_{\alpha\beta} x^\alpha_j\, dx^\beta\right\}\\ &= y^i_k \cdot dx^k_j + \varphi^*(y^i_k) \cdot u^k_{\alpha\beta} x^\alpha_j\, dx^\beta .\end{aligned}$$

On the other hand

$$\begin{aligned}&\varphi^*(y^i_k\, x^k_{jm}\, \theta^m)\\ &= \varphi^*(y^i_k) \left\{u^k_\alpha x^\alpha_{jm} + u^k_{\alpha\beta} x^\alpha_j\, x^\beta_m \quad \theta^m\right\}\\ &= y^i_k\, x^k_{jm}\, \theta^m + \varphi^*(y^i_k)\, u^k_{\alpha\beta} x^\alpha_j\, x^\beta_m\, y^m_\delta\, dx^\delta\\ &= y^i_k\, x^k_{jm}\, \theta^m + \varphi^*(y^i_k)\, u^k_{\alpha\beta} x^\alpha_j\, dx^\beta .\end{aligned}$$

Hence both the expressions considered failed to be invariant by an equal error and so their difference is invariant. Q. E. D.

I will leave it as an instructive exercise to the reader to identify the θ^i θ^i_j of this section with our earlier ones θ in A_p .

Let me finally discuss the relation of these formulas to those one finds in the text on connections. We have already remarked that J_2M is naturally "fibered" over J_1M and J_1M over M

$$M \xleftarrow{\pi_1} J_1M \xleftarrow{\pi_2} J_2M \xleftarrow{\pi_3} J_3M \longleftarrow \quad .$$

the fibers being diffeomorphic to $GL(n,\mathbb{R})$ for π_1 and to a Euclidean space of dim $n\binom{n}{2}$ for π_2 .

Actually each of the spaces $J_k(M)$ should be thought of as a "Lie group over M", in the sense that there is a Lie group $GL(n,\mathbb{R})^{(k)}$ which naturally acts on the right in $J_k(M)$; Furthermore locally the map

$$M \xleftarrow{\pi^k} J_k(M)$$

always has a section, and if

$$s : U \longrightarrow J_k(M)$$

is such a section over u then the map

$$\psi_s : U \times GL(n,\mathbb{R})^{(k)} \longrightarrow \pi^{-1}(u)$$

given by

$$\psi_s(u, g) = s(u) \cdot g$$

in a diffeomorphism. Such "Lie groups over M" are technically called principal bundles over M and occur in many places in geometry. Here let me just observe that the group $GL(n, \mathbb{R})^{(1)} = GL(n, \mathbb{R})$ and that the right action of $\|a^i_j\|$ on $J_1(\mathbb{R}^n)$ is simply given by:

(5.25) $$a^* \cdot x^i = x^i \ ; \ a^* x^i_j = x^i_k a^k_j$$

The higher groups $GL(n, \mathbb{R})^{(0)}$ $k > 2$ are the groups formed by the k - jets at 0 of invertible diffeomorphisms of $\mathbb{R}^n$.

Now with this in mind, consider an arbitrary smooth section

$$s : J_1M \longrightarrow J_2M \ .$$

Topologically there is no obstruction to such a section because these fibers are contractible. In any case we have the following:

PROPOSITION 4.2. Let $s : J_1M \longrightarrow J_2M$ be any smooth section. Then the forms $s^* \theta^i_j$ restrict to the left invariant forms denoted by θ^i_j in section 2, on any fiber $GL(n, \mathbb{R})$ of J_1M.

It follows that the θ^i together with the θ^i_j determine a framing of TJ_1M, and that the field of planes $\mathcal{H}$ determined by the kernel of the

$s^* \theta^i_j$:

$$\mathcal{H}(s) = \text{Ker}\, s^* \theta^i_j$$

is transversal to the fibers of J_1M .

Proof : If one applies s^* to the local expression (5. 23) for θ^i_j one obtains

$$\breve{s}^* \theta^i_j = y^i_k \left\{ dx^k_j - s^* (x^k_{jm}) \theta^m \right\}$$

because $s^* x^i = x^i$ and $s^* x^i_j = x^i_j$ in our notation convention and so s^* pulls the θ^m on J_2M to the θ^m on J_1M . Now restriction to a fiber annihilates the θ^m and the first term clearly restricts to our left invariant forms of section 2.

At this stage we are very close to the theory of connections as it is treated in most contemporary texts. Indeed one definition of "connection" for a manifold M is the following.

Definition. A connection on the manifold M is a smooth plane field $\mathcal{H}$ of dimension n^2 on J_1M , which is:

(5. 25) transversal to the fibers ,

and

(5. 26) invariant under the action of $GL(n, \mathbb{R})$ on $J_1(M)$.

Hence the horizontal space $\mathcal{H}(s)$ of a section $s : J_1(M) \longrightarrow J_2(M)$

nearly constitutes a connection on M, in that it satisfies (5.25), but need not satisfy (5.26). If $\mathcal{H}(s)$ does satisfy (5.26) as well, then we call s equivariant, and these now yield all "torsion free" connections on M as we will see in a moment. In the present context the torsion of a connection $\mathcal{H}$ is defined as follows. The horizontal space $\mathcal{H}$ enables us to extend the left invariant forms on $\theta^i_j \in \Omega\, GL(n, \mathbb{R})$ to forms $\tilde{\theta}^i_j$ on all of J_1M by the requirement that

$\tilde{\theta}^i_j$ restrict to θ^i_j on each fiber and $\tilde{\theta}^i_j$ annihilates $\mathcal{H}$.

It follows that the expressions

$$T^i = d\,\theta^i + \tilde{\theta}^i_j \wedge \theta^j$$

are "horizontal" forms in the sense that they are in the span of the θ^i, and these are called the torsion forms of the connection M. Now in view of (5.24) we have

$$d\,\theta^i + s^*\,\theta^i_j \wedge \theta^j \equiv 0 ,$$

so that $\mathcal{H}(s)$, if it is invariant, is automatically torsion free.

Let us now look in detail at what the equivariant sections of J_2M look like. Under the transformation $a^* x^i_j = x^i_k a^k_j$, we have

$$a^*\,\theta^i = A^i_k\,\theta^k$$

where

$$A^i_k a^k_j = \delta^i_j$$

is the inverse of a . Similarly

$$a^* (y^i_k \, dx^k_j) = A^i_\alpha (y^\alpha_k \, dx^k_\beta) \, a^\beta_j \quad .$$

Hence $\mathcal{K}(s)$ will be equivariant if and only if

$$a^* (s^* \theta^i_j) = A^i_\alpha (s^* \theta^\alpha_\beta) \, a^\beta_j \tag{5.27}$$

and this condition implies a restriction on the functions $s^* x^i_{jk}$, which so far were arbitrary smooth functions of the x^i and x^i_j .

In fact substitution (5.23) and (5.27) yields the relation :

$$a^* \cdot s^* (x^i_{jk}) = s^* (x^i_{\beta\lambda}) \, a^\beta_j \, a^\lambda_k \quad , \tag{5.28}$$

whence $s^* (x^i_{jk})$ must have the form

$$s^* (x^i_{jk}) = f^i_{\alpha\beta} (x) \cdot x^\alpha_j \cdot x^\beta_k \tag{5.29}$$

where now the $f^i_{\alpha\beta}$ are arbitrary smooth functions of the x's symmetric in the lower indices.

In any case it is clear that locally such sections can be found and they can then be patched together over M by a partition of unity without trouble. This development is now completed by the following

PROPOSITION 5.3. <u>Every torsion free connection</u> $\mathcal{K}$ <u>on</u> J_1M <u>is of the form</u> $\mathcal{K}(s)$ <u>where</u> s <u>is an equivariant section of</u> J_2M <u>over</u> J_1M .

<u>Proof</u>: This follows directly from the following

LEMMA. <u>Let</u> s_1 <u>and</u> s_2 <u>be two sections of</u> $J_2(M)$. <u>Then</u>

$$s_1^* \theta^i_j - s_2^* \theta^i_j = \Gamma^i_{jk}\, \theta^k \tag{5.30}$$

<u>where the</u> Γ^i_{jk} <u>are smooth function on</u> J_1M , <u>symmetric in</u> jk . <u>Conversely given, a section</u> s_1 <u>and a system of functions</u> Γ^i_{jk} <u>with</u> $\Gamma^i_{jk} = \Gamma^i_{kj}$, <u>then there is a unique section</u> s_2 <u>with</u>

$$s_1^* \theta^i_j - s_2^* \theta^i_j = \Gamma^i_{jk}\, \theta^k \ . \tag{5.31}$$

This in turn follows directly from the local representation (5.23), and the fact that the $s^*(x^k_{jm})$ are quite arbitrary except for the symmetry in j and m .

To prove the proposition, consider a torsion free connection $\mathcal{K}$ on J_1M , and let $\tilde{\theta}^i_j$ be the corresponding extension of the invariant forms on the fiber. Also let s be a section of $J_2(M)$. Then

$$\tilde{\theta}^i_j - s^* \theta^i_j$$

is horizontal and hence there exist C^∞ functions Γ^i_{jk} such that

$$\tilde{\theta}^i_j - s\, \theta^i_j = \Gamma^i_{jk}\, \theta^k$$

Now because $\mathcal{H}$ is torsion free

$$d\,\theta^i + \tilde{\theta}^i_j \wedge \theta^j = 0 \quad ,$$

and in any case

$$d\,\theta^i + s^*\,\theta^i_j \wedge \theta^j = 0 \quad .$$

Hence subtracting these two we obtain the condition

$$\Gamma^i_{jk}\,\theta^k \wedge \theta^j = 0 \quad .$$

Therefore Γ^i_{jk} is symmetric in j and k, and so by the lemma, there is and s_1 with

$$s_1{}^*\,\theta^i_j = \theta^i_j + \Gamma^i_{jk}\,\theta^k = \tilde{\theta}^i_j \quad .$$

<u>Remarks.</u> Connections are often introduced as differential operators on tensor fields over M and a word concerning the relationship of that definition and our present one is appropriate. The essential point linking them is that the J_1 construction enables one to <u>define all tensor-fields on</u> M <u>as functions on</u> J_1M <u>subject to an equivariance condition which is characteristic of the tensor field.</u> Precisely, given a representation

$$\rho : GL(n, \mathbb{R}) \longrightarrow \text{End } V$$

the tensors of type ρ on M, can be defined to be the function

$$f : J_1(M) \longrightarrow V$$

subject to

$$g^* f = \rho^{-1}(g) \cdot f \qquad g \in GL(n, \mathbb{R}) \tag{5.33}$$

With this understood, let $\mathcal{H}$ be a connection on J_1M , X a vector field on M . Clearly X lifts to a univque horizontal vector field $X_{\mathcal{H}}$ on J_1M which projects to X . Hence if f is any tensor field on M ,

$$X_{\mathcal{H}} \cdot f : J_1 M \longrightarrow V$$

is a new function which because of the equivariance of $\mathcal{H}$ again satisfies 5.33. This is the <u>covariant derivative</u> $\nabla_X f$, of f in the direction X vis-à-vis the connection $\mathcal{H}$.

In conclusion let me now interpret the curvature of connection $\mathcal{H}$ as the obstruction to the integrability of $\mathcal{H}$.

PROPOSITION 5.4 . *Let $\mathcal{H}$ be a connection on J_1M and let $\tilde{\theta}^i_j$ be the corresponding extension of the invariant form on the fiber. Then $\mathcal{H}$ is integrable if and only if the forms*

$$R^i_j(\mathcal{H}) = d\,\tilde{\theta}^i_j + \tilde{\theta}^i_k \wedge \tilde{\theta}^k_j \tag{5.34}$$

vanish identically.

Proof : Integrability of $\mathcal{H}$ is equivalent to the requirement that $d\,\tilde{\theta}^i_j$ be in the ideal $\mathcal{J}$ generated by $\tilde{\theta}^i_j$ as we learned in chapter 2. Now

$d\tilde{\theta}^i_j$ and $R^i_j(\mathcal{K})$ differ by an element in $\mathcal{J}$ so the integrability criterion is equally expressed by the requirement that $R^i_j(\mathcal{K})$ be in $\mathcal{J}$. However, $R^i_j(\mathcal{K})$ has the virtue of being a horizontal form: That is, there exist functions R^i_{jke} skew in k and e, such that

$$R^i_j(\mathcal{K}) = R^i_{jke}\, \theta^k \wedge \theta^e \tag{5.35}$$

granting (5.35) the proposition is immediate, because the θ's are complementary to the $\tilde{\theta}^i_j$'s : so that

$$R^i_j(\mathcal{K}) \subset \mathcal{J} \Leftrightarrow R^i_{jke} \equiv 0 \ . \tag{5.36}$$

To prove (5.35) it is sufficient to show that

$$\iota(V)\, R^i_j \equiv 0$$

for any vertical vector-field on $J_1(M)$, and as these are spanned by the infinitesimal right translations, of $GL(n, \mathbb{R})$ it is sufficient to prove

$$\iota(X)\, R^i_j = 0 \ . \tag{5.37}$$

Now the equivariance of $\mathcal{K}$ under $GL(n, \mathbb{R})$ implies that $\mathcal{L}(X)\, \tilde{\theta}^i_j \subset \mathcal{J}$. But then by the structure equations on each fiber $\mathcal{L}(X)\, \tilde{\theta}^i_j = \tilde{\iota}(X)\, d\,\tilde{\theta}^i_j =$ $= -\iota(X)\, \tilde{\theta}^i_k \wedge \tilde{\theta}^k_j$, yielding (5.37) . Q. E. D.

As an immediate corollary we now have:

COROLLARY. The invariant forms

$$R^i_j = d\,\theta^i_j + \theta^i_k \wedge \theta^k_j$$

in $J_2(M)$ play the role of a universal obstruction to the integrability of a torsion free connection on J_1M in the sense that if

$$s : J_1M \longrightarrow J_2M$$

is the section defining the torsion free connection $\mathcal{H}$ then $\mathcal{H}$ is integrable if and only if

$$s^* R^i_j \equiv 0.$$

6. The natural thickenings of M .

Let me from now on take the position that $J_k M$ is defined as the space of k - jets at 0 of imbeddings

$$f : U \longrightarrow \mathbb{R}^n \quad U \subset \mathbb{R}^n \tag{6.1}$$

where U is some neighborhood of 0 in a fixed $\mathbb{R}^n$.

This corresponds to chosing fixed coordinate p in our model Diff M / Diff $(M; p^{k+1})$, and in the "honest" approach this means that we associate to f the point of $J_k(M)$ with coordinates

$$\frac{\partial^{|\alpha|} f^i}{\partial x^\alpha}(0)$$

relative to local coordinates at f(o) .

The present version has the advantage of clearly indicating both the left action of Diff M and the right action of $GL(n, \mathbb{R})$ on the $J_k M$'s . Indeed they are both given by composition:

$$f \longrightarrow \varphi \circ f \circ g \qquad \varphi \in \text{Diff}(M), \quad g \in GL(n, \mathbb{R}) \tag{6.2}$$

It follows that we can naturally form the spaces

$$J_k M / GL(n, \mathbb{R})$$

where we identify points under the right action of $GL(n, \mathbb{R})$, and more generally the spaces

$$J_k M / G$$

for G a subgroup of $GL(n, \mathbb{R})$. In particular if $O(n) \subset GL(n, \mathbb{R})$ denotes the orthogonal subgroup of $\mathbb{R}^n$, then the space

$$J_k M / O(n)$$

is of especial interest, because $O(n)$ is a maximal compact subgroup of $GL(n, \mathbb{R})$. For $k = 1$ the space

$$J_1 M / GL(n, \mathbb{R})$$

reduces to M as $GL(n, \mathbb{R})$ is the fiber of the principal bundle $J_1 M$ over M. On the other hand the space

$$J_1 M / O(n)$$

is already interesting. First of all it fits into the sequence of maps

$$M \xleftarrow{\pi} J_1 M / O(n) \xleftarrow{\pi'} J_1(M)$$

where π' is a principal fibering with group $O(n)$, and π is a "fibering over M with fiber $GL(n, \mathbb{R}) / O(n)$ " .

Recall here that a map $E \xrightarrow{\pi} M$ is called a fibering with fiber F

if M admits a covering by open sets $\mathcal{U}_\alpha$ such that over $\mathcal{U}_\alpha$ E admits a product decomposition into $\mathcal{U}_\alpha \times F$ compatible with π : That is, there exist diffeomorphism

$$\varphi_\alpha : \mathcal{U}_\alpha \times F \longrightarrow \pi^{-1}(\mathcal{U}_\alpha)$$

which map the slices $p \times F$ to $\pi^{-1}(p)$; $p \in \mathcal{U}_\alpha$. In the present instance the fiber of π is the homogeneous space $GL(n, \mathbb{R}) / O(n)$ which is contractible. Indeed the well known Gramm Schmidt construction of an orthogonal base from an arbitrary one can be interpreted as a product decomposition

$$GL(n, \mathbb{R}) = T(n) \times O(n) \tag{6.3}$$

where $T(n)$ denotes the triangular matrices with positive diagonal entries. Hence

$$GL(n, \mathbb{R}) / O(n) = T(n)$$

which is clearly contractible - one simply lets the off-diagonal entries go to 0, and the diagonal ones to 1 .

Another proof of this fact can be based on the spectral theorem from which one deduces that

$$GL(n, \mathbb{R}) = S^{+}(n) \times O(n) \tag{6.4}$$

where $S^{+}(n)$ denotes the convex set of positive definite symmetric matrices, and using this point of view it is easy to see that

$$\text{a section} \quad s : M \longrightarrow J_1(M) / O(n) \quad ,$$

corresponds to describing a Riemannian structure on M, with $(\pi')^{-1} s \subset J_1(M)$ playing the role of the corresponding sub-bundle

$$O(M) \subset J_1(M)$$

furnished by the orthogonal frames of the structure.

Note here that from this point of view the existence of a Riemannian structure on M follows from the quite general proposition that a fibering with contractible fibers admit sections.

Note also that such a section can be used to construct a homotopy equivalence of $J_1 M / O(n)$ with M, and using the fact that the fibers of J_k over J_{k-1} $k \geq 2$, are in any case contractible we see that for all k,

PROPOSITION 6.1. The manifolds $J_k M / O(n)$ are naturally acted on by Diff (M) and their projections:

$$M \longleftarrow J_k M / O(n)$$

are homotopy equivalences.

In this sense then we should think of the $J_k M / O(n)$ as k-th order thickening of M.

In particular the cohomology groups of all these space are canonically isomorphic to the cohomology of M, and it makes good sense to look for the characteristic classes of M in the complex of $\mathrm{Diff}(M)$ - invariant forms on $J_k(M) / O(n)$, and in particular one would expect the most complete answer in the complex

$$(6.5) \qquad \Omega(J_\infty M / O(n) = \lim_{k \to \infty} \Omega(J_k M / O(n) \quad .$$

If we now compare this with our discussion of the Gelfand-Fuks complex, we see, with little difficulty, that the complex of invariant forms in (6.5) is furnished by the complex

$$(6.6) \qquad \Omega_c(a_n ; O(n))$$

of the continuous forms on the Lie algebra of formal vector-fields modulo the orthogonal group. Thus we have a natural map

$$(6.7) \qquad H^* \Omega_c(a_n , O(n)) \longrightarrow \lim_k H^* (J_k M / O(n)) \cong H^*(M) \quad .$$

To sum up, the Gelfand-Fuks point of view, leads us to the

THEOREM : <u>The inclusion of the</u> Diff M - <u>invariant forms the DeRham theory of the k-th order thickenings of</u> M , <u>leads to a natural map</u>

$$\chi : H^* \Omega_c(a_n , On) \longrightarrow H^* (M)$$

<u>and therefore to natural cohomology invariants on M</u> .

The next question is, of course, to compute the universal object

$$H^* \Omega_c(\alpha_n, On)$$

and in the next section I will outline how such computations proceed with the aid of some classical theorems of Invariant theory.

7. The "first main theorem" of invariant theory and some of its applications.

In his remarkable book on the classical groups, Herman Weyl explicitly constructs a basis for the invariants of certain representation of the classical groups.

For $GL(n, \mathbb{R}) = \mathrm{Aut}(V)$ his "First Main theorem" then asserts that all the invariants of $\mathrm{Aut}(V)$ acting on the tensor spaces

$$\overset{k}{\otimes} V \otimes \overset{\ell}{\otimes} V^* \tag{7.1}$$

are generated by contractions -- that is -- by the natural pairing of $V \otimes V^* \to \mathbb{R}$.

Explicitly one can state this result in the following form:

THEOREM 7.1 : <u>There are invariant maps</u>

$$\varphi : \overset{k}{\otimes} V \otimes \overset{\ell}{\otimes} V^* \longrightarrow \mathbb{R}$$

<u>only if</u> $k = \ell$, <u>and then these are generated by the map</u>

$$\varphi(v_1 \otimes \cdots v_k \otimes \theta^1 \otimes \cdots \otimes \theta^k) = \Pi\, \theta^i(v_i)$$

<u>or</u> φ <u>composed with an arbitrary permutation</u> σ, <u>of the factors in</u> $\overset{k}{\otimes} V$; <u>that is, by maps</u> $\varphi \circ \sigma$ <u>given by</u>

$$\varphi \circ \sigma(v_1 \otimes \cdots \otimes v_k \otimes \theta^1 \otimes \cdots \otimes \theta^k) = \prod_{i=1}^{k} \theta^i(v_{\sigma(i)}) \tag{7.2}$$

A dual and quite equivalent formulation reads:

THEOREM 7.2 : The Aut(V) module

$$\overset{k}{\otimes} V \otimes \overset{\ell}{\otimes} V^*$$

has no invariant subspaces when $k \neq \ell$.

If $k = \ell$, and $\{v_i , \theta^j\}$ is a set of dual bases for V and V^* , then the subspace of invariants in

$$\overset{k}{\otimes} V \otimes \overset{k}{\otimes} V^*$$

is generated by elementary invariants of the form

$$\sigma \left\{ v_{i_1} \otimes v_{i_2} \cdots \otimes v_{i_k} \right\} \otimes \theta^{i_1} \otimes \theta^{i_2} \otimes \cdots \theta^{i_k} \tag{7.3}$$

where σ is some permutation of the factors in $\overset{k}{\otimes} V$.

These theorems are also valid if Aut(V) is replaced by $\mathrm{Aut}(V)^+$, the automorphisms of $\det > 0$ of V but are not correct for the Special linear orthongonal groups, where additional determinental factors appear. On the other hand if $G \subset \mathrm{Aut}(V)$ is the subgroup preserving a symmetric or skew symmetric form, then Herman Weyl also shows that

THEOREM 7.3 : If G is the subgroup preserving the form $\eta : V \times V \to \mathbb{R}$, then G - invariants linear forms

$$\varphi : \overset{k}{\otimes} V \longrightarrow \mathbb{R}$$

exist only if k is even, and they are then generated by the elementary forms:

(7.4) $$\varphi(v_1 \otimes \cdots \otimes v_{2n}) = \Pi\eta(v_{2i}, v_{2i+1}) \quad ,$$

and forms obtained from φ by permutating the factors arbitrarily.

Note that in this case V and V^* are isomorphic G - modules , so that (7.4) may be thought of as an elementary invariant coming from $GL(n; \mathbb{R})$ via an identification of half the V's with V^*'s .

Essentially both these theorems are easy to comprehend in that they assert that the obvious invariants generate all invariants. However, the proofs are nontrivial and I will not have time to discuss them here. By and large, Herman Weyl's book still seems to me the best reference. See also [2].

Armed with these results one can already make a large number of interesting computations. Let me illustrate by computing

$$H^* \{\Omega(g\ell_n)\}$$

where $g\ell_n$ the Lie algebra of $GL(n; \mathbb{R})$. From our present point of view it is best to identify $GL(n; \mathbb{R})$ with $Aut(V)$, $\dim V = n$, and then correspondingly

$$g\ell_n \simeq End(V) \simeq V \otimes V^* \quad .$$

Hence - from our discussions in section 3 -

$$H^q(gl_n) \approx \mathrm{Inv}_{\mathrm{Aut}(V)} \wedge^q (V \otimes V) \quad .$$

Furthermore our forms $\theta^i_j \in \Omega^1(gl_n)$ are a basis for $V^* \otimes V$ of the form $\theta^i \otimes v_j$, so that the invariants of Aut(V) on $\overset{q}{\otimes}(V^* \otimes V)$ are generated by expressions

$$\theta^{i_1}_{j_1} \otimes \theta^{i_2}_{j_2} \dots \otimes \theta^{i_n}_{j_n}$$

where the j'th are a permutation of the i's . Now the natural projection

$$\begin{array}{c} \overset{q}{\otimes} (V^* \otimes V) \\ \big\downarrow \\ \wedge^q (V^* \otimes V) \end{array}$$

admits an Aut(V) - invariant section given by the skew symmetric tensors. Hence the invariants below are all images of invariants above.

The skew symmetry of the wedge now makes it clear that the basic "potential" invariants are given by the cyclic contractions:

$$\theta^i_i \; ; \sim \theta^i_j \wedge \theta^j_i \; ; \sim \theta^i_j \wedge \theta^j_k \wedge \theta^k_i \; ; \quad \text{etc} .$$

Further observe that the ones with an even number of constituents vanish, again because $\wedge$ is skew: For example

$$\theta^i_j \wedge \theta^j_i = - \theta^j_i \wedge \theta^i_j \equiv - \theta^i_j \wedge \theta^j_i$$

It follows then that <u>our invariants are generated by the classes</u>

$$(7.6)\qquad h_i = \operatorname{trace}\{\theta \wedge \cdots \wedge \theta\} \quad (2i - 1 \text{ factors})$$

<u>with</u> $i = 1, 2, \cdots$.

This is about as far as the first main theorem takes one. What is needed to complete the computation of $H^*(g\ell_n)$ is the following.

LEMMA. <u>For</u> $i > n$ <u>the forms</u> h_i <u>can be expressed as in terms of the</u> h_i, $i \le n$. <u>Further</u>

$$h_1 \wedge h_2 \wedge \cdots \wedge h_n \neq 0$$

<u>and so represent the volume form on</u> $GL(n ; \mathbb{R})$.

Unfortunately I know of no simple proof of this fact purely in the framework of representation theory. Clearly this is a result in the framework of the "second fundamental theorem" of Herman Weyl dealing with relations between the basic invariants, and I refer you to his book for details. Actually a detailed proof - going back to the original computations of R. Brauer are found on page 140 of Herman Weyl's I. A. S. Lectures of 1934-35, purely in the context of invariant theory.

As another illustration let me now turn to the cohomology of $(a_n, g\ell_n)$, where the "First main theorem" enables one to eliminate all the jets of order greater than two.

Here of course $gl_n \subset a_n$ denotes the linear vector-fields. This subalgebra acts reductively on $\Omega(a_n)$ and considered as the corresponding $\mathrm{Aut}(V)^*$ module we clearly have

(7.7) $$\Omega^1(a_n) \simeq V^* \oplus (V^* \otimes V) \oplus (V^* \otimes S_2 V) \oplus \cdots \simeq V^* \otimes SV$$

where S_k denotes the kth Symmetric product of V, and $S(V)$ the symmetric algebra. Further our old classes $\{\theta^i_\alpha\}$ are a basis for the space.

Hence the elementary invariants under the action of $\mathrm{Aut}(V)^+$ are again given by <u>total contractions</u> of our θ^i_α. Thus for instance

$$\theta^i_i \quad \text{and} \quad \theta^i \wedge \theta^j_{ij}$$

are invariants in $\Omega(a_n)$. Now recall that <u>by definition</u> $\Omega(a_n, gl_n)$ is generated by <u>only those invariants which do not involve any</u> θ^i_j's. Hence our first example is not in this complex while the second one is.

This restriction on the θ^i_α makes it impossible to create invariants in $\Omega(a_n; gl_n)$ involving a θ^i_α with $l(\alpha) > 2$. Indeed if we try, say

$$\theta^i \wedge \theta^j \wedge \theta^k_{ijk}$$

the answer is zero, because the ij in θ^k_{ijk} is symmetric, while $\theta^i \wedge \theta^j$ is of course skew, and this phenomenon is easily seen to persist in any elementary contraction of θ^i_α's in which some α has length >2 .

Thus our complex consists entirely of forms created by contraction out of the θ^i and the θ^i_{jk} . Further the symmetry of the lower indices in θ^i_{jk} seriously restricts the non-zero invariants. For example

$$\theta^i_{jk} \wedge \theta^\ell_{mi} \wedge \theta^j \wedge \theta^k \wedge \theta^m$$

is out. In fact using this symmetric it is now easy to come to the following conclusion:

THEOREM 7.6 . The complex $\Omega(a_n, g\ell_n)$ vanishes in $\dim > 2n$, has only even dimensional elements, and is generated by the $2k$ - forms

$$\Sigma_k = \text{Trace}(R^k) \qquad R = \|R^i_j\| \quad k \le n$$

where R^i_j denotes the universal curvature - form

$$R^i_j = -\theta^i_{jk} \wedge \theta^k \quad .$$

of section 4.

COROLLARY .

(7.8) $$H^*\{\Omega(a_n, g\ell_n)\} = \Omega(a_n, g\ell_n) = \mathbb{R}[c_1, \cdots, c_n] \quad .$$

where $\mathbb{R}[c_1, \cdots, c_n]$ denote the truncated polynomial ring of proposition (4.4).

The first equality is immediate because $\Omega(\mathfrak{a}_n ; \mathfrak{gl}_n)$ has only even forms - whence $d \equiv 0$. To see that the c_1's generate as well as the Σ_i recall that

$$\det(1 + R) = \sum_{i=0}^{n} c_i(R)$$

defined our Chern-forms while

$$\Sigma_k = \text{Trace } R^k .$$

Hence these are related by the Newton formulae:

$$(7.9) \quad \begin{cases} \Sigma_1 - c_1 = 0 \\ \Sigma_2 - c_1 \Sigma_1 + 2c_2 = 0 \\ \Sigma_k - \Sigma_{k-1} c_1 + \cdots \pm kc_k = 0 \end{cases} .$$

which show that over $\mathbb{Q}$ these forms span the same space. Over the integers the c's ,are preferable because they generate the Σ's over $\mathbb{Z}$, but not vice versa.

Geometrically $H^*\{\Omega(\mathfrak{a}_n, \mathfrak{gl}_n)\}$ corresponds to the cohomology of the Diff M - invariant forms on

$$J_\infty M / GL(n, \mathbb{R})$$

so that (7.8) evaluates this ring and also implies that the map

$$J_\infty M \longrightarrow J_2 M$$

induces are isomorphism in the cohomology of the $GL(n, \mathbb{R})$ - basic Diff M - invariant forms.

At this stage then, one could expect M to have "Chern-classes" $c_i(M) \in H^{2i}(J_2 M / GL(n, \mathbb{R}) \approx H^{2i}(M)$. However, half of these - the odd ones - disappear under the natural map:

$$J_\infty M / O(n) \longrightarrow J_\infty M / GL(n, \mathbb{R}) \quad ,$$

whose algebraic counterpart is the homomorphism

(7.10) $$H^* \Omega(a_n ; gl_n) \xrightarrow{\pi^*} H^*(a_n ; O(n))$$

To study this arrow one uses a standard spectral sequence argument, which then easily leads to the formula (4.32) of section 4 .

8. Characteristic classes of Foliations.

At this stage we have seen that every M naturally gives rise to a map

(8.1) $$H^*(\mathfrak{a}_n, O(n)) \xrightarrow{\varphi(M)} H^*\{J_2M/O(n)\} \approx H^*(M)$$

and that in the dimensions $\le \dim n$, this map takes the elements c_{2i} of the L.H.S. to the Pontrajgin classes of M on the right hand side. Thus our construction yields all the classical characteristic classes. On the other hand $H^*(\mathfrak{a}_n, O(n))$ also has classes in much higher dimensions and they seem to be completely extraneous at this stage. For instance for $n = 1$, how is the class

$$h_1 c_1 \in H^3\{\mathfrak{a}_n, O(n)\}$$

to be interpreted?

The correct answer here seems to be that one should consider an n - manifold M to be a special case of a foliation $\mathfrak{F}$ of codimension n.

Indeed for such a foliation all our constructions go through and one ends up with a natural map

$$H^*(\mathfrak{a}_n, O(n)) \xrightarrow{\varphi(\mathfrak{F})} H^*\{M(\mathfrak{F})\}$$

of our universal L.H.S. into the cohomology of the manifold $M(\mathfrak{F})$ on which $\mathfrak{F}$ is defined. As the dim of $M(\mathfrak{F})$ can be any integer $\ge n$ all the class of $H^*(\mathfrak{a}_n; O(n))$ become potential characteristic classes, and for instance $h_1 c_1$

becomes the Godbillon-Vey class for foliation of codim 1 . This is, of course, the class that sparked much of the recent work in this subject.

To obtain this development recall that a foliation $\mathfrak{F}$ of codim q on a manifold $M(\mathfrak{F})$ can be defined as an integrable sub-bundle $E_{\mathfrak{F}}$ of the tangent bundle TM of M , which has codim q .

Thus taking $E = 0$, defines the (trivial) foliation of M by points and that is why every n - manifold trivially defines a foliation of codimension n .

On the other hand, by the Frobenius theorem, any such an E , defines a whole class of submersions

$$U \xrightarrow{\varphi} \mathbb{R}^n$$

of open sets which $U \subset M$ to $\mathbb{R}^n$, with the property that

$$\text{Ker } df = E|U \quad .$$

Hence $\mathfrak{F}$ can also be described by a collection $\{\mathfrak{U}_\alpha, \varphi_\alpha\}$ consisting of a cover $\mathfrak{U}_\alpha$ of M , and submersions

$$\mathfrak{U}_\alpha \xrightarrow{\varphi_\alpha} \mathbb{R}^n \tag{8.3}$$

such that on $\mathfrak{U}_\alpha \cap \mathfrak{U}_\beta$

$$\text{Ker } d\varphi_\alpha = \text{Ker } d\varphi_\beta \quad . \tag{8.4}$$

From this point of view a codimension - n foliation $\mathfrak{F}$, looks very much like an n - manifold with the φ_α playing the role of the coordinate charts.

Following this line of reasoning one defines $J_k(\mathfrak{F})$ to be the space of k - jets of maps φ with source in M and target 0 in $\mathbb{R}^n$ which are in the submersions of $\mathfrak{F}$ in the sense that near the source of φ Ker $d\varphi$ defines the foliation $\mathfrak{F}$.

Note that if $\mathfrak{F}$ is the trivial foliation of M, then

$$J_k(\mathfrak{F}) \simeq J_k(M)$$

and one merely assigns to each φ its inverse to pass from one side to the other.

In short then, replacing $J_k(M)$ by $J_k(\mathfrak{F})$ one obtains a "thickening"

$$J_k(\mathfrak{F}) \;/\; O(n)$$

of M, and the invariant forms on this thickening give rise to the natural extension of our map (8. 1) :

$$H^*(a_n, O_n) \longrightarrow H^*\{J_k(\mathfrak{F})/O(n)\} \approx H^*(M) \quad .$$

Furthermore, note that $J_1(\mathfrak{F})$ corresponds to the frame-bundle of the dual of the quotient bundle Q defined by $\mathfrak{F}$: Thus if

$$O \longrightarrow E \longrightarrow TM \longrightarrow Q \longrightarrow O$$

defines Q then

$$J_1(\mathfrak{F}) = \text{Frame-bundle of } Q^* ,$$

i. e., the fibers of $J_1(\mathfrak{F})$ over $p \in M$ correspond to frames $\theta_p^1, \cdots, \theta_p^n$ of 1 - forms at p, which are independent and vanish on E_p.

Hence under $\varphi(\mathfrak{F})$ the c_{21} of the L. H. S. go over into the characteristic classes of Q^* on M, and in particular, the truncation of the c's in $H^*(\mathfrak{a}_n, O(n))$ none implies the Vanishing theorem which I noted many years ago.

The Pontriagin ring of the normal bundle to a foliation of dimension n, vanishes in dim > 2n.

At present the map (8.5) essentially summarizes the general state of affairs concerning the characteristic classes of foliations. It expresses the vanishing condition just described and at the same time predicts new characteristic classes.

However, there is of course no assurance that all the classes of $H^*(\mathfrak{a}_n, O(n))$ can be distinguished by foliations, or that no other characteristic classes of foliations exist, and by and large these two aspects of the subject are much harder to treat.

To answer the first question one needs a large arsenal of examples of foliations and here the first step was taken by Godbillon-Vey-Rousarie, who considered the following very simple, and beautiful homogeneous foliation.

On $SL(2,\mathbb{R})$, let $\mathfrak{F}$ be the foliation given by

$$\mathfrak{F} = \{\text{Ker } \theta_2^1\} \quad .$$

where θ_2^1 is the left invariant form of that name from section 3, restricted to $SL(2,\mathbb{R})$. We have

$$d\theta_2^1 + \theta_1^1 \wedge \theta_2^1 + \theta_2^1 \wedge \theta_2^2 = 0$$

by our structure equations, whence

(8.6) $$d\theta_2^1 = \theta_2^1 \wedge (\theta_1^1 - \theta_2^2) \quad ,$$

so that $\mathfrak{F}$ is integrable, and indeed the "leaves" of $\mathfrak{F}$ are, of course, the cosets of the subgroup

$$H = \begin{pmatrix} * & 0 \\ * & * \end{pmatrix}$$

of $SL(2,\mathbb{R})$ given by the condition $x_2^1 = 0$.

Now this foliation in itself is not yet interesting in the sense that $h_1 c_1(\mathfrak{F})$ is necessarily zero because $SL(2,\mathbb{R})$ is a non-compact 3 manifold and so has no cohomology in dimension 3. However, if we divide the whole example on the left by a discrete subgroup $\Gamma \subset SL(2,\mathbb{R})$ with $\Gamma \backslash SL(2,\mathbb{R})$ compact, then the example becomes interesting and in particular

$$h_1 c_1(\Gamma\backslash\mathfrak{F}) \neq 0 \quad .$$

Let me close by showing you why this is so. What we need is the following algorithm for computing $h_1 c_1(\mathfrak{F})$, which - read the other way around - was the original definition of this characteristic class in Godbillon-Vey.

PROPOSITION . Let $\mathfrak{F}$ be the foliation on M given by

$$\text{Ker } \theta$$

where θ is a non-vanishing 1 - form on $\mathfrak{F}$. Then from the integrability condition there exists 1 - form η on M with

$$d\theta + \eta \wedge \theta = 0 \quad .$$

In terms η the class $h_1 c_1(\mathfrak{F})$ is then represented by $\eta \wedge d\eta$:

(8.7) $$h_1 c_1(\mathfrak{F}) = \{\eta \cdot d\eta\} \quad .$$

To see this, recall that in $\Omega(\mathfrak{a}_1, O_1)$ we have

$$d\theta^1 + \theta^1_1 \wedge \theta^1 = 0$$

whence $h_1 = \theta^1_1$, and $c_1 = d\theta^1_1 + \theta^1_1 \wedge \theta^1_1 = d\theta^1_1$.

It follows that

(8.8) $$h_1 c_1 = \theta^1_1 \wedge d\theta^1_1$$

in $\Omega(\mathfrak{a}_1, O_1)$.

Now one simply interprets θ as defining a section

$$s_\theta : M \longrightarrow J^1(\mathcal{F})$$

with

$$s_\theta{}^* \theta^1 = \theta$$

and η as a section

$$s_\eta : (M) \longrightarrow J_2(M)$$

with $s_\eta{}^* \theta_2^1 = \eta$, to identify (8. 8) with (8. 7) . Q. E. D.

Note here that in higher codimension we would not be able to describe a foliation by a <u>single</u> θ , nor would we have a section s_η and that is why it took some time to unravel the proper generalizations of the class $h_1 c_1$.

Armed with this algorithm we return to the example $\mathcal{F}$ on $SL(2, \mathbb{R})$. Clearly

$$\eta = -(\theta_1^1 - \theta_2^2) = -2\theta_1^1$$

in view of the identity

$$\theta_1^1 + \theta_2^2 = 0$$

on $SL(2, \mathbb{R})$. Hence, computing again by our structure equations, yields

$$d\eta = -2\,\theta_2^1 \wedge \theta_1^2$$

and

$$\eta d\eta = -2\,\theta_1^1 \wedge \theta_2^1 \wedge \theta_1^2 \quad .$$

Now this 3 - form is clearly $\neq 0$ on $SL(2, \mathbb{R})$ and being left invariant comes from the corresponding form on $\Gamma \backslash SL(2, \mathbb{R})$, which clearly represents $h_1 c_1$ of $\Gamma \backslash \mathfrak{F}$. Integrating over the compact manifold $\Gamma \backslash SL(2, \mathbb{R})$ now implies that

$$h_1 c_1 (\mathfrak{F}) \neq 0 \quad .$$

This construction can clearly be extended to higher dimensions and this yields corresponding nontriviality results for certain charactertistic classes, see for instance [10], and the work of Kamber-Tondeur [44]. However, these are far from sufficient to settle the matter and what is really needed at this stage is an arsenal of nonhomogeneous examples. The master par excellence of such foliations at present is W. Thurston who first of all built examples of foliations of codimension 1, whose invariant $\int_M h_1 c_1 (\mathfrak{F})$ takes on any real value, and more recently proved that every manifold with Euler number 0, admits a foliation of codimension 1. For these developments see [66], [67], also [48].

Bibliography

Anderson, D. W.

[1] 1972 A generalization of the Eilenberg-Moore spectral sequences. Bull. AMS, 78, 784-786.

Atiyah, M., Bott, R., and Patodi, V.

[2] 1973 On the heat equation and the index theorem. Invent. Math., 19, 279-330.

Borel, A. and Hirzebruch, F.

[3] 1959 Characteristic classes and homogeneous spaces, II. Am. Journal of Mathematics, 81, 315-382.

Bott, R.

[4] 1971 The Lefschetz formula and exotic characteristic classes. Proc. Diff. Geom. Conference, Rome 1971.

[5] 1971 On a topological obstruction to integrability. Proc. International Congress Math. (Nice, 1970) Vol. 1, 27-36. Gauthier-Villars, Paris.

[6] 1972 Lectures on characteristic classes and foliations (Notes by Lawrence Conlon). Lecture Notes in Mathematics # 279, 1-94. Springer-Verlag, New York.

[7] 1973 On the Gelfand-Fuks cohomology. Proc. Diff. Geom. Conference, Stanford, 1973.

[8] 1973 On the Chern-Weil homomorphism and the continuous cohomology of Lie groups. Advances in Math., 11, 289-303.

[9] 1973 Some remarks on continuous cohomology. Proc. of International Conference on Topology and Related Topics, Univ. of Tokyo, Japan

Bott, R. and Haefliger, A.

[10] 1972 On characteristic classes of Γ-foliations. Bull. AMS, 78, 1039-1044.

Brown, E. H.

[11] 1965 Abstract homotopy theory. Trans. AMS, 119, 79-85.

Cartan, H.

[12] 1950 Notions d'algèbre différentielle. La transgression dan une groupe de Lie. Colloque de Topologie, Bruxelles, 15-27, 57-71.

Chevalley, C.

[13] 1946 "Theory of Lie Groups", Princeton University Press, Princeton.

Chevalley, C., and Eilenberg, S.

[14] 1948 Cohomology theory of Lie groups and Lie algebras. Trans. AMS, 63, 85-124.

Chern, S. and Simons, J.

[15] 1971 Some cohomology classes in principal fiber bundles and their applications to Riemannian geometry. Proc. Nat. Acad. Sci, U.S.A., 68, 791-794.

de Rham, G.

[16] 1932 Sur la théorie des intersections et les integrales multiples. Commet. Math. Helv., 4, 151-157.

Dold, A.

[17] 1972 "Lectures on Algebraic Topology." Springer, New York.

Dold, A., and Puppe, D.

[18] 1961 Homologie Nicht-Additiver Funktoren. Anwendungen. Ann. Inst. Fourier, Grenoble, 11, 201-312.

Eilenberg, S.

[19] 1950/1951 Homologie des groupes. Séminaire Henri Cartan, Ecole Normal Superieure, Paris.

Eilenberg, S. and MacLane, S.

[20] 1945 Relation between homology and homotopy groups of spheres. Ann. of Math. (2), 46, 480-509.

Eilenberg, S. and Zilber, J.

[21] 1950 Semi-simplicial complexes and singular homology. Ann. of Math, 51, 499-513.

Fogarty, J.

[22] 1969 "Invariant Theory." Benjamin, New York.

Gelfand, I.

[23] 1964 "Generalized Functions." Academic Press, New York.

Gelfand, I., and Fuks, D.

[24] 1968 Cohomologies of the Lie algebra of vector fields in a circle. Funct. Anal., 2, 342-3.

[25] 1969 Cohomologies of the Lie algebra of vector fields on a manifold. Funct. Anal., 3, 155.

[26] 1969 The cohomology of the Lie algebra of tangent vector fields on a smooth manifold, I. J. Functional Anal., 3, 194-210.

[27] 1970 The cohomology of the Lie algebra of tangent vector fields on a smooth manifold, II. J. Functional Anal., 4, 110-116.

[28] 1970 The cohomology of the Lie algebra of formal vector fields. Izv. Ann. SSSR, 34, 327-342.

Gelfand, I., Fuks, D., and Kalinin, D.

[29] 1972 Cohomology of the Lie algebra of formal hamiltonian vector fields. Funct. Anal., 6, 25-29.

Godbillon, C.

[30] 1972 Cohomologie d'algèbres de Lie de champs de vecteurs formels. Sém. Bourbaki, 421-01 to 421-19

Godbillon, C., and Vey, J.

[31] 1971 Un invariant des deuilletages de codimension un. C. R. Acad. Sci., Paris, 273, 92.

Greenberg, M. J.

[32] 1967 "Lectures on Algebraic Topology." Benjamin, New York.

Guillemin, V.

[33] 1973 Cohomology of vector fields on a manifold. Advances in Math., 10, 192-220.

Haefliger, A.

[34] 1970 Feuilletages sur les variétés ouvertes. Topology, 9, 183-194.

[35] 1971 Homotopy and integrability. Lecture Notes in Math # 197, 133-163. Springer-Verlag, New York.

[36] 1972 Sur les classes caractéristiques des feuilletages. Sém. Bourbaki, 412-01 to 412-21.

Heitsch, J.

[37] 1973 Deformations of secondary characteristic classes. Topology, 12, 381-388.

Hochschild, G., and Mostow, G. D.

[38] 1962 Cohomology of Lie groups. Ill. J. Of Math., 6, 367-401.

Hochschild, G., and Serre, J-P.

[39] 1953 Cohomology of Lie algebras. Ann. of Math., 57, 591-603.

Hu, S. -T.

[40] 1964 "Elements of General Topology." Holden-Day, San Francisco.

Husemoller, D.

[41] 1966 "Fiber Bundles." McGraw-Hill, New York.

Kamber, F., and Tondeur, P.

[42] 1973 Cohomologie des algebres de Weil relatives tronquées. Algèbres de Weil semi-simpliciales. Homomorphisme caracteristique d'un fibré principal feuilleté. Comptes Rendu Acad. Sci. Paris, t. 276.

[43] 1974 Characteristic invariants of foliated bundles. Manuscripta Math., 11, 51-89.

[44] Foliated Bundles and characteristic classes. Springer notes to be published.

Kobayashi, S.

[45] 1961 Canonical forms on frame bundles of higher order contact. Diff. Geometry, Proc. Symp. III, AMS, Tuscon.

Koszul, J. -L.

[46] 1950 Homologie et cohomologie des algèbres de Lie. Bulletin de Soc. Math. de France, 78, 5-127.

Lamotke, K.

[47] 1968 "Semisimpliziale algebraische Topologie." Springer, Berlin.

Lawson, H. B., Jr.

[48] 1974 Foliations. Bull. AMS, 80, 369-418.

Mather, J.

[49] 1973 Loops and foliations. Proc. of Int. Conf. on Topology and Related Topics, Univ. of Tokyo, Japan, 1973.

May, J. P.

[50] 1967 "Simplicial Objects in Algebraic Topology." Van Nostrand, Princeton.

Milnor, J.

[51] 1956 Construction of universal bundles, I, II. Ann. Math., 63, 272-284 and 430-436.

[52] 1957 The realization of a semi-simplicial complex.' Ann. Math., 65, 357-362.

[53] 1957 "Notes on Characteristic Classes." Princeton Univ. Press, Princeton.

Mosher, R., and Tangora, M.

[54] 1968 "Cohomology Operations." Harper and Row.

Schwartz, L.

[55] 1950 "Théorie des Distributions." Hermann, Paris.

Segal, G.

[56] 1968 Classifying spaces and spectral sequences. Inst. des Hautes Etudes Scientifiques, Publ. Math., 34, 105-112.

[57] 1974 "Categories and cohomology theories." Topology, 13, 293-312.

Serre, J-P.

[58] 1951 Homologie singulière des espaces fibrés. Ann. of Math, 54, 425-505.

[59] 1965 "Lie Algebras and Lie Groups." Benjamin, New York.

Shulman, H.

[60] 1972 "On characteristic classes." Thesis, Univ. of Cal., Berkeley.

Spanier, E.

[61] 1966 "Algebraic Topology." McGraw Hill, New York.

Stasheff, J.

[62] 1963 Homotopy associativity of H-spaces II. Trans. AMS, 108, 293-312.

Steenrod, N.

[63] 1943 Homology with local coefficients. Annals of Math., 44, 610-627.

[64] 1951 "The topology of fibre bundles." Princeton U. Press, Princeton.

tom Dieck, T.

[65] 1974 "On the homotopy type of classifying spaces." Manu. Math., 11, 41-49.

Thurston, W.

[66] 1972 Noncobordant foliations of S^3. Bull. Am. Math. Soc., 78, 511-514.

[67] 1973 The theory of foliations of codimension greater than one. Proc. of Stanford U. Conf. on Diff. Geom. 1973.

[68] 1974 Foliations and groups of diffeomorphisms. Bull. AMS, 80, 304-312.

Trauber, P.

[69] 1973 "The continuous cohomology of the Lie algebra of vector fields on a manifold." Ph. D. thesis, Princeton Univ.

Van Est, W.

[70] 1955 Une application d'une méthode de Cartan-Leray. Indag. Math., 17, 542-544.

Weil, A.

[71] 1952 Sur les théorèmes de de Rham. Comment. Math. Helv., 26, 119-145.

Weyl, H.

[72] 1946 "The Classical Groups." Princeton Univ. Press, Princeton.

Whitehead, J. H. C.

[73] 1949 Combinatorial Homotopy, I. Bull. AMS, 55, 213-245.

Reprinted from ADVANCES IN MATHEMATICS Vol. 16, No. 2, May 1975

On the Shape of a Curve

RAOUL BOTT

Department of Mathematics, Harvard University, Cambridge, Massachusetts 02138

DEDICATED TO STAN ULAM

The dubious honor of being your anti-keynote speaker[1] has no doubt fallen to me because I am an engineer and applied mathematician gone astray—a victim of the seductive wiles of pure mathematics.

Of course the days of penitence in matters of chastity are long gone, so do not expect me to be contrite. Rather my theme will be that the temptation was simply too great. However let me say this concerning the various branches of our subject. Subjectively I find no difference among any of them. When I worked on networks with Duffin I found them as fascinating as I found Lie groups later on in my work with Samelson, fixed point theory in my work with Atiyah, and as I find foliations in my work with Haefliger at the moment.

But to get on with it, let me cite an instance of the "Pure Temptation" in mathematics which was eminently successful:

We start from the "practical" question: How many solutions does a real polynomial equation of degree n have?

A little experimentation then shows that such an equation, say,

$$a_n x^n + \cdots + a_0 = 0, \qquad a_n \neq 0,$$

might well have no solutions, that it has at most n of them, and that there seems to be something going on behind the scenes which is hard to get at. On the other hand if we abandon the practical world of the real numbers and succumb to the "pure" by introducing the *complex numbers* $\mathbb{C}$, we find an infinitely more satisfying state of affairs which sheds light not only on our original question but eventually illuminates nearly every aspect of analysis. Over $\mathbb{C}$ the situation is of course the following.

[1] First Los Alamos Workshop on Mathematics in the Natural Sciences, June 12–18, 1974.

THEOREM 1. *The general polynomial equation of degree n has precisely n solutions.*

Here and in the following, "general" or "generic" is used to mean all polynomials except possibly those whose *coefficients* satisfy a finite number of polynomial equations.

For example, the general polynomials of degree two are all

$$p(x) \equiv ax^2 + bx + c,$$

with $a \neq 0$ and $b^2 - 4ac \neq 0$.

In particular the general polynomials are open and dense in the set of all polynomials and their complement is of measure zero. Thus the "special" polynomials in their complement are always limits of the generic ones so that even their properties can often be best understood as the limiting property of the generic ones. Thus Theorem 1 illuminates all polynomial equations.

Note finally that the cardinality is really the only invariant of a finite set, so that Theorem 1 does describe the shape of the generic solution set completely.

At this stage the pure temptation is of course to proceed further and ask the corresponding question for polynomials in more variables, and the moment one succumbs to it, one is introduced into the heart not only of topology and algebraic geometry but also of much of modern mathematics.

Let me pursue this path in the case of two variables, and to keep the discussion as concrete as possible, let us try first to determine the "shape" of the set S of points $(x, y) \in \mathbb{C}^2$ satisfying the equation

$$x^3 + y^3 = 1. \qquad (*)$$

A first result is the

PROPOSITION. *The set $S \subset \mathbb{C}^2$ is a smooth surface of real dimension two in $\mathbb{C}^2 = \mathbb{R}^4$.*

To prove this, proceed just as over the reals. These *real* points, $S_{\mathbb{R}}$, of S, are clearly the level sets where the function $f(x, y) = x^3 + y^3$ equals unity. It follows that

$$xf_x + yf_y = 3 \quad \text{on} \quad S_{\mathbb{R}}.$$

Hence the gradient of $f \neq 0$ along $S_{\mathbb{R}}$ and therefore, by the implicit function theorem, $S_{\mathbb{R}}$ is a smooth *curve* in $\mathbb{R}^2$. Over $\mathbb{C}$ this argument is equally valid, and that also explains why the algebraic geometer calls S a *curve* over $\mathbb{C}$. Geometrically, i.e., over $\mathbb{R}$, S is however given by two independent real conditions in $\mathbb{R}^4$ and hence has dimension two.

We next try to determine the "shape" of S. In particular, let us see how S fits into the classification of compact orientable surfaces which Lipman Bers already discussed earlier during our Conference. You will recall that the basic shapes of such objects are the sphere with g-handles (Fig. 1).

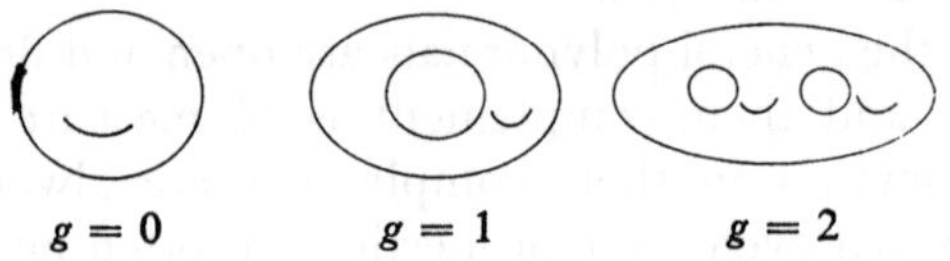

FIGURE 1

To investigate this problem, let us project S onto its first variable x. Clearly this projection, say π, maps S *onto* $\mathbb{C}$ and the inverse image of a point $x \in \mathbb{C}$ consists of all $y = (1 - x^3)^{1/3}$. Thus two types of situations arise:

(a) If $x^3 = 1$, then $\pi^{-1}(x)$ consists of one point.

(b) If $x^3 \neq 1$ then $\pi^{-1}(x)$ consists precisely of the three cube roots of $1 - x^3$.

In Fig. 2 I have tried to illustrate the situation.

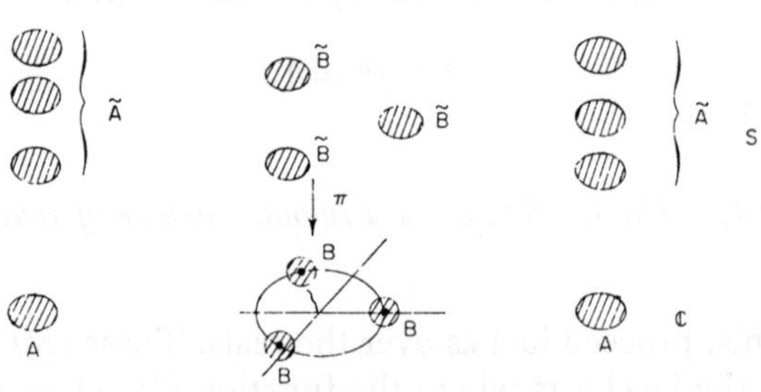

FIGURE 2

Over a region such as A, S has three disjoint copies of A, while over a region such as B, S has essentially just *one* copy of B. How this all fits together is admittedly a little mind-boggling but locally there is no difficulty. Near the branch points (the cube roots of 1) one argues as follows. Let ρ be such a cube root so that $\rho^3 = 1$. Then near ρ write $x = \rho + \epsilon$, so that

$$1 - x^3 = 1 - (\rho + \epsilon)^3 = -3\rho^2\epsilon \cdot \{1 + (1/\rho)\epsilon + (1/3\rho^2)\epsilon^2\}.$$

Hence near ρ, y is given by

$$y = \sqrt[3]{\epsilon} \cdot \{a_0 + a_1\epsilon + \cdots\}$$

where $a_0 = (-3\rho^2)^{1/3}\rho \neq 0$ and the power series converges for small ϵ. Thus near $\pi^{-1}\rho$, the projection π looks essentially like the map $y \to y^3$ which sends the unit disk into itself. (In other respects the point $\pi^{-1}\rho$ is however quite undistinguished on S, for we already know that S has *no* singular points!)

Before assembling S out of the little disks of Fig. 1, we should first come to grips with the fact that S is not compact. Clearly as we march to ∞ in x, say along the integers $x = 1, 2, 3, \ldots$, every one of the three inverse images under π in S will also march to ∞ on S.

However a remedy is again close at hand. Just as the Gauss Sphere, that is the surface of genus zero, is obtained by adding "one point at ∞" to $\mathbb{C}$, so in order to obtain a closed surface $\bar{S}$ we expect to add three points to S. This turns out to be correct, but care has to be taken to establish this fact. Essentially we have to check that our point at ∞ is not "ramified relative to π" the way the cube roots $1, \rho, \rho^2$ were. For instance consider the equation

$$y^2 + x^3 = 1 \qquad (*')$$

instead of (*). Here the solution set S' clearly looks like *two* sheets spread out over the $\mathbb{C}$-plane, ramified at $1, \rho, \rho^2$ as before, but in this case I claim that one should add only *one* point at ∞, to obtain a smooth closed surface $\bar{S}'$.

Now how is all this to be understood and not only by the inspired but also by us, the simple minded?

The technique that reduces these questions to a routine computation again starts with an abstraction, the so-called projective space $\mathbb{C}P_2$, which plays the role of the Gauss Sphere in dimension two.

If in $\mathbb{C}^3$ *we remove the origin and then identify two triples* (x, y, z) *and* (x', y', z') *of* $\mathbb{C}^3 - 0$, *if and only if there exists a* $\lambda \in \mathbb{C} - 0$ *with*

$$(\lambda x', \lambda y', \lambda z') = (x, y, z),$$

the resulting space is called the complex projective space of dimension two and is notated $\mathbb{C}P_2$.

Notice that the map

$$(x, y) \to (x, y, 1) \qquad (^{\dagger})$$

imbeds $\mathbb{C} \times \mathbb{C}$ into $\mathbb{C}P_2$ and that the complement of this $\mathbb{C} \times \mathbb{C}$ in $\mathbb{C}P_2$ is given by the class of $(x, y, 0)$, that is, by a $\mathbb{C}P_1$, which in turn is precisely $\mathbb{C} \cup \infty$, i.e., our "Gauss Sphere." Thus the spaces $\mathbb{C}P_2$ etc., which are easily seen to be compact, are indeed the *natural higher dimensional analogs of the Gauss Sphere.*

Once we are aquainted with this concept it naturally suggests that we consider the set PS of points in $\mathbb{C}P_2$ on which the *homogenous equation*

$$x^3 + y^3 = z^3$$

is satisfied. Indeed this set is again easily seen to *be compact* and *obviously agrees with our* S *on* $\mathbb{C} \times \mathbb{C} \subset \mathbb{C}P_2$, *because there,* z *can be taken to be equal to unity.*

Thus our PS is certainly a "compactification" of S, but now the question arises in what sense PS is to be understood as a surface, whether it is smooth etc. In short the question one encounters here is, to *what extent and in what sense can one extend the calculus from* $\mathbb{C}^2$ *to* $\mathbb{C}P_2$.

Actually the answer is quite simple: Let

$$U_x = \{(x, y, z) \in \mathbb{C}P_2 \text{ with } x \neq 0\},$$
$$U_y = \{(x, y, z) \in \mathbb{C}P_2 \text{ with } y \neq 0\}, \text{ and}$$
$$U_z = \{(x, y, z) \in \mathbb{C}P_2 \text{ with } z \neq 0\}.$$

Then, by our earlier remark U_x, U_y, U_z are each isomorphic to $\mathbb{C} \times \mathbb{C}$ and *between them they cover all points of* $\mathbb{C}P_2$. *Thus to do the calculus on* $\mathbb{C}P_2$ *one simply has to work three times as hard as in* $\mathbb{C} \times \mathbb{C}$; one has to study all phenomena in *each* of the three coordinate patches:

U_x, U_y, and U_z. To illustrate: Using the plausible names for our variables, e.g. (y, z in U_x, etc.), the equation

$$x^3 + y^3 = z^3$$

is satisfied precisely on the sets

$$\begin{aligned} x^3 + y^3 &= 1 \quad \text{in} \quad U_z, \\ x^3 + 1 &= z^3 \quad \text{in} \quad U_y, \\ 1 + y^3 &= z^3 \quad \text{in} \quad U_x, \end{aligned}$$

and, since each of these defines a smooth curve (by our old method of proof in $\mathbb{C} \times \mathbb{C}$), PS is indeed a smooth object. Furthermore the *points at* ∞ *of* $x^3 + y^3 = 1$ *now appear plainly in both* U_y *and* U_x as the points where $z = 0$, i.e., where x, respectively y, range over -1, $-\rho$, $-\rho^2$.

We test this procedure next for the equation

$$y^2 + x^3 = 1.$$

Here the homogeneous version is

$$y^2 z + x^3 = z^3$$

and its manifestations on U_x, U_y, U_z are

$$\begin{aligned} y^2 + x^3 &= 1 \quad \text{on} \quad U_z, \\ z + x^3 &= z^3 \quad \text{on} \quad U_y, \\ y^2 z + 1 &= z^3 \quad \text{on} \quad U_x. \end{aligned}$$

Hence the point at ∞ of $y^2 + x^3 = 1$ appears in U_y *alone* as the point $z = 0$, $x = 0$.

After this long detour, we may return to our curve S. What we have learned is that $PS \in \mathbb{C}P_2$ is a smooth compact surface of $\mathbb{C}P_2$ and that PS is obtained from S by adding three *points* at ∞. Thus:

$$PS = S \cup p_1^\infty \cup p_2^\infty \cup p_3^\infty.$$

In short, PS is $\bar{S}$, our closed version of S, and we can finally turn to our main global problem, which now can be formulated in the form:

What is the genus (number of handles) of $\bar{S}$?

Alas, we aren't quite there yet. First we need to check that $\bar{S}$ is orientable, that is, that $\bar{S}$ can be covered by little *real* coordinate disks so that on the overlap the Jacobian is positive. At this stage another magnificent quality of the complex numbers (and complex analytic functions) makes its appearance. The point is that, if

$$\omega = f(z)$$

is an analytic function of the complex variable z, then the Jacobian $=$ $\det(\partial f_i/\partial x_j)$ of the coordinate transformation of real and imaginary parts of ω to the real and imaginary parts of f, is always $\geqslant 0$, and >0 when $f'(z) \neq 0$.

I leave the proof of this fact to you, and, armed with it, we can now easily prove the orientability, by first finding a *good* complex coordinate near each point of S. Indeed x will do near all points which are not over branch points for π, and at these y will do. Finally, at ∞ in PS, z will do, but here one could easily argue that orientability is not affected by removing a few points, so that S orientable $\Rightarrow$ $\bar{S}$ orientable. So then, $\bar{S}$ is orientable and our question concerning its genus is legitimate. Note further that along the way of proving this, we have actually noted that: (1) not only the calculus but really the whole theory of complex variables makes sense on PS, and (2) that orientability is automatic for such complex analytic beings.

But let me now really turn to the problem of identifying the shape, i.e., genus of $\bar{S}$, and here I will again take the most "uninspired" way out. I will not try to make you imagine *how* PS really looks, although it would be a good exercise to do so; rather I will apply to this problem the oldest *invariant* of topology, and as you will see this invariant will reduce the question to the simplest arithmetic. In a sense then this is actually the *most inspired* method, only it is not our inspiration that is needed, only that of Euler.

The invariant of a surface I want to use goes back to Euler, is called its "Euler number," and is computed as follows:

Consider a "triangulation" of a smooth surface: into α_2 curvilinear triangles, and let α_1 denote the number of edges and α_0 the number of vertexes.

For instance, for the Gauss Sphere (Fig. 3) we could take the triangulation with $\alpha_2 = 6$, $\alpha_1 = 9$, and $\alpha_0 = 5$. The number

$$e = \alpha_0 - \alpha_1 + \alpha_2$$

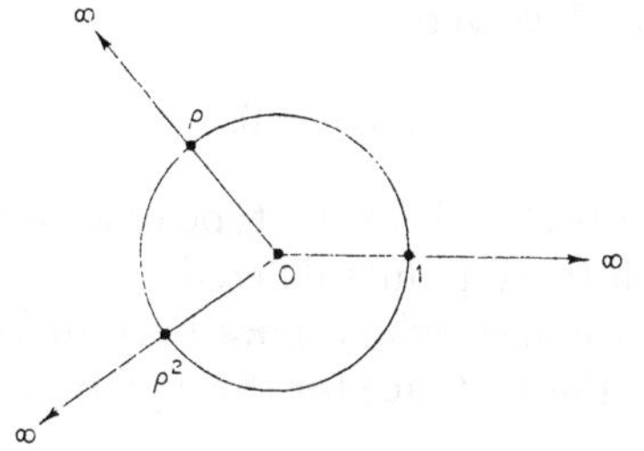

FIGURE 3

is then called the Euler number of the triangulation, and, as was already essentially known to Euler, the following theorem is valid:

THEOREM 2. *The Euler number of a triangulation of a surface X is independent of the triangulation. Thus e(X) is well defined, and this invariant of X is related to the genus of X by the formula*

$$e(X) = 2 - 2g(x).$$

For instance for the Gauss Sphere we find by Fig. 3 that

$$e(\mathbb{C}P_1) = 5 - 9 + 6 = 2.$$

Therefore to determine the shape of $\bar{S}$ it is sufficient to compute its Euler number, and this is now easy to do in view of our geometric understanding of $\bar{S}$ spread over the Gauss Sphere (Fig. 2).

We triangulate $\mathbb{C}P_1 = \mathbb{C} \cup \infty$, say as indicated in Fig. 3, but any refinement of that triangulation will also do. Now since our branch points are vertexes in the triangulation, it is easy to see that the *inverse image of the triangulation under π is a triangulation of $\bar{S}$* and simply counting we see that the number $\tilde{\alpha}_i$ (of that triangulation) is related to the α_i of the initial triangulation of $\mathbb{C}P_1$ by the relation:

$$\tilde{\alpha}_2 = 3\alpha_2, \qquad \tilde{\alpha}_1 = 3\alpha_1, \qquad \tilde{\alpha}_0 = 3\alpha_0 - 3 \times 2.$$

The 3×2 of course corrects for the fact that the three branch points were covered only once rather than three times. These equations lead to

$$e(\bar{S}) = 3e(\mathbb{C}P_1) - 3 \times 2,$$

so that from $e(\mathbb{C}P_1) = 2$ we get

$$e(\bar{S}) = 0.$$

Thus we have shown that: $\bar{S}$ is of the type of a torus, and our original S looks like a torus with three points deleted.

Let us test this procedure, which goes back to Hurwitz, I believe, on a few more examples. For instance for the equation

$$y^2 + x^3 = 1, \tag{*'}$$

a triangulation count gives:

$$\tilde{\alpha}_2 = 2\alpha_2, \quad \tilde{\alpha}_1 = 2\alpha_1, \quad \tilde{\alpha}_0 = 2\alpha_0 - 4 \times 1$$

because there are now four branch points. Hence

$$\begin{aligned} e(\bar{S}') &= 2e(\mathbb{C}P_1) - 4 \times 1 \\ &= 0. \end{aligned}$$

Thus $\bar{S}'$ is again a torus while S is now a torus with *one point* removed. Actually much more can be said; namely, *the torus* is the shape of the *generic polynomial* of degree three in the *following sense.*

THEOREM 3. *Consider the polynomials in two variables $p(x, y)$ of degree three, whose associated homogeneous polynomial in three variables $\tilde{p}(x, y, z)$ vanishes on a nonsingular surface in $\bar{S}(p) \subset \mathbb{C}P_2$. These polynomials then form a generic set, and $\bar{S}(p)$ is of the shape of a torus.*

To prove this, one just has to show that (1) the nonsingularity of $\bar{S}(p)$ can be written down in terms of polynomials in the coefficients of p, and (2) that nearly all polynomials have *topologically* isomorphic shapes. Then it is easy to see that these polynomials form a *connected set*, and of course on a connected set the Euler number cannot change so that it is sufficient to compute it for *one* example.

In the same spirit we can now believe the following solution of our initial question.

THEOREM 4. *The generic polynomial in two variables of degree n, has for its compactified solution set in $\mathbb{C}P_2$, a surface of Euler number*

$$\begin{aligned} e(\bar{S}) &= ne(\mathbb{C}P_1) - n(n-1) \\ &= -n(n-3). \end{aligned}$$

Indeed, all we really need to do is to apply our procedure to the equation

$$x^n + y^n = 1$$

which is only ramified over the nth roots of one, and therefore clearly leads to the equation for $e(\bar{S})$ given above. Q.E.D.

So much for my main example. I wanted to cover it in great detail, so that you would not only be exposed to some of the concepts and techniques of topology and algebraic geometry, but also learn a useful fact which isn't as generally known as it should be.

Let me point out, for instance, that the moment one tackles higher order partial differential equations, our theorem becomes very pertinent, and this is especially so in the study of hyperbolic equations à la Petrovsky [1].

But I will not try to propagandize for this fledgling result in topology here. Rather, I would like to show you an *application* of the full grown topology to number theory which is maybe the most spectacular achievement of recent mathematics.

To explain this development let me sketch in the extensions of those concepts we have already met which will be needed later on.

First of all the notion of a smooth surface has its natural extension in the notion of smooth n-manifold. For our purposes, let us make do with this definition of such an object.

A subset M_n of $\mathbb{R}^N$ is called a smooth n manifold, if in the vicinity of each point $p \in M_n$, M_n is the level set of $N-n$, smooth functions $\{f_\alpha\}$ whose gradients are independent at p in the sense that the matrix $\| \partial f_\alpha / \partial x_j \| j = 1, \ldots, N$ has rank, $N - n$ at p.

On such M_n's one can locally describe points by n parameters, i.e., they are locally isomorphic to $\mathbb{R}^n$, but usually not globally so. Hence the calculus can be extended from $\mathbb{R}^n$ to M^n, but with care and usually more work, as global conditions have to be checked in many coordinate patches. These manifolds are also the natural habitat, in the view of the pure mathematician, of objects like tensor-fields, densities etc.

Now then, once we understand these n-manifolds, what about their "shape." Unfortunately already for $n = 3$ no complete list of shapes, such as in dim 2, can be constructed. We therefore have to get by with partial results, and it becomes even more vital than in dimension two to find a fairly good set of invariants of these shapes. A first candidate is the Euler number, and it is clear how to extend its definition. Namely, we cut M_n into α_n (*curvilinear triangles of* dim n) and let α_i, $i \leqslant n$ be the number of i dimensional curvilinear triangles (called i-simplexes in the jargon of topology) which occur in such a triangulation. Then we set

$$e(M) = \alpha_0 - \alpha_1 + \alpha_2 \pm \alpha_n$$

and hope that $e(M)$ turns our to be independent of the triangulation. Of course here it becomes even highly nontrivial whether such a triangulation exists, but it does, and indeed the Euler number is an invariant of M_n.

Next we might hope for more invariants as the dimension increases; and this is also true, though none of them are ever quite as tractable as Euler's. Let me just describe one of these invariants for it points the way for all of "cohomology theory."

I already remarked that the calculus carries over to M_n. In particular one can speak of differentials or line integrals, if you prefer, on M_n. Thus a differential ω on M_n induces a differential

$$\sum a_i(x)\, dx_i$$

on each coordinate patch.

Now just as in $\mathbb{R}^n$, we can also speak of *exact* differentials, i.e., one whose integral $\int_c \omega$ along the curve c depends only on the endpoints of c when c is perturbed a little bit. They satisfy the condition that for every one of their local manifestations

$$(\partial a_i/\partial x_j) - (\partial a_j/\partial x_i) = 0.$$

Again, just as in $\mathbb{R}^n$, any C^∞ function f on M_n has a differential df which is clearly *exact*. On the other hand, unlike the case in $\mathbb{R}^n$, it is not any more true that every exact form is df for some function. This fact leads to the vector space of *interesting line integrals*,

$$H^1(M^n) = \text{exact differentials/gradients of functions},$$

and a very instructive exercise for all of you not steeped in these matters would be to convince yourselves that on a surface

$$\dim H^1(M) = 2 \text{ genus } (M).$$

Thus for surfaces dim $H^1(M)$ is just as good an invariant as the Euler number was, except of course, *a priori*, harder to compute. Now returning to M_n, I think you can see how the land lies—one might hope to define for every $k \leqslant n$ vector spaces

$$H^k(M) = \text{exact } k\text{-dim volume integrals/trivially exact } k\text{-dim volume integrals}$$

and hope for finite dimensionality when M is compact. Well all this can be done, as was suggested by Eli Cartan and leads one to the so-called *de Rham Cohomology of* M_n. In terms of these slightly more elaborate invariants of shape, our old friend the Euler number is expressed as follows:

$$e(M_n) = \sum (-1)^i \dim H^i(M_n),$$

and this expression proves its invariance as no choices are involved in the right-hand side.

Let me next show you a marvelous phenomenon discovered by Lefschetz in the twenties which will play a fundamental role in the application to algebra and number theory I want to explain.

First we note that the change-of-variable formula in the calculus leads to an *induced homomorphism*

$$H^k(N) \xleftarrow{H^k(f)} H^k(M)$$

for every smooth function

$$N \xrightarrow{f} M.$$

(Note that the arrows are reversed, the characteristic contravariance of cohomology.) In particular then if $f: M \to M$ is a map of M into itself, then

$$H^k(f): H^k(M) \to H^k(M)$$

is a linear endomorphism and its trace is well defined whenever $H^k(M)$ is finite-dimensional, and, therefore in particular, when M is compact. Now what Lefschetz observed is the following remarkable fact:

$$\text{number of fixed points of } f = \sum (-1)^k \operatorname{trace} H^k(f)$$

Here of course $p \in M$ is fixed under f if $f(p) = p$, and the formula above is literally true only if the fixed points are "orientable" and "nondegenerate", in the sense that, at each fixed point p, we have

$$\det(1 - (\partial f_i/\partial x_j)(p)) > 0.$$

$\{x_\alpha\}$ being focal coordinates centered at p and f being locally expressed by the f_i. (Thus $x_i(f(p)) = f_i(x)$.)

Now this formula is a joy in itself but I really don't have time to explore it with you. Rather let me stop here with "topology" and return, in a sense, to our original problem of looking at the shape of the solution set to a polynomial, or a number of polynomial equations, *but this time more from the point of view of number theory.*

The number theorist is in the final analysis interested in the integer solutions of a set of equations, but over the years he has developed great insight into how to localize such problems to each prime and then put the local information together again. For him the field $\mathbb{Z}_p$, of the integers mod p, is therefore of profound interest, and he has by now a large body of algebraic information about it. Furthermore, much of our previous discussion makes sense over any field k; e.g., we can define the projective space kP_n of dim n for every field k, just as we did before by taking as points of kP_n the rays in k^{n+1}. Also, given a finite number of homogenous polynomials $p_1, \ldots, p_r$ in $(n + 1)$ variables we can study the solution points over k:

$$S_k \subset kP_n \qquad \textit{where all} \quad p_i = 0;$$

and in algebraic geometry one tries to study this situation by analogy with what was found over the field $\mathbb{C}$. Of course the matter is delicate. Simple phenomena over $\mathbb{C}$ turn out to split into many cases in characteristic p. In fact the situation is reminiscent of the step from classical to quantum mechanics, so that what was found in the "real world of $\mathbb{C}$," is only a hint of what happens in characteristic p.

With this preamble we are about ready to explain the "Weil Conjectures" which have pointed the way for so much of this development in the last 30 years, and which were solved only last year by a young Belgian mathematician, Deligne, using all the machinery built over the past 20 years by Grothendieck and his school of algebraic geometry.

The problem deals with a set of homogenous polynomial equations

$$p_i(x_0, \ldots, x_n) = 0 \qquad i = 1, \ldots, k \tag{**}$$

with coefficients in $\mathbb{Z}_p$, and the problem is again to describe the solutions of (**) over $\mathbb{Z}_p$ and its various extension fields.

Now $\mathbb{Z}_p$ has up to isomorphisms precisely one extension field $\mathbb{F}_k$ of order p^k for each k. Let us therefore set

$$S_k = \text{number of solutions of (**) in } \mathbb{F}_k P_n ,$$

and combine these integers into the formal counting series,

$$Z(t) = \sum_{k=0}^{\infty} \frac{S_k t^k}{k} . \qquad \text{(R)}$$

At first sight (and even at second!) there is very little that can be said about (R). After all, each S_k is computed by a difficult combinatorial procedure.

On the other hand if one knows a little algebra and a little topology and has a great imagination, as A. Weil had, then it is not hard to guess at some quite explicit properties of $Z(t)$.

The algebra we need to know is this:

(1) The role of $\mathbb{C}$ in the classical theory is always played by the *algebraic closure* of the field in which the equations have their coefficients.

(2) The algebraic closure, $\bar{\mathbb{Z}}_p$, has an automorphism $F: \bar{\mathbb{Z}}_p \to \bar{\mathbb{Z}}_p$, given by sending $x \to x^p$, which has the property, that the fixed set of F^k acting on $\bar{\mathbb{Z}}_p$ is isomorphic to the field $I\mathbb{F}_k$.

By the way, here is a nice example of the richness of characteristic p: The rather dull complex conjugation, taking us from $\mathbb{C}$ to $\mathbb{C}$ and having $\mathbb{R}$ as its fixed set is, in this framework, replaced by the much more flamboyant "Frobenius automorphism" F.

With this information at hand we can trivially reinterpret the number S_k in the following manner: Let $\hat{S}$ denote the set of solutions of (**) in $\bar{\mathbb{Z}}_p P_n$. Since the coefficients of the defining set of equations are in $\mathbb{Z}_p$ and hence fixed under F, this transformation acts on $\hat{S}$, and we clearly have:

$$S_k = \text{number of fixed points of } F^{(k)} \text{ acting on } \hat{S}.$$

Well, the cat is out of the bag now. In this formulation the topologist of course is reminded of the Lefschetz fixed point formula.

Indeed if F *were* an endomorphism of a *compact manifold* X, and if we *could* assume that all fixed points of all iterates are transverse and orientable, then by the Lefschetz formula we would have:

$$\text{number of fixed points } F^k = \sum (-1)^i \text{ trace } \{H^i(F)\}^k.$$

Hence if $\{\alpha_e\}$ and $\{\alpha_0\}$ run over the eigenvalues of $H^i(F)$: $H^i(X) \to H^i(X)$ with i even and i odd, respectively, then the above gives

$$\text{number of fixed points } F^k = \sum_e \alpha_e{}^k - \sum_0 \alpha_0{}^k,$$

and hence for the formal series in question:

$$Z(F, t) \equiv \sum \frac{(\text{number of fixed points } F^k)t^k}{k} = \log \frac{\prod(1 - t\alpha_e)}{\prod(1 - t\alpha_0)}.$$

It follows that in this topological situation we would have the relation:

$$e^{Z(F,t)} = \frac{\prod(1 - t\alpha_e)}{\prod(1 - t\alpha_0)}; \qquad (***)$$

i.e., the exponential of Z would be a rational function of t and the number of factors on top and on bottom correspond to the dimension of the even and odd cohomology of X, respectively.

Well, André Weil saw all this many years ago (see [4]) and conjectured that there should be finite dimensional vector spaces $H^i(\hat{S})$ associated to the set $\hat{S}$ with the Lefschetz property etc., and therefore that finally the counting series should have the desired property (***). In fact he went further. Using an even deeper analogy with the algebraic situation over $\mathbb{C}$, where X is a complex submanifold of $\mathbb{C}P_n$ and therefore has very special properties, he conjectured that the Eigenvalues α of F on $H^i(\hat{S})$ would have the bound

$$|\alpha| = p^{i/2}.$$

Well as I said, this assertion was proved only last year by P. Deligne in a tour de force of combined techniques. The rationality of $e^{Z(t)}$ had been proved earlier, and I refer you to [2] for a very short history of the subject.

Finally let me point out, again by earlier work of Deligne, that this solution of the Weil conjecture also implies the truth of the celebrated

"Ramanujan Conjecture" in number theory. This conjecture asserts that in the formal power series

$$\Delta = x \prod_{n=1}^{\infty} (1 - x^n)^{24}$$

the coefficient $\tau(p)$ of x^p (p a prime) satisfies the inequality

$$|\tau(p)| \leqslant 2p^{11/2}.$$

Of course it is a nontrivial and very beautiful story to see how this completely *elementary* conjecture is related to our main discussion, but I have neither the time nor really the expertise to make that connection for you. Rather I herewith rest my case for the fascination of pure mathematics, and, although all of this is far removed from turbulence, field theory and the many questions we have been hearing about, I am deeply convinced that uncanny relationships, such as the path from the Ramanujan Conjecture to the Lefschetz Formula, are still to be found in our subject and some of these may yet prove very useful in the more applied world. In fact, I would not be too surprised if discrete mod p mathematics and the p-adic numbers would eventually be of use in the building of models for very small phenomena.

References[1]

1. I. G. Petrovsky, On the diffusion of waves and the lacunas for hyperbolic equations, *Math. USSR-Sb.* **17** (59)(1945), 289–370.
2. J.-P. Serre, Valeurs propres des endomorphismes de Frobenius, Séminaire Bourbaki, No. 446, Feb. 1974.
3. C. L. Siegel, "Topics in Complex Function Theory," Vol. I, Wiley, New York, 1969.
4. A. Weil, Number of solutions of equations in finite fields, *Bull. Amer. Math. Soc.* **55** (1949), 497–508.

[1] I cite here only four papers. C. L. Siegel's book on function theory is for those interested in following up the early part of my lecture. The A. Weil paper contains the original conjectures, and Serre's article gives an account of Deligne's arguments as well as an intelligent bibliography.

Printed by the St Catherine Press Ltd., Tempelhof 37, Bruges, Belgium

Proceedings of Symposia in Pure Mathematics
Volume 27, 1975

ON THE GELFAND-FUKS COHOMOLOGY*

RAOUL BOTT**

Introduction. Roughly four years ago Gelfand and Fuks had the inspired idea of considering the *continuous* cohomology of the Lie-algebra of C^∞-vector fields on a manifold ([3], [4], [5]). They showed that the answer was finite dimensional for all compact manifolds and that for S^1 this cohomology is a tensor product of an exterior algebra in dimension 3 and a polynomial ring with generator in dimension 2:

$$H_c(\mathscr{L}S^1) = E(\omega) \otimes \boldsymbol{R}[y].$$

In this report I would like to present a new point of view towards their work which arose, on the one hand, out of the new insights into this cohomology due to Guillemin and Trauber, and on the other out of conversations between G. Segal, A. Haefliger and myself during the past year. I will finally formulate a conjecture which would completely settle the homotopy theoretic nature of this cohomology. Although one finds evidence for its validity on every side, its proof has so far eluded us. I should also mention that when I recently showed this conjecture to Gelfand he told me that it had also, quite independently, been formulated by Fuks.

To set the stage, recall that if L is any Lie algebra, then its cohomology can be defined as the cohomology of the complex

$$AL = \sum A^q L$$

of alternating forms on L, under the differential operator:

AMS (MOS) subject classifications (1970). Primary 17B65, 18H25, 17B55; Secondary 53C10, 55B25, 57D35, 58H05.

*General lecture given at the Institute.

**This research was partially supported by NSF grant GP 31359X-1.

$$d\varphi(x_1, \cdots, x_{q+1}) = \sum(-1)^{i+j}\varphi([x_i\, x_j], x_1 \cdots \hat{x}_i \cdots \hat{x}_j \cdots x_q).$$

Now if L carries a topology, the continuous alternating forms clearly form a subcomplex $A_c L$ of AL and one may then call its cohomology the continuous cohomology of L:

$$H_c^*(L) \equiv H^*(A_c L).$$

With this understood, the Gelfand-Fuks cohomology of M is simply the continuous cohomology of its Lie algebra of vector fields, $\mathscr{L}M$, considered with the C^∞-topology:

$$\text{Gelfand-Fuks}\,(M) = H_c^*(\mathscr{L}M).$$

In the next section I will outline a new proof that $H_c^k(\mathscr{L}M)$ is finite dimensional for every k, provided that M admits a covering $\mathscr{U} = \{U_\alpha\}_{\alpha\in J}$ by a *finite* collection of open sets such that all nonvacuous intersections of the U_α are diffeomorphic to $\boldsymbol{R}^n$.

1. The case of $\boldsymbol{R}^n$. Let $\mathscr{L} = \mathscr{L}\boldsymbol{R}^n$, $n > 0$, and $A = A_c\mathscr{L}$. Consider the subcomplex $A_{pt} = A_{pt}\mathscr{L}$ consisting of all forms in A whose support is at the origin of $\boldsymbol{R}^n$. The aim of this section is to sketch in a proof of the following.

THEOREM 1. *The inclusion ι: $A_{pt} \to A$ induces an isomorphism in cohomology*

$$\iota_* : H^*(A_{pt}) \cong H^*(\mathscr{L}\boldsymbol{R}^n). \tag{1.1}$$

Now as we will see in a moment, the complex $A_{pt}\mathscr{L}$ is much more tractable and it is not hard to show that it is finite dimensional. Hence Theorem 1 implies finite-dimensionality of the Gelfand-Fuks theory on $\boldsymbol{R}^n$. Note also that the cohomology $H^*(A_{pt}\mathscr{L})$ figures in the Gelfand-Fuks approach, as the *continuous* cohomology of the Lie algebra $\mathfrak{a}_n$ of formal vector fields on $\boldsymbol{R}^n$:

$$H^*(A_{pt}) \cong H_c^*(\mathfrak{A}_n),$$

but we will proceed independently of this assumption here.

To prove (1. 1), consider the action of $\boldsymbol{R}^*$ on $\boldsymbol{R}^n$ given by

$$f_t \cdot x = tx, \qquad x \in \boldsymbol{R},\ t \in \boldsymbol{R}^*.$$

Now each f_t is a diffeomorphism, and therefore acts on the vector fields and therefore acts also on A. Note that if $X = \sum a_i\,(\partial/\partial x^i)$ is a vector field, then

$$(f_t^* X)_x = t \sum a_i(x/t)\,\frac{\partial}{\partial x_i}. \tag{1.2}$$

Hence f_t^* acting on A shrinks supports towards 0 as $t \to \infty$. In particular, if $\delta_p^k \in A^1$, $k = 1, \cdots, n$, is given by

$$\delta_p^k\Big(\sum a_i \frac{\partial}{\partial x_i}\Big) = a_k(p), \qquad p \in \boldsymbol{R}^n,$$

then (1.2) implies that

$$f_t^* \delta_p^k = t\delta_{(p/t)}^k. \tag{1.3}$$

More generally, if $D^\alpha = \partial^{\alpha_1}/\partial x_1^{\alpha_1} \cdots \partial^{\alpha_n}/\partial x_n^{\alpha_n}$ is a differential operator monomial, then one checks that

$$f_t^*(D^\alpha \delta_p^k) = t^{1-|\alpha|} D^\alpha \delta_{(p/t)}^k, \qquad |\alpha| = \sum \alpha_i. \tag{1.4}$$

Now because every $\omega \in A^q$ can be represented as a finite sum of elements of the type

$$\omega = \int \mu(x_1, \cdots, x_q)\, D^{\alpha_1} \delta_{x_1}^{i_1} \wedge D^{\alpha_2} \delta_{x_2}^{i_2} \wedge \cdots \wedge D^{\alpha_n} \delta_{x_q}^{i_q}\, dx_1 \cdots dx_q$$

where $\mu(x_1, \cdots, x_q)$ has compact support and is continuous, the formula (1.4) explains how f_t^* acts on all of A. In particular we see that the action of $\boldsymbol{R}^*$ on A_{pt} is semisimple and decomposes A_{pt} according to its "weights":

$$A_{pt} = A_{pt}^{(n)} + A_{pt}^{(n-1)} + \cdots \tag{1.5}$$

with $f_t^* | A_{pt}^{(k)} = t^k$, $k \leqq n$.

The crucial lemma for the proof (1.1) is the following:

Lemma. *For every $\omega \in A$, $\lim_{t\to\infty} (1/t^n)\omega$ exists and is a well-defined element $l_n(\omega)$ of $A_{pt}^{(n)}$. Furthermore l_n defines a surjection of A onto $A_{pt}^{(n)}$*:

$$A \xrightarrow{\ l_n\ } A_{pt}^{(n)} \longrightarrow 0. \tag{1.6}$$

Explanation. By (1.4) all that has to be explained is why f_t^* never blows up to order greater than n. For instance what about

$$\omega = \int \mu(x_1, \cdots, x_q)\, \delta_{x_1}^1 \wedge \delta_{x_2}^1 \wedge \cdots \wedge \delta_{x_q}^1\, dx_1 \cdots dx_q$$

with $q > n$? Of course here it is the *alternation* that saves the day, so that the "worst" term is of the form $\delta_0^1 \wedge \delta_0^2 \wedge \cdots \wedge \delta_0^n$ in $A_{pt}^{(n)}$.

Granting this lemma, one extends it as follows:

Lemma. *Let $F^n A \supset F^{n-1} A \supset \cdots$ be the filtration on A defined by*

$$\omega \in F^k A \Leftrightarrow \lim_{t\to\infty} \frac{1}{t^k}\omega$$

exists, and let $l_k\, \omega$ be that limit. Then one has the exact sequence

$$0 \to F^{k-1}A \to F^k A \xrightarrow{\ l_k\ } A_{pt}^{(k)} \to 0, \tag{1.7}$$

which is in fact split by the inclusion $A_{pt}^{(k)} \subset F^k A$.

Corollary. *If A_{pt} is graded by "weight" then the graded group GA associated to the filtration $\{F^k A\}$ is isomorphic to A_{pt}*:

$$GA \cong A_{pt}. \tag{1.8}$$

To complete the proof of the theorem one still needs a homotopy formula. For this purpose let R be the radical vector field generating the action of $\boldsymbol{R}^n$ on $\boldsymbol{R}^n$: $R = \sum x^i (\partial/\partial x^i)$. Then the crucial homotopy formula in A is

$$\frac{d}{dt} f_t^* = \frac{1}{t}\{i(R)d + d\,i(R)\} f_t^*. \tag{1.9}$$

Applied with $t = 1$ to $A_{pt}^{(k)}$ we see that in $A_{pt}^{(k)}$ multiplication by k is homotopic to 0. Hence

COROLLARY. $H(A_{pt}^{(k)}) = 0$ *for* $k \neq 0$.

Next integrate (1.9) from 1 to T to obtain

$$f_T^* - 1 = K_T d + dK_T \tag{1.10}$$

with

$$K_T \omega = \int_1^T i(R) f_t^* \omega \frac{dt}{t}. \tag{1.11}$$

In general, the limit at $T \to \infty$ of K_T will not exist, but for $F^k A$ with $k < 0$, it will and this limit will also lie in $F^k A$. Hence, in particular, $H(F^{-1}A) = 0$. This granted, we are finished for then $H(A/F^{-1}A) \cong H(A)$. But (1.8) implies that

$$G(A/F^{-1}A) \cong A_{pt}/F^{-1}A_{pt}, \tag{1.12}$$

whence by the corollary (just above) $H(A/F^{-1}A) \cong H\{A_{pt}^{(0)}\}$. Q.E.D.

It remains to understand why the limit K_T as $T \to \infty$ exists on and preserves $F^{-1}A$. For this purpose write

$$K_T = \int_1^T i(R)\, t f_t^* \frac{dt}{t^2}.$$

Now changing variables to $\lambda = 1/t$, we get $K_T = \int_{1/T}^1 \{i(R)(1/\lambda) f_{1/\lambda}^*\}\, d\lambda$ so that on $F^{-1}A$ this integral converges to

$$K_\infty = \int_0^1 \left\{ i(R) \frac{1}{\lambda} f_{1/\lambda}^* \right\} d\lambda.$$

To see that $K_\infty F^{-1}A \subset F^{-1}A$, note that $i(R)$ commutes with f_t^* and furthermore $f_t^* f_u^* = f_{tu}^*$. Hence

$$f_t^* K_\infty = \int_0^1 \left\{ i(R) \frac{1}{\lambda} f_{t/\lambda}^* \right\} d\lambda = \int_0^{1/t} \left\{ i(R) \frac{1}{\lambda} f_{1/\lambda}^* \right\} d\lambda$$

which tends to 0 as $t \to \infty$.

2. The Künneth formula. If L_1 and L_2 are two finite-dimensional Lie algebras and $L_1 \oplus L_2$ denotes their direct sum then it is of course well known that

(2.1) $$H^*(L_1 \oplus L_2) \cong H^*(L_1) \otimes H^*(L_2),$$

and this relation follows directly from the fact that on the level of complexes

(2.2) $$A(L_1 \oplus L_2) \cong A(L_1) \otimes A(L_2).$$

These considerations immediately lead one to suspect that (2.1) is also valid in the Gelfand-Fuks theory, but as in this context (2.2) has to be replaced by a completed tensor-product,

$$A_c(L_1 \oplus L_2) \cong A_c(L_1) \hat{\otimes} A_c(L_2),$$

it is difficult to prove directly the Künneth relation. However, the techniques of the last section can be extended without difficulty to prove the Künneth formula for the disjoint unions of $\boldsymbol{R}^n$'s: Thus one has the formula

(2.3) $$H^*\left(\bigoplus_{\alpha=1}^{k} \mathscr{L}\boldsymbol{R}^n_\alpha\right) = \bigotimes_{1}^{k} H^*(\mathscr{L}\boldsymbol{R}^n_\alpha).$$

At this stage one is ready to carry out a program suggested by G. Segal to get at the Gelfand-Fuks theory of a general manifold M.

For this purpose, let $\mathscr{U} = \{U_\alpha\}_{\alpha \in J}$ be a finite ordered open covering of M with the property that all intersections $U_{\alpha_1} \cap \cdots \cap U_{\alpha_k}$ are either empty or diffeomorphic to $\boldsymbol{R}^n$. (Such coverings always exist on compact manifolds.) Corresponding to $\mathscr{U}$, consider the *semisimplicial* space

(2.4) $$\mathscr{U}M\colon \mathscr{U}_0 M \leftleftarrows \mathscr{U}_1 M \overset{\leftarrow}{\leftleftarrows} \mathscr{U}_2 M \overset{\leftarrow}{\underset{\leftarrow}{\leftleftarrows}} \mathscr{U}_3 M$$

where $\mathscr{U}_k M$ is the disjoint union of the k-fold intersections of the $\{U_\alpha\}$, and the arrows indicate the inclusions

$$\mathscr{U}_{\alpha_0} \cap \cdots \cap \mathscr{U}_{\alpha_k} \subset \mathscr{U}_{\alpha_1} \cap \cdots \cap \hat{\mathscr{U}}_{\alpha_i} \cap \cdots \cap \mathscr{U}_{\alpha_k}.$$

Now then the functor $\mathscr{L}$ clearly behaves contravariantly vis-à-vis these maps (one can restrict vector fields to open subsets) so that applying $\mathscr{L}$ to $\mathscr{U}M$ gives rise to a "cosemisimplicial topological Lie algebra":

$$\mathscr{L}\circ\mathscr{U}M : \mathscr{L}\mathscr{U}_0 M \rightrightarrows \mathscr{L}\mathscr{U}_1 M \overset{\rightarrow}{\rightrightarrows}.$$

As usual, the alternating sum of the arrows now defines a differential operator δ, and the classical proof of the fact that the Čech cohomology of a fine sheaf vanishes here leads to the result that

(2.5) $$H_\delta\{\mathscr{L}\mathscr{U}M\} \cong \mathscr{L}M.$$

Now this same argument extends further to yield a corresponding formula for the simplicial *module* $A_c\circ\mathscr{L}\circ\mathscr{U}M$:

(2.6) $$H_\delta\{A_c\circ\mathscr{L}\circ\mathscr{U}M\} = A_c\mathscr{L}M.$$

The module $A_c\circ\mathscr{L}\circ\mathscr{U}M$ of course also carries the differential operator d of the

Gelfand-Fuks theory, which combined with δ makes $A_c \circ \mathscr{L} \circ \mathscr{U} M$ into a double complex, whose total cohomology can, by (2.6), be identified with $H(A_c \mathscr{L} M)$.

On the other hand, filtering this double complex "the other" way produces a spectral sequence with E_2 terms:

$$E_2 = H_\delta \{H_d(A_c \circ \mathscr{L} \circ M)\} = H_\delta \{H^*(\mathscr{L} \mathscr{U} M)\}$$

and converging to the same total cohomology. In view of our Künneth formula however, the semisimplicial module $H^*(\mathscr{L}\mathscr{U}M)$ takes the form:

$$\overset{\alpha}{\bigotimes} H^*(\mathfrak{a}_n) \rightrightarrows \overset{\alpha\beta}{\bigotimes} H^*(\mathfrak{a}_n) \overset{\rightarrow}{\underset{\rightarrow}{\rightarrow}} \tag{2.7}$$

whose δ-cohomology is now seen to be finite dimensional by virtue of known facts about $H^*(\mathfrak{a}_n)$. Actually with a little care (and the Eilenberg-Zilber theorem) one can push (2.7) to obtain the original spectral sequence of Gelfand-Fuks. Indeed it is not hard to evaluate our E_2 term to be

$$\tilde{H}_\delta\{H^*(\mathscr{L}\mathscr{U}M)\} = \sum_k H_*^a\left(M^{(k)}; M_{k-1}^{(k)}; \overset{k}{\bigotimes} \tilde{H}^*(\mathfrak{a}_n)\right) \tag{2.8}$$

where on the right M^k denotes $M \times \cdots \times M$ k times, and M_{k-1}^k denotes the subset consisting of points for which two or more of the coordinates are equal, H_* denotes homology (!) and the a indicates that we consider the homology which behaves alternatingly relative to the permutation group on k letters. Finally the ~ denotes augmented cohomology.

But rather than indicate how this is done, I will show in the next section how the recent work of Anderson [1] leads to a similar sequence in homology theory pure and simple.

3. The conjecture. We have already remarked on a likely Künneth formula:

$$H^*(\mathscr{L} M \oplus \mathscr{L} N) \cong H^*(\mathscr{L} M) \oplus H^*(\mathscr{L} N) \tag{3.1}$$

for disjoint unions, and using the spectral sequence just developed it is not hard to actually prove this result for the sort of manifolds under consideration. Now in topology this type of functional behavior is exhibited by the homology of *function spaces*. That is, if we define a functor $F(X)$ on spaces by setting

$$F(X) = H^*\{\text{Maps } (X; Y)\}$$

(real coefficients, say) then, under some finiteness conditions

$$F(X_1 \cup X_2) = F(X_1) \otimes F(X_2). \tag{3.2}$$

This suggests that the Gelfand-Fuks theory of an n-manifold M should be the standard cohomology of a function space: Maps $(M; Y_n)$. Furthermore, if such a formula held, then

$$H^*(Y_n) = H^*\{\text{Maps}(\mathbf{R}^n; Y_n)\}$$

would have to be equal to $H^*(\mathscr{L}\mathbf{R}^n) = H^*(\mathfrak{a}_n)$.

Now already in the original work of Gelfand-Fuks there occurs a space with this property and it is defined as follows:

Let $E_n \to^{\pi} BGL_n$ be the universal bundle over $BGL_n(\boldsymbol{C})$. Then the $2n$-skeleton of $BGL_n(\boldsymbol{C})$ is a well-defined space $BGL_n(2n)$, and Y_n is simply $\pi^{-1}\, BGL_n(2n)$. Note in particular that $GL(n, \boldsymbol{C})$ acts naturally on Y_n. In any case, this is then a natural candidate for our *universal space* vis à vis the Gelfand-Fuks theory. However, this hypothesis is not compatible with the work of Guillemin [6] and its extension by Trauber [8].

To explain these results one considers the *diagonal subcomplex*

$$A_\Delta \mathscr{L} M \subset A_c \mathscr{L} M$$

whose q-forms are precisely those with support on the diagonal in M^q. Now in [6] Guillemin evaluated $H(A_\Delta \mathscr{L} M)$ as follows:

Let PM be the principal $GL(n, \boldsymbol{R})$ bundle of M, and let $Y_n M$ be the associated bundle

$$Y_n M = PM \times_{GL(n,\boldsymbol{R})} Y_n,^1$$

so that $Y_n M \to M$ is a bundle with fiber Y_n.

Then Guillemin shows that

$$H^r(A_\Delta \mathscr{L} M) \cong H^{r+n}(Y_n M). \tag{3.3}$$

It follows therefore that in any case our sought-after homotopy functor for Gelfand-Fuks must admit a natural map from $H^*(Y_n M)$ decreasing dimension by n.

Now the construction which most naturally satisfies these conditions is the space $\Gamma(Y_n M)$ of continuous sections of $Y_n M$ over M. Indeed if PM is trivial, then

$$\Gamma(Y_n M) = \mathrm{Map}(M; Y_n),$$

while, in all cases, there is the natural evaluation map

$$\Gamma(Y_n M) \times M \xrightarrow{e} Y_n M$$

which evaluates (s, m) at m to yield $s(m)$. In cohomology, this map induces a homomorphism

$$e^*: H^*(Y_n M) \to H^*(\Gamma\, Y_n M) \otimes H^*(M)$$

which after integration over the fiber produces the desired homomorphism of

$$H^*(Y_n M) \to H^*(\Gamma\, Y_n M).$$

Evaluation at several points similarly fits the results of [9]. All this leads us to the main

(3.4) *Conjecture.* $H^*(\mathscr{L} M) \cong H^*(\Gamma\, Y_n\, M)$.

Now apart from the heuristic arguments leading up to (3.4) there is considerable

[1] $GL(n, \boldsymbol{R}) \subset GL(n, \boldsymbol{C})$ acts on Y_n.

independent evidence. First of all, the Anderson spectral sequence [1] is applicable to the right-hand functor and *provides the same E_2-term as we found for the left-hand one.* Secondly, as Andre Haefliger has pointed out, (3.4) also fits the recent work of Mather and Thurston. Where they show an isomorphism of cohomology for two topological categories C_1 and C_2, our Conjecture (3.4) can be interpreted as an isomorphism between the corresponding *continuous* cohomologies.

But this connection would take me too far afield to explain here (see [7], [2] for these concepts).

Naturally the question now arises why such a reasonable conjecture once made is not immediately demonstrable. In view of the functional similarities of the two functors one clearly only needs a map between them which induces an isomorphism on $\boldsymbol{R}^n$. But unfortunately such a map seems hard to come by.

Let me conclude by observing that for spheres (3.4) rather easily leads to the result that

$$H^*(\mathscr{L} S^n) = H^*(\operatorname{Map}(S^n; Y_n))$$

which for $n = 1$ yields the desired $E(\omega) \otimes \boldsymbol{R}[y]$ but for $n \geqq 2$ gives a rather larger answer than indicated by Gelfand-Fuks. Furthermore, these results can be checked independently by yet another method, due to A. Haefliger.

BIBLIOGRAPHY

1. D. W. Anderson, *A generalization of the Eilenberg-Moore spectral sequence*, Bull. Amer. Math. Soc. **78** (1972), 784–786. MR **46** #9987.

2. R. Bott, *Some remarks on continuous cohomology*, Proc. Internat. Conference on Topology and Related Topics, University of Tokyo, Japan, April 1973.

3. I. M. Gel'fand and D. B. Fuks, *Cohomologies of the Lie algebra of tangent vector fields on a smooth manifold.* I, Funkcional. Anal. i Priložen **3** (1969), no. 3, 32–52. (Russian) MR **41** #1067.

4. ———, *Cohomologies of the Lie algebra of tangent vector fields on a smooth manifold.* II, Funkcional. Anal. i Priložen. **4** (1970), no. 2, 23–31 = Functional Anal. Appl. **4** (1970), 23–32. MR **44** #2248.

5. ———, *Cohomology of the Lie algebra of formal vector fields*, Izv. Akad. Nauk SSSR Ser. Mat. **34** (1970), 322–337 = Math. USSR Izv. **4** (1970), 327–342. MR **42** #1103.

6. V. Guillemin, *Cohomology of vector fields on a manifold*, Advances in Math. **10** (1973), no. 2, 192–220.

7. J. Mather, *Loops and foliations*, Proc. Internat. Conference on Topology and Related Topics, University of Tokyo, Japan, April 1973.

8. P. Trauber, *Gelfand-Fuks cohomology*, Ph. D. Thesis, Princeton University, Princeton, N. J., 1973.

HARVARD UNIVERSITY

Advances in Mathematics—395

ADVANCES IN MATHEMATICS 20, 555–555 (1976)

On the de Rham Theory of Certain Classifying Spaces*

R. BOTT

Harvard University, Cambridge, Massachusetts 02138

H. SHULMAN

University of Pennsylvania, Philadelphia, Pennsylvania 19174

AND

J. STASHEFF

Temple University, Philadelphia, Pennsylvania 19122

In his thesis, H. Shulman proved the vanishing phenomena for characteristic classes of foliations without the usual geometric constructions of connections, curvature, etc. In this note, we present a setting for this, which was noted independently by the senior authors about two years ago.

Since that time, this subject has grown considerably, so that our account here is to a large extent anachronistic. In particular, the work of Kamber and Tondeur, who combine simplicial and curvature techniques but avoid classifying spaces, as well as the work of Vey, Bott, Haefliger and others has progressed far beyond the results outlined here. Nevertheless, our ideas are on the one hand very simple, but on the other, involve technicalities that have been resolved only recently, that a short account at this time still seems worthwhile.

The basic concept, which has been in the air for quite some time, and notably in the work of Deligne on mixed Hodge structures, is that the de Rham theory can be profitably employed as a tool for studying certain nonmanifolds, namely, those that are obtained as the geometric realization of a *simplicial manifold.*

* The authors gratefully acknowledge support from the NSF under grants GP 43613X, GP 22222X, and GP 11111X.

Indeed, if

$$M: M_0 \leftleftarrows M_1 \leftleftarrows M_2 \cdots$$

is such an object (that is, M is a contravariant functor from the category of finite nonempty ordered sets to C^∞-manifolds), then applying the de Rham functor Ω, produces a cosimplicial module

$$\Omega M: \Omega M_0 \rightrightarrows \Omega M_1 \rightrightarrows \Omega M_2 \cdots,$$

whose associated double complex, also denoted by ΩM, should be thought of as the *de Rham complex* of the *simplicial manifold* M. On the other hand, the object M also has a natural space associated to it, called its geometric realization and denoted by $|M|$, and a folk theorem has it that

EXTENDED DE RHAM THEOREM.

$$H^*(|M|; \mathbb{R}) \simeq H^*(\Omega M).$$

provided de Rham is valid on each of the constituent manifolds M_i of M.

Now, this very plausible extension of de Rham, can be immediately applied for instance to the study of $H^*(BG)$, where G is a Lie group and BG its classifying space. Indeed, as is pointed out most succinctly in Segal's paper [11], BG can be taken to be the geometric realization of a semisimplicial manifold

$$NG: * \leftleftarrows G \leftleftarrows G \times G \leftleftarrows G \times G \times G \cdots,$$

which is really the manifold version of the old Eilenberg–McLane construction.

Thus, applying our Folk theorem in this instance yields a double complex for the computation of $H^*(BG)$, which is of the form

$$\Omega NG: * \rightrightarrows \Omega G \rightrightarrows \Omega(G \times G) \rightrightarrows \Omega(G \times G \times G),$$

and so can be interpreted as a *completed version* of the usual Bar construction. (The classical Bar construction, of course, fails for de Rham, because the multiplication $G \times G \to G$ does not induce a coalgebra structure on ΩG.)

In view of the generalized de Rham theorem, one can seek representatives for the characteristic ring of G in the complex ΩNG, and this was first done by Shulman in his thesis, where he extended and reinterpreted a construction given in [14] for GL_q.

In particular, Shulman obtained the following positioning of the real characteristic classes of G in the de Rham complex ΩNG:

THEOREM II (Shulman). *A real characteristic class Φ associated to an invariant polynomial of degree q on the Lie-algebra of G, has a representative in ΩNG that involves only forms of degree $\geqslant q$. Thus*:

$$\Phi = \Phi_{q-1} + \Phi_{q-1} + \cdots + \Phi_0 ,$$

where $\Phi_i \in \Omega^{q+i} N_{q+i} G$.

This result also follows from the work of Kamber and Tondeur [6], and can also be deduced from a theorem of Bott–Hochshild [2], which computes the simplicial homology of each $\Omega^q NG$ in terms of the continuous cohomology of G acting on the qth symmetric power of the dual of the Lie algebra $\mathfrak{g}$ of G. The actual result is:

THEOREM (Bott, Hochshild).

$$H_\delta{}^p(\Omega^q NG) \simeq H^{p-q}_{\mathrm{cont}}(G;\, S^q\mathfrak{g}^*).$$

(Here, H_{cont} denotes continuous Eilenberg–McLane cohomology of G.) The earlier statement now follows from the fact that $H^0_{\mathrm{cont}}(G;\, S^q\mathfrak{g}^*) = \mathrm{Inv}_G \cdot (S^q\mathfrak{g}^*)$. Thus, these invariants begin a cocycle Φ_0 in ΩNG, which can be completed by adding correction terms Φ_i, $i > 0$, because there, the δ-cohomology vanishes.

Note that a result corresponding to Theorem II would not hold if Ω is replaced, say, by the integral singular theory. Thus, here, the very special nature of the de Rham theory comes to the fore.

We come now to an account of the vanishing and exotic class phenomena in this setting.

Our first observation is that the construction of ΩNG goes through essentially word for word for the "smooth categories" Γ_q encountered in the theory of foliations à la Haefliger. That is, the classifying space $B\Gamma_q$ for Haefliger structures is again the geometric realization of a semisimplicial *manifold* $N\Gamma_q$ (see Section 2 for details).

$$N\Gamma_q : \mathbb{R}^q \underset{\longleftarrow}{\overset{\longleftarrow}{}} m_1(\Gamma_q) \underset{\longleftarrow}{\overset{\longleftarrow}{\longleftarrow}} m_2(\Gamma_q) \cdots .$$

Thus, one may speak of the de Rham complex $\Omega N\Gamma_q$, and use it to study the behavior of the usual characteristic classes of the normal bundle to a foliation. Indeed, in our frame work, this amounts to studying the map that the functor

$$\Gamma_q \xrightarrow{\nu} GL_q(\mathbb{R}),$$

associating to a germ of a diffeomorphism $\gamma \in m(\Gamma_q)$ its differential, induces in de Rham:

$$\Omega N\Gamma_q \xleftarrow{\Omega(\nu)} \Omega NG, \qquad G = GL(q, \mathbb{R}).$$

Now, however, the striking feature of $N\Gamma_q$ is that, *for every q, the manifolds $m_r(\Gamma_q)$ of $N\Gamma_q$ are q-dimensional.* As a consequence, $\Omega(\nu)$ annihilates all forms of dim $> q$ for trivial reasons, and therefore, by Theorem II, ν^* annihilates all real characteristic classes of dim $> 2q$.

In this way then, we have a very transparent proof of the following

VANISHING THEOREM (Bott). *Let*

$$B\nu: B\Gamma_q \to BGL_q$$

be the map induced by the Jacobian map $\nu: \Gamma_q \to GL_q$. Then, over the reals, $(B\nu)$ induces the zero homomorphism in dim $> 2q$.

Actually, this argument also immediately suggests how to construct potentially new characteristic classes in $B\Gamma_q$. Namely, let $\mathscr{F}_q\Omega NGL_q$ denote the complex of ΩNGL_q consisting of forms of degree $> q$. Then, because dim $N\Gamma_q = q$, $\Omega\nu^*$ annihilates this submodule and so induces a map

$$\bar{\Omega}\nu^*: \Omega NGL_q/\mathscr{F}_q\Omega GL_q \to \Omega N\Gamma_q.$$

Hence, $H^*(\Omega NGL_q/\mathscr{F}_q\Omega GL_q)$ has a natural map to $H^*(B\Gamma_q)$. On the other hand, this cohomology is relatively computable and leads to all the potential exotic classes in $B\Gamma_q$, as determined, say, in [3, or 6].

This essentially surveys the content of this note and in the various sections, we will be concerned mainly with the technicalities that arise when one seeks to carry out the above program.

1. The Generalized de Rham Theorem

A simplicial space

$$X: X_0 \leftleftarrows X_1 \overset{\longleftarrow}{\leftleftarrows} X_2 \cdots \tag{1.1}$$

has several geometric realizations associated to it, and they are discussed and compared in detail by Segal in [12].

On the one hand, one has the fat realization $\| X \|$ of X, obtained from the disjoint union

$$\coprod X_i \times \Delta^i, \qquad \Delta^i \text{ the } i \text{ simplex},$$

by identifying only with respect to the "boundaries" in X. On the other hand, one has the more geometrically intuitive lean realization $| X |$, which also is obtained from $\| X \|$ by identifying with respect to the degeneracies in X. The natural map

$$\| X \| \rightarrow | X |$$

is, in general, not a weak homotopy equivalence, but is one if all the degeneracies in X are cofibrations. Furthermore, it is only the fat realization that behaves relative to maps. More precisely, it has the property:

(*) *If $X_i \rightarrow Y_i$ is a map of simplicial spaces that is a weak homotopy equivalence for each i, then so is the induced map*

$$\| X_i \| \rightarrow \| Y_i \|.$$

This assertion is essentially proved by Segal [12, Appendix]. However, he there states the result for homotopy equivalences, and alas, operates generally in the category of compactly generated Hausdorf spaces. On the other hand, the argument he outlines also goes through in the present context.

We turn now to a singular analog of our Folk theorem.

Theorem (Folk). *Let S_* denote the functor of singular chains, so that S_* applied to the simplicial space X, gives rise to a simplicial chain complex $i \rightarrow S_*(X_i)$, whose associated double complex will be denoted by $S_*(X)$. With this understood, one has a functorial isomorphism*

$$H\{S_*(X)\} = H(S_* \| X \|). \tag{1.2}$$

Remarks. (1) When X is a simplicial set, the formula (1.2) reduces to the well-known fact that the singular theory of the realization is computable combinatorially. (2) The usual spectral sequences of a double complex of course now yield spectral sequences converging to $H(S_* \| X \|)$. In particular, one has an E_2 term of the form

$$E_2^{p,q} = H_p(H_q(X)),$$

(where the internal H is singular and the external one derived from the simplicial structure) which is familiar in the literature. (3) A special instance of (1.2) occurs in the work of Kozul [7, 8] on fiber spaces, and there are hints at (1.2) in many contexts, e.g., Stasheff [15], etc. But we have not been able to find an explicit reference in the literature, and therefore, will outline a proof of (1.2) along lines suggested by Segal. First, consider the singular functors $\mathscr{S}_q$ associating to a space its singular q simplexes. Applying $\mathscr{S}_q$ to the X_i of X, produces a doubly simplicial set $(i, j) \to S_j(X_i)$, which we will denote by $X_{i,j}$. Now, any double simplicial set $Y_{i,j}$ has fat and lean geometric realizations $\| Y_{i,j} \|$ and $| Y_{i,j} |$ obtained from

$$\coprod X_{i,j} \times \Delta^i \times \Delta^j$$

in the evident manner, and as the cofibration conditions are here eminently satisfied, the two realizations have the same weak homotopy type. Furthermore, $H_*(\mathscr{S} \| Y_{i,j} \|)$ can be computed combinatorially, that is, from the free chain complex $\mathscr{F} Y_{i,j}$ generated by the $Y_{i,j}$: Thus

$$H(\mathscr{F} Y_{*,*}) \simeq H(\| Y_{i,j} \|).$$

Note now that in our situation, i.e., $X_{i,j} = \mathscr{S}_j(X_i)$, the left-hand side is precisely $H\{S_*(X)\}$. Hence, it suffices to show that $\| X_{i,j} \|$ has the same weak homotopy type as $\| X \|$. Now, the realization of $\| X_{i,j} \|$ can be factored through realizing in the vertical direction (i.e., along j, and then in the horizontal direction.) On the other hand, the vertical realization $\| X_{i,j} \|_j$ creates the fat geometric realization of the singular complex of X_i. Thus, we have a natural map

$$\| X_{i,j} \|_j \to X_i ,$$

which is a weak homotopy equivalence for each i.

Hence, the induced map in the fat realization

$$\| X_{i,j} \| = \| \, \| X_{i,j} \|_j \|_i \to \| X \|$$

is also a weak homotopy equivalence. Q.E.D.

We are now ready to consider the de Rham version of (1.2). For this purpose, recall first of all that the de Rham complex ΩM of a smooth manifold is mapped into the singular $\mathbb{R}$-cochains $S^*(M)$ on M, only via the subcomplex $\tilde{S}_*(M)$, generated by the smooth singular simplexes in $S_*(M)$. Precisely, one has the natural inclusion:

$$i_* : \tilde{S}_*(M) \hookrightarrow S_*(M),$$

which by a smoothing argument is shown to be a chain equivalence, so that the obvious map

$$\Omega^* M \to \tilde{S}^*(M)$$

given by integration, composed with $(i_*)^{-1}$, yields the de Rham homomorphism

$$H^*(\Omega^* M) \to H^*(SM) = H^*(M).$$

Now, the classical smoothing argument for the chain equivalence of i_* is by and large only carried out for Hausdorf manifolds in the literature (see [17] for instance) and for certain models of the Haefliger classifying spaces it would be convenient to have it in general.

One technique for extending α to the non-Hausdorf situation is to use the strong excision property of the singular theory, which allows one to compute $H\{S_*(M)\}$ by means of the subcomplex $H\{S_*^{\mathcal{U}}(M)$ generated $\mathcal{U}$-small simplexes, with $\mathcal{U}$ any open cover of M. On the other hand, M is *locally Hausdorf* so that $\tilde{S}\mathcal{U}(M) \subset S^{\mathcal{U}}(M)$ is a chain equivalence by the classical argument for a cover consisting of Hausdorf sets. In detail, this argument takes the following form:

Let $\mathcal{U} = \{\mathcal{U}_\alpha\}_{\alpha \in J}$ be an open cover of a space X, which for simplicity we assume indexed by an order index set J.

This situation naturally defines a *simplicial space* $X_{\mathcal{U}}$, whose ith components consist of the disjoint union of the $(i+1)$-fold intersections of the $\mathcal{U}_\alpha$. Thus

$$X_{\mathcal{U}} : \coprod \mathcal{U}_\alpha \leftleftarrows \coprod_{\alpha<\beta} \mathcal{U}_\alpha \cap \mathcal{U}_\beta \; \substack{\leftarrow \\ \leftarrow \\ \leftarrow}, \qquad \text{etc.} \tag{1.3}$$

Further, the inclusion $\mathcal{U}_\alpha \subset X_J$ combine to define a natural map ϵ of X_J into the trivial semisimplicial object defined by X (and again denoted by X) which plays the role of an augmentation. In fact

$$|X| \xleftarrow{|\epsilon|} |X_U| \tag{1.4}$$

is a homotopy equivalent if X is paracompact as is pointed out in Segal [11].

Now, let $\mathcal{S}_q$ be the qth singular functor. Then, $\mathcal{S}_q$ applied to (1.3) yields a homotopy equivalence:

$$|\mathcal{S}_q X| \xleftarrow{|\mathcal{S}_q{}'|} |\mathcal{S}_q X_{\mathcal{U}}|.$$

Applying the free-abelian group functor, this translates into the following well-known (Folk) resolution for the $\mathcal{U}$-small-simplexes of X, which is valid without any conditions on the space X:

THE SINGULAR MEYER VIETORIS RESOLUTION. *Let $S_*^{\mathcal{U}}X$ denote the group of singular $\mathcal{U}$-small chains on X. Then, ϵ induces a resolution of $S_*^{\mathcal{U}}$ in the sense that for each q:*

$$0 \longleftarrow S_q^{\mathcal{U}}(X) \xleftarrow{\epsilon^*} S_q(\coprod \mathcal{U}_\alpha) \longleftarrow S_q(\coprod_{\alpha<\beta} \mathcal{U}_\alpha \cap \mathcal{U}_\beta) \longleftarrow \tag{1.5}$$

is an exact sequence.

In short then, ϵ induces chain equivalences of the total complexes

$$S_*^{\mathcal{U}} \xleftarrow[\epsilon_*']{} S_*(X_{\mathcal{U}}),$$

and by the combinatorial nature of the argument, also of the differentiable singular chains when X is a manifold M.

Thus, in that case, one has a natural diagram:

$$\begin{array}{ccccc} S_*(M) & \xleftarrow{i_*} & S_*^{\mathcal{U}}(M) & \xleftarrow{\epsilon} & S_*(M_J) \\ & & \uparrow s_*' & & \uparrow s_* \\ & & \tilde{S}_*^{\mathcal{U}}(M) & \xleftarrow{\tilde{\epsilon}} & \tilde{S}_*(M_J). \end{array} \tag{1.6}$$

Now, s_* induces a chain equivalence if we apply the smoothing to a Hausdorf cover $\{\mathcal{U}_\alpha\}_{\alpha \in J}$ of M. Hence, s_*' is an equivalence, and of course, i_* is one by the excision property of S_*.

At this stage, we have the following straightforward corollary of the Vietoris sequence:

Corollary. *There is a natural de Rham map*

$$H^*(\Omega M) \longrightarrow H^*(M;\mathbb{R}), \tag{1.7}$$

whether M is Hausdorf or not. Further, if $\{\mathcal{U}_\alpha\}_{\alpha\in J}$ *is any open cover of M by paracompact sets, then*

$$H^*(\Omega M_{\mathcal{U}}) \simeq H^*(M;\mathbb{R}). \tag{1.8}$$

In short, the de Rham–Čech construction always gives the singular theory provided only that the cover is by paracompacts.

On the other hand, the map

$$H^*(\Omega M) \xrightarrow{\epsilon^*} H^*(\Omega M_{\mathcal{U}})$$

will in general not be an isomorphism. Of course, if M is paracompact itself, a smooth partition of one can be used in the usual way to prove that ϵ^* is an isomorphism.

If we now combine (1.2) with (1.7), we obtain the thought after

Generalized de Rham Theorem. *For a simplicial manifold*

$$M\colon n \to M_n\,,$$

there is a functorial map of

$$H^*(\Omega M) \to H^*(\|M\|). \tag{1.9}$$

Furthermore, if each M_i *is paracompact, then* (1.9) *is an isomorphism.*

2. The de Rham Theory of $B\Gamma_q$

To apply our de Rham theorem, we have to recall the basic steps in Haefliger's classification theory for foliations. In this theory, which proceeds in close analogy to Lie group Bundle theory, the structure group is replaced by a C^∞ category Γ_q defined as follows:

The objects $o(\Gamma_q)$ consist of $\mathbb{R}^q$ in its usual topology and C^∞ structure.

The morphisms $m(\Gamma_q)$ are the germs of diffeomorphisms of $\mathbb{R}^q$, in the sheaf topology over $\mathbb{R}^q$. Thus, under the source and target maps

$$\mathbb{R}^q \overset{s}{\underset{t}{\leftleftarrows}} m(\Gamma_q), \tag{2.1}$$

$m(\Gamma_q)$ is a sheaf over $\mathbb{R}^q$, and as such, inherits a C^∞ structure of a q-manifold from $\mathbb{R}^q$. The resulting manifold is, of course, highly non-Hausdorf.

Now, the classification theory of Haefliger starts with numerable Γ_q-cocycles $\{\gamma_{\alpha\beta}\}$ over a space X defines equivalences of such cocycles to obtain the notion of a Haefliger structure on X, and then shows that the *concordance classes of such structures* are classified by a classifying space. The most universally applicable model for such a classifying space seems to be the "Milnor construction applied to Γ_q" and in Segal's terminology, this is equivalent to applying our realization functor to the nerve $N\hat{\Gamma}_q$ of the unwinding of Γ_q over $\mathbb{Z}^+$. We recall this terminology: First, for any category $\mathscr{C}$, the nerve of $\mathscr{C}$ is a simplicial object:

$$N\mathscr{C}\colon o(\mathscr{C}) \leftleftarrows m_1(\mathscr{C}) \leftleftarrows\!\!\leftarrow m_2(\mathscr{C}) \leftleftarrows\!\!\leftleftarrows, \tag{2.2}$$

where $o(\mathscr{C})$ are the objects and $m_i(\mathscr{C})$ denotes diagrams

$$\cdot \xleftarrow{\gamma_1} \cdot \xleftarrow{\gamma_2} \cdots \xleftarrow{\gamma_i}$$

of morphisms in $m(\mathscr{C})$, the structure maps being attained by dropping the end arrows and composing adjacent ones. If $\mathscr{C}$ carries a topology (C^∞ structure, etc.) $N(\mathscr{C})$, then $N(\mathscr{C})$ clearly carries a topology (C^∞ structure, etc.). Second, we recall that if $\mathbb{Z}^+$ denotes the category with objects the positive integers and one morphism for every $i \leqslant j$, then the unwinding of $\mathscr{C}$ over $\mathbb{Z}^+$ is defined by Segal to be the subcategory $\hat{\mathscr{C}}$ of the product category $\mathscr{C} \times \mathbb{Z}^+$, obtained by deleting all morphisms $(j, i \leqslant i)$ with j not an identity. The essential virtue of this construction is that whereas in $N\mathscr{C}$, the degeneracies well may not be cofibrations, they always will be in $N\hat{\mathscr{C}}$.

Furthermore, the natural map

$$\hat{\mathscr{C}} \to \mathbb{Z}^+$$

induces a map

$$| N\hat{\mathscr{C}} | \to | N\mathbb{Z}^+ |,$$

and as the right-hand space is the infinite simplex, the star cover of $N\mathbb{Z}^+$ pulls back naturally to $| N\hat{\mathscr{C}} |$, and it is this cover that plays an essential role in the classification theory.

In any case, it is then seen rather easily that $\mathscr{C}$-valued cocyle classes on open numerable covers on a space X and are classified by maps into

$| N\mathscr{C} |$, see [4, 5, 10] for details. In the situation at hand, this classification implies the following

THEOREM (Buffet–Lor). *The concordance classes of Γ_q-structures are classified by the space* $| N\hat{\Gamma}_q |$.

Remarks. (1) Actually the use of $\hat{\Gamma}_q$ rather than Γ_q is only a matter of convenience as tom Dieck [16] has shown recently that for any topological category, $\mathscr{C}$ the unwound lean realization is of the same homotopy type as the fat realization of $\mathscr{C}$:

THEOREM (tom Dieck). *The natural map $\hat{\mathscr{C}} \to \mathscr{C}$ induces a homotopy equivalence*

$$| N\mathscr{C} | \sim \| N\mathscr{C} \|.$$

(2) Quite recently, Segal has suggested other models for the classifying space of foliations, which work well on paracompact spaces. In particular, he shows that the realization of the topological category $\mathscr{E}_q$ described below is such a classifying space, and as will be immediately apparent, $N\mathscr{E}_q$ is definitely a Hausdorff simplicial manifold with paracompact constituents, whose degeneracies are cofibrations.

Segal's result is the following:

THEOREM (Segal). *Let $\mathscr{E}_q$ be the topological category whose objects are all pairs (x, U), whose U is an open subset of $\mathbb{R}^q$ and $x \in U$, and whose morphisms $(x, U) \xrightarrow{f} (x', U')$ are all smooth embeddings $f\colon U \to U'$ such that $f(x) = x'$. Then, the map sending f to its germ at x defines a natural map $| \mathscr{E}_q | \to | N(\Gamma_q)|$, which is a paracompact equivalence.*

In short then, any one of the three spaces $\|N\Gamma_q\|$, $|N\hat{\Gamma}_q|$, $|N\mathcal{E}_q|$ can be considered as a classifying space for foliations, and they all have the following nearly self-evident property.

PROPOSITION. *The simplicial spaces $N\Gamma_q$, $N\hat{\Gamma}_q$, and $N\mathscr{E}_q$ are all q-dimensional.*

To explain this phenomenon, consider the case $N\Gamma_q$, and in particular, of $m_2(\Gamma_q)$. One has the diagram:

$$\begin{array}{ccc} m_2(\Gamma_q) & \longrightarrow & m(\Gamma_q) \\ {\scriptstyle s^{-1}(t)}\downarrow & & \downarrow{\scriptstyle t} \\ m(\Gamma_q) & \xrightarrow{s} & o(\Gamma_q) = \mathbb{R}^q, \end{array}$$

which exhibits $m_2(\Gamma_q)$ as the pull-back of t by s. Both s and t are local homeomorphisms. Hence, $s^{-1}(t)$, and the composition $s \circ s^{-1}(t)$ will also be local homeomorphisms. This shows that $m_2(\Gamma_q)$ is a C^∞ q-manifold.

Finally, because the target of a composition of two diffeomorphisms varies smoothly with the source of the first one, it follows that composition $m_2(\Gamma_q) \to m_1(\Gamma_q)$ is smooth, and also a local homeomorphism. The general case is quite analogous. The argument of course works equally for $\hat{\Gamma}_q$ because $\mathbb{Z}^+$ is a discrete category. In $\mathscr{E}_q$, the remark is even more immediate.

The Vanishing Theorem of the Introduction is now also immediate for any one of our three models. For example, the composition

$$\nu: \hat{\Gamma}_q \to \Gamma_q \to GL_q,$$

of the forgetful functor, together with sending a germ into its Jacobian, induces a map

$$N\nu: N\hat{\Gamma}_q \to NGL_q,$$

of smooth simplicial manifolds, and the induced maps of the de Rham complex

$$(N\nu)^*: \Omega^*NGL_q \to \Omega^*N\hat{\Gamma}_q,$$

clearly annihilates the subcomplex $\mathscr{F}_q\Omega^*NGL_q$ of forms of degree $> q$. The de Rham map, therefore, factors through the quotient to yield the diagram:

$$\begin{array}{ccccc} H(\Omega^*NGL_q) & \xrightarrow{r^*} & H(\Omega^*NGL_q/\mathscr{F}_q\Omega^*NGL_q) & \xrightarrow{\pi^*} & H(\Omega^*N\hat{\Gamma}_q) \\ \downarrow \wr & & & & \downarrow \\ H(|NGL_q|) & & \xrightarrow{\quad |\nu|^* \quad} & & H^*(|N\hat{\Gamma}_q|), \end{array} \tag{2.3}$$

with the vertical maps indicating our generalized de Rham homomorphism. On the left-hand side, this is an isomorphism because we are dealing with paracompact constituents in $N(GL_q)$. (In fact, here in the group case, m_r is simply isomorphic to r copies of the group under consideration.)

From the position of the characteristic ring in Ω^*NGL_q, it is now clear that already, r^* annihilates all characteristic classes of dim $2q$, so that (2.3) is a natural strengthening of the vanishing phenomena with the algebra $H(\Omega^*NGL_q/\mathscr{F}_q\Omega^*NGL_q)$ playing the role of potentially new characteristic classes. This algebra will be computed in a subsequent note to show that it agrees with the exotic classes found by a variety of people, in particular, one finds here all the smooth classes of $|N\hat{\Gamma}_q|$ in the sense of Bott–Haefliger [3].

We close this section with a few additional comments concerning this whole argument.

(1) Clearly, the procedure we outlined here is valid whenever one deals with a category Γ of "Lie-type." Rather than defining these, let us just consider two examples:

EXAMPLE 1. Let $\Gamma = \Gamma_q\mathbb{C} \subset \Gamma_{2q}$ consisting of germs preserving a holomorphic structure.

EXAMPLE 2. Let $\Gamma = \Gamma_q S \subset \Gamma_{2q}$ consist of the germs preserving a simplectic form on $\mathbb{R}^{2q}$.

In these cases, $N\hat{\Gamma}$ will not only be a simplicial manifold, but will be a complex one in Example 1 and a symplectic one in Example 2. We will write G for the image group of the jacobian map ν for Γ. Thus, in our first case, $G = GL_q(\mathbb{C}) \subset GL_{2q}$; in the second, $G = Spl(q) \subset GL_{2q}$. In each of these cases, therefore, the normal map

$$\nu\colon \Gamma \to G$$

can be studied by de Rham methods, and in particular, we get a natural map of

$$H(\Omega^*NG/\mathscr{F}_r\Omega^*NG) \to H^*(|N\hat{\Gamma}|),$$

where r is the dimension of the real representation of Γ involved ($2q$ in both our examples). On the other hand, it will not be true that this procedure always yields the "continuous cohomology of $N\hat{\Gamma}$." For instance, it fails in the symplectic case. However, it is likely that this procedure would work in the limit as $\nu\colon \Gamma \to G$ is replaced by its kth prolongations: $\nu^{(k)}\colon \Gamma \to G^{(k)}$.

REFERENCES

1. R. BOTT, Lectures on characteristic classes and foliations (Notes by Lawrence Conlon), *in* "Lecture Notes in Mathematics," No. 279, pp. 1–94. Springer-Verlag, New York, 1972.
2. R. BOTT, On the Chern–Weil homomorphism and the continuous cohomology of Lie groups, *Advances in Math.* 11 (1973), 289–303.

3. R. Bott and A. Haefliger, On characteristic classes of Γ-foliations, *Bull. Amer. Math. Soc.* **78** (1972), 1039–1044.
4. A. Haefliger, Feuilletages sur les variétés ouvertes, *Topology* **9** (1970), 183–194.
5. A. Haefliger, Homotopy and integrability, *in* "Lecture Notes in Mathematics," No. 197, Springer–Verlag, New York, 1970.
6. F. Kamber and P. Tondeur, Characteristic invariants of foliated bundles, *Manuscripta Math.* **11** (1974), 51–89.
7. J. L. Koszul, Multiplicateurs et classes caractéristiques, *Trans. Amer. Math. Soc.* **89** (1958), 256–266.
8. J. L. Koszul, Espaces fibrés associés et pré-associés, *Nagoya Math. J.* **15** (1959), 155–169.
9. J. P. May, "Simplicial Objects in Algebraic Topology," Van Nostrand, Princeton, 1967.
10. M. Mostow, J. Perchik, and R. Bott, Notes on Gel'fand Fuks cohomology and characteristic classes, Eleventh Holiday Symposium, New Mexico State University, Las Cruces, New Mexico, 1973.
11. G. Segal, Classifying spaces and spectral sequences, *Inst. Hautes Études Sci., Publ. Math.* **34** (1968), 105–112.
12. G. Segal, Categories and cohomology theories, *Topology* **13** (1974), 293–312.
13. G. Segal, Classifying spaces for foliations, to appear.
14. H. Shulman, On characteristic classes, Thesis, Univ. of California, Berkeley, 1972.
15. J. Stasheff, Homotopy associativity of H-spaces II, *Trans. Amer. Math. Soc.* **108** (1963), 293–312.
16. T. tom Dieck, On the homotopy type of classifying spaces, *Manuscripta Math.* **11** (1974), 41–49.
17. F. W. Warner, "Foundations of Differential Manifolds and Lie Groups," Scott Forsman, Glenview, Illinois, 1971.

Société Mathématique de France
Astérisque 32-33 (1976)

On Characteristic Classes in the Framework of Gelfand-Fuks Cohomology

by Raoul Bott

1. Introduction.

It is a great privilege to address this colloquium in honour of H. Cartan. Every mathematician of my generation, be he a topologist or geometer or for that matter an algebraist or analyst, has been profoundly influenced by H. Cartan's research exposition and general point of view. In my own case, my first introduction to topology in 1949, were the lectures of Steenrod on fiberbundles on the one hand, and the Cartan Harvard Lectures of the year before on which K. Reidemeister lectured at the I. A. S., to a small private seminar (I believe only E. Specker and I were the audience), on the other. Of course in Professor Reidemeister's rather rambling and philosophical presentation some of the famed French lucidity was lost and that may account for the fact that, in spite of my great admiration for the French school, I have never been quite able to imitate it in my own work.

Nevertheless, the ideas I learned then (largely thanks to much remedial tutoring by my friend Specker), and later, through the years, from the famed Cartan Seminar, are still the foundation of my knowledge; and I would therefore like, on this occasion, above all else to thank Henri Cartan in the name of my whole generation for the selfless way in which he has always worked to make mathematics, on all levels, accessible and intelligible to all of us. He has truly been our teacher.

For my topic here I have chosen to speak on a subject which has fascinated me for the past three years or so and which is especially appropriate here also, because it complements the work of Chern-Weil and H. Cartan on characteristic classes in a manner which I think would have pleased E. Cartan. I am referring to the work of Gelfand-Fuks on the cohomology of the Lie-algebra LM of vector fields on a C^∞ manifold M.

Recall that these "infinitesimal motions" X on M correspond to homogeneous linear first order C^∞ differential operators on M. In view of the fact that quite generally, the commutator of an operator of order n and one of order m is of order $n + m - 1$, it follows that under the law

$$[X, Y] = X \circ Y - Y \circ X$$

these vector fields form a Lie algebra over $\mathbb{R}$; that is, a vector space over $\mathbb{R}$ endowed with a skew symmetric bilinear multiplication $X, Y \to [X, Y]$, subject to the Jacobi identity

$$\Sigma [X, [Y, Z]] = 0$$

where the sum extends over cyclic permutations of the variables.

Now this Lie algebra LM is of course the basic tool in all aspects of modern differential geometry. However, traditionally the emphasis has been on the structure of LM as a $C^\infty(M)$ - module. Thus, for instance, the famous DeRham complex

$$\Omega(M) = \Sigma\, \Omega^q(M)$$

consists of the complex of $C^\infty(M)$ - alternating $C^\infty(M)$ - multilinear maps form

$$LM \times \cdots \times LM \text{ to } C^\infty(M),$$

and the exterior derivative

$$d : \Omega^q(M) \longrightarrow \Omega^{q+1}(M)$$

is miraculous and unique in so far as it preserves the $C^\infty(M)$ linearity.

Finally, I think most of us also attributed the finite dimensionality of the DeRham cohomology

$$H_{DR}(M) = \text{Ker } d/\text{Im } d$$

to this same $C^\infty(M)$ - linearity.

Now Gelfand and Fuks had the simple but inspired idea to replace this linearity condition by continuity in the C^∞ topology: Precisely then, they define a complex

$$A(M) = \oplus A^q(M) \tag{1.1}$$

where $A^q(M)$ consists of the multi - $\mathbb{R}$ - linear maps

$$\omega : \underbrace{LM \times \cdots \times LM}_{q} \longrightarrow \mathbb{R}$$

which are continuous in the C^∞ topology on LM, endow $A(M)$ with a differential operator d, given by

$$d\omega(x_1, \cdots, x_{q+1}) = \sum_{i<j} (-1)^{i+j+1} \omega([x_i, x_j], x_1, \cdots, \hat{x}_i, \cdots, \hat{x}_j, \cdots, x_{q+1}), \tag{1.2}$$

and ask for the cohomology of the resulting differential complex.

I will write GF(M) for its cohomology

$$GF(M) = HA(M) = \mathrm{Ker}\ d/\mathrm{Im}\ d \quad \text{in } (1.1),$$

but before discussing it any further, a word about (1.1) is in order. This formula is best first understood in the context of Lie groups. Given such an object, that is, a group in the category of C^∞ Manifolds, we may naturally single out the subcomplex

$$\Omega\mathfrak{g} \equiv \mathrm{Inv}_G \cdot \Omega(G)$$

of forms on G invariant under left translation. As the DeRham d commutes with translations, $\Omega\mathfrak{g}$ comes equipped with a differential operator, so that $H(\Omega\mathfrak{g})$ is a well defined object, and we further have a natural map

$$H(\Omega\mathfrak{g}) \longrightarrow H(\Omega G) = H_{DR}(G) \quad .$$

Now here $\mathfrak{g}$ really stands for the Lie algebra of Left invariant vector-fields on G, and the d of $\Omega\mathfrak{g}$ is then easily seen to be given by the formula (1.1) if the $X_i \in \mathfrak{g}$, $\omega \in \Omega^q(\mathfrak{g})$. Of course finally (1.1) defines cohomology for any abstract Lie algebra.

In short then, the Gelfand-Fuks complex deals with "the continuous subcomplex of the cochain complex of the abstract Lie algebra LM".

Now LM is clearly in some sense the Lie algebra of the group Diff M of Diffeomorphisms of M, hence in some sense A(M) bears the same relation to Diff(M) as $\Omega(\mathfrak{g})$ bears to G. If one thinks along these lines the continuity assumption also turns out to be essential and natural. Indeed for a Lie group G, the complex $\Omega(\mathfrak{g})$ can easily be characterized by a universal property: To wit: the element $\omega \in \Omega^a(\mathfrak{g})$

assigns to every C^{∞} map

$$f : M \longrightarrow G ,$$

the element $\omega(f) = f^{*}\omega$ in $\Omega^{q}(M)$, and this function clearly has the following two properties:

a.) Functoriality. For a commuting diagram of C^{∞} maps as below,

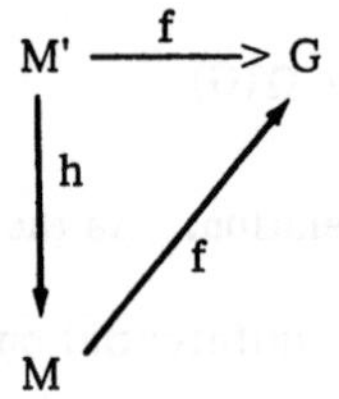

we have $\omega(f') = h^{*}\omega(f)$.

b.) Invariance. Given the diagram

$$M \xrightarrow{\ f\ } G \xrightarrow{\ \ell_g\ } G$$

where ℓ_g denotes left translation by g, then

$$\omega(\ell_g \circ f) = \omega(f) .$$

Conversely given a function $f \mapsto \omega(f)$ subject to a.) and b.) there is a unique $\underline{\omega} \in \Omega(\mathfrak{g})$ such that

$$\omega(f) = f^{*}\underline{\omega}$$

(Indeed just choose $1 : G \longrightarrow G$ to be the identity and set $\underline{\omega} = \omega(1)$.)

Now if we apply this procedure to Diff(M), where we declare a map

$$W \longrightarrow \mathrm{Diff}(M)$$

to be C^∞ if and only if the transpose map

$$W \times M \xrightarrow{f^t} M$$

given by

$$f^t(w, m) = f(w)(m)$$

is C^∞, then one sees without trouble that every such map,

$$f : W \longrightarrow \mathrm{Diff}(M)$$

induces a homomorphism

$$f^! : A(M) \longrightarrow \Omega(W)$$

which is natural and invariant, and as I said before, for this $f^!$ just to exist, the continuity assumption is already essential.

But all this is really a little beside the point - I remark on it as much to show off some familiarity with arrows as to justify the concept - once asked, the problem to compute $H^*\{A(M)\}$ is clearly interesting and in a series of papers Gelfand and Fuks proceeded to investigate it. Among the results they proved the following:

THEOREM (Gelfand-Fuks). The ring $GF^q(M)$ is of finite dimension for each q if M is compact, and in particular, for the circle S^1,

$$GF(S^1) = E(\omega) \otimes \mathbb{R}[y]$$

is the tensor product of an exterior algebra with a 3-dimensional generator ω and a polynomial ring $\mathbb{R}[y]$ with a 2-dimensional generator.

Furthermore, they gave the following explicit representatives for ω and y: Let S^1 be represented as $\mathbb{R}^1/\mathbb{Z}$ so that vector-fields $f\frac{\partial}{\partial x}$ can be identified with functions f of period 1. With this understood

$$\mu(f,g) = \int_0^1 \det \begin{vmatrix} f' & g' \\ f'' & g'' \end{vmatrix} dx$$

$$\omega(f,g,h) = \det \begin{vmatrix} f & g & h \\ f' & g' & h' \\ f'' & g'' & h'' \end{vmatrix}_0$$

where the 0 in the second formula denotes evaluation at 0. Of course, any other point would do just as well, and in fact one could also integrate this expression for ω. On the other hand μ can never be represented by a formula that does not involve the behavior of f and g on all of S^1.

2. The local complex.

I do not have time here to describe how Gelfand and Fuks arrived at their results. Rather let me explore only the first step in their computation, for as it turned out later, it furnishes the addendum to the theory of characteristic classes I alluded to earlier.

The first step in question is based on the remark that $\theta \in A(M)$ has a natural support in M, so that it makes sense to speak of the subcomplex $A_K(M)$ of elements whose support lies in K, and therefore in particular of the minimal subcomplexes $A_p(M)$ with support at a point $p \in M$. To study $A_p(M)$ choose a coordinate system $x_1, \cdots, x_n$ centered at p, so that every $X \in L(M)$ is locally given by

$$X = \sum_i a^i \frac{\partial}{\partial x^i}$$

where the a^i are C^∞ functions near p.

From the most elementary properties of distributions it then follows that all the 1-forms of $A_p(M)$ are generated by the finite partial derivatives of the δ - function applied to the components of X. Hence $A_p(M)$ is really independent of M, and therefore a purely local and universal complex which we will often simply denote by A_p:

$$A_p = A_p(M) \qquad \text{for any } M \ni p \ .$$

Gelfand and Fuks take the position that A_p is again the continuous analogue of $\Omega\mathfrak{g}$. Indeed if $\mathfrak{a}_n$ denotes the Lie algebra of formal vector-fields

$$X = \Sigma a_i \frac{\partial}{\partial x_i}$$

where the a_i are formal power series in the x's, and if we topologise $\mathfrak{a}_n$ in the usual power series topology, - x^α is small if $|\alpha|$ is large - then it is obvious that

$$A_p = \Omega_c(\mathfrak{a}_n),$$

is the complex of continuous alternating forms on $\mathfrak{a}_n$.

Well, in any case, to get on with an explicit computation of $H^* A_p = H^*_c(\mathfrak{a}_n)$ we proceed as follows:

In terms of our local coordinates x^i at p, define 1-forms

$$\theta^i_\alpha \in A^1_p = \Omega^1_c(\mathfrak{a}_n),\ i = 1, \cdots, n,$$
$$\alpha = (\alpha_1, \alpha_2, \cdots, \alpha_k)$$

by:

$$\theta^i_\alpha(X) = (-1)^{|\alpha|} \frac{\partial^{|\alpha|}}{\partial x^\alpha} a^i \Big|_p .$$

Here the α's are k-tuples of non negative integers and $|\alpha| = \Sigma\alpha_i$.

Our earlier statement then translates to the assertion that

(2.1) <u>The θ^i_α span A^1_p and hence generate all of</u> A_p.

Note that of course the θ's are symmetric in their lower indices so that $\theta^i_{12} \equiv \theta^i_{21}$, etc.

The sign has been chosen so that the lower indices correspond to Lie - derivatives of θ^i in the direction $\frac{\partial}{\partial x^{\alpha_1}}, \cdots, \frac{\partial}{\partial x^{\alpha_k}}$. That is, we have

$$\mathcal{L}\left(\frac{\partial}{\partial x^j}\right) \theta^i = \theta^i_j$$

$$\mathcal{L}\left(\frac{\partial}{\partial x^j}\right) \cdot \mathcal{L}\left(\frac{\partial}{\partial x^k}\right) \theta^i = \theta^i_{jk}, \text{ etc,}$$

where $\mathcal{L}$ is still given by (   ) and so commutes with d.

Now because the $\left(\frac{\partial}{\partial x^i}\right)$ commute with one another, the $\mathcal{L}\left(\frac{\partial}{\partial x^i}\right)$ also commute.

Hence the map

$$\frac{\partial}{\partial x^{\alpha_1}} \otimes \cdots \otimes \frac{\partial}{\partial x^{\alpha_n}} \longrightarrow \mathcal{L}\left(\frac{\partial}{\partial x^{\alpha}}\right) \cdots \mathcal{L}\left(\frac{\partial}{\partial x^{\alpha_n}}\right)$$

factors through the symmetric, or polynomial, ring $\mathbb{R}\left[\frac{\partial}{\partial x^i}, \cdots, \frac{\partial}{\partial x^n}\right]$ generated by the $\frac{\partial}{\partial x^i}$ over $\mathbb{R}$, and it follows that under $\mathcal{L}$, the complex A^1_p is naturally an $\mathbb{R}\left[\frac{\partial}{\partial x^i}\right]$ module.

From this point of view (2. 1) can therefore be stated in the form:

Under $\mathcal{L}$ the 1-forms A^1_p constitute a free module with n generators $\theta^1, \cdots, \theta^n$ over the polynomial ring $\mathbb{R}\left[\frac{\partial}{\partial x^i}\right]$:

$$A^1_p \simeq \mathbb{R}\left[\frac{\partial}{\partial x_i}, \cdots, \frac{\partial}{\partial x_n}\right] (\theta^1, \cdots, \theta^2)$$

It follows now that structure equations of A_p are completely determined once they are written down for θ^i, and hence by the following, easily verified,

PROPOSITION 2. 1.

$$d\theta^i + \theta^i_j \wedge \theta^j = 0\,,^\dagger \tag{2. 2}$$

† We use the usual convention that repeated indices are summed.

To see how the formula (2.2) determines all other structure equations, one need only apply $\mathcal{L}\left(\frac{\partial}{\partial x^k}\right)$ etc., to both sides of (2.2) and recall that $\mathcal{L}$ acts by derivations. Explicitly one obtains:

(2.3) $$d\theta^i_k + \theta^i_{jk} \wedge \theta^j + \theta^i_j \wedge \theta^j_k = 0$$

(2.4) $$d\theta^i_{k\ell} + \theta^i_{jk\ell} \wedge \theta^j + \theta^i_{jk} \wedge \theta^j_\ell + \theta^i_{j\ell} \wedge \theta^j_k + \theta^i_j \wedge \theta^j_{k\ell} = 0$$

in the first two cases.

Clearly now (2.3) bears a considerable similarity to the structure equations of $GL(n, \mathbb{R})$ and to bring this out further let us define the 2-forms $R^i_j \in A^2_p$ by

(2.5) $$R^i_j \equiv d\theta^i_j + \theta^i_k \wedge \theta^k_j \ .$$

Then (2.3) takes the form

(2.6) $$R^i_j \equiv - \theta^i_{jk} \wedge \theta^k$$

while the structure equations of $GL(n, \mathbb{R})$ of course were $R^i_j \equiv 0$. In fact these forms behave just as the curvature tensor of a "torsion free connection" does, and as we will see later the R^i_j do play a role of the <u>universal</u> or <u>formal</u> curvature forms of a manifold. Here let us just as an exercise in the calculus of $A_p(M)$ derive the classical curvature identities:

PROPOSITION 2.2. <u>The forms $R^i_j \in A^2_p$ have the following properties:</u>

(2.7) $$R^i_j \wedge \theta^j = 0 \ , \ dR^i_j = R^i_k \wedge \theta^k_j - \theta^i_k \wedge R^k_j$$

<u>which correspond to the two Bianchi Identities in Torsion free Geometry.</u>

Proof: By (2.6)

$$R^i_j \wedge \theta^j = - \theta^i_{jk} \wedge \theta^k \wedge \theta^j \equiv 0$$

because θ^i_{jk} is symmetric in its lower indices.

To obtain the second identity consider the equation (2.4) and multiply it with θ^ℓ summing over the repeated index ℓ. On obtains

$$(2.8) \qquad d\theta^i_{k\ell} \wedge \theta^\ell + \theta^i_{jk} \wedge \theta^j_\ell \wedge \theta^\ell + \theta^i_{j\ell} \wedge \theta^j_k \wedge \theta^\ell + \theta^i_j \wedge \theta^j_{k\ell} \wedge \theta^\ell = 0$$

where the second form of (2.4) has cancelled out by the symmetry of $\theta_{jk\ell}$ in j, k, and ℓ. But (2.8) simply restates (2.7) in the equivalent form

$$- dR^i_k + \theta^i_j \wedge R^i_k - R^i_j \wedge \theta^j_k = 0 \ .$$

On the cohomological side (2.6) and (2.7) imply that the subalgebra

$$\underline{W} = \left\{\theta^i_j \, , \, R^i_j\right\}$$

generated by the indicated forms in A_p is, first of all, finite dimensional and secondly closed under d. Indeed by (2.6) any monomial in the R's of length > n vanishes, while (2.7) together with (2.5) expresses d of the θ's and R's in terms of the θ's and R's. Actually this subcomplex already carries all the cohomology of A_p.

THEOREM: (Gelfand-Fuks) The inclusion

$$\underline{W} \subset A_p(U)$$

induces an isomorphism in cohomology.

The proof of this theorem is nontrivial, and involves some nice elementary invariance theory, to eliminate the forms θ^i_α with $|\alpha| > 2$; but here I have time only to describe the final outcome.

For this purpose let us define the ith Chern polynomial as a function on $GL(n, \mathbb{R})$ by the formula:

$$\Sigma t^i c_i(A) = \det(1 + tA) . \tag{2.9}$$

It is then clear that each c_i is a polynomial in the entries A^i_j of A, for example

$$\begin{cases} c_1(A) = A^i_i \\ c_2(A) = (-A^i_j A^j_i + (A^i_i)^2)/2 \\ c_n(A) = \det A \end{cases} \tag{2.10}$$

Now as the R^i_j are two-forms and so commute we can evaluate these polynomials c_i on the R^i_j's to obtain forms

$$c_i(R) \in A^{2i}_p . \tag{2.11}$$

We call these the Chern classes in A_p and in terms of them the Pontrjagin forms $p_i(R) \in A^{4i}_p$ are defined simply by

$$p_i(R) = c_{2i}(R) . \tag{2.12}$$

Of course in the literature one often encounters these forms with factors of $(-1)^i\left(\frac{\sqrt{-1}}{2\pi}\right)^i$, but this is not to the point here, where they are, for the moment, only elements of $\underline{W} \subset A^{4i}_p$. In any case in terms of these $c_i(R)$ one can reduce $\underline{W}$ further to a complex WU_n defined as follows.

Consider the polynomial ring $\mathbb{R}[c_1, \cdots, c_n]$ with (weight of c_i) = 2i , and let $\underline{\mathbb{R}}[c_1, \cdots, c_n]$ be the quotient of $\mathbb{R}[c_1, \cdots, c_n]$ by the ideal of elements with weight $> 2n$. Here of course the weight w of a monomial $c_1^{\alpha_1} \cdots c_n^{\alpha_n}$ in $\alpha_1 \cdot 1 + \alpha_2 \cdot 2 + \cdots + \alpha_n \cdot n$.

With this understood one sets

$$WU_n = \underline{\mathbb{R}}[c_1, \cdots, c_n] \otimes E(h_1, \cdots, h_n)$$

where $E(h_1, \cdots, h_n)$ denotes an exterior algebra with generators h_i in dimensions $2i - 1$. Finally a differential operator d is introduced in WU_n by setting

(2.13)
$$d(c_i \otimes 1) = 0$$
$$d(h_i \otimes 1) = 1 \otimes c_i \ .$$

The cohomology of A_p is now made rather explicit by:

PROPOSITION 2.3. <u>The map $1 \otimes c_i \to c_i(R)$ has an extension to a map of complexes</u>

$$WU_n \longrightarrow \underline{W} \subset A_p$$

<u>and any such extension induces an isomorphism</u>

(2.14)
$$H^*(WU_n) \xrightarrow{\simeq} H^*(A_p) \simeq H^*(\mathfrak{a}_n)$$

As an example, consider the case of $n = 1$. Then $WU_1 \subset \underline{W}$ can be taken to be

$$c_1 = R_1^1 , \ h_1 = \theta_1^1 ,$$

for we clearly have $dh_1 = c_1$. In this case $H(WU_1)$ is generated by 1 , and

$c_1 h_1$ in dim 3. Thus:

(2.15) $$H^* A_p(\mathbb{R}^1) = \begin{cases} \mathbb{R} \text{ in dim } 0 \\ \mathbb{R} \text{ in dim } 3 \end{cases}.$$

The computations of course become more complicated in higher dimensional cases but were carried out by Vey [11] who produced an explicit base for $H^* A_p(\mathbb{R}^n)$ in terms of the monomials in the h's and c's of WU_n.

So much then for our brief excursion into the A_p. We have learned first of all, that this "subcomplex of minimal support" has structure equations which are easy to write down, and which are very analogous to the equations one meets in differential geometry. Secondly we have seen that its cohomology is finite dimensional and computable.

Now in the computation of $H^* A(M)$, the cohomology of $A_p(M)$ plays a fundamental role, which I will have not time to explore here. Let me just give those of you familiar with some topology the barest outlines of the final outcome.

Already Gelfand-Fuks observed that every manifold M carries a bundle EM, which is associated to its principal frame-bundle and whose fibers $E_p M$, have cohomology isomorphic to that of $A_p(M)$:

$$H^*(E_p M) \simeq H^* A_p(M).$$

Thus E is in a sense a geometric realization of the family of complexes $A_p(M)$, $p \in M$, and in terms of E, the final result on $H^* AM$ is given by the following theorem which was conjectured independently two years ago by Fuks and myself and for which two independent proofs have recently been given by Haefliger and Segal.

THEOREM (Haefliger, Segal). Let ΓE denote the space of continuous sections of E. Then

(2.16) $$H^*(AM) \simeq H^*(\Gamma E) .$$

This beautiful result is, of course, nevertheless a negative one, as it reduces the mysterious $H^*(AM)$ to the complicated but still rather well understood $H^*(\Gamma E)$.

Observe that (2.16) implies, in particular, that

$$H^*(A\mathbb{R}^n) = H^*(\Gamma E) = H^*(E_p) = H^*(A_p\mathbb{R}^n)$$

the second equality being true because E is contractible to E_p over p. Finally let us check (2.16) for the circle, where Gelfand and Fuks made their computations. In this case E_p is the three sphere S^3, which conforms with (2.15). Further as the tangent bundle of S^1 is trivial, we have

$$E = S^3 \times S^1$$

whence $\Gamma E = \mathrm{Map}(S^1, S^3)$, the space of maps from S^1 to S^3. Now this function space has well known homology, to wit:

(2.17) $$H^* \mathrm{Map}(S^1, S^3) = \mathbb{R}[y] \otimes E(\omega)$$

is a polynomial ring with generator μ in dimension 2, tensored with an exterior algebra with generator ω in dimension 3.

3. On the Relations with Characteristic classes.

I have already remarked on the fact that the basic structure equations of $A_p(M)$ bear a great similarity to the structure equations of torsion free geometry, and I would now like to explain this déjà vu phenomenon.

For this purpose let us think of M as the homogeneous space

$$M = \mathrm{Diff}(M) / \mathrm{Diff}(M;p)$$

where Diff(M;p) denotes the subgroup keeping p fixed. Now the group Diff(M;p) has a natural filtration by the subgroups

$$\mathrm{Diff}(M;p) \supset \mathrm{Diff}(M;p^2) \supset \mathrm{Diff}(M;p^3) \supset \cdots$$

keeping p fixed to higher and higher order. Thus

$$\mathrm{Diff}(M;p^k)$$

here denotes those diffeomorphisms whose Taylor expansion at p differ from the identity by terms of order k or higher. Dividing successively by these subgroups and setting

$$J_k(M) = \mathrm{Diff}(M) / \mathrm{Diff}(M;p^{k+1})$$

one obtains a tower of finite dimensional manifolds over M

$$M \longleftarrow J_1(M) \longleftarrow J_2(M) \longleftarrow . \tag{3.1}$$

on all of which Diff(M) acts naturally, and it therefore makes sense to speak of the Diff(M) invariant forms on $J_k(M)$.

The manifolds $J_k(M)$ are of course the higher frame bundles or jet bundles over M, and well known to all geometers, and I claim on these spaces the complex $A_p(M)$ can be naturally identified with the complex of $\mathrm{Diff}(M)$ invariant forms.

Precisely, one has a natural isomorphism :

$$A_p(M) \simeq \lim_{n\to\infty} \mathrm{Inv}_{\mathrm{Diff}(M)} \; \Omega J_n(M) \; . \tag{3.2}$$

From this point of view, our forms θ^i, θ^i_j, θ^i_{jk}, etc., now appear successively in $J_1(M)$, $J_2(M)$, etc., and although the $d\theta^i$ are of course invariant already in $J_1(M)$, they are not decomposable there. Thus the formula

$$d\theta^i + \theta^i_j \wedge \theta^j = 0 \tag{3.3}$$

only becomes valid in $J_2(M)$, etc.

In short then, the complex $A_p(M)$ provides one with a complete description of the "Tautological" forms on the $J_k(M)$. These of course were known to the geometers all along since E. Cartan I expect, and now in retrospect I recall having learned about them from Kobayashi years ago at the Tucson seminar on Diff. Geometry [8]. However he never pursued the cohomology of this complex, and its significance in the framework of characteristic classes.

Before I explain that significance note that (3. 2) corresponds to the well known formula for the invariant forms on a homogeneous space G/H of a Lie group

$$\mathrm{Inv}_G \, \Omega(G/H) \simeq \mathrm{Inv}_H \, \Omega(\mathfrak{g}/\mathfrak{h}) \; , \tag{3.4}$$

in our infinite dimensional case. Indeed in this situation, the Lie algebra of $\mathrm{Diff}(M; p^k)$

clearly corresponds to the vector fields vanishing to order k at p, so that finally as $k \to \infty$,

(3.5)
$$\mathrm{Inv}_G\, \Omega J_\infty M = \Omega \alpha_n .$$

In any case the isomorphism (3.2) is natural and therefore immediately furnishes us with some sort of characteristic ring in $\lim_k H^*\Omega(J_k M)$ which I simply denote by $H^*(J_\infty M)$.

Now observe that in our tower $J_1(M)$, $J_2(M)$, etc., are all of the same homotopy type. Indeed the fibers of all these projections are all contractible groups. Hence our natural map

(3.6)
$$\begin{array}{ccc} H^*(\alpha_n) & \longrightarrow & H^*(J_\infty M) \\ & & \wr\wr \\ & & H^*(J_1 M) \end{array}$$

is not quite as far removed from where we want it - that is in $H^*(M)$ - as might at first appear.

In fact at this stage only the following small modification is needed to come up with genuine characteristic classes.

Let us choose coordinates $x^1, x^2, \cdots, x^n$ centered at p, and in terms of these define the groups

$$\mathrm{Diff}_0(M; p^k)$$

as the subgroups of those $f \in \mathrm{Diff}(M)$ whose jet at p takes the form

$$j_p(f)\, x^i = a^i_j x^j + \text{terms of order} \geq (k+1)$$

with $\| a^i_j \|$ in the group $O(n)$.

Then for $k \geq 2$, we have the exact sequence

$$0 \longrightarrow \mathrm{Diff}(M; p^k) \longrightarrow \mathrm{Diff}_0(M; p^k) \longrightarrow O(n) \longrightarrow 0$$

which gives rise to a corresponding fibering

$$J_k(M) \xrightarrow{\pi} J_k(M)/O(n)$$

where the right hand term is $\mathrm{Diff}(M)/\mathrm{Diff}_0(M; p^k)$.

This construction now induces a natural arrow

$$H^*_c(\mathfrak{a}_n; O_n) \longrightarrow H^*\{J_\infty M/O(n)\} \tag{3.7}$$

where the left hand side denotes the usual relative Lie algebra cohomology with O_n the subalgebra of formal orthogonal vector fields, and the R. H. S. is short for

$$\lim_{k\to\infty} H^*\Omega\{J_k M/O(n)\} \quad .$$

On the other hand $J_k(M)/O(n)$ has the homotopy type of M for every k, so that (3.7) real is really an arrow

$$H^*_c(\mathfrak{a}_n; O_n) \longrightarrow H^*(M) \quad . \tag{3.8}$$

Thus from this point of view the natural characteristic classes of an n - manifold, are given by $H^*_c(\mathfrak{a}_n; O_n)$.

The computations of Gelfand-Fuks can now be modified without too much trouble and one then finds the following result (see [3]).

<u>Proposition.</u> $H^*(\mathfrak{a}_n, O_n) = H^*(WO_n)$ <u>where</u> WO_n <u>is the complex</u>

$$E(h_1, h_3, \cdots) \otimes \underline{\mathbb{R}}[c_1, \cdots, c_n]$$

with

$$d(1 \otimes h_1) = 1 \otimes c_1$$

$$d(1 \otimes h_3) = 1 \otimes c_3, \text{ etc.}$$

In particular in $\dim \leq n$, $H^*(\mathfrak{a}_n, O_n)$ is isomorphic to the ring generated by the c_{2i}, $4i \leq n$, and, up to a constant factor, these of course correspond to the well known Pontrjagin classes of M

$$c_{2i}(M) = p_i(M) .$$

On the other hand, just as in WU_n there are other classes in $H(\mathfrak{a}_n, O_n)$ of much higher dimension, which therefore seem spurious in the present context.

Nevertheless, as I will explain in a moment, these phantom classes come into their own once we extend our point of view from manifolds to Foliations.

But first a final remark on how and why the formulas (3.3) were reminiscent of curvature formulas; which will also explain why the c_{2i} represent the usual Pontrjagin classes - say of a Riemann structure on M. The point is, of course, that a Riemann structure on M corresponds to a section s of $J_1(M)/O_n$ over M, and the corresponding Levi-Civita connection then defines an O_n - equivariant section

$$J_1(M) \xrightarrow{\nabla} J_2(M) ,$$

under which the R^i_j pull back to the curvature of the Riemann structure which the

$c_{2i}(R^i_j)$, being $O(n)$ basic, correspond to the Pontrjagin classes on M. Once this relation is made, one can link up the computation of $H^*(\mathfrak{a}_n; O_n)$ with the usual machinery of connections and curvature, as well as the Weil algebra of $g\ell(n)$. In fact, the $\underline{W}$ of the previous section is clearly a truncated form of that Weil algebra. All in all the last proposition therefore is a refined version of the classical results.

4. The Characteristic classes of Foliations.

Let me briefly expalin now how the higher classes of WO_n serve as potential characteristic classes of foliations. Roughly speaking the idea is that one may think of a manifold M, as a special case of a foliation - i.e., we think of M as foliated by points.

Thus extended the constructions of the last section are seen to define a natural arrow

(4.1) $$H^*(\alpha_n, O(n)) \xrightarrow{\varphi(\mathfrak{F})} H^*\{M(\mathfrak{F})\}$$

to the cohomology of the manifold on which the foliation $\mathfrak{F}$ of codimension n is defined. As now dim (M) is unrestricted, the total cohomology of the L.H.S. comes into play.

To see how such an arrow comes about, recall that a foliation $\mathfrak{F}$ on M of codimension n can be defined as a subbundle $E_{\mathfrak{F}}$ of the tangent bundle of M which is locally the kernel of submersions of M to $\mathbb{R}^n$:

Thus near each $p \in M$,

(4.2) $$E \simeq \ker d f$$

where $f : V \longrightarrow \mathbb{R}^n$, $p \in V$, is a smooth map whose differential is onto at each point. With this understood define $J_k(\mathfrak{F})$ to be the space of k-jets of submersions f, from M to $\mathbb{R}^n$, with target $0 \in \mathbb{R}^n$, and which are compatible with $\mathfrak{F}$ in the sense that $E_{\mathfrak{F}}$ is Ker df.

In the case of the trivial foliation, $(E = 0)$, $J_k(\mathfrak{F})$ is then easily seen to be a local description of our old $J_k(M)$ and in this framework it is not difficult to generalize (3.6) to (4.1). Furthermore under $\varphi(\mathfrak{F})$ the Pontrjagin classes of $H^*(\mathfrak{a}_n, O_n)$ go over into the Pontrjagin classes of the normal bundle TM/E of $\mathfrak{F}$. Thus in particular the truncation of $\underline{W}$ therefore implies the vanishing condition for the characteristic classes of such a normal bundle which I noted some years ago. (At the same time (4.1) also brings us the secondary classes, occasioned by this vanishing phenomenon.)

Thus for $n = 1$, the class $h_1c_1 \in H^3(\mathfrak{a}_n; O_n)$ corresponds to the Godbillon, Vey class, which of course started this whole development. The general setting (4.1) for its generalization was noted later by many of us: Bernstein,Rosenfeld in the U.S.S.R., and by Malgrange, and Bott-Haefliger in the West. Of course there are also other points of view which explain these phenomena. In particular the work of Chern-Simons, Cheeger deals with secondary classes quite generally, and Kamber and Tondeur have their own approach to all manner of foliated bundles and their characteristic classes.

Unfortunately however, the most difficult question concerning these classes remains unsolved. We do not know whether all the classes of $H^*(\mathfrak{a}_n; O_n)$ can be distinguished by actual foliations.

Now the master par excellence for constructing foliations is Thurston, but even he has only been able to detect a few of them. On the other hand of course he has detected these with a vengeance. Indeed, he can forge a foliation $\mathfrak{F}$ on which the class $h_1c_1{}^n \in H^{2n+1}(\mathfrak{a}_n, O_n)$ takes on any preassigned real value!

But this really leads us to another topic, beyond the scope of these remarks; all I wanted to do here, apart from saluting Henri Cartan, is to show how a natural question, such as Gelfand-Fuks asked themselves, has borne fruit in many interesting directions and has taught us to rethink old phenomena in new ways.

Bibliography

I have not attempted a complete bibliography here, but the following papers should serve as an introduction into most phases of the subject.

Bott, R.

[1] 1971 On a topological obstruction to integrability. Proc. International Congress Math. (Nice, 1970) Vol. 1, 27-36. Gauthier-Villars, Paris.

[2] 1972 Lectures on characteristic classes and foliations (Notes by Lawrence Conlon). Lecture Notes in Mathematics #279, 1-94. Springer-Verlag, New York.

Bott, R. and Haefliger, A.

[3] 1972 On characteristic classes of Γ-foliations. Bull. AMS, 78, 1039-1044.

Cartan, H.

[4] 1950 Notions d'algèbre différentielle. La transgression dans un groupe de Lie. Colloque de Topologie, Bruxelles, 15-27, 57-71.

Chern, S. and Simons, J.

[5] 1971 Some cohomology classes in principal fiber bundles and their applications to Riemannian geometry. Proc. Nat. Acad. Sci., U. S. A., 68, 791-794.

Gelfand, I., and Fuks, D.

[6] 1968 Cohomologies of the Lie algebra of vector fieldson a circle. Funct. Anal., 2, 342-3.

[7] 1969 Cohomologies of the Lie algebra of vector fields on a manifold. Funct. Anal., 3, 155.

[8] 1969 The cohomology of the Lie algebra of tangent vector fields on a smooth manifold, I. J. Functional Anal., 3, 194-210.

[9] 1970 The cohomology of the Lie algebra of tangent vector fields on a smooth manifold, II. J. Functional Anal., 4, 110-116.

[10] 1970 The cohomology of the Lie algebra of formal vector fields. Izv. Ann. SSSR, 34, 327-342.

Godbillon, C.

[11] 1972 Cohomologie d'algèbres de Lie de champs de vecteurs formels. Sém. Bourbaki, 421-01 to 421-19.

Godbillon, C., and Vey, J.

[12] 1971 Un invariant des feuilletages de codimension un. C. R. Acad. Sci., Paris, 273, 92.

Guillemin, V.

[13] 1973 Cohomology of vector fields on a manifold. Advances in Math., 10, 192-220.

Haefliger, A.

[14] 1970 Feuilletages sur les variétés ouvertes. Topology, 9, 183-194.

[15] 1971 Homotopy and integrability. Lecture Notes in Mathematics #197, 133-163, Springer-Verlag, New York.

[16] 1972 Sur les classes caractéristiques des feuilletages. Sém. Bourbaki, 412-01 to 412-21.

Kamber, F., and Tondeur, P.

[17] Foliated Bundles and characteristic classes. Springer notes to be published.

Kobayashi, S.

[18] 1961 Canonical forms on frame bundles of higher order contact. Diff. Geometry, Proc. Symp. III, AMS, Tuscon.

Thurston, W.

[19] 1972 Noncobordant foliations of S^3. Bull. Am. Math. Soc., 78, 511-514.

[20] 1973 The theory of foliations of codimension greater than one. Proc. of Stanford U. Conf. on Diff. Geom. 1973.

[21] 1974 Foliations and groups of diffeomorphisms. Bull. AMS, 80, 304-312.

Raoul BOTT
Harvard University
CAMBRIDGE, Mass. 02138, USA

Lecture Notes in Mathematics

Edited by A. Dold and B. Eckmann

652

Differential Topology, Foliations and Gelfand-Fuks Cohomology

Proceedings of the Symposium held at the Pontifíca Universidade Católica do Rio de Janeiro, 5–24 January, 1976

Edited by Paul A. Schweitzer, s.j.

Springer-Verlag
Berlin Heidelberg New York 1978

On Some Formulas for the Characteristic Classes of Group-actions

Raoul Bott[†]

1. Introduction. This is an account only of the material of my last two lectures of the Rio conference, as the earlier lectures dealt with well known matters.

My emphasis here then is the study of our foliation invariants in the context of group actions on a manifold, and I will start by showing you the naturality of our basic construction now pays off by extending directly to the equivariant situation.

Recall then the main conclusions of our earlier discussions. Essentially they amounted to this (see also [1],[2] for details):

If we let JM denote the space of jets - based at $0 \in \mathbb{R}^n$ - of diffeomorphisms of $\mathbb{R}^n$ to M, then JM carries an algebra of natural forms which is isomorphic to the cochain algebra $C^*(\mathfrak{a}_n)$ where $\mathfrak{a}_n$ is the Lie algebra of formal vector-fields on $\mathbb{R}^n$.

In short there is a natural arrow:

(1.1) $$C^*(\mathfrak{a}_n) \longrightarrow \operatorname{Inv}_{\operatorname{Diff} M} \Omega^* JM$$

of $C^*(\mathfrak{a}_n)$ onto the algebra of Diff(M) - invariant forms on JM. Further we saw that although (1.1) induces the 0 - homomorphism in cohomology, it induces an interesting one once we divide JM by the natural action of O_n = the orthogonal group of $\mathbb{R}^n$.

Indeed then (1.1) induces a "basic" homomorphism:

(1.2) $$C^*(\mathfrak{a}_n; O_n) \longrightarrow \operatorname{Inv}_{\operatorname{Diff} M} \Omega^*(JM/O_n)$$

and this map certainly recaptures all the usual characteristic classes of M.

[†]The author gratefully acknowledges with thanks partial support from the NSF under Grant MPS 74-11896.

More precisely we saw that:

(1.3) $$H^*(\mathfrak{o}_n, O_n) \simeq H^*(WO_n)$$

where WO_n is the differential algebra given by:

(1.4) $$WO_n = \underline{\mathbb{R}}[c_1, \cdots, c_n] \otimes E(h_1, h_3, \cdots h_k) \quad k \text{ odd and } = n \text{ or } n-1$$

with differential:

$$d\, c_i \otimes 1 = 0 \quad , \quad d(1 \otimes h_i) = c_i$$

and $\underline{\mathbb{R}}$ denoting the quotient of the indicated polynomial ring by the elements of $\dim > 2n$, the c_i having dimension $2i$.

Further we saw - and this is of course quite standard - that

$$H^*(\Omega^*(JM/O_n)) \simeq H^*(M)$$

so that (1.2) induces a natural map

$$H^*(\mathfrak{o}_n, O_n) \longrightarrow H^*(M) \quad ,$$

and finally we identified the image of c_{2i} under this arrow with the Pontryagin classes of M.

The main virtue of this point of view is then, that we see that a manifold determines its own Pontryagin forms naturally on the space

$$\widetilde{M} = JM/O_n \quad ,$$

which I think of as a naturally thickened version of M, and that furthermore these forms are invariant under any diffeomorphism of M.

Now let us put this construction to use when an abstract (i.e., discrete) group Γ, acts on M via diffeomorphism. These data naturally define two semisimplicial manifolds

$$M\Gamma : M \leftleftarrows \Gamma \times M \, \overleftarrow{\overleftarrow{\leftarrow}} \, \Gamma \times \Gamma \times M \tag{1.5}$$

and

$$B\Gamma : * \leftleftarrows \Gamma \, \overleftarrow{\overleftarrow{\leftarrow}} \, \Gamma \times \Gamma \tag{1.6}$$

whose geometric realizations then respectively correspond to, the M - bundle, $|M\Gamma|$, induced by the Γ action over the classifying space $|B\Gamma|$ of Γ, and $|B\Gamma|$ itself.

Now if

$$X : X_0 \leftleftarrows X_1 \, \overleftarrow{\overleftarrow{\leftarrow}} \, X_2 \cdots \tag{1.7}$$

is any semisimplicial manifold the cohomology of its geometric realization $|X|$ can be computed in various ways. First of all the double complex $\Omega^{**}X$

$$\Omega^{**}X : \Omega^*(X_0) \xrightarrow{\delta} \Omega^*(X_1) \xrightarrow{\delta} \cdots \tag{1.8}$$

obtained from (1.7) by applying the de Rham functor Ω^* to each (X_i) and then giving $\Omega^{**}X$ the sum of the de Rham differential and the differential operator δ derived from the simplicial structure, computes $H^*(|X|)$ thus:

$$H^*(|X|) = H\{\Omega^{**}(X)\} \tag{1.9}$$

On the other hand one also has the complex of "compatible forms" on the realization of X:

$$|X| = X_0 \cup X_1 \times \Delta_1 \cup X_2 \times \Delta_2 \cup \cdots \tag{1.10}$$

in the sense of Whitney-Thom-Sullivan. This complex is denoted by $\Omega^*|X|$, and once again one has:

(1.11) $$H^*(|X|) = H^*(\Omega^*|X|) .$$

For a fine account of all this I refer you to Dupont, [5].
The formula (1.9) is also to be found in [4].

In particular then, we can compute the cohomology of $M\Gamma$ from the double complex

(1.12) $$\Omega^{**}M\Gamma \ : \ \Omega^*(M) \xrightarrow{\delta} \Omega^*(M \times \Gamma) \longrightarrow$$

Furthermore as we will see later, on the first constitutent δ takes the form

(1.13) $$\delta\omega \mid M \times j = \omega \otimes 1 - j^*\omega \otimes 1 .$$

Hence the natural linear map

(1.14) $$\Omega^* M \longrightarrow \Omega^{**}M\Gamma$$

<u>becomes a cochain map only on the Γ - invariant forms on M.</u>

In view of this state of affairs it suggests itself to replace $M\Gamma$ by $\widetilde{M\Gamma}$.

Then the arrows (1.2) and (1.14) combine to yield a chain map

(1.15) $$C^*(a_n, O_n) \longrightarrow \Omega^*(\widetilde{M}) \longrightarrow \Omega^{**}\widetilde{M\Gamma} .$$

On the other hand the homotopy equivalence of M with $\widetilde{M}$ easily implies that of $M\Gamma$ and $\widetilde{M\Gamma}$. Hence in homology (1.15) induces a homomorphism

(1.16) $$H^*(a_n, O_n) \longrightarrow H^*(|M\Gamma|) .$$

Thus we see that the invariance of our construction immediately yields "equivariant characteristic classes" for a Γ - manifold in $|M\Gamma|$. And here of course, as the cohomology of $M\Gamma$ has no a priori bound, all the classes of WO_n potentially come into play.

Finally if M is a compact Γ - orientable manifold we can define the equivariant characteristic numbers of M by following (1. 16) with integration over the fiber in the fibering

$$\begin{array}{c} M\Gamma \\ \downarrow \pi \\ B\Gamma \end{array} .$$

There results an additive homomorphism

$$H^*(\mathfrak{a}_n;O_n) \longrightarrow H^*(B\Gamma) \quad , \tag{1. 17}$$

and my aim in the next sections will be to derive some explicit recipes for (1. 17) , and to review some of Thurston's examples in this framework.

2. <u>Formulas for the Godbillon-Vey Class in $H^*(B\Gamma)$</u> . Note that (1. 16) can also be thought of in this manner.

The discreteness of Γ naturally defines a folitation $\mathfrak{F}_\Gamma$ on $M\Gamma$ which is transversal to the fibers in the projection $M\Gamma \to B\Gamma$, and of codimension n . The characteristic classes of $\mathfrak{F}_\Gamma$ integrated over the fiber induce (1. 16) . With this interpretation it suggests itself that the constructions which are known to represent the characteristic classes of folitations should extend to the semisimplicial situation provided only that one has a suitable de Rham theory at hand. Now the double complex $\Omega^{**}X$ does not fit the bill because its multiplication is not anticommutative. On the other hand the compatible complex of Dupont is perfectly suitable and therefore yields explicit recipes quite readily.

Let me start with the Godbillon-Vey class ω, itself. This invariant - corresponding to a generator of $H^3(\mathfrak{a}_1, O_1)$ - is defined on oriented foliations $\mathfrak{F}$ of codimension one, and can be computed according to the algorithm:

Let $\mathfrak{F}$ be described as the kernel of a non-degenerate 1 - form φ. Then integrability implies that there exists a 1 - form η with

$$d\varphi = \eta \wedge \varphi .$$

(2.1) Now set

$$\omega = \eta \wedge d\eta .$$

Then $d\omega = 0$. Further the cohomology class of ω is independent of the choices involved, and represents $\omega(\mathfrak{F})$.

The extension of (2.1) to the semisimplicial case is now immediate.

Given a s. s. manifold

$$X : X_0 \leftleftarrows X_1 \Lleftarrow X_2 \cdots$$

a foliation $\mathfrak{F}$ on X is simply a foliation $\mathfrak{F}_k$ on each X_k, such that, all the structure maps:

$$X(\alpha) : X_k \longrightarrow X_{k'}$$

are transversal to $\mathfrak{F}_{k'}$ and induce isomorphisms

$$\mathfrak{F}(\alpha) : \mathfrak{F}_k \longrightarrow X(\alpha)^{-1} \mathfrak{F}_{k'} .$$

Such data then define a foliation $|\mathfrak{F}|$ on the geometric realization $|X|$ of X, in the following manner:

On $X_k \times \Delta^k$, the natural projection

$$X_k \xleftarrow{\pi_L} X_k \times \Delta^k \tag{2.2}$$

induces the foliation $\pi_L^{-1} \circ \mathfrak{F}_k$, and these foliations are compatible under the identifications which assemble the $X_k \times \Delta^k$ to form $|X|$. Precisely we have in mind here the so-called "Fat realization" of $|X|$, which therefore only identifies by the boundary maps. Thus if α is such a boundary map, and

$$\Delta(\alpha) : \Delta^{k-1} \longrightarrow \Delta^k$$

the corresponding map of the $k-1$ simplex Δ^{k-1} onto a face of Δ^k, then the identification corresponding to α is described by the diagram:

$$\begin{array}{ccc} X_k \times \Delta^{k-1} & \xrightarrow{1\times\Delta(\alpha)} & X_k \times \Delta^k \\ {\scriptstyle X(\alpha)\times 1}\downarrow & & \\ X_{k-1} \times \Delta^{k-1} & & \end{array} \tag{2.3}$$

That is, the two images of a point $(p,q) \in X_k \times \Delta^{k-1}$ under the horizontal and vertical arrows are to be identified in $|X|$.

It is clear then that the $\pi_L^{-1} \circ \mathfrak{F}_k$ do define a compatible collection of foliations on $|X|$, and that is precisely what one means by a foliation on $|X|$.

Similarly, one defines the de Rham complex $\Omega^*|X|$ of "compatible forms" on $|X|$ in terms of the diagram (2.3): Thus:

(2.4) A q - form φ on $|X|$ is a collection $\{\varphi_k\}$ of q - forms on $X_k \times \Delta^k$ such that

$$\{1 \times \Delta(\alpha)\}^* \varphi_k = \{X(\alpha) \times 1\}^* \varphi_{k-1} .$$

Finally one has the Dupont extension of the Whitney-Thom-Sullivan Theorem to the effect that

$$H^*(\Omega X) \sim H(\Omega^{**} X)$$

the isomorphism being induced by "integration over the simplexes"

$$\varphi \mapsto \sum_k \pi_*^L \varphi_k \tag{2.5}$$

where π_L is the projection (2.2) and π_*^L denotes integration over the fiber of π_L. Note that the sum has only q nonzero terms, when φ is of dimension q.

Now then, with all this understood our first remark is:

PROPOSITION 2.6. Let $\mathfrak{F}$ be an oriented codimension 1 foliation on the paracompact simplicial manifold X. Then $\omega(\mathfrak{F}) \in H^*(\Omega|X|)$ can be computed by the Godbillon-Vey algorithm (2.1) provided only we interpret "form" to mean "compatible form" on $|X|$.

The proof of this is quite straightforward, so let me only start the argument and in the process derive an explicit algorithm for $\omega(\mathfrak{F})$ in terms of the structure maps of X.

Consider then an $\mathfrak{F}$ as envisaged in the proposition, and let $\mathfrak{F}_0$ on X_0 be represented as kernel of φ with

$$d\varphi = \eta\varphi .$$

We next try to extend φ and η to $X_1 \times \Delta^1$ in a compatible manner. To fix the notation let 0 and 1 be the vertices of Δ^1 and let x^0 and x^1 be the corresponding barycentric coordinates on Δ^1. Also let

$$\sigma^0 : \Delta^0 \longrightarrow \Delta^1, \quad \text{and} \quad \sigma^1 : \Delta^0 \longrightarrow \Delta^1$$

be the inclusions sending Δ^0 to 0 and 1 respectively, and let σ_0 and σ_1 be the corresponding maps:

$$X_0 \underset{\sigma_1}{\overset{\sigma_0}{\leftleftarrows}} X_1 .$$

Now by our hypothesis on $\mathfrak{F}$ the forms

$$\sigma_0^* \varphi \quad \text{and} \quad \sigma_1^* \varphi$$

both represent $\mathfrak{F}_1$ in the same orientation. Hence there exists a smooth positive function μ_1 on X_1 such that

(2.6) $$\sigma_1^* \varphi = \mu_1 \, \sigma_0^* \varphi .$$

Using μ_1 we now construct the form

(2.7) $$\varphi_1 = \sigma_0^* \varphi \cdot \mu_1^{x^1} \qquad \text{on} \quad X_1 \times \Delta^1 .$$

Here we have identified the forms on X_1 with their pullback to $X_1 \times \Delta^1$ under π_L and $\mu_1^{x^1}$ is defined by

$$\mu^{x^1}(p, a) = \mu_1(p)^{x^1(a)} .$$

Hence φ_1 restricts to $\sigma_i^*\varphi$ on $\partial(X_1 \times \Delta^1)$ and is compatible. Furthermore the kernel of φ_1 clearly represents $\pi_L^{-1}\mathfrak{F}_1$. Indeed any nonzero multiple of $\sigma_0^*\varphi$ would.

Next we wish to extend η compatibly to $X_1 \times \Delta^1$. For this purpose differentiate (2.7). One obtains

(2.8) $$d\varphi_1 = (\log \mu_1 \cdot dx^1 + x^1 \, d \log \mu_1 + \sigma_0^*\eta) \wedge \varphi_1 \quad .$$

Now the term in the bracket is not compatible with η as it stands. However, it can be modified to become so in the following manner:

By differentiating (2.6) we obtain

$$d\,\sigma_1^*\varphi = \sigma_1^*\eta \cdot \sigma_1^*\varphi = \{d \log \mu_1 + \sigma_0^*\eta\}\,\sigma_1^*\varphi$$

whence

(2.9) $$\sigma_1^*\eta \equiv d \log \mu_1 + \sigma_0^*\eta \mod (\sigma_1^*\varphi) \quad .$$

Hence we may replace $d \log \mu_1$ in (2.8) by $\sigma_1^*\eta - \sigma_0^*\eta$ to obtain:

$$d\varphi_1 = \{\log \mu_1 \, dx^1 + x^1 \, \sigma_1^*\eta_1 + x^0 \, \sigma_0^*\eta\} \wedge \varphi_1$$

and this time the term

(2.10) $$\eta_1 = \{\log \mu_1 \, dx^1 + x^1 \, \sigma_1^*\eta + x^0 \, \sigma_0^*\eta\}$$

is clearly compatible with η.

This construction now extends to all of $|X|$ to yield the following compatible collection φ_k and η_k on $X_k \times \Delta^k$.

For each k, we let $\sigma_0, \sigma_1, \cdots, \sigma_k$ be the maps of $X_k \to X_0$ corresponding to the inclusion of Δ^0 into Δ^k as the $k+1$ vertexes.

Then we define μ_i, $i = 1, \cdots, k$, by

$$\sigma_i^* \varphi = \mu_i \sigma_0^* \varphi$$

and finally set

$$\varphi_k = \sigma_0^* \varphi \left(\prod_{i=0}^{k} \mu_i^{x^i} \right) \tag{2.11}$$

and correspondingly set

$$\eta_k = \sum_{i=0}^{k} \{\sigma_i^* \eta \cdot x^i + \log \mu_i \cdot dx^i\} \quad . \tag{2.12}$$

<u>The forms</u> $\eta_k \wedge d\eta_k = \omega_k(\mathfrak{F})$ <u>are therefore again compatible, closed, and give an algorithm for computing</u> $\omega(\mathfrak{F})$ <u>in</u> $H^*(|X|)$.

At this stage one may of course return to the more economical complex $\Omega^{*,*}(X)$ by integrating over the simplices. Then

$$\pi_*^L \omega(\mathfrak{F}) = \omega^{3,0} + \omega^{2,1} + \omega^{1,2} + \omega^{0,3} \tag{2.13}$$

with $\omega^{i,j} \in \Omega^i(X_j)$ and using (2.13) one obtains explicit formulae. For example:

$$\omega^{3,0} = \eta \wedge \varphi, \tag{2.14}$$

while $\omega^{2,1}$ is obtained by integrating $\eta_1 \wedge d\eta_1$ over the 1 - simplex Δ^1. Thus this component is given by:

$$\int_{\Delta^1} (x^0 \sigma_0^*\eta + x^1 \sigma_1^*\eta + \log\mu_1\, dx^1) \wedge (dx^1(\sigma_1^*\eta - \sigma_0^*\eta) + d\log\mu_1\, dx^1 + x^0\, d\sigma_0^* + x^1\, d\sigma_1^*\eta)\ .$$

Only the terms involving a dx survive and hence with a little algebra one arrives at

$$\omega^{2,1} = \sigma_1^*\eta \wedge \sigma_0^*\eta - d\{\log\mu_1(\frac{\sigma_1^*\eta + \sigma_0^*\eta}{2})\} \tag{2.15}$$

To obtain the next term we proceed similarly with $\eta_2 \wedge d\eta_2$. The result is

$$\omega^{1,2} = \sum_{i,j=0}^{2} \int_{\Delta^2} \log\mu_i\,(d\log\mu_j - \sigma_j^*\eta)\, dx^i\, dx^j \qquad 0 \leq i,\ j \leq 2$$

(2.16)

$$= \sum_{0\leq i<j\leq 2} (-1)^{i-j+1} \log u_i(d\log\mu_j - \sigma_j^*\eta)$$

Finally $\omega^{0,3}$ vanishes because there is no term involving three dx's in $\eta_3 \wedge d\eta_3$.

So far I have discussed only the original Godbillon-Vey class ω for foliations of codimension 1. If $\mathfrak{F}$ is orientable but has higher codimension say q - their class ω is generalized to $\mathfrak{F}$, by again describing $\mathfrak{F}$ as the kernel of a decomposable form φ, this time of dimension q. Then integrability again implies that

$$d\varphi = \eta \wedge \varphi$$

but now it is the form

$$\omega(\mathfrak{F}) = \eta \wedge (d\eta)^q$$

which is an invariant of $\mathfrak{F}$ and thus gives the extended Godbillon-Vey class in $H^{2q+1}(M)$. In WO_q this class then represents $h_1 c_1{}^q$.

In the semisimplicial situation this extension works equally well and (2. 12) again yields an explicit algorithm for the computations of $\omega(\mathfrak{F})$. Note by the way that in codimension q, only components of the type:

$$\pi_*^L\, \omega(\mathfrak{F}) = \omega^{2q+1,0} + \cdots + \omega^{q,q+1} \tag{2.17}$$

survive because $\eta_k \wedge d\eta_k{}^q$ can contain no more than $(q+1)$ dx's for any k. Furthermore the extremal terms are easy to compute from (2. 12):

$$\omega^{2q+1,0} = \eta \wedge (d\eta)^q$$

$$\omega^{q,q+1} = \int_{\Delta^q} (\log \mu_i\, dx^i)\{(d \log \mu_j - \sigma_j^*\eta)\, dx^j)^q\} \;. \tag{2.18}$$

There are corresponding formulae for the other classes in $H^*(WO_n)$ but they are harder to write down explicitly, because, as in the case of just a single foliation $\mathfrak{F}$ on M, one has to bring in a comparison of torsion-free and Riemannian connections.

However the compatible complex again enables one to extend the single-space recipes to the semisimplicial case in a more or less straightforward manner, but rather than discussing these let me now specialize our formulae to the case of a group action. We therefore have to recall briefly how $M\Gamma$ and $B\Gamma$ are constructed and why $M\Gamma$ carries a natural foliation.

Recall first, that if C is a a category, then it defines a natural semisimplicial object

$$BC : O(C) \leftleftarrows \mathfrak{m}_1(C) \Lleftarrow \mathfrak{m}_2(C) \tag{2.19}$$

starting with the objects $O(C)$ of C, and $\mathfrak{m}_i(C)$ being the i-tuples of composable morphisms in C. Thus a typical element in $\mathfrak{m}_k(C)$ is given by the diagram:

(2.20)
$$\cdot \overset{\gamma_1}{\longleftarrow} \cdot \overset{\gamma_2}{\longleftarrow} \cdot \ldots \cdot \overset{\gamma_k}{\longleftarrow} \cdot$$

the γ 's being arrows in C. The boundary maps in BC are then given by sending such a diagram into one with either an end arrow deleted, or two consecutive one composed. For example

(2.21)
$$\partial_0(\gamma_1, \gamma_2) = \gamma_2$$
$$\partial_1(\gamma_1, \gamma_2) = \gamma_1 \circ \gamma_2$$
$$\partial_2(\gamma_1, \gamma_2) = \gamma_1$$

This construction applied to the category determined by the group Γ yields $B\Gamma$. When applied to the category Γ_M determined by the action of Γ on M, it yields $M\Gamma$: Thus

$$M\Gamma \equiv B\Gamma_M$$

where the category Γ_M has as objects the manifold M, and as morphisms pairs (γ, m) with $\gamma \in \Gamma$, $m \in M$. Finally the composition

(2.22)
$$(\gamma_1, m_1) \circ (\gamma_2, m_2) = (\gamma_1 \circ \gamma_2, m_2)$$

exists only if $\gamma_2(m_2) = m_1$ and is then given by the RHS of (2.22).

Pictorially these are then best described by arrows of the form

(2.23)
$$\gamma(m) \overset{\gamma}{\longleftarrow} m$$

In any case, all of this granted it is easy to see that the following proposition holds.

PROPOSITION 2. 24. Let $\mathfrak{F}$ be a foliation on M, such that Γ acts transversally to $\mathfrak{F}$ on M. Then $\mathfrak{F}$ extends to a foliation $\mathfrak{F}\Gamma$ on $M\Gamma$.

In particular the foliation of M by points induces a natural foliation $\mathfrak{F}_M$ on $M\Gamma$.

Note of course that $\mathfrak{F}_M$ then corresponds to the foliation by points of $M\Gamma$ considered as a simplicial manifold in the obvious sense.

When we specialize our formulae to $\mathfrak{F}_M$, matters simplify to some extent as now η can clearly be taken to be zero. If we furthermore follow our recipe, in the case when M is oriented and compact, and integrate $\omega(\mathfrak{F}_M)$ over the fiber in the projection

$$M\Gamma \xrightarrow{\pi} B\Gamma$$

then - M being q - dimensional - only $\omega^{q,q+1}$ enters into the computation. In this manner we obtain the following algorithm for

$$\pi_* \, \omega(\mathfrak{F}_M) \; .$$

PROPOSITION 2. 25. The characteristic "number" $\pi_* \, \omega(\mathfrak{F}_M) \in H^{q+1}(B\Gamma)$ is represented by the cocycle:

$$(2.26) \quad \omega(\gamma_1, \cdots, \gamma_{q+1}) = \int_{\Delta^{q+1} \times M} (\log \mu_i \, dx^i)(d \log \mu_j \, dx^j)^q$$

where the μ_i are functions on M, determined in the following manner:

Choose a volume φ on M, and then define μ_i by:

$$(2.27) \qquad (\gamma_i \gamma_{i+1} \cdots \gamma_{q+1})^* \varphi = \mu_i \varphi \; .$$

Remark. (1) Thurston has been aware of this formula all along although his conventions and his setting might have differed somewhat. For those of us less geometrically inspired it was however not so obvious. On the other hand - and this is the main point of my remarks - in the context of the compatible complex the formula does become easily accessible.

(2) The following geometrical interpretation of 2.25 was pointed out to me by Thurston quite recently. If $y_1, \cdots, y_{q+1}$ are coordinates in $\mathbb{R}^{q+1}$ then the form

(2.28) $$\Psi = \Sigma(-1)^i y_i \, dy_1 \wedge \cdots \widehat{dy_i} \wedge \cdots dy_n$$

clearly satisfies the identity

$$d\Psi = n \text{ volume in } \mathbb{R}^n .$$

On the other hand, up to a universal constant (2.26) clearly gives the integral of Ψ pulled back to M under the map

$$f_\Gamma : M \longrightarrow \mathbb{R}^{q+1}$$

defined by

(2.29) $$y^i \circ f_\Gamma = \log(\mu_i)$$

Thus our formula for the Eilenberg MacLane cocycle $\omega(\mathfrak{F})$ can be written;

(2.30) $$\omega(\mathfrak{F})(\gamma_1, \cdots, \gamma_{q+1}) = \text{volume enclosed by } f_\Gamma M \text{ in } \mathbb{R}^{q+1} .$$

3. <u>The Infinitesimal Case.</u> When the action of Γ on a foliation $\mathfrak{F}$ over M comes by restriction of an action of a connected Lie group G containing Γ, the characteristic classes of the Γ - action can be described in Lie-algebra terms. Such a description is often more easy to handle, and our aim in this section is therefore to describe an infinitesimal analogue of (2.18) in this framework. The technique involves the lifting of $\mathfrak{F}$ to to $(G/K) \times M$, with K a maximal compact subgroup of G, and is thus typical of the Van-Est theorems and theorems on continuous cohomology, as well as of the whole approach of Kamber and Tondeur to foliated bundles. See [7].

However I will not pursue these aspects, but rather head as quickly as possible towards the infinitesimal formula for $\omega(\mathfrak{F})$.

Recall then, first of all, that G/K is contractible:

$$G/K \simeq pt \quad . \tag{3.1}$$

Hence if we consider the natural Γ action on G/K then (3.1) extends to a homology equivalence of the quotient of G/K by the action of Γ (denoted $[G/K]\Gamma$) with $B\Gamma$,"

$$[G/K]\Gamma \simeq B\Gamma \tag{3.2}$$

Thus $[G/K]\Gamma$ is a thickened version of $B\Gamma$. Now consider the edge homomorphism in $[G/K]\Gamma$,

$$\operatorname*{inv}_{\Gamma} \Omega^*(G/K) \to \Omega^{**}([G/K]\Gamma) \quad . \tag{3.3}$$

Because $\Gamma \subset G$, the left hand side certainly contains the subcomplex of G - invariant forms on G/K, which is finally identified with the relative Lie algebra complex $\Omega^*(g; K)$. Thus we have

$$\Omega^*(\underline{g};K) \equiv \operatorname{Inv}_G \Omega^*(G/K) \subset \operatorname{Inv}_\Gamma \Omega^*(G/K) \tag{3.4}$$

combining (3.4) with (3.3) and (3.2) gives rise to an arrow[†]

$$\iota_* : H^*(\underline{g};K) \to H^*(B\Gamma) \quad , \tag{3.5}$$

and our first aim is then to show that:

PROPOSITION 3.6. <u>In the situation envisaged the characteristic classes of the</u> Γ <u>action lift naturally to</u> $H^*(\underline{g};K)$.

To see this result one first has to understand the following diagram of G spaces:

$$G \times_K M \xrightarrow[\approx]{t} (G/K) \times M \xrightarrow{\pi} M \quad , \tag{3.7}$$

where π is the natural product projection, and the action of G on $(G/K) \times M$ is the product action. On the other hand G <u>acts on</u> $G \times_K M$ <u>purely on the left, and</u> $G \times_K M$ <u>is</u> defined as the quotient

$$G \times_K M = (G \times M)/K \quad , \tag{3.8}$$

with K acting by

$$(g, m) \cdot k = (gk, k^{-1}m) \quad . \tag{3.9}$$

Finally the twist map t is induced by

$$\tilde{t} : G \times M \to G \times M \tag{3.10}$$

sending

$$(g, m) \text{ to } (g, g \cdot m) \quad . \tag{3.11}$$

[†] In [5] and also independently in a recent paper of Shulman and Tischler, this arrow is explicitly described on the chain level.

It is then clear that $\tilde{t}$ is a diffeomorphism which sends the K action (3.9) into the K action $(g, m) \to (gk, m)$ and so induces the equivalence t of (3.7). The virtue of this twisting map t, is of course that the G - action on $G \times_K M$ is given by the left translation of G. It follows that the G - invariant forms on $G \times_K M$ can be identified with the forms on $G \times M$, which are

(1) invariant under left translation of G by G, and

(2) K - basic under the action (3.8) of K on $G \times M$. Thus:

$$\text{(3.12)} \qquad \operatorname{Inv}_G \Omega^*(G \times_K M) \sim K \text{ basic forms in } \Omega^*(\mathfrak{g}) \otimes \Omega^* M \ .$$

In any case, the plan of procedure is now suggested by the diagram:

$$\text{(3.13)} \qquad \begin{array}{ccccc} \operatorname{Inv}_G \Omega^*(G \times_K M) & \to & \Omega^{**}(G \times_K M\Gamma) & \approx & \Omega^{**}(M\Gamma) \\ \downarrow \pi_* & & & & \downarrow \pi_* \\ \operatorname{Inv}_G \Omega^*(G/K) & \to & \Omega^*(G/K\Gamma) & \approx & \Omega^{**}(B\Gamma) \end{array}$$

where the lower line induces (3.5), and π_* denotes integration over the fiber. The homotopy equivalence in the upper line is of course again a consequence of the contractibility of G/K. In view of (3.13) our lifting problem clearly amounts to realizing the characteristic classes of $\mathfrak{F}$ lifted to $G \times M$, by G - invariant forms. Thus we need the following.

PROPOSITION 3.14. The characteristic classes of $\mathfrak{F}$ admit a natural lifting to the complex (3.12).

Let me carry out the proof, but again only for our Godbillon-Vey class $\omega(\mathfrak{F})$. Consider then the pull-back $\tilde{\mathfrak{F}}$ of $\mathfrak{F}$ to $G \times M$, under the map

$$\text{(3.15)} \qquad G \times M \xrightarrow{\pi \circ \tilde{t}} M \ .$$

To describe $\tilde{\mathfrak{F}}$ let us identify the tangent space of $G \times M$, at (g, m) with $\mathfrak{g} \oplus T_m M$, using the left invariant vector-fields on G to identify G_g with $\mathfrak{g}$, and using the projections for the direct sum decomposition. Also if $x \in \mathfrak{g}$ is an element of the Lie algebra of G, let $\dot{x} \in \Gamma(TM)$ be the corresponding infinitesimal motion on M, induced by the action of G on M.

Precisely if e^{tx} is the one-parameter subgroup generated by $x \in \mathfrak{g}$, then

(3.16) $$\dot{x}_m = \text{tangent of } e^{tx}m \text{ at } m \ .$$

With this understood, the kernel of $\tilde{\mathfrak{F}}$ is described by:

$x + y \in T_{(g,m)}(G \times M)$ _is in_ $\tilde{\mathfrak{F}}$ _if and only if_

(3.17) $$\dot{x}_m + y_m \in \mathfrak{F} \ .$$

Indeed the curve (ge^{tx}, m) goes over into $ge^{tx}m$ under our map, and hence its tangent at $t = 0$ goes to $g_*\dot{x}_m$. On the other hand y goes to $g_* y_m$. Hence $g_*(\dot{x}_m + y_m) \in \mathfrak{F}$. But G preserves the foliation so that (3.17) follows.

An immediate corollary of (3.17) is the following:

PROPOSITION 3.18. _Let the foliation_ $\mathfrak{F}$ _be described as the kernel of the decomposable_ q _form_ φ. _Then_ $\tilde{\mathfrak{F}}$ _admits a natural representation as the kernel of a decomposable form_ $\tilde{\varphi} \in \Omega^*(\mathfrak{g}) \otimes \Omega^*(M)$. _Furthermore, in the natural double grading of this complex_ $\tilde{\varphi}$ _has components:_

$$\tilde{\varphi} = \tilde{\varphi}^{q,0} + \cdots + \tilde{\varphi}^{0,q}$$

with:

$$\tilde{\varphi}^{0,q} = \varphi \ .$$

(3.19) $$\tilde{\varphi}^{1,q-1} = \sum_\alpha x'_\alpha \wedge \iota(\dot{x}_\alpha)\varphi$$

where x_α runs over a basis of g, and x'_α over a dual base in g^*.

Proof. This is purely a linear algebra matter. If θ is a 1-form with $\mathfrak{F}$ in its kernel, then

(3.20) $$\tilde{\theta} = \theta + \sum_\alpha x'_\alpha\, \iota(\dot{x}_\alpha)\,\theta$$

will have $\tilde{\mathfrak{F}}$ in its kernel. Indeed for any $x + y$ subject to (3.17) we then have $\iota(x+y)\tilde{\theta} = \theta(y) + \theta(\dot{x}) = 0$.

Hence if $\theta^1 \wedge \cdots \wedge \theta^q = \varphi$ locally, then $\tilde{\theta}^1 \wedge \cdots \wedge \tilde{\theta}^q$ describes $\tilde{\mathfrak{F}}$ locally, and expanding this product clearly yields (3.19).

To proceed further we need to compute $d\tilde{\varphi}$ and express it as $\tilde{\eta} \wedge \tilde{\varphi}$ in the double complex $\Omega^*(g) \otimes \Omega^*(M)$.

For this purpose, let us set

$$\iota(\dot{x}_\alpha)\varphi = \varphi_\alpha$$

and use the double index summation convention, so that

(3.21) $$\tilde{\varphi} = \varphi + x'_\alpha \wedge \varphi_\alpha + \cdots$$

describes the "beginning" of $\tilde{\varphi}$.

Also, let us assume that on M the integrability of $\mathfrak{F}$ is expressed by

(3.22) $$d\varphi = \eta \wedge \varphi$$

where η is a global 1-form. Then I claim that:

PROPOSITION 3.23. The η of (3.22) lifts naturally to one $\tilde{\eta}$ in $\Omega^*\underline{g}\otimes\Omega^*M$ such that

$$d\tilde{\varphi} = \tilde{\eta}\wedge\tilde{\varphi} .$$

Further the $\tilde{\eta}$ is given by

(3.24) $$\tilde{\eta} = \eta - x'_\alpha\{\mu(x_\alpha) - \eta(\dot{x}_\alpha)\}$$ where $\mu(x)$ is defined by (3.27).

Proof. Differentiating (3.21) yields

(3.25) $$d\tilde{\varphi} = d\varphi - x'_\alpha\wedge d\varphi_\alpha + \cdots .$$

Further

(3.26) $$d\varphi_\alpha = d\iota(\dot{x}_\alpha)\varphi = \mathfrak{L}(\dot{x}_\alpha)\varphi - \iota(\dot{x}_\alpha)d\varphi.$$

where $\mathfrak{L}(\dot{x}_\alpha)$ is the Lie derivative in the direction $\dot{x}$. Because G preserves $\mathfrak{F}$, $\mathfrak{L}(\dot{x})$, $x\in\underline{g}$ must preserve φ up to multiples, whence

(3.27) $$\mathfrak{L}(\dot{x})\varphi = \mu(x)\varphi\ ,\ x\in\underline{g}\ ,\ \text{with}\ \mu(x)\in\Omega^0(M) .$$

This $\mu(x)$ is, of course, the infinitesimal analogue of the $\mu(\sigma)$ in Section 2.

In any case, combining (3.25), (3.26), (3.27) with (3.22) one obtains the formula

$$d\tilde{\varphi} = \eta\wedge\varphi - \mu(x_\alpha)x'_\alpha\wedge\varphi$$
$$+ x'_\alpha\wedge\{\eta(\dot{x}_\alpha)\varphi - \eta\wedge\varphi_\alpha\} + \cdots$$

which, up to terms of order ≥ 2 in the $\underline{g}$ direction, is given by

(3.28) $$d\tilde{\varphi} = \{\eta - x'_\alpha(\mu(x_\alpha) - \eta(\dot{x}_\alpha)\}\wedge\tilde{\varphi} .$$

But as $\tilde{\eta}$ exists and clearly is the sum of forms of type $(1,0)$ and $(0,1)$ this equation fixes $\tilde{\eta}$. Q. E. D.

To assemble the pieces, we shall have to determine whether $\tilde{\varphi}$ and $\tilde{\eta}$ are K basic in our complex.

In general this will, of course, not be the case. However, by averaging over K, we can arrange it that both the φ and the η of our discussion are invariant under the action of K, i.e., that infinitesimally

$$\mathcal{L}(\dot{x})\varphi = 0 \quad ; \quad \mathcal{L}(\dot{x})\eta = 0 \quad \text{for} \quad x \in \underline{k} \tag{3.29}$$

and under this hypothesis we have the following.

PROPOSITION 3.30. The condition (3.29) implies that $\tilde{\varphi}$ and $\tilde{\eta}$ are K basic[†] relative to the action of K on G x M.

The proof of this fact is a straightforward check (though not quite trivial) which I will take up in greater generality at another time.

At this stage, we are ready to give an infinitesimal recipe for $\omega(\mathfrak{Z})$.

Indeed, expanding $\tilde{\eta}(d\tilde{\eta})^q$, will have all possible type of components:

$$\omega(\mathfrak{Z}) = \omega^{2q+1,0} + \cdots + \omega^{0,2q+1} ,$$

of which the simplest are given by:

$$\omega^{0,2q+1} = \eta \cdot d\eta^q , \tag{3.31}$$

$$\omega^{q+1,q} = - x'_{\alpha_1} \wedge \cdots x'_{\alpha_{q+1}} \nu_{\alpha_1} d\nu_{\alpha_2} \wedge \cdots d\nu_{\alpha_{q+1}} \tag{3.32}$$

where we have now set

[†] The K basic forms are those annihilated by the vector fields along the orbit of the K action and invariant under that action.

(3.33) $$\nu(x) = \mu(x) - \eta(\dot{x}) \qquad x \in \underline{g}$$

and have abbreviated $\nu(x_\alpha)$ to ν_α.

If we think of $\Omega^*(\underline{g}) \otimes \Omega^*(M)$ as the complex $\Omega^*(\underline{g}; \Omega^*(M))$ of forms on $\underline{g}$ with values in $\Omega^* M$, then (3.33) can be thought of as a 1-form on $\underline{g}$ with values in $\Omega^0(M)$:

$$\nu \in \Omega^1(\underline{g}; \Omega^0(M)) \quad ,$$

and there (3.32) takes the form

(3.34) $$\omega^{q+1,q}(x_1, \cdots, x_{q+1}) = \frac{1}{(q+1)} \Sigma (-1)^i \nu(x_i)\, d\nu(x_1) \cdots \widehat{d\nu(x_i)} \cdots d\nu(x_{q+1}) \quad .$$

Hence we get the following corollary, which is an infinitesimal analogue of 3.25.

PROPOSITION 3.35. Suppose $Z \subset M$ is a cycle of dimension q on M. Then

(3.36) $$\int_Z \omega(\mathfrak{F}) \in H^{q+1}(\underline{g}; K)$$

is represented by the cocycle

(3.37) $$\omega_Z(\mathfrak{F})(x_1, \cdots, x_q) = \int_Z \nu(x_1) d\nu(x_2) \cdots d\nu(x_{q+1})$$

where $\nu(x)$ is defined by

(3.38) $$\{\mathfrak{L}(x) - \eta(\dot{x})\}\varphi = \nu(x)\varphi \quad .$$

4. On the Examples of Thurston and Heitsch. The history of examples of foliations with varying classes $\omega(\mathfrak{F})$ is roughly as follows.

In the complex analytic case, I had observed already before 1970 that the foliation:

(4. 1) $$\mathcal{F}_\lambda = \{\lambda_1 z_1 \frac{\partial}{\partial z_1} + \lambda_2 z_2 \frac{\partial}{\partial z_2}\}$$

on $\mathbb{C}^2 - 0$, had for its g. v. invariant:

$$\int_{S^3} \omega(\mathcal{F}_\lambda) = c\,\{\frac{\lambda_1}{\lambda_2} + \frac{\lambda_2}{\lambda_1} - 2\} ,$$

and thus varied continuously with λ.

At that time, I thought that the corresponding real invariant would always vanish. However, soon thereafter in 1971, the paper of Godbillon-Vey appeared with the Roussarie example of the foliation $\mathcal{F}$ on $\Gamma\backslash SL(2;\mathbb{R})$ induced by the Lie algebra of triangular matrices in $SL(2;\mathbb{R})$, Γ being a discrete subgroup with compact quotient space.

Thereafter, Thurston produced his examples. In particular he constructed examples of a family of actions of a group Γ acting on S^1, such that the corresponding g. v. number $\in H^2(B\Gamma)$ varied continuously. In an appendix, Robert Brooks has written up the details treating this example with the formula (2. 25) much like Dupont in [5] treated the Euler class of flat bundles. Here let me just outline and comment on the plan of this very ingenious example.

We start by observing that $SL(2;\mathbb{R})$ acts on S^1 in the classical manner

(4. 1) $$z \longrightarrow \frac{az+b}{cz+d} , \quad |z| = 1, \begin{pmatrix} ab \\ cd \end{pmatrix} \in SL(2,\mathbb{R}) .$$

Hence for any $\Gamma \subset SL(2,\mathbb{R})$ there is a natural action of Γ on S^1.
Where $\Gamma\backslash SL(2,\mathbb{R})$ is compact, $H^2(\Gamma;\mathbb{R}) \neq 0$ and the g. v. class of the action will also be nonzero. On the other hand, as we let Γ vary in a continuous family of such subgroups,

the corresponding characteristic class does not vary. Thus, the moduli of Riemann surfaces do not furnish varying examples.

Thurston therefore twisted these homogeneous actions on S^1 in the following manner.

Consider the double cover

$$S^1 \xrightarrow{\pi} S^1$$

given by sending z to z^2; $|z| = 1$. Then every diffeomorphism f of S^1, admits precisely two liftings $\tilde{f}$ relative to π. Thus,

$$\tilde{f} : S^1 \longrightarrow S^1$$

is a diffeomorphism of S^1, with

$$\pi \circ \tilde{f} = f \circ \pi .$$

On the double cover $\mathrm{Diff}^{(2)} S^1$ of $\mathrm{Diff}\, S^1$ the function $f \longrightarrow \tilde{f}$ now becomes single valued and defines a homomorphism:

$$\mathrm{Diff}^{(2)} S^1 \xrightarrow{\Psi_2} \mathrm{Diff}\,(S^1) \tag{4.2}$$

Note that if f is lifted to an "equivariant map" : $\underline{f}(x + 2\pi) = \underline{f}(x) + 2\pi$,

$$\underline{f} : \mathbb{R} \longrightarrow \mathbb{R}$$

then $\tilde{f}$ is represented by:

$$\tilde{f}(x) = 1/2\, \underline{f}\,(2x) \quad \text{or} \quad 1/2\, \{\underline{f}(2x) + \pi\} \ .$$

Hence, in particular, if f is a rotation by α and hence represented by

$$x \longrightarrow x + \alpha$$

then $\widetilde{f}$ is represented by:

(4.3) $$x \longrightarrow x + \alpha/2 , \quad \text{and} \quad x \longrightarrow x + \frac{\alpha}{2} + \pi .$$

It follows that if the map

$$SL(2, \mathbb{R}) \longrightarrow PSL(2, \mathbb{R}) \longrightarrow \mathrm{Diff}\, S^1$$

given by (4.1), is lifted to a map of $SL(2, \mathbb{R})$ to $\mathrm{Diff}^{(2)}(S^1)$ and then followed by Ψ_2, there results a homomorphism of

(4.4) $$SL(2, \mathbb{R}) \underset{SO(2)}{*} SL(2, \mathbb{R}) \xrightarrow{1 \times \Psi_2} \mathrm{Diff}\,(S^1)$$

where on the left we have in mind the free product amalgamated along the rotations according to (4.3), and it is this action which gives rise to Thurston's example. More precisely, he takes the Γ represented by generators

$$\{X, Y, Z, W\}$$

and with the relation

(4.5) $$[X, Y] = [Z, W] ,$$

chooses $X, Y \in SL(2; \mathbb{R})$ so that $[X, Y]$ is a rotation by $0 < \alpha < \pi$, (see Appendix for details), and also $Z', W' \in SL(2, \mathbb{R})$ so that $[Z', W']$ is a rotation by 2α. Then lifting Z' and W' and applying Ψ_2, one obtains Z, W in $\mathrm{Diff}\, S^1$, which clearly satisfy the relation (4.5).

Now varying α, Thurston obtains his example. Note that as this example is obtained by amalgamation of two infinitesimal situations, it cannot be directly treated by our infinitesimal method, although, as Bob Brooks point out our global and infinitesimal cocycles agree where they should.

More recently Thurston has found examples of varying the higher g.v. classes by actions of certain Γ's on the spheres. Thus, his examples actually vary the characteristic numbers of certain Γ actions.

During the Rio conference, James Heitsch suggested a different approach to these examples, which quite recently has enabled him to vary a large number of characteristic classes independently (see [6]).

In our terminology, Heitsch passes from the sphere, where Thurston worked, to a foliation $\mathfrak{F}_\lambda$ on $\mathbb{R}^n - 0$ of the type I used in the complex case, but here he starts out with a sufficiently special λ so that $\mathfrak{F}_\lambda$ admits a large group of automorphism. Let me conclude by taking up the first instance of his construction to vary $h_1 c_1^3 \in H^4(B\Gamma)$.

We will use our infinitesimal recipe for this purpose, so recall first of all that the homomorphism

$$H^*(g;K) \longrightarrow H^*(B\Gamma) \tag{4.6}$$

is injective for any semi-simple Lie group and any discrete subgroup Γ with $\Gamma\backslash G$ compact. Indeed Γ will then have a subgroup of finite index $\Gamma' \subset \Gamma$ such that Γ' acts freely on G/K. The natural map

$$H^*(g;K) \longrightarrow H^*(\Gamma'\backslash G/K) \tag{4.7}$$

is then injective in the top dimension and both sides satisfy Poincaré duality. Hence (4.7) is injective, but $\Gamma'\backslash G/K \simeq B\Gamma'$, and $B\Gamma'$ and $B\Gamma$ have equal rational cohomology. Q. E. D.

Finally recall that by a theorem of Borel's any semi-simple G admits a Γ with $\Gamma\backslash G$ compact. Thus for our purposes, varying the infinitesimal class with a semi-simple group G also varies the class in some $B\Gamma$.

With these remarks we are ready to take up the Heitsch example.

Let then $\mathfrak{F}_\lambda$ be generated by X_λ in $\mathbb{R}^4 - 0$ where

$$X_\lambda = \sum_{i=1}^{2} \lambda_i\left(x_i \frac{\partial}{\partial x_i} + y_i \frac{\partial}{\partial y_i}\right). \tag{4.8}$$

Then the natural action of $SL(2;\mathbb{R})$ on the (x_i, y_i) space defines an action of

$$G = SL(2,\mathbb{R}) \times SL(2,\mathbb{R})$$

on $\mathbb{R}^4$, which obviously preserves X_λ and hence $\mathfrak{F}_\lambda$.

The infinitesimal action of $\mathfrak{g}$ on $\mathbb{R}^4$ is therefore generated by

$$u_i = x_i \frac{\partial}{\partial x_i} - y_i \frac{\partial}{\partial y_i} \quad ; \quad v_i = x_i \frac{\partial}{\partial y_i} + y_i \frac{\partial}{\partial x_i} \tag{4.9}$$

and

$$h_i = x_i \frac{\partial}{\partial y_i} - y_i \frac{\partial}{\partial x_i} .$$

Thus the h_i generate the action of K, and a class in $H^4(\mathfrak{g};K)$ is determined by its value on $u_1 \wedge v_1 \wedge u_2 \wedge v_2$.

Let us now apply the infinitesimal recipe to $\mathfrak{F}_\lambda$. For φ we may choose

$$\varphi = \iota(X_\lambda)\nu \quad , \quad \nu = dx_1\, dy_1\, dx_2\, dy_2 . \tag{4.10}$$

Then

$$d\varphi = \mathcal{L}(X_\lambda)\nu = 2(\lambda_1 + \lambda_2)\nu \quad .$$

Hence an admissible η is given by

$$\eta = (\lambda_1 + \lambda_2) \cdot \frac{\Sigma \lambda_i^2 \, dr_i^2}{\Sigma \lambda_i^2 r_i^2} \quad , \quad i = 1, 2 \tag{4.11}$$

where we have set

$$r_i^2 = x_i^2 + y_i^2 \quad . \tag{4.12}$$

Indeed η is clearly invariant under K, and satisfies the relation

$$d\varphi = \eta \wedge \varphi \quad .$$

Further note that φ is invariant under all of G, so that the ν of (3.33) is entirely given by the formula

$$\nu(x) = \eta(x) \quad .$$

Now by direct computation

$$\begin{aligned} u_i \cdot r_i^2 &= 2(x_i^2 - y_i^2) \\ v_i \cdot r_i^2 &= 4\, x_i\, y_i \end{aligned} \tag{4.13}$$

hence

(4.14)
$$\nu(u_i) = -\tau(u_i) = \frac{\left(\lambda_1+\lambda_2\right)2\lambda_i\left(x_i^2-y_i^2\right)}{\Sigma\,\lambda_i^2\,r_i^2}$$

$$\nu(v_i) = -\tau(v_i) = \frac{\left(\lambda_1+\lambda_2\right)4\lambda_i\,x_i\,y_i}{\Sigma\,\lambda_i^2\,r_i^2}$$

Now (3.37) together with (2.28) yield the formula:

$\omega(\mathfrak{J}_\lambda)(u_1,v_1,u_2,v_2)$ = <u>volume enclosed by the map</u> $S^3 \longrightarrow R^4$ <u>given by the four functions</u>

$\nu(u_1),\ \nu(v_1),\ \nu(u_2),\ \nu(v_2)$

<u>on</u> $S^3 \subset R^4$.

Clearly this map is homogeneous of degree zero, hence we may change coordinates from x_i to x_i/λ_i, and setting

(4.15)
$$z_i = x_i + \sqrt{-1}\,y_i$$

we see that the volume enclosed by the map in question will be proportional to $(\lambda_1+\lambda_2)^4/(\lambda_1\lambda_2)^2$ times the volume enclosed by the unit sphere under the map,

(4.16)
$$\{z_1,z_2\} \longrightarrow \{z_1^2, z_2^2\}/\,|z_1|^2 + |z_2|^2 ,$$

which is easily seen to be nonzero. <u>Hence</u>

$$\int \omega(\mathfrak{J}_\lambda)(u_1,v_1,u_2,v_2) = \text{const.}\ \frac{\left(\lambda_1+\lambda_2\right)^4}{\left(\lambda_1\lambda_2\right)^2}$$

<u>and therefore varies with</u> λ.

In higher even dimensions this method of Heitsch's works equally well and leads to an independent variation of all the $h_1 c^\alpha$ classes. In odd dimensions the argument is a little more subtle. However, he can also treat this case by construction which - on the sphere - goes back to Thurston.

All in all then Thurston and Heitsch have set us well on the way of showing that all the potential classes of $H^*(a_n;o_n)$ are non-trivial, independent and, in the appropriate cases, variable. For classes involving many h's, corresponding independence theorems were first obtained by Kamber and Tondeur [7].

Appendix

by

Robert Brooks

In this appendix, we will evaluate the Godbillon-Vey class in the case of some specific actions of groups on the circle. We will then show how these calculations lead to an example due to Thurston showing how one can vary the Godbillon-Vey "number."

Given an action of G on S^1, recall that on BG, ω is given by the 2-cocycle

$$\omega(g,f) = \int_{S^1} \log(\mu_f)\, d\log(\mu_{gf}) \quad .$$

If G is a discrete subgroup of $PSL(2,\mathbb{R})$, then we have a natural action of G on S^1, which we view as the boundary of the upper half plane, given by the linear fractional transformations -

$$\begin{pmatrix} a & b \\ c & d \end{pmatrix}(z) = \frac{az+b}{cz+d}$$

For ease in computation, we can conjugate this action by the linear fractional transformation $z \to \frac{z-i}{z+i}$ taking the upper half plane into the disk $|z| \leq 1$ - we will want to pass back and forth freely between the two pictures.

Viewing f and g as acting on $|z| = 1$, we may now write

$$f(z) = \theta_1\left(\frac{z-\eta_1}{\overline{\eta}_1 z - 1}\right); \quad g \circ f(z) = \theta_2\left(\frac{z-\eta_2}{\overline{\eta}_2 z - 1}\right)$$

with $|\theta_1| = |\theta_2| = 1$, $|\eta_1|, |\eta_2| < 1$

and so viewing S^1 as $\mathbb{R}/\mathbb{Z}$, then f and g become

$$f(x) = \frac{1}{2\pi i} \log(\theta_1(\frac{e^{2\pi ix} - \eta_1}{\bar{\eta}_1 e^{2\pi ix} - 1}))\ ; \quad g \circ f(x) = \frac{1}{2\pi i} \log(\theta_2(\frac{e^{2\pi ix} - \eta_2}{\bar{\eta}_2 e^{2\pi ix} - 1})) \ .$$

Choosing the standard volume on S^1, $\mu_f = f'$ and $\mu_g = g'$, we can now easily compute by residues to get

$$\omega(f, g) = \int_{S^1} \log(\mu_f)\, d\log(\mu_{gf})$$

$$= \int_0^1 \log(1 - e^{-2\pi ix}\eta_1)(1 - e^{2\pi ix}\bar{\eta}_1)[\,2\pi i\,](\frac{e^{-2\pi ix}\eta_2}{1 - e^{-2\pi ix}\eta_2} - \frac{\bar{\eta}_2}{1 - e^{2\pi ix}\bar{\eta}_2})\, dx$$

$$= 2\pi i \log(\frac{1 - \eta_2\bar{\eta}_1}{1 - \bar{\eta}_2\eta_1}) \ .$$

We can give this number a somewhat more geometric flavor by now passing to the upper half plane - writing $\eta_1 = \frac{a - i}{a + i}$, $\eta_2 = \frac{b - i}{b + i}$, we have

$$\omega(f, g) = 2\pi i \log(\frac{a + i}{\bar{a} - i})(\frac{\bar{b} - i}{b + i})(\frac{\bar{a} - b}{\bar{b} - a}) \ .$$

Now in the upper half plane, a geodesic is of the form $re^{i\theta} + k$, where r, k and θ are real. So given $z_1 = re^{i\theta_1} + k$, $z_2 = re^{i\theta_2} + k$, the expression $\log(\frac{z_1 - \bar{z}_2}{z_2 - \bar{z}_1}) = i(\theta_1 - \theta_2)$. Hence well-known formulas in non-Euclidean geometry give us

(*) $$\omega(g, g) = (-4\pi^2) \text{ area } (\Delta) \ ,$$

where Δ is the geodesic triangle with vertices i, a, and b. Here, of course, a and b satisfy $f(a) = i$, $(g \circ f)(b) = i$.

The formula (*) relates well to the formulas of [5] in the following manner: If a Lie group G operates on a manifold M of dimension q, then we get a map from $\mathfrak{g} = \text{Lie}(G)$ to the Lie algebra of vector fields on M in the following way: given $X \in \mathfrak{g}$, then it defines a one-parameter group f_t of diffeomorphisms of M, and the infinitesimal generator of this flow is a vector field on M.

Let K be a maximal compact subgroup of G, with Lie algebra $\mathfrak{k}$, and choose a K-invariant volume θ on M. Then $\mathcal{L}_X(\theta) = \mu_X \cdot \theta$ defines a function μ_X on M, which we may think of as the "divergence" of X. One may define an "infinitesimal" Godbillon-Vey class by the formula

$$\omega(X_0, X_1, \cdots, X_q) = \int \mu_{X_0}\, d\mu_{X_1} \wedge \cdots \wedge d\mu_{X_q}$$

and one sees easily that this defines an element of $H^{q+1}(\mathfrak{g}/\mathfrak{k})$, independent of θ.

In [5] is constructed, given a discrete subgroup Γ of G, a map $H^*(\mathfrak{g}/\mathfrak{k}) \to H^*(B\Gamma)$. In the specific case of our standard $PSL(2,\mathbb{R})$ action on S^1, one can check easily that this "infinitesimal" Godbillon-Vey class agrees with our "global" Godbillon-Vey class.

Using the formula $(*)$, we can construct foliated circle bundles having arbitrary Godbillon-Vey number, according to the following scheme due to Thurston. Let a and b be any two points in the upper half plane. Then there is a unique parallelogram (in the sense that opposite sides are of equal length) having as three vertices i, a, and b. Labelling the fourth vertex c, then there are unique elements f and g satisfying

$$f(c) = a \qquad f(b) = i$$

$$g(c) = b \qquad g(a) = i$$

Then $fgf^{-1}g^{-1}(i) = i$, and so $fgf^{-1}g^{-1}$ is a rotation. A little non-Euclidean geometry shows us that the rotation is the non-Euclidean area of the parallelogram. Since the parallelogram may have an arbitrary area, we have shown

LEMMA. <u>For any</u> $0 < \alpha < 2\pi$, <u>there are</u> $f, g \in PSL(2, \mathbb{R})$ <u>whose commutator is a rotation through angle</u> α.

Now let G be the fundamental group of the two-holed torus. G is given as $\{X, Y, Z, W : XYX^{-1}Y^{-1} = ZWZ^{-1}W^{-1}\}$, and BG has a fundamental 2 - cycle

$$[BG] = (X, Y) - (XYX^{-1}, X) - (XYX^{-1}Y^{-1}, Y) - (Z, W) + (ZWZ^{-1}, Z) + (ZWZ^{-1}W^{-1}, W)$$

and this is the object on which, for careful choices of X, Y, Z, and W, we want to evaluate our cocycle ω.

Fixing a positive integer n, and some $\alpha < \frac{2\pi}{n}$, choose f and g with commutator α, and f' and g' with commutator $n\alpha$.

Set $X = g$, and $Y = f$. We set $\widetilde{f} = (x) = \frac{1}{n} f'(nx)$, and similarly for $\widetilde{g}$ - we may think of $\widetilde{f}$ and $\widetilde{g}$ as liftings of f' and g' to an n - fold covering of the circle. Of course it is clear that the commutator of $\widetilde{f}$ and $\widetilde{g}$ is now α.

Now $\omega(g, f) - \omega(gfg^{-1}, g)$ is simply $(-4\pi^2) \cdot \alpha$, as can be seen by applying (*) to the parallelogram with vertices i, a, b, and c. $\omega(gfg^{-1}f^{-1}, g) = 0$, since $gfg^{-1}f^{-1}$ is a rotation. Similarly, $\omega(g', g') - \omega(g'f'g'^{-1}, g') = (-4\pi^2) \cdot n\alpha$, and it is easy to check by the original integral formula that passing to an n - fold covering simply multiplies the result by n. Hence choosing $X = g$, $Y = f$, $Z = \widetilde{g}$, $W = \widetilde{f}$, we get $\omega([BG]) = 4\pi^2(\eta^2 - 1)\alpha$.

Varying α between 0 and $\frac{2\pi}{n}$, and taking n arbitrarily large, we see that ω may take any real value.

Bibliography

1. R. Bott, M. Mostow, J. Perchik, Gelfand-Fuks cohomology and foliations, Proceedings of the Eleventh Symposium New Mexico State University, 1973.

2. R. Bott, Some aspects of Invariant theory, C.I.M.E. Lectures delivered at Varenna, August 1975, to be published.

3. R. Bott, Lectures on characteristic classes and foliations (Notes by Lawrence Conlon), Lecture Notes in Mathematics vol. 279, 1-94. Springer-Verlag, New York.

4. R. Bott, H. Shulman, J. Stasheff, On the de Rham theory of certain classifying spaces, to be published in Advances of Math.

5. J. L. Dupont, Semisimplicial de Rham cohomology and characteristic classes of flat bundles, to be published in Top.

6. J. Heitsch, Independant variation of secondary classes, to be published.

7. F. W. Kamber, P. Tondeur, Foliated bundles and characteristic classes, Lecture Notes in Mathematics vol. 493, Springer-Verlag, New York.

8. W. Thurston, Foliations and groups of diffeomorphisms, Bull. AMS, 80, 304-312.

Reprinted from "Representation Theory of Lie Groups," London Mathematical Society Lecture Notes Series, No. 34, pp. 65-90.

4 · The geometry and representation theory of compact Lie groups

RAOUL BOTT (Harvard University)
(Notes by G.L. Luke)

1. On the topology of compact Lie groups

A Lie group G is a C^∞-manifold which is also a group and such that the group laws:

multiplication $m : G \times G \to G$

inverse $i : G \to G$

are C^∞-maps.

Examples

(i) $\mathbb{R}^n, +$; the additive group of n-dimensional Euclidean space,

(ii) $R^*, \cdot$; the multiplicative group of the real line,

(iii) $GL(V)$; the automorphisms of a finite dimensional real vector space V. Choosing a basis for V gives an isomorphism of $GL(V)$ and $GL(n, \mathbb{R})$, the group of $n \times n$ invertible matrices with real coefficients.

If $\mathbb{C}$ denotes the complex numbers and $\mathbb{H}$ the quaternions then $GL(n, \mathbb{C})$ and $GL(n, \mathbb{H})$ are defined similarly.

(iv) compact examples are:

(a) $\mathbb{R}/\mathbb{Z}, +$, the quotient of the additive reals by the integers which is isomorphic to the circle group S^1 of complex numbers of unit norm.

(b) A cartesian product of Lie groups is also a Lie group. A product of circles

$$T^n = S^1 \times \ldots \times S^1 \quad , \quad n \text{ copies}$$

is called a torus.

(c) $SU(n+1)$ = the group of $(n+1)\times(n+1)$ matrices with complex coefficients and satisfying $X\cdot \bar{X}^t = 1$, $\det X = 1$. In particular, if $X \in SU(2)$ then $X = \begin{pmatrix} a & \alpha \\ -\bar{\alpha} & \bar{a} \end{pmatrix}$ when $|a|^2 + |\alpha|^2 = 1$. Hence $SU(2)$ is homeomorphic to S^3, the 3-sphere of unit vectors in $\mathbb{C}^2$.

The fact that the topological spaces S^1 and S^3 can be given the structure of a Lie group suggests the question: 'Which spheres are Lie groups?'. Perhaps the most accessible theorem in this area is the following.

Theorem. *None of the even dimensional spheres, S^{2n}, are Lie groups.*

Proof. Suppose S^{2n} is a Lie group. Consider the multiplication map $m : S^{2n}\times S^{2n} \to S^{2n}$. The restrictions $m : \{e\}\times S^{2n} \to S^{2n}$ and $m : S^{2n}\times\{e\} \to S^{2n}$ both correspond to the identity map.

Let $H^*(S^{2n},\mathbb{R})$ denote the cohomology ring of S^{2n} with real coefficients (the deRham cohomology computed from differential forms). Then $H^0(S^{2n},\mathbb{R}) \cong \mathbb{R}$, $H^{2n}(S^{2n},\mathbb{R}) \cong \mathbb{R}$, $H^m(S^{2n},\mathbb{R}) = 0$ in other degrees.

The multiplication map m induces a ring homomorphism

$$m^* : H^*(S^{2n},\mathbb{R}) \to H^*(S^{2n}\times S^{2n},\mathbb{R})$$

and it is well known that

$$H^*(S^{2n}\times S^{2n},\mathbb{R}) \cong H^*(S^{2n},\mathbb{R}) \otimes H^*(S^{2n},\mathbb{R}) \ .$$

Using this identification and 1 for the generator in $H^0(S^{2n},\mathbb{R})$, λ for the generator in $H^{2n}(S^{2n},\mathbb{R})$ we see that $m^*(1) = 1\otimes 1$ and $m^*(\lambda)$ must have the form $a(\lambda\otimes 1) + b(1\otimes\lambda)$ for some real numbers a and b. Restricting the multiplication map to $m^* : H^*(S^{2n},\mathbb{R}) \to H^*(S^{2n}\times\{e\},\mathbb{R}) \cong H^*(S^{2n},\mathbb{R})$

where it equals the identity, we see that $a = 1$ and similarly, $b = 1$. Hence $m^*(\lambda) = \lambda\otimes 1 + 1\otimes\lambda$.

However, m^* is also a ring homomorphism and hence

$$m^*(\lambda^2) = m^*(\lambda)^2 = (\lambda\otimes 1 + 1\otimes\lambda)^2$$
$$= \lambda^2\otimes 1 + (\lambda\otimes 1)(1\otimes\lambda) + (1\otimes\lambda)(\lambda\otimes 1) + 1\otimes\lambda^2 \quad .$$

But $\lambda^2 \in H^{4n}(S^{2n},\mathbb{R})$ and hence $\lambda^2 = 0$. Further, as λ may be represented by a form of even degree, we see that

$$(\lambda\otimes 1)(1\otimes\lambda) = (1\otimes\lambda)(\lambda\otimes 1) = \lambda\otimes\lambda$$

and hence $2(\lambda\otimes\lambda) = 0$. Finally, it is clear that $\lambda\otimes\lambda \in H^{2n}(S^{2n},\mathbb{R}) \otimes H^{2n}(S^{2n},\mathbb{R})$ is not zero. Hence no such multiplication exists.

Note: In the case of an odd dimensional sphere, the generator $\lambda \in H^{2n+1}(S^{2n+1},\mathbb{R})$ may be represented by a form of odd degree and so

$$(\lambda\otimes 1)(1\otimes\lambda) = -(1\otimes\lambda)(\lambda\otimes 1)$$

and the argument above fails.

This theorem has the following generalization.

Theorem (H. Hopf). *Let* G *be a compact connected Lie group, then* G *has the cohomology (real coefficients) of a product of odd dimensional spheres,*

$$H^*(G,\mathbb{R}) \cong H^*(\prod_\alpha S^{2k_\alpha-1},\mathbb{R}) \quad .$$

The numbers k_α where $2k_\alpha-1$ is the dimension of a component sphere, are called the exponents of G . The cohomology ring of a sphere of dimension $2k_\alpha-1$ is an exterior algebra on one generator $x_{2k_\alpha-1}$ in degree $2k_\alpha-1$ and hence

$$H^*(G;\mathbb{R}) \cong \bigotimes_i E(x_i); i = 2k_\alpha - 1$$

which raises the problem: 'Compute the exponents of G.'

This leads us to the main structure theorem for compact Lie groups.

Lemma. *Let G be a compact Lie group. Then G admits a Riemannian metric which is both left and right invariant.*

For simplicity, we will say that G is primitive if there exists only one bi-invariant Riemannian structure (up to a multiplicative constant).

Theorem. *Let G be a compact, connected, simply-connected primitive Lie group. Then G is one of the following types.*

$A_n \sim SU(n+1)$, $n = 0,1,\ldots$

$B_n \sim \widetilde{SO}(2n+1)$, *the simply connected covering of the special orthogonal group,* $n = 1,2,\ldots$

$C_n \sim Sp(n)$, *the group of* $n\times n$ *matrices with coefficients in* $\mathbb{H}$ *such that* $X\bar{X}^t = 1$, $n = 1,2,\ldots$

$D_n \sim \widetilde{SO}(2n)$, $n = 1,2,\ldots$

These groups are known as the classical groups, and the following, as the exceptional groups:

G_2 , F_4 , E_6 , E_7 , E_8 .

To find the exponents of a group we might proceed by a case-by-case attack. The enumeration of the classical groups suggests using induction. This works perfectly for $SU(n)$ and $Sp(n)$.

The group $SU(n)$ acts naturally on the sphere S^{2n-1} of unit vectors in $\mathbb{C}^n$ and the stability group of a point on the sphere is SU(n-1) . Hence he have the fibering

$$SU(n-1) \to SU(n) \to SU(n)/SU(n-1) \simeq S^{2n-1} .$$

Passing to cohomology the twisting disappears,

$$H^*(SU(n),\mathbb{R}) \cong H^*(SU(n-1),\mathbb{R}) \otimes H^*(S^{2n-1},\mathbb{R})$$

and hence, by induction,

$$H^*(SU(n),\mathbb{R}) \cong H^*(S^3,\mathbb{R}) \otimes H^*(S^5,\mathbb{R}) \otimes \ldots \otimes H^*(S^{2n-1},\mathbb{R}) .$$

A similar argument works for Sp(n) . The group acts naturally on the sphere S^{4n-1} of unit vectors in quaternionic space $\mathbb{H}^n$ with stability group Sp(n-1) which leads to

$$H^*(Sp(n),\mathbb{R}) \cong H^*(S^3,\mathbb{R}) \otimes H^*(S^7,\mathbb{R}) \otimes \ldots \otimes H^*(S^{4n-1},\mathbb{R}) .$$

It is interesting to note that the Betti numbers for these groups were obtained by R. Brauer in the 1920's by a quite different method.

For the special orthogonal groups, the corresponding fibering is $SO(n-1) \to SO(n) \to S^{n-1}$. The numbers n-1 and n are of different parity and the groups SO(n-1) and SO(n) belong to different series. A straightforward induction fails.

We pass now to other methods of computing cohomology. The cohomology of the group-manifold G may be computed via the deRham complex of differential forms, $(\Omega^*(G),d)$ where $\Omega^*(G)$ denotes the exterior algebra of differential forms with real coefficients and d the exterior derivative.

Let $Inv_{left}\Omega^*(G)$ or $\Omega^*(\mathfrak{g})$ denote the subcomplex of left invariant forms. Such a form is determined by its value at the identity $e \in G$ and hence $\Omega^*(\mathfrak{g})$ is a finite dimensional space. Let $\mathcal{H}^*(\mathfrak{g})$ denote the subcomplex of bi-invariant differential forms. We will show later that $d|\mathcal{H}^*(\mathfrak{g}) \equiv 0$.

Theorem. *Let* G *be a compact, connected Lie group, then*

$$H^*(G,\mathbb{R}) \cong H^*(\Omega^*(\mathfrak{g}),d) \cong \mathcal{H}^*(\mathfrak{g}) \quad .$$

Note that this isomorphism reduces the problem of computing the cohomology of G to a problem in invariant theory. It was by this method that R. Brauer determined the exponents of the classical groups. The method was also known to Cartan although the actual isomorphism theorem first appeared in a paper of Eilenberg and Chevalley in the 1940's.

Proof of the Theorem. A left invariant differential form is determined by its value at the identity and evaluation yields the isomorphism

$$\mathrm{Inv}_{\mathrm{left}}\Omega^*(G) \cong \Lambda^*(T_e^*(G)) \quad .$$

The exterior algebra $\Lambda^*(T_e^*(G))$ may be interpreted as the multilinear alternating forms on the tangent space $T_e(G)$ of G at the identity. Further, $T_e(G)$ is isomorphic to the Lie algebra $\mathfrak{g}$ of left invariant vector fields on G and an invariant metric gives us an isomorphism between $\mathfrak{g}$ and $\mathfrak{g}^*$ which explains our notation $\Omega^*(\mathfrak{g})$.

Consider the inclusion $\Omega^*(\mathfrak{g}) \hookrightarrow \Omega^*(G)$. Averaging the left translations gives a projection

$$\int L_g^* dg : \Omega^*(G) \to \Omega^*(\mathfrak{g})$$

and it follows that both $\Omega^*(\mathfrak{g})$ and $H^*(\Omega^*(\mathfrak{g}))$ are direct summands in $\Omega^*(G)$ and $H^*(\Omega^*(G))$. Note that if w is a closed form then so is dL_g^*w . The connectedness of G implies that e and g may be joined by a path and hence w and $L_g^* w$ are cohomologous. Finally $\int (L_g^* w - w)dg$ is an average of coboundaries and hence a coboundary, so w and $\int L_g^* w$ are cohomologous. This shows that

$$H^*(G,\mathbb{R}) \cong H^*(\Omega^*(\mathfrak{g})) \quad .$$

We may use the same argument and right translation restricted to $\Omega^*(\mathfrak{g})$ to show that $\Omega^*(\mathfrak{g})$ and $H^*(\mathfrak{g})$ have the same cohomology and it remains only to show that $d|H^*(\mathfrak{g}) \equiv 0$.

The map induced on $T_e^*(G)$ by the inverse map i is simply -1 . Hence the induced map on $\Lambda^p T_e^*(G)$ is $(-1)^p$. Notice also that i^* preserves bi-invariant forms. Hence if w is a bi-invariant p-form then

$$(-1)^{p+1}dw = i^* dw = di^* w = (-1)^p dw$$

and hence $dw = 0$ as required.

2. The adjoint action and its orbits

Throughout this section we will assume that G is compact and connected. We commence by describing our notation.

Let $g \in G$ and $X \in T_e(G)$. Then gX denotes the image of X under the map induced on the tangent space by left translation by g , similarly Xg for right translation and gXg^{-1} for conjugation. We have already defined the adjoint representation of G on $T_e(G)$ by $Ad(g)X = gXg^{-1}$. This is the canonical finite dimensional representation possessed by every Lie group. The derivative of Ad is denoted by

$$ad : T_e(G) \to End(T_e(G))$$

and, in fact, $ad(X)(Y) = [X,Y]$ for $X,Y \in T_e(G)$.

This is easy to see when G is a matrix group. Then

$Ad(g)(X) = gXg^{-1}$ and $\exp(tX) = 1 + tX + \frac{t^2X^2}{2!} + \ldots$ (matrix multiplication). Hence

$$\begin{aligned} ad(X)(Y) &= \frac{d}{dt} Ad(tX)(Y)\Big|_{t=0} \\ &= \frac{d}{dt}\{e^{tX}Y\, e^{-tX}\}\Big|_{t=0} \\ &= [X,Y]\ , \text{ the coefficient of } t\ . \end{aligned}$$

For example $G = SO(3)$. The tangent space $T_e(G)$ consists of skew symmetric matrices and taking the basis

$$e_1 = \begin{pmatrix} 0 & 1 & 0 \\ -1 & 0 & 0 \\ 0 & 0 & 0 \end{pmatrix},\quad e_2 = \begin{pmatrix} 0 & 0 & 1 \\ 0 & 0 & 0 \\ -1 & 0 & 0 \end{pmatrix},\quad e_3 = \begin{pmatrix} 0 & 0 & 0 \\ 0 & 0 & 1 \\ 0 & -1 & 0 \end{pmatrix},$$

the adjoint representation of $SO(3)$ becomes the usual representation on $\mathbb{R}^3$. The (non-trivial) orbits of the Ad action are the 2-spheres centered at 0 and every sphere cuts a line through 0 in two points. We will generalize this fact later and hence we give it a technical proof.

Fix $X \neq 0$ on a given line and consider the function defined on a sphere by $Y \to \langle X,Y\rangle$ where $\langle\ ,\ \rangle$ is the Euclidean inner product on $\mathbb{R}^3$. The sphere intersects the line precisely at the maximum and minimum of the function.

Returning to the general compact group G with Lie algebra $T_e(G) = \mathfrak{g}$ and $\langle\ ,\ \rangle$ any Ad-invariant inner product on $\mathfrak{g}$. The infinitesimal version of the invariance $\langle Ad(g)X, Ad(g)Y\rangle = \langle X,Y\rangle$ is computed by putting $g = e^{tZ}$ and differentiating at $t = 0$. This yields

$$\langle [Z,X],Y\rangle + \langle X,[Z,Y]\rangle = 0\ .$$

For $X \in \mathfrak{g}$, let $\mathfrak{g}_X = \{Y \mid [X,Y] = 0\}$, the centralizer of X and $\mathfrak{g}^X = [X,\mathfrak{g}]$, the image of adX. Then $\mathfrak{g}_X$ is a subalgebra of $\mathfrak{g}$, the Lie algebra of $G_X = \{g \in G \mid Ad(g)X = X\}$ the stabilizer of X. These spaces fit together in an exact sequence

$$0 \to \mathfrak{g}_X \to \mathfrak{g} \to \mathfrak{g}^X \to 0 .$$

Lemma. *The spaces $\mathfrak{g}_X$ and $\mathfrak{g}^X$ are perpendicular complements.*

Proof. Let $Y \in (\mathfrak{g}^X)^\perp$. Then $\langle [Y,X],\mathfrak{g}\rangle = \langle Y,[X,\mathfrak{g}]\rangle = 0$ and so $[Y,X] = 0$, $Y \in \mathfrak{g}_X$ and $(\mathfrak{g}^X)^\perp \subseteq \mathfrak{g}_X$. The argument is completed by noting that $\dim(\mathfrak{g}^X)^\perp = \dim \mathfrak{g}_X$.

An element $X \in \mathfrak{g}$ is said to be regular if $\dim \mathfrak{g}_X \leq \dim \mathfrak{g}_Y$ for all $Y \in \mathfrak{g}$.

Lemma. *If X is regular then $\mathfrak{g}_X$ is abelian.*

Proof. Suppose $\mathfrak{g}_X$ is not regular. Then there exist $Y,Z \in \mathfrak{g}_X$ such that $[Y,Z] \neq 0$.

Consider $\mathfrak{g}_{X+tZ}$ for t small. Then $[X+tY,X] = 0$ implies that $ad(X+tY)$ preserves the decomposition $\mathfrak{g} = \mathfrak{g}_X \oplus \mathfrak{g}^X$. Further, since $ad(X)$ acts isomorphically on $\mathfrak{g}^X$, so does $ad(X+tY)$ for t sufficiently small. Hence $\mathfrak{g}^X \subseteq \mathfrak{g}^{X+tY}$ and $\mathfrak{g}_{X+tY} \subseteq \mathfrak{g}_X$ by the previous lemma. However, $Z \notin \mathfrak{g}_{X+tY}$ and hence $\dim_{X+tY} < \dim \mathfrak{g}_X$ which is a contradiction. When X is regular, the abelian Lie algebra $\mathfrak{g}_X$ is called a Cartan subalgebra of $\mathfrak{g}$. It is not hard to show that the Cartan subalgebras are the maximal abelian subalgebras of $\mathfrak{g}$. These generalize the lines through the origin in the case of $SO(3)$.

First Conjugacy Theorem. *Let G be a compact connected Lie group with Lie algebra $\mathfrak{g}$ and H a Cartan subalgebra of $\mathfrak{g}$. Then for each $Y \in \mathfrak{g}$, the adjoint orbit*

$$O(Y) = Ad(G)(Y)$$

intersects H *in a finite non-empty set of points.*

Proof. Let $X \in \mathfrak{g}$ be such that $H = \mathfrak{g}_X$, $\langle\ ,\ \rangle$ be an Ad-invariant inner product on $\mathfrak{g}$ and $f : O(Y) \to \mathbb{R}$ be defined by $f(Z) = \langle Z,X\rangle$. The orbit $O(Y)$ is compact and hence $f\ O(Y)$ achieves its minimum. We can assume $Y \in O(Y)$ is the minimum.

For each $Z \in \mathfrak{g}$, $\frac{d}{dt} f(\mathrm{Ad}(\exp tZ)(Y))\big|_{t=0}$ since $f(Y)$ is the minimum of $f|O(Y)$. Hence $\langle [Z,Y],X\rangle = 0$ for all $Z \in \mathfrak{g}$,

$$\langle[\mathfrak{g},Y],X\rangle = \langle Y,[X,\mathfrak{g}]\rangle = \langle Y,\mathfrak{g}^X\rangle = 0$$

and hence $Y \in \mathfrak{g}_X = H$.

Notice that if $Y \in H$ then $O(Y)$ meets H perpendicularly since $\mathfrak{g}_Y$ and $\mathfrak{g}^Y$ are perpendicular. This implies that $O(Y)$ intersects H in a discrete and hence a finite set of points. Also, the argument actually shows that all the critical points of $f|O(Y)$ belong to $O(Y) \cap H$.

Corollary. *Let* H_1 *and* H_2 *be Cartan subalgebras of* $\mathfrak{g}$. *Then there exists* $g \in G$ *such that* $\mathrm{Ad}(g)H_1 = H_2$.

Proof. Let $H_1 = \mathfrak{g}_{X_1}$ and $H_2 = \mathfrak{g}_{X_2}$. Then there exists $g \in G$ such that $\mathrm{Ad}(g)X_2 \in H_1$. Hence $\mathrm{Ad}(g)H_2 = H_1$.

Let H be a Cartan subalgebra of $\mathfrak{g}$ and $W(H)$ denote the group of automorphisms of H induced by the adjoint automorphism of G. Then $W(H)$ is called the Weyl group of H and is easily seen to be finite.

Proposition. *For each* $Y \in \mathfrak{g}$ *and Cartan subalgebra* H, *the intersection* $O(Y) \cap H$ *is an orbit of the Weyl group.*

Proof. Let $X,Y \in H$ be such that $X = \mathrm{Ad}(g)(Y)$ for some $g \in G$. Then $\mathfrak{g}_X = \mathfrak{g}_{\mathrm{Ad}(g)Y} = \mathrm{Ad}(g)\mathfrak{g}_Y$. But $H \subseteq \mathfrak{g}_Y$ and hence

$Ad(g)H \subseteq \mathfrak{g}_X$. Therefore H and $Ad(g)H$ are both Cartan subalgebras of $\mathfrak{g}_X$.

Let G_X be the stabilizers of X and G_X^o the connected component of the identity. Then $\mathfrak{g}_X$ is the Lie algebra of the compact group G_X^o and there exists $Z \in \mathfrak{g}_X$ such that $Ad(\exp Z)Ad(g)H = H$. But $Ad(\exp Z)(X) = X$ and so

$$X = Ad(\exp Z)(X) = Ad\{(\exp Z) \cdot g\}(Y)$$

and $Ad\{(\exp Z) \cdot g\}$ induces a Weyl group transformation of H .

We illustrate these notions for the group $U(n)$. Then the Lie algebra consists of skew Hermitian matrices which we identify with the Hermitian matrices by multiplication by $\sqrt{-1}$. The adjoint representation is equivalent to the representation of $U(n)$ acting on Hermitian matrices by conjugation.

A diagonal Hermitian matrix corresponds to a regular element of $\mathfrak{g}$ if and only if all the diagonal elements are distinct and the corresponding Cartan subalgebra consists of all the diagonal Hermitian matrices. The Weyl group is the set of permutations of the diagonal elements. Finally, the First Conjugation Theorem becomes the well known fact that any Hermitian matrix can be put into diagonal form.

For a general compact Lie group G and Cartan subalgebra $H \subseteq \mathfrak{g}$, the set of singular elements (non-regular) of H is a finite union of hyperplanes. This configuration is called the Infinitesimal Diagram of G .

For example, let $G = SU(3)$ and H be the set of real diagonal matrices of trace 0 . Let $Y \in H$ be given by

$$Y = \begin{pmatrix} \lambda_1 & 0 & 0 \\ 0 & \lambda_2 & 0 \\ 0 & 0 & \lambda_3 \end{pmatrix} , \qquad \lambda_1 + \lambda_2 + \lambda_3 = 0 .$$

Then the diagram of $SU(3)$ consists of the 2-plane H and hyperplanes $\lambda_1 = \lambda_2$, $\lambda_2 = \lambda_3$, $\lambda_1 = \lambda_3$.

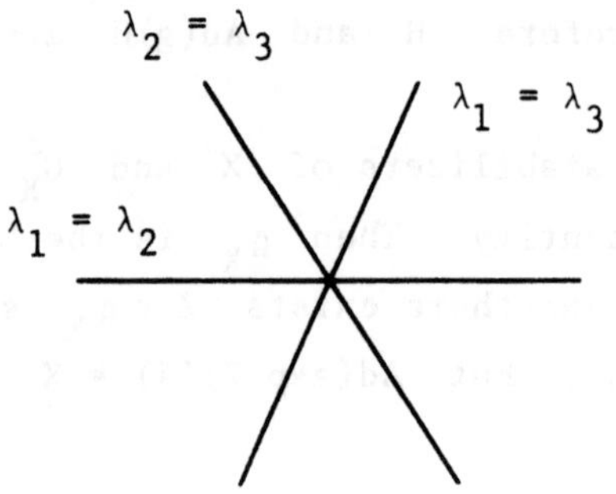

The Weyl group is generated by reflecting across the hyperplanes in the diagram and the orbits have the following form:

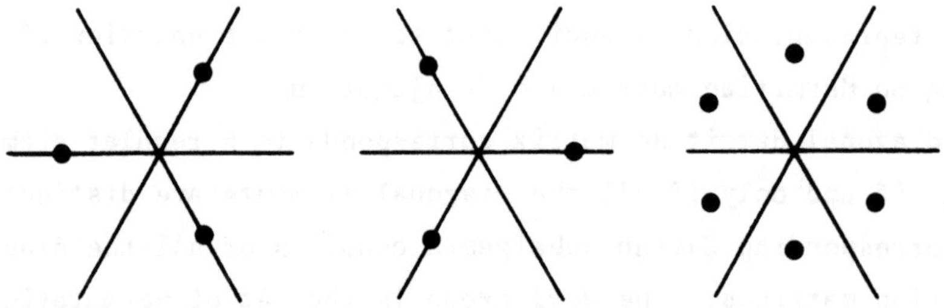

In the general orbit there are 6 critical points, not simply a maximum and minimum.

Theorem (Bott). *The critical points of* f *are all non-degenerate and an index of a critical point is equal to twice the number of hyperplanes crossed by a line joining* X *to the critical point.*

The proof of this theorem appears in [2] and will not be given here. We content ourselves with illustrating the result for SU(3) .

For the general orbit, the critical points have indices as illustrated on the right below and the orbit in $\mathfrak{g}$ is the flag manifold

$U(3)/U(1) \times U(1) \times U(1)$.

For the first degenerate case the orbit is the complex projective space $U(3)/U(2)\times U(1)$ and the critical points have indices 0,2,4

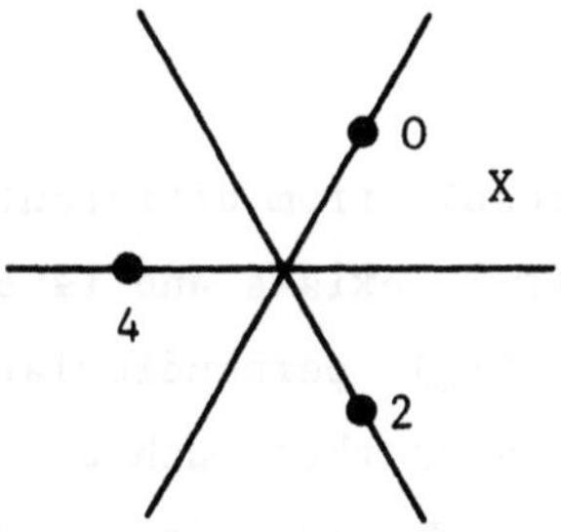

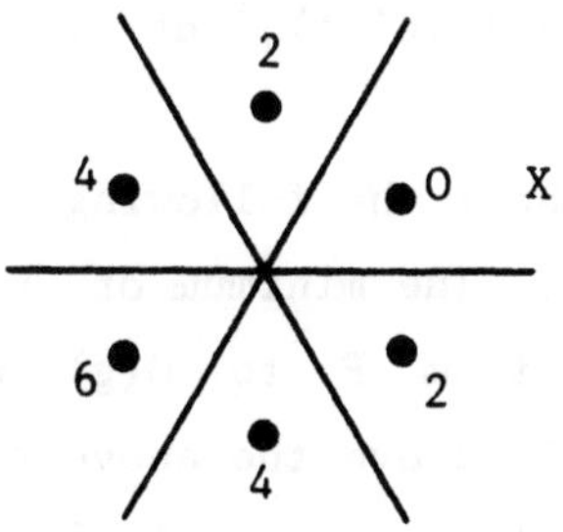

Returning to the theorem, we see that as a consequence of general Morse theory, each adjoint orbit is a union of even dimensional cells, one for each critical point. It follows that $H^*(O(Y),\mathbb{Z})$ is torsion-free and the i-th Betti number is the number of critical points of index i . Hence the cohomology of an orbit can be computed from the infinitesimal diagram of the group.

3. The Poincaré series of the loop space of a compact group

The exponential map $\exp : \mathfrak{g} \to G$ restricts to a diffeomorphism between a neighbourhood of $0 \in \mathfrak{g}$ and $e \in G$. If $H \subseteq \mathfrak{g}$ is a Cartan subalgebra and $T = \exp H$, the corresponding maximal torus, then $\exp : H \to T$ may be thought of as the universal covering map. As in the case of the algebra H , we define the Weyl group of T to be the group of automorphisms of T induced by conjugation in G .

Theorem. *Each orbit of the conjugation action of* G *intersects the maximal torus* T *precisely in the orbit of the Weyl group.*

Let <u>Ad</u> denote the conjugation action and let G have a bi-invariant Riemannian metric. Then let $P \in T$ be a generic point - not in the image of the singular set of $H = T_e(T)$. By $\Omega(P,O(g))$ we denote the space of smooth paths from P to $O(g)$, the Ad-orbit through g.

We define the energy path u by

$$E(u) = \int |u'(t)|^2 \, dt \quad .$$

Then we have the following fundamental result from differential geometry. The minimum of E on $\Omega(P,O(g))$ exists and is a geodesic from P to $O(g)$ which meets $O(g)$ perpendicularly.

We will prove the above theorem by showing that such a geodesic lies completely in T and hence T and $O(g)$ actually intersect.

Lemma. *Let G be a Lie group which acts on a Riemannian manifold M by isometries. Then a geodesic of M which intersects one of the G-orbits perpendicularly intersects them all perpendicularly.*

Proof. Let $u:[0,1] \to M$ be a geodesic (parametrized by path length) which meets an orbit $O(m)$, $m \in M$ perpendicularly at $u(1)$.

Each $X \in \mathfrak{g}$, the Lie algebra of G, induces a Killing field ξ_X on M and a 1-parameter family of isometries corresponding to e^{sX}. There results a perturbation:

$$u(\, , \,) : [0,1] \times (-\varepsilon,\varepsilon) \to M$$

of the geodesic $u = u(\cdot,0)$ by geodesics $u(\cdot,s)$ with the property that $u(1,s)$ meets $O(m)$ perpendicularly.

Consider $g(t) = (\xi_X(u(t)) \, , \, u'(t))$, the inner product of the Killing field and the velocity vector field along the geodesic. Then if ∇ denotes the Laplace-Beltrami connection operator, we have

$$u'(t)g = (\nabla_{u'(t)}\xi_X, u'(t))$$

$$= (\nabla_{\xi_X} u', u')(u(t)) \quad \text{since} \quad [\xi_X, \frac{d}{dt} u(t,s)] = 0$$

$$= \frac{1}{2}\, \xi_X(u(t))(u', u')$$

$$= 0 \quad \text{since } u \text{ is parametrized by path length.}$$

Hence $g(t)$ is constant along the geodesic u and since it vanishes at $u(1)$ it vanishes everywhere along u.

It follows that u is perpendicular to all the orbits it intersects since ξ_X gives a typical tangent vector to an orbit.

Returning to the theorem, for a general point $P \in T$, $T_P(O(P))$ is the orthogonal complement to $T_P(T)$. Hence a geodesic from P to $O(g)$ which is perpendicular to $O(g)$ is also perpendicular to $O(P)$ and hence lies in T. Therefore $O(g) \cap T \neq \phi$.

The proof that the intersection is an orbit of the Weyl group is similar to that given for the Cartan subalgebra.

We now study the indices of the critical points of the energy function E. In particular, we study the critical points of E on the paths from $P \in T$ to the identity $e \in G$. We know that the critical points are geodesics.

When $G = SU(3)$ and T is the maximal torus of special unitary diagonal matrices with Lie algebra denoted by $\mathfrak{t}$, then the Lie algebra $\mathfrak{g}$ decomposes under the adjoint action restricted to T as

$$\mathfrak{g} = \mathfrak{t} \oplus E_{\alpha_1} \oplus E_{\alpha_2} \oplus E_{\alpha_3}$$

with T acting trivially on t and where $E_{\alpha_1}, E_{\alpha_2}, E_{\alpha_3}$ are distinct two dimensional representations of T

$$\alpha_i : T \to SO(2) \ , \quad i = 1,2,3 \ .$$

Let $U_{\alpha_i} \subseteq T$ be the kernel of α_i. Then the configuration

$$\bigcup_{i=1}^{3} \exp^{-1}(U_{\alpha_i}) \subseteq t$$

is called the Diagram of G. We illustrate this with the Infinitesimal Diagram marked in thicker lines.

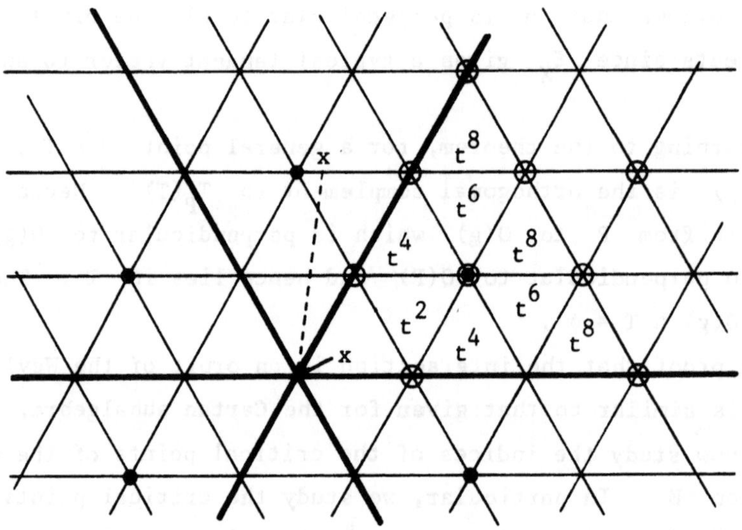

The vertices correspond to $\exp^{-1}(Z(G))$ where $Z(G)$ denotes the centre of G, and those distinguished by a dot to $\exp^{-1}(e)$ The geodesics joining a point P to e, lift in the covering to the straight lines joining the points of $\exp^{-1}(P)$ to 0.

We fix a fundamental Weyl chamber, that is, a component of the infinitesimal diagram. Then each simplex containing a point of $\exp^{-1}(P)$ corresponds, under the Weyl group, to a unique simplex in the fundamental chamber (this fact depends on

$\pi_1(SU(3)) = 0$). Hence, by using the Weyl group, we may move our geodesics into the fundamental chamber so that they are in one-to-one correspondence with the simplices.

As before, the index of the energy function on $\Omega(P,e)$ corresponding to a particular geodesic, is twice the number of lines crossed by the geodesic in the Diagram. This is also illustrated.

Of course, this generalizes to an arbitrary compact, connected, simply connected Lie group G.

We define the Poincaré series of a space X to be

$$P_t(X) = \sum_{i=0}^{\infty} t^i \dim H^i(X,\mathbb{R})$$

and ΩX to be the set of continuous loops in X from a fixed base point. We note that ΩG is homotopy equivalent to $\Omega(P,e)$. Then as a corollary of the previous theorem, since all the indices of E on $\Omega(P,e)$ are even, $H^*(\Omega G,\mathbb{Z})$ has no torsion and

$$P_t(\Omega G) = \sum t^{\lambda(\Delta)}$$

where the sum is taken over the simplices on the fundamental chamber and $\lambda(\Delta)$ is twice the number of hyperplanes crossed by a line joining an interior point of the simplex to 0.

When G is a simple group, then there is a unique leading simplex in the fundamental chamber with index 2. Hence $\pi_2(\Omega G) = \mathbb{Z}$ and so $\pi_3(G) = \mathbb{Z}$, $\pi_2(G) = 0$.

These results give us a neat method for computing the Betti numbers from the Diagram. We illustrate it for $G = SU(3)$.

Consider the following strip of the diagram.

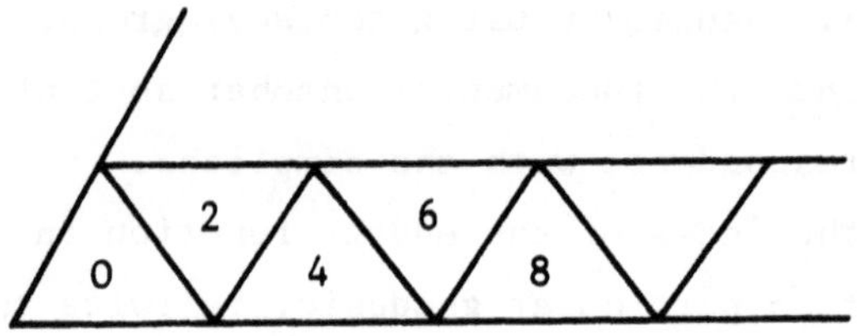

In each rhombus the indices are 4 greater than those to the immediate left and hence the contribution of this strip to the Poincaré series is

$$(1+t^2) + t^4(1+t^2) + (t^4)^2(1+t^2) + \dots = \frac{(1+t^2)}{1-t^4} .$$

Applying the same technique to the series of strips

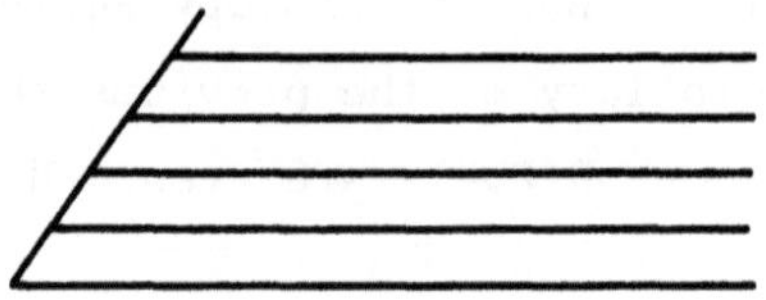

we see that

$$P_t(\Omega SU(3)) = \frac{1+t^2}{(1-t^4)^2} = \frac{1}{(1-t^2)(1-t^4)} .$$

Recall that $H^*(SU(3),\mathbb{R}) = H^*(S^3),\mathbb{R}) \otimes H^*(S^5,\mathbb{R})$ and so $\frac{1}{1-t^2}$ corresponds to S^3 and $\frac{1}{1-t^4}$ to S^5.

By comparing this Morse theory approach with the standard constructions in topology one obtains a relation between the exponents - which is easily computable and for E_8 - actually gives all the exponents.

The idea is this:

The general rule for computing $P_t(\Omega G)$ from the correspondin spheres is

$$P_t(\Omega G) = \Pi \frac{1}{1-t^{\ell_i-1}}$$

where ℓ_i is the dimension of a component sphere. The formula holds for G simple, connected and simply connected. Hence we have: If

$$P_t(G) = \Pi(1+t^{2k_i-1}) \text{ , then } P_t(\Omega G) = \Pi \frac{1}{(1-t^{2k_i-2})} \text{ .}$$

On the other hand using the Diagram we find that

$$P_t(\Omega G) = \frac{Q(t)}{\pi(1-t^{2a_i})}$$

where Q is a polynomial such that $Q(1) = 1$ and $\Sigma a_i \alpha_i$ *is the sum of the positive roots* expressed in terms of the *simple roots* α_i . By comparing the behaviour of these series near $t = 1$, it can be shown that

$$\ell ! \Pi d_i \Pi(d_i-1) = \Pi a_i$$

where ℓ is the rank of G and the d_i come from expressing the highest root μ in terms of the simple roots $\alpha_i, \mu = \Sigma d_i \alpha_i$ [2] .

In the case of E_8 , this formula actually gives an expression for $\Pi(k_i-1)$ as a product of ℓ primes

$$\Pi(k_i-1) = 1.7.11.13.17.19.23.29$$

and hence

$$P_t(E_8)$$

$$= (1+t^3)(1+t^{15})(1+t^{23})(1+t^{27})(1+t^{35})(1+t^{39})(1+t^{47})(1+t^{59})$$

4. The Weyl character formula

We commence by reviewing the theory of characters of representations of compact groups.

Let G be a compact group and W a representation of G. Then W is a vector space equipped with a continuous homomorphism from G into the group of automorphisms of W, denoted $x \to W(x)$, $x \in G$. The character χ_W of W is defined by

$$\chi_W(x) = \text{Trace } W(x) , \quad x \in G$$

and characters have the properties:

1) $\chi_{W_1 \oplus W_2} = \chi_{W_1} + \chi_{W_2}$

2) $\chi_{W_1 \otimes W_2} = \chi_{W_1} \cdot \chi_{W_2}$

3) $\int_G \chi_W(x)\,dx = \dim W^G$ where W^G denotes the space of vectors fixed by G. (This property follows from applying the trace homomorphism to the projection $P_W = \int_G W(x)\,dx$ onto W^G.)

4) If W^* denotes the contragredient or dual representation to W, then

$$\chi_{W^*}(x) = \chi(x^{-1}) = \overline{\chi(x)}$$

Next we describe the Schur orthogonality relations. Suppose W and W' are nonequivalent, irreducible representations of G. Then $\text{Hom}(W,W')^G = 0$ by Schur's lemma, that is $(W^* \otimes W')^G$ has zero invariant subspace. Hence $\int_G \overline{\chi_W}\chi_{W'} = 0$. If W and W' are equivalent and irreducible then $\dim \text{Hom}(W,W')^G = 1$ and hence $\int_G \overline{\chi_W}\chi_{W'} = 1$. Therefore, the characters form an orthonormal set of functions on G.

Notice that the characters are class functions (constant on conjugacy classes) and the Peter-Weyl theorem gives us that the characters of irreducible representations form an orthonormal basis of class functions. Hence, W can be expressed as a direct sum of irreducible representations

$$W \simeq a_1W_1 \oplus a_2W_2 \oplus \dots \oplus a_nW_n$$

where the $a_i = \dim \operatorname{Hom}(W,W_i)^G = \int_G \overline{\chi_W}\chi_{W_i}$.

Suppose now that G is a compact, connected Lie group. The representation theory reduces to finding the characters of the irreducible representations. We have shown that every conjugacy class in G intersects a fixed maximal torus T and hence a character of a representation is determined by its restriction to T. Of course, each representation W of G restricts to a representation $W|T$ and $\chi_W|T = \chi_{W|T}$. Hence an initial step in calculating the irreducible representations of G is to find the representations and characters of T.

Let $\hat{T}$ denote the set of continuous homomorphisms from T into the circle group. Hence $\hat{T}$ is the set of irreducible representations of T. Then if $\mathbb{Z}[\hat{T}]$ is the ring of formal, finite linear combinations of elements of $\hat{T}$, with integer coefficients, the restriction $W|T$ corresponds, via its character, to an element

$$\sum a_i\lambda_i \in \mathbb{Z}[\hat{T}] \quad , \quad a_i = 0,1,2,\dots,\lambda_i \in \hat{T} \ .$$

The simplest example occurs when $T = S^1$, the circle group. Then $\hat{S}^1 \cong \mathbb{Z}$. If we let z be a generator of $\hat{S}^1$ so $\hat{S}^1 = \{\dots,z^{-2},z^{-1},1,z,z^2,\dots\}$, then

$$\mathbb{Z}[\hat{S}^1] = \mathbb{Z}[z,z^{-1}] \ .$$

When $G = SU(2)$, we have a natural representation on $\mathbb{C}^2$. The torus

$$T = \left\{ \begin{pmatrix} z & 0 \\ 0 & z^{-1} \end{pmatrix} \middle| \; |z| = 1 \right\}$$

is maximal in $SU(2)$ and acts naturally on the symmetric powers $S^k(\mathbb{C}^2)$. Thinking of $S^k(\mathbb{C}^2)$ as the space of polynomials in two variables, x and y say, homogeneous of degree k, the element $\begin{pmatrix} z & 0 \\ 0 & z^{-1} \end{pmatrix}$ of the torus acts by

$$\begin{pmatrix} z & 0 \\ 0 & z^{-1} \end{pmatrix} x^{k-i} y^i = z^{k-2i} x^{k-i} y^i \quad .$$

Hence the character of this representation of T on $S^k(\mathbb{C}^2)$ is

$$\left(z^{-k} + z^{-(k-2)} + \ldots + z^{(k-2)} + z^k \right)$$

and this formula also gives the corresponding element of $\mathbb{Z}[\hat{T}]$ where now z^ℓ denotes the homomorphism

$$\begin{pmatrix} z & 0 \\ 0 & z^{-1} \end{pmatrix} \to z^\ell \quad .$$

Note that these elements of $\mathbb{Z}[\hat{T}]$ are symmetric under the substitution $z \to z^{-1}$. This follows from the fact that the Weyl group acts on $\mathbb{Z}[\hat{T}]$ fixing those elements which come from restriction of representations of G.

Consider the product

$$(z^{-k} + z^{-(k-2)} + \ldots + z^{k-2} + z^k)(z - z^{-1}) = z^{k+1} - z^{-(k+1)}$$

Such an expression $z^r - z^{-r}$ is called an elementary alternating expression and the product shows us that the character of $S^k(\mathbb{C}^2)$ has the form

$$\frac{\text{elementary alternating expression}}{\text{smallest elementary alternating expression}} .$$

This is the Weyl Character Formula for SU(2) .

For a general compact connected group G with maximal torus T , let W(T) be the Weyl group and $\lambda \in \hat{T}$. Then

$$\sum_{w \in W(T)} (-1)^w \lambda^w$$

denotes the element of $\mathbb{Z}[\hat{T}]$ given by the sum of λ^w , the image of λ under the action of $w \in W(T)$ on $\hat{T}$ with coefficient $(-1)^w$, the determinant of the action of w on the Lie algebra of T . For an irreducible representation V of G with highest weight λ , the Weyl Character Formula is

$$\chi_V | T = \frac{\Sigma(-1)^w (\lambda+\rho)^w}{\Sigma(-1)^w \rho^w}$$

where ρ denotes half the sum of the positive roots. Here we have used the same symbol to denote both a character and its derivative. Also, although $\Sigma(-1)^w \rho^w$ may not actually give a character unless we pass to a covering of G , the quotient certainly makes sense on G . This formula shows how to find the complete set of characters of irreducible representations of G by alternating vertices of the Diagram.

To construct a representation of G from a given $\lambda \in \hat{T}$ in the fundamental chamber, we may proceed as follows. Let L_λ be the line bundle over G/T constructed from the representation λ , $L_\lambda = G \times_\lambda \mathbb{C}$. It can be shown that G/T is a complex manifold and that L_λ is a complex analytic line bundle. The natural representation of G on the holomorphic sections of L_λ is an irreducible representation with highest weight λ . This result is known as the Borel-Weil Theorem.

Appendix: We describe the Weyl Character Formula in terms of the Diagram of G (assuming G simple and simply connected).

By using the canonical bilinear form defined on the Lie algebra of T in terms of the roots, the weights may be interpreted as vertices of the Diagram. For a weight λ , we have the following alternating expression $\Sigma(-1)^w\lambda^w$ and the smallest alternating expression comes from alternating the vertex nearest to the origin in the fundamental chamber, namely ρ .

For $G = SU(3)$, the Diagram with ρ is as illustrated

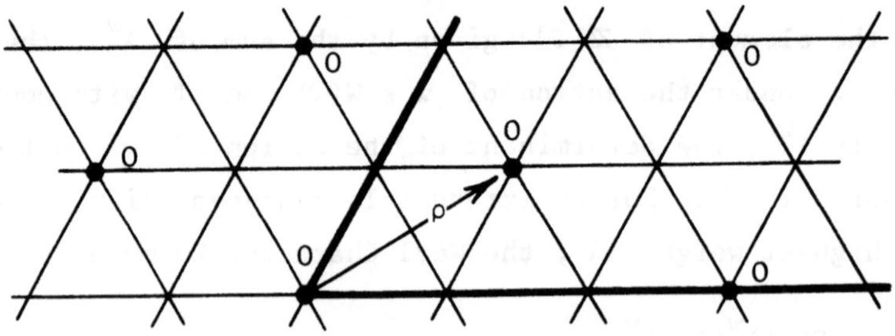

and the formula $\chi_V.\Sigma(-1)^w\rho^w = \Sigma(-1)^w(\lambda+\rho)^w$ may be interpreted geometrically as follows. The hexagon

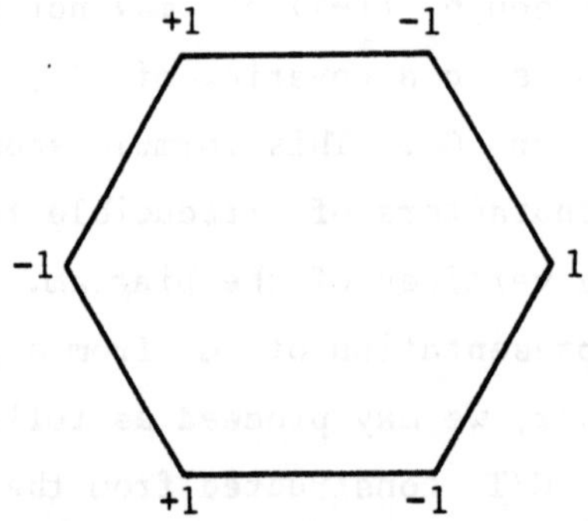

corresponds to the smallest alternating expression $\Sigma(-1)^w\rho^w$.

In terms of the Diagram, with vertices labelled according to the multiplicity of a weight occurring in the representation V , multiplication of a weight τ with multiplicity k by $\Sigma(-1)^w\rho^w$, corresponds to translating the hexagon on the Diagram to centre at the vertex τ and then replacing the label k at τ

by $\pm k$ at the vertices of the translated hexagon. Finally, on summing all the new labels, we obtain $\Sigma(-1)^{w}(\lambda+\rho)^{w}$.

In this way the Hermann-Weyl formula yields the character of $V\lambda$ as the solution of a difference equation; and one can solve it inductively by the following procedure.

1) Start with putting a 1 under some lattice point λ of the fundamental chamber.

2) Now for our hexagon centered anywhere but at $\lambda+\rho$, in the fundamental chamber, the alternating sum of the multiplicities of V must add up to 0.

3) Use the fact that the multiplicities occur symmetrically relative to the Weyl group. For instance in the last figure I have indicated the result of this procedure when applied to the representation with maximal weight ρ^{2}, and I recommend that after cutting out the above hexagon the reader compute the character of the representation with ρ^{3} as maximal weight.

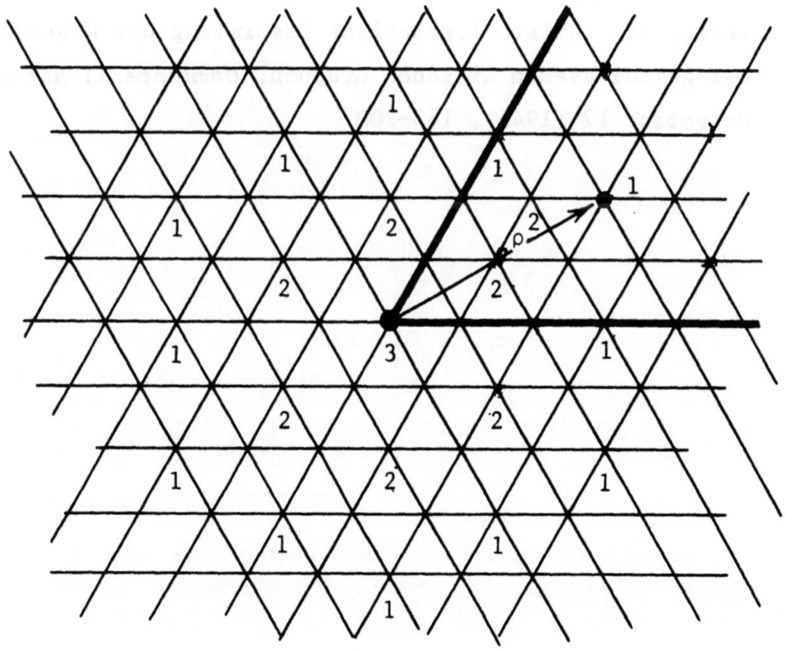

References

[1] Bott, R. and Samelson, H. 'Applications of the theory of Morse to symmetric spaces', *American Journal of Mathematics* 80 (1958), 964-1029.

[2] Bott, R., 'An application of the Morse theory to the topology of Lie groups', *Bulletin de la Société Mathématique de France* 84 (1956), 251-282.

[3] Bott, R. and Samelson, H., 'The cohomology ring of G/T', *Proceedings of the National Academy of Sciences* 41 (1955), 490-493.

[4] Serre, J.-P., 'Représentations linéaires et espaces homegènes kähleriens des groupes de Lie compacts', Séminaire Bourbaki, No.100, Paris, 1954.

[5] Stiefel, E., 'Uber eine Beziehung zwischen geschlossenen Lie'schen Gruppen und diskontinuierlichen Bewegungsgruppen euklidischer Räume und ihre Anwendung auf die Aufzählung der einfachen Lie'schen Gruppen', *Commentarii Mathematici Helvetici* 14 (1941/2), 350-380.

[6] Stiefel, E., 'Kristellographishe Bestimmung der Characktere der geschlossenen Lieschen Gruppen, *Commentarii Mathematici Helvetici* 17 (1945), 165-200.

Extrait de *L'Enseignement mathématique*, T. XXIII, fasc. 3-4, 1977

ON THE CHARACTERISTIC CLASSES OF GROUPS OF DIFFEOMORPHISMS [1]

by Raoul BOTT [2]

It gives me great pleasure to join in this celebration of Beno Eckmann's sixtieth birthday. I had the good fortune to meet Beno at the very outset of my mathematical career in 1950 at Princeton. Of course at that time I knew much too little of the subject to have other than neighbourly conversations with him—they lived next to us in the housing project—but over the years I have, like all of us, learned much from his superb lectures, expositions and clarifying point of view. I have always been especially fond of his paper on the Radon Hurwitz numbers, and it played an important part in Shapiro and my thinking about Clifford Algebras.

But let me turn to my talk, I am not allowed to reminisce indefinitely. The question I would like to discuss here has fascinated me for the past six or seven years and is related to the work of Gelfand Fuks on the one hand, and the work of Haefliger on Foliation and that of Chern, Simons and Cheeger on the other. Here I will however launch into the subject by presenting it from the point of view of group cohomology, which I take to be the common denominator of all of us here.

Consider then an abstract group Γ acting (on the right) in C^{∞} manner on the smooth manifold M:

$$M \times \Gamma \to M \tag{1}$$

If one the chooses a volume θ for M, then every $f \in \Gamma$ maps θ into a multiple of itself:

$$f^* \theta = \mu(f) \theta ,$$

and out of this magnification factor we may manufacture a function of $(n+1)$ arguments, $\{f_1, ..., f_{n+1}\{$ in Γ, $n = \dim M$, by setting

$$\omega(f_1, \ldots, f_{n+1}) = \int_M \log \mu_1(\bar{f}_1) \cdot d \log \mu(\bar{f}_2) \ldots d \log \mu(\bar{f}_{n+1}) , \tag{2}$$

[1]) Presented at the Colloquium on Topology and Algebra, Zurich, April 1977.

[2]) My thanks are due to the Guggenheim Foundation, the National Science Foundation and the Forschungsinstitut for partial support of the work reported here.

where

$$\bar{f}_i = f_1 \circ f_2 \circ \ldots \circ f_i .$$

More generally let us choose a Riemann structure on M, and let ∇ denote the associated Levicivita connection on M.

Then $f \in \Gamma$ acts on ∇ and the difference

$$f^* \nabla - \nabla = \xi(f)$$

is a well-defined 1-form on M with values in the bundle of endomorphisms of T, the tangent bundle of M:

$$\xi(f) \in \Omega^1(M; \text{End } T) .$$

Then for any partition $\alpha_1 + \ldots + \alpha_k = n$ of n we define functions $\omega^\alpha(f_1, \ldots, f_{n+1})$ by the recepee:

$$(3) \quad \omega^\alpha(f_1, f_2 \ldots f_{n+1}) = \int_M \Sigma \log \mu_1 \operatorname{Trace}(\xi_2 \ldots \xi_{\alpha_1}) \ldots \operatorname{Trace}(\xi_{\alpha_k+1}, \ldots \xi_{\alpha_{n+1}}) ,$$

where μ_i stands for $\mu(\bar{f}_i)$ computed for the volume of the Riemann structure, and ξ_i stands for $\xi(\bar{f}_i)$. Finally the sum is taken alternatingly.

Concerning these expressions one now has the following

PROPOSITION. *The $(n+1)$-cochains ω and ω^α just defined are cocycles and their classes*

$$(4) \qquad [\omega^\alpha] \in H^{n+1}(\Gamma, \mathbf{R})$$

are independent of the choices involved.

By the way the formula (1.1) for ω, which goes back to Thurston, agrees with $\omega^{1, \ldots 1}$, up to a constant, because if θ is taken as volume of the Riemannian structure, then one easily finds that

$$\text{trace } \xi(f) = -d \log \mu(f) .$$

The classes ω^α are therefore certain invariants of the action of Γ on M, and I propose to call them characteristic numbers because they are constructed in analogy with the usual ones.

Several questions arise now naturally. First of all, where did these formulas come from. Next, how many similar formulas could one expect, and finally are these classes independent in the sense that for any linear nontrivial combination

$$\psi = \Sigma a_\alpha \omega^\alpha ,$$

there is some action of some Γ on some M so that ψ is nonzero in $H^{n+1}(\Gamma)$.

The first question we can answer quite precisely. The second one really only when $M = S^1$, although the work that André Haefliger is reporting on reduces the problem to a purely computational one whenever the action of the group of Diffeomorphisms on the principal bundle of M is known. Thus for surfaces and in view of the recent result of A. Hatcher that Diff $(S^3) \sim O(4)$ also for the three sphere, this question is essentially answered. On the other hand the third question is not even decided for S^1.

Here I will address myself mainly to the first one and will also try to indicate my personal recepee for deriving the explicit cocycles I have written down. The first step in this direction is then to explain that the classes ω^α have their antecedents in the characteristic classes of foliations, and appear in $H^*(\Gamma)$ by virtue of the fact that the associated bundle

$$M\Gamma \quad \downarrow \pi \quad *\Gamma \tag{5}$$

over the classifying space of Γ given by the action of Γ, admits a canonical foliation $F\Gamma$, transversal to the fiber.

Let me start therefore with a quick reveiw of the characteristic classes of a Foliation. There are really two quite different ways of doing this, sort of a high and a low road to the same goal. The high road is closer to Gelfand Fuks theory. while the low road is closer to the considerations of Chern, Simons, Cheeger. For our purposes this is the best way to start.

Recall first that a foliation F on M is defined by a sub-bundle $E \in T$ whose sections $S(E)$ are closed under the lie bracket:

$$[S(E), S(E)[\in S(E) . \tag{6}$$

I often denote a foliation by the associated exact sequences

$$F: 0 \to E \to T \xrightarrow{\pi} Q \to 0$$

induced by the inclusion of E in T.

Now in the usual theory of characteristic classes every connection ∇ on Q, induces a well defined homomorphism

$$\varphi(\nabla)^*: \mathbf{R}[c_1, \dots, c_q] \to \Omega^*(M)$$

of the polynomial ring in the universal Chern-classes, c_i in dim $2i$, to the forms on M. Quite a long time ago I noticed that for a special class of connections on Q—called the *basic* ones—this homomorphism factors through the quotient ring

(7) $$\underline{\mathbf{R}}[c_1, \ldots, c_q] = \mathbf{R}[c_1, \ldots, c_q] / (\dim > 2q)$$

in which the ideal of classes in dim $> 2q$ has been set equal to zero. These basic connections are characterised by the condition that,

(8) $$\Delta_X^B \cdot \tilde{Y} = \pi[X, \tilde{Y}], \text{ for } X \in S(E),\ Y \in S(Q), \quad \tilde{Y} \in S(T),\ \pi \tilde{Y} = Y,$$

and the integrability enters here, because the right hand side is independent of the choice of $\tilde{Y}$ if and only if E is integrable.

In any case then, from such a ∇^B, we have the commutative diagram:

(9) $$\begin{array}{ccc} \mathbf{R}[c_1 \ldots c_q] & \longrightarrow & \Omega^*(M) \\ \downarrow & \nearrow & \\ \underline{\mathbf{R}}[c_1 \ldots c_q] & & \end{array}$$

expressing a vanishing phenomenon which yields the only known topological obstruction to integrability.

Now it is hardly new in topological circles that if something vanishes then something else should simultaneously appear somewhere. Nevertheless it took many years and the work of Godbillon-Vey-Roussarie, and for them the work of Gelfand Fuks was the inspiration, as well as ideas of Chern-Simons, conversation with Milnor, letters with A. Haefliger until we discovered the classes which "appear" as a result of (9).

The answer is as follows. One first constructs a complex

$$WO_q = \underline{\mathbf{R}}[c_1, \ldots, c_q] \otimes E(h_1, h_3, \ldots, h_p),\ p \text{ odd} \leqslant q$$

with

$$d h_i = c_i,\ i \text{ odd}$$

and hence analogous to the complex introduced in the Gelfand Fuks work, describing $H^*(a_n)$ the cohomology of formal vector-fields in $\mathbf{R}^n$. Then in terms of this complex the correct extension of (9) yields a *cannonical homotopy class of maps*:

(10) $$\varphi^*(F): WO_q \to \Omega^*(M)$$

for any foliation F.

A representative of $\varphi^*(F)$ is obtained by choosing two connections on Q, namely a basic one ∇^B and a Riemannian one ∇^R, that is some connection preserving some Riemann structure on Q. Indeed the standard comparison procedure between connections then yields difference forms,

$$h_i(\nabla^B, \nabla^R) \in \Omega^*(M)$$

with

$$c_i(\nabla^B) - c_i(\nabla^R) = d\, h_i(\nabla^B, \nabla^R),$$

and $\varphi^*(F)$ is then induced by the map

$$c_i \to c_i(\nabla^B)$$
$$h_i \to h_i(\nabla^B, \nabla^R) \quad i \text{ odd}.$$

Because $c_i(\nabla^R) \equiv 0$ for odd i, in view of the special symmetiies of a Riemannian connection, this $\varphi^*(F)$ is then well defined. Finally the homotopy class of $\varphi^*(F)$ is seen to be independent of all choices because both the basic and the Riemann connections form convex sets.

The cohomology of the algebras WO_q can be computed explicitly without too much trouble. For instance for $n = 1$, H^* has classes only in dim 1 and 3:

$$H^*(WO_1) = \{1; h_1 c_1\}$$

while

$$H^*(WO_2) = \{1; c_2; h_1 c_1^2, h_1 c_2\}.$$

In general a basis for $H^*(WO_n)$ in terms of monomials $h_\alpha \cdot c_\beta$ $h_{\alpha 1} \ldots h_{\alpha_k} c_{\beta 1} \ldots c_{\beta_n}$. with the α, β subject to certain inequalities is known, and all "exotic classes" i.e. ones involving an h, are annihilated by multiplication with any element of positive dimension. The c_{2i} of course correspond to the usual Pontryagin classes.

Finally note that there is a map

$$WO_{q+1} \to WO_q$$

defined by sending c_i to c_i and h_i to h_i for $i \leqslant q$, and sending c_{q+1} to 0 as well as h_{q+1} — if $(q+1)$ is odd. The image of *this map constitutes the stable classes* of WO_q. The simplest of these are of course the Pontryagin classes of Q.

So much then for a quick review of the characteristic classes of foliations. As mentioned before, they are pertinant to our discussion because the bundle

$$\begin{array}{c} M\Gamma \\ \pi \downarrow \\ *\Gamma \end{array}$$

has a canonical foliation $F\Gamma$, transversal to the fibers M, of π. Hence in particular the classes $h_1 c^\alpha \in H^{2n+1}(WO_n)$, $n = \dim M$, can be applied to $F\Gamma$. Finally integrating over the fiber in π, we obtain classes $\pi_* h_1 c^\alpha (F\Gamma)$ in $H^{n+1}(*\Gamma) = H^{n+1}(\Gamma)$ and I claim that the formulae I wrote down at the start, are explicit recepees for these classes. Precisely we have:

Let the s_i and c_i be related as the power sums are to the elementary functions, that is by the Newton formulae. Thus

$$s_1 - c_1 = 0$$
$$s_2 - s_1 c_1 + 2c_2 = 0$$
$$s_n - s_{n-1} c_1 + \dots \pm n c_n = 0,$$

and let $s^\alpha = s_1^{\alpha 1} \dots s_n^{\alpha n}$. *Then, the form* $\omega^\alpha(f_1 \dots f_{n+1})$ *represents a constant multiple of* $\pi_* h_1 s^\alpha (F\Gamma)$:

(11) $$[\omega^\alpha] \cong \pi_* h_1 s^\alpha (F\Gamma).$$

To prove such a result in any particular case one could in principle start with any free action of Γ on a manifold E which is highly connected so that E/Γ is a high approximation to the classifying space $B\Gamma \equiv *\Gamma$, of Γ, and then compute our classes in

$$\frac{E \times M}{\Gamma}$$

with $F(\Gamma)$ being the foliation, which, when lifted to $E \times M$, is given by the "horizontal space" TE of tangents along E:

$$\pi^{-1}(F\Gamma) = TE \subset T(E \times M).$$

However for explicit formulas which are valid in general it is best to extend the concepts of *foliation* of *de Rham theory*, and of *characteristic classes* to a larger category in which explicit models of the classifying space $B\Gamma$ and its associated bundle $M \cdot \Gamma$ can be constructed.

The appropriate category for such a program is the category of *simplicial manifolds* where such explicit models are by now well known. Thus $M\Gamma$ can be represented by:

(12) $$M\Gamma: M \leftleftarrows M \times \Gamma \leftleftarrows M \times \Gamma \times \Gamma \dots,$$

with the bundary maps given by

$$\partial_0 (m; \gamma_1, \dots, \gamma_n) = (m\gamma_1; \gamma_2 \dots \gamma_n)$$
$$\partial_i (m; \gamma_1, \dots, \gamma_n) = (m; \gamma_1 \dots, \gamma_i \circ \gamma_{i+1}, \dots, \gamma_n); \ i > 0$$

and $*\Gamma$ by the corresponding simplicial set, where M is equal to a single point $*$:

(13) $$*\Gamma:\ * \leftleftarrows \Gamma \leftleftarrows\!\!\leftarrow \Gamma \times \Gamma \ldots,$$

Both of these are examples of simplicial manifolds in the sense that each is a collection of manifolds $\{X_k\}\ k = 0, \ldots$ between which the usual semisimplicial structure maps are given by C^∞ maps. To any such "semi-simplicial" manifold

$$X:\ X_0 \leftleftarrows X_1 \leftleftarrows\!\!\leftarrow X_2$$

one associates a geometric realisation:

$$|X| = (\amalg X_k \times \Delta^k) / \sim ,$$

obtained from the disjoint union of the $X_k \times \Delta^k$ where Δ^k is the k-simplex, by identifying $X_k \times \partial\Delta^k$ with the appropriate subsets of $X_{k-1} \times \Delta^{k-1}$.

Applied to (12) and (13) this construction then yields a map

(14) $$\begin{array}{c} |M\Gamma| \\ \downarrow \pi \\ |*\Gamma| \end{array}$$

which represents the associated M-bundle over the classifying space of Γ.

Now the point I wish to make is that the notions of Foliation, de Rham Theory, basic and Riemann connections all extend to this category, and provided one uses the Thom-Whitney-Sullivan-Dupont model for this extended de Rham theory so does the characteristic homomorphism $\varphi_*(F)$. To make this plausible let me say a few words concerning each one of these steps.

A foliation F on X, is simply a foliation F_k on each X_k which is preserved by the structure maps. For example the foliation $F\Gamma$ on $M \cdot \Gamma$ is represented by the trivial foliation

$$F_k\colon 0 \to T\{M \times \Gamma^{(k)}\} \to T\{M \times \Gamma^{(k)}\} \to 0 .$$

of M by points, so that the normal bundle to F_k corresponds to $T\{M \times \Gamma^{(k)}\}$.

For the De Rahm theory of $|X|$ one uses the ring $\Omega^*(|X|)$ of *compatible forms on* $\amalg X_k \times \Delta^k$. Thus an r-form in $\Omega^*(|X|)$ is a function, which assigns to each k an r-form $\theta(k) \in \Omega^*(X_k \times \Delta^k)$ so that in the diagram

$$\begin{array}{ccc} \Omega^*(X_k \times \Delta^{k-1}) & \xleftarrow{1 \otimes i^*} & \Omega^*(X_k \times \Delta^k) \\ {\scriptstyle \partial_i^* \otimes 1}\downarrow & & \\ \Omega^*(X_{k-1} \times \Delta^{k-1}) & & \end{array}$$

one has the relation

(15) $$(1 \times i^*)\, \theta(k) = (\partial_i{}^* \times 1)\, \theta(k-1).$$

Here of course i is any inclusion of a face in Δ_k and ∂_i the corresponding structure map of X_k to X_{k-1}.

This de Rahm theory is clearly anticommutative and computes the cohomology of $|X|$ under mild hypotheses on X:

$$H^*(|X|) = H\{\Omega^*(|X|)\},$$

as was shown recently by Dupont [8]. Actually there is a more economical model for $H^*(|X|)$, namely the double complex

$$\Omega^{**}(X) = \oplus\, \Omega^p(X_q)$$

with differential $d \pm \delta$ where δ in the alternating sum of the structure maps, but this complex is not anticommutative and therefore not so suitable for our purposes. (See [8] or [4] for the De Rahm theorem in this context).

The final step is now to define basic and Riemann connections in this extended situation, and this is again all done by a compatability condition. Thus if

$$F_k\colon 0 \to E_k \to TX_k \to Q_k \to 0$$

represents F on X_k then a basic connection for Q, is a compatible collection of basic connections for the pull-back of Q_k to $X_k \times \Delta^k$.

For instance in our situation, such a compatible collection can be constructed as follows:

Let ∇ be any connection on M. On the component $M \times f_1 \times \ldots \times f_k \times \Delta^k$ define $\nabla(k)$ by

(16) $$\nabla(k) = \nabla + x_i\, \xi_i,$$

where the ξ_i are defined as in (2) and the x_i are the usual baricentric coordinates of Δ^k. This is then a compatible collection as easily checked.

One proceeds similarly with the Riemann connections on Q although it is harder to give explicit formulas in that case and for this reason I know of no really attractive recepee for the classes involving the higher h_i's. For the $h_i\, S^\alpha$ one gets by because of the following stratagem:

Let θ be a volume on M. Then θ admits a natural compatible extension to $\Omega^1(|M \cdot \Gamma|)$ by setting

(17) $$\theta(f_1, \ldots, f_k) = u(\bar{f}_1)^{x1} \cdot u(\bar{f}_2)^{x2} \ldots u(\bar{f}_k) \cdot {}^{xk}\, \theta$$

where the x^i are again the baricentric coordinates on Δ^k, and the $\mu(f_i)$ are defined as in (3). (See [2] for some of the details and examples.)

So much then for the first question I raised. Concerning the second one, André Haefliger is reporting in detail so let me here just summarise the situation.

The proper formulation of the question is really in the frame work of Gelfand-Fuks. One seeks the smooth cohomology of the group of diffeomorphisms of M. Thus one sets $D = \text{Diff}(M)$ and tries to compute the cohomology of the complex of "smooth" Eilenberg cochains on D. Now the final answer for this problem, on which André Haefliger and I are working at the moment, seems to be the following one: There is a natural bundle $E(M)$ over M, on which D acts. This group therefore also acts on the space $S(E)$ of sections of E. Hence we may finally construct the homotopy quotient.

(18) $$S(E)/D = S(E) \underset{D}{\times} U(D)$$

that is, the bundle associated to the action of D on $S(E)$ over the classifying space of D. The cohomology of this space we believe gives the smooth cohomology of D.

(19) $$H^*_{\text{smooth}}(D) \simeq H^*(S(E)/D).$$

Of course this formula is of use only if D and its action on $S(E)$ are well understood. For instance for $M = S^1$, E is given by $S^3 \times S^1$, D by S^1 and the action of D on E by the translation. Hence

$$S(E) = \text{Map}(S^1, S^3)$$

and for $S(E)/D$ one finds—as André just tought me last week:

(20) $$H^*(S(E)/D) \simeq \mathbf{R}[\omega, e] \quad \omega \cdot e = 0$$

where $\omega, e \in H^2$.

Thus for the circle every smooth class of Diff (S^1) is a power of ω—and this is just the $h_1 c_1$ of our discussion, or of one other class, e, which is the usual Euler class of the circle bundle $S^1 \cdot \Gamma$ over $*\Gamma$; Γ = Diff (S^1) in the discrete topology. By a different argument Gelfand and Fuks had already determined this case a long time ago, but they made an error in the ring structure, and it is in any case interesting to verify the answer in this manner.

For the 2-sphere (18) gives an algorithm as I mentioned earlier and which I hope André is discussing in greater detail in his lecture. In any case one finds many classes besides those generated by the "foliation classes" and most likely the smooth cohomology of Diff (S^2), though finitely generated in each dimension is not finitely generated as a ring.

Finally a word concerning the last question I raised. The first example of a nontrivial foliation class, was given by Roussarie and Godbillon-Vey [7]. In our frame work, their example is as follows: Let $\Gamma \subset SL(2, \mathbf{R})$ be a discrete subgroup with compact fundamental domain. These exist—indeed every fundamental group of a Riemann surface of genus $\geqslant 2$, can be realised in this manner. Now let P be the subgroup of triangular matrixes in $G = SL(2, \mathbf{R})$. Then

$$G / P = S^1 ,$$

and hence Γ acts naturally on S^1. Now it turns out that both ω and $e \in H^2(\Gamma)$ are nontrivial for this action, on the other hand they are proportional, for any such "homogenous" example.

Since the paper of G.-V., this homogenous case has been studied by many people, notably Kamber and Tondeur [10], and quite recently (Unpublished) Baker. André Haefliger and I in our original note [3] were most probably the first to note some extensions of the example, however one of our assertions there, our independence theorem, was based on a misconception.

Let me try to rectify the matter by stating here the true fact, from which our independence assertion was incorrectly deduced. This theorem gives a sort of proportionality result for the classes $h_1 c_\alpha$.

Consider then the general case of a connected semi-simple Lie group G and let P be a parabolic subgroup of G, with dim $G/P = n$. Then after complexification $G^{\mathbf{C}}/P^{\mathbf{C}}$ is a *compact complex analytic* n-manifold and hence has well defined Chern classes and Chern numbers.

Our proportionality theorem then asserts that:

Proposition. *For any* $\Gamma \subset G$, *the classes* $h_1 c^\alpha (F\Gamma) \in H^{2n+1}(|G/P \cdot \Gamma|)$ *are proportional to the Chern numbers* c^α *of* $G^{\mathbf{C}}/P^{\mathbf{C}}$. *Thus for any two multi-indexes* α *and* β

$$h_1 c^\alpha (F\,\Gamma) \cdot c^\beta = h_1 c^\beta (F\,\Gamma) \cdot c^\alpha . \tag{21}$$

Unfortunately this of course does not imply that the proportionality factor does not vanish, and that is alas, what happens most of the time in these examples.

Kamber and Tondeur, and as I said quite recently Baker, have explored this homogenous case in great detail and have found quite a few independent classes by this means, see [10] and [9]. On the other hand no "stable class" has ever been detected by a homogenous example.

Using the Lefschetz-Hyperplane Theorem for $G^{\mathbf{C}}/P^{\mathbf{C}}$, Haefliger and I have quite recently been able to explain this. Indeed one has the following quite general fact:

THEOREM. *All stable classes of* $F\Gamma$ *in* $H^*(|G/P \cdot \Gamma|)$ *vanish.*

Here P is parabolic in G, and $\Gamma \subset G$. Presumably the same is true in all homogenous cases but this is not clear to us so far.

The master of non homogenous actions has been first and foremost Thurston, and last year Heitch has been able to considerably extend Thurston's computations [9]. I will not be able to report on this work here, except to state what Thurston's constructions imply for the smooth cohomology of Diff (S^1). He shows first of all that e and ω are independent, and in fact he constructs a smooth family of actions of the group

$$\Pi = \{X, Y, U, V \mid [X, Y] = [U, V]\}$$

on S^1 for which the class ω varies continuously. See [2] for details in this context.) He also constructed examples to show that $\omega^n \neq 0$. On the other hand we still have no example for which e^n, $n \geqslant 2$ is nonzero.

BIBLIOGRAPHY

[1] BOTT, R. Lectures on characteristic classes and foliations (Notes by Lawrence Conlon). *Lecture Notes in Mathematics 279*, pp. 1-94, Springer Verlag, New York.

[2] —— On some formulas for the characteristic classes of group actions. *Proceedings of the Conference on Foliations*, Rio de Janeiro, 1976.

[3] BOTT, R. and A. HAEFLIGER. On characteristic classes of Γ-foliations. *Bull. AMS 78* (1972), pp. 1039-1044.

[4] BOTT, R., H. SHULMAN and J. STASHEFF. On the de Rham theory of certain classifying spaces. (To be published in *Advances of Math.*)

[5] CHERN, S. and J. SIMONS. Somo cohomology classes in principal fiberbundles and their applications to Riemannian geometry. *Proc. Nat. Acad. Sci. U.S.A. 68* (1971), pp. 791-794.

[6] GELFAND, I. M. and D. B. FUKS. Cohomology of the Lie algebra of tangent vector fields of a smooth manifold. I, II. *Funkcional Anal. i Prilozen. 3* (1969), No. 3, pp. 32-52; ibid. *4* (1970), pp. 23-32 (Russian), MR *41*, 1067.

[7] GODBILLON-VEY. Un invariant des feuilletages de codimension 1. *C. R. Acad.Sci. Paris 273* (1971), p. 92.

[8] Dupont, T. L. Semisimplicial de Rham cohomology and characteristic classes of flat bundles. (To be published in *Topology*.)
[9] Heitch, J. Residue formulae for characteristic classes. (To be published.)
[10] Kamber, F. W. and P. Tondeur. Foliated bundles and characteristic classes. *Lecture Notes in Mathematics 493*, Springer Verlag, New York.
[11] Thurston, W. Foliations and groups of diffeomorphisms. *Bull. AMS 80*, (1974), pp. 304-307.

(Reçu le 5 mai 1977)

Raoul Bott

Harvard University
Cambridge
Mass. 02138
USA

Topology Vol. 16, pp. 285–298. Pergamon Press, 1977.

THE COHOMOLOGY OF THE VECTOR FIELDS ON A MANIFOLD

R. BOTT and G. SEGAL

(Received 1 May 1977)

§1. INTRODUCTION

THE SMOOTH vector fields on a smooth manifold M form a Lie algebra Vect(M) under the bracket. Gelfand and Fuchs[7–9] have studied the Lie algebra cohomology of Vect(M), which they define by means of a cochain algebra $A(M)$, where $A^k(M)$ is the vector space of continuous **R**-multilinear maps

$$\underbrace{\mathrm{Vect}(M)\times\cdots\times\mathrm{Vect}(M)}_{\leftarrow k \rightarrow}\rightarrow \mathbf{C},$$

and the differential d: $A^k(M)\rightarrow A^{k+1}(M)$ is defined by the formula

$$\mathrm{d}\alpha(\xi_1,\ldots,\xi_{k+1})=\sum_{i<j}(-1)^{i+j-1}\alpha([\xi_i,\xi_j],\xi_1,\ldots,\hat{\xi}_i,\ldots,\hat{\xi}_j,\ldots,\xi_{k+1}).$$

'Continuous' refers to the usual C^∞ topology on Vect(M). (Actually Gelfand and Fuchs considered the cohomology with real coefficients, but we have found it convenient to change from **R** to **C**.)

In this paper we shall prove that when M is either a compact manifold or the interior of a compact manifold with boundary the cohomology of Vect(M) is the same as that of the space of continuous cross-sections of a certain natural fibre bundle E_M on M associated to its tangent bundle. The fibre of E_M is an open manifold F whose cohomology is that of Vect($\mathbf{R}^n$). The result was conjectured independently by Fuchs and the first author, and has also been proved by Haefliger[11] and Trauber by different methods.

The history of the present proof is roughly as follows. Gelfand and Fuchs in their first highly original papers considered only vector fields with compact supports. They observed that the cochains of the Lie algebra can be regarded as distributions, and they obtained their results by studying the natural filtration of the cochains by the dimension of their supports. Our plan has been to work with general vector fields on non-compact manifolds, so that they can be restricted to open submanifolds. This enables us to prove the theorem by first considering the case of $\mathbf{R}^n$ and then using a patching-together process. We treat $\mathbf{R}^n$ by a contraction argument which we think illuminates the essential homotopy-invariance of the Lie algebra cohomology (cf. Prop. (2.1)). It turns out that the Lie algebra of all vector fields on $\mathbf{R}^n$ has the same cohomology as the jets of vector fields at a single point, i.e., the "formal vector fields" at the origin in $\mathbf{R}^n$. Given the validity of the local result it seemed, especially in view of Anderson's calculation of the cohomology of mapping spaces[1], that a fairly simple patching argument should prove the conjecture. But there were two difficulties: first the technical difficulty that the patching involved surprisingly delicate questions of convergence, and secondly the fundamental difficulty that there seemed to be no map relating the Lie algebra cohomology to that of a space of sections. A solution of these difficulties along the lines presented here was devised, more or less, by the second author in 1974, and has been gradually improved by the authors jointly since then. It is still not very satisfying.

It may be worth commenting on the relationship between our approach and others. Haefliger's framework is Sullivan's explicit algebraic model of the rational homotopy category in terms of minimal differential graded algebras. Roughly one can say that he constructs on the one hand a minimal model for the Lie algebra cochains and on the other hand a minimal model for the homotopy type of the space of sections, and proves that the two models are actually isomorphic. An advantage of his approach is that it incorporates an explicit calculation of the cohomology, at least for manifolds whose Pontrjagin classes vanish. Our method, on the other hand, is more self-contained, and relates the Gelfand–Fuchs cochains comparatively directly to the cochain algebra of the space of sections. It has the advantage of incorporating a proof that the cohomology is finite dimensional, a result due to Gelfand and Fuchs in the case of vector fields with compact supports.

Trauber's method is to use the Gelfand–Fuchs filtration. Its relation to Haefliger's approach is roughly that of Quillen's[13] to Sullivan's[16] treatment of the rational homotopy category.

Concerning the general nature of the problem one can say the following. If it were possible to associate a classifying space $B\mathfrak{g}$ naturally to a Lie algebra $\mathfrak{g}$ so that $H^*(\mathfrak{g}) = H^*(B\mathfrak{g})$ then the theorem would be rather easy. For the points of the principal bundle P of a Riemannian manifold M define exponential maps $\mathbf{R}^n \to M$ which induce homomorphisms of Lie algebras $\mathrm{Vect}(M) \to V_n$, where V_n is the formal vector fields at O in $\mathbf{R}^n$. Applying the functor B would give a map $P \to \mathrm{Map}(B\mathrm{Vect}(M); BV_n)$ equivariant with respect to the action of O_n on P and BV_n. By adjunction we should have $B\mathrm{Vect}(M) \to \mathrm{Map}_{O_n}(P; BV_n)$. But $\mathrm{Map}_{O_n}(P; BV_n)$ is the space of sections of a bundle on M with fibre BV_n, associated to P. One could hope that the last map was a cohomology equivalence.

But unfortunately no such classifying space functor $\mathfrak{g} \to B\mathfrak{g}$ can possibly exist. If it did there would be a natural integral lattice in the cohomology of a Lie algebra. In fact $B\mathfrak{g}$ exists as a real homotopy type, but then it is its "continuous" cohomology that is relevant for us. To make clear the significance of this we should point out that our result is closely related to the fact that the cohomology of the Lie algebra $\mathfrak{g}$ of a reductive Lie group G is (with complex coefficients) that of the *complexification* $G_{\mathbf{C}}$ of G as a space, by the maps

$$\begin{aligned}(\text{cochains of } \mathfrak{g}) &= (\text{invariant forms on } G)\\ &= (\text{holomorphic invariant forms on } G_{\mathbf{C}}) \subset (\text{forms on } G_{\mathbf{C}}).\end{aligned}$$

On the other hand if one associates a simplicial set abstractly to the cochains of $\mathfrak{g}$ following Sullivan's general prescription[16] one gets the quotient by G of the usual singular complex of G. The homotopy type of this is $(EG_\delta \times G)/G_\delta$, where G_δ is G with the discrete topology and EG_δ is its universal space. This is the homotopical fibre of the map $BG^\delta \to BG$, and there seems to be no obvious relation between it and $G_{\mathbf{C}}$.

The plan of the paper is as follows. In §2 we study $A(\mathbf{R}^n)$, proving in particular that its cohomology is finite dimensional. In §3 we construct the manifold F whose cohomology is that of $A(\mathbf{R}^n)$. In §4 we construct our fundamental map. In §5 we describe without proof how to construct a cochain algebra representing a mapping space or a space of sections. The result here is essentially due to Anderson[1], but we give a self-contained treatment which we hope is of some interest in its own right. The main theorem is stated in §6, and proved modulo a convergence lemma whose proof is postponed to §8. The result of §5 is proved in §7.

§2. THE CASE OF EUCLIDEAN SPACE

The radial contractions $T_t: \mathbf{R}^n \to \mathbf{R}^n$ defined by $T_t(x) = tx$ (for $0 < t \leq 1$) induce endomorphisms of $A(\mathbf{R}^n)$ which induce the identity on cohomology. For inner automorphisms of a Lie algebra induce the identity on its cohomology, and as the one-parameter group of diffeomorphisms $\{T_t\}$ is generated by a vector field ρ the derivative with respect to t of the action of T_t on $A(\mathbf{R}^n)$ comes from the inner automorphism of $\mathrm{Vect}(\mathbf{R}^n)$ defined by ρ. (This will be made more explicit in the proof of (2.2) below.)

Now for any manifold M an element of $A^q(M)$, being a multilinear function on smooth vector fields on M, can be thought of as a distribution with compact support on the product of q copies of M, with values in some appropriate vector bundle. The contractions T_t replace the cocycles of $A(\mathbf{R}^n)$ by cohomologous ones with smaller support.

It follows from this that,

PROPOSITION 2.1. *If U is a star-shaped open set of $\mathbf{R}^n$ then $A(U) \to A(\mathbf{R}^n)$ is a cohomology equivalence.*

But it is natural to ask whether one can pass to the limit as $t \to 0$ and obtain cocycles whose support is a single point. We shall prove that that is so.

PROPOSITION 2.2. *Let $A_0(\mathbf{R}^n)$ be the subalgebra of $A(\mathbf{R}^n)$ consisting of cochains whose support is the origin. Then the inclusion $A_0(\mathbf{R}^n)\to A(\mathbf{R}^n)$ is a cohomology equivalence.*

Remark. Associating to a vector field on $\mathbf{R}^n$ its Taylor expansion at the origin is a continuous surjection of Lie algebras from Vect($\mathbf{R}^n$) to the algebra $\hat{V}_n$ of formal vector fields at the origin in $\mathbf{R}^n$. The subalgebra $A_0(\mathbf{R}^n)$ is exactly the continuous cochain algebra of $\hat{V}_n$. It has a purely algebraic description, as we shall see in §3.

Proof of 2.2. We shall actually prove considerably more, namely that the cohomology class of a cochain α is determined by its values on a certain finite dimensional space of polynomial vector fields.

Let us write V for $\mathbf{R}^n$. A vector field on V is an element of $\mathscr{E}(V)\otimes V$, where $\mathscr{E}(V)$ is the smooth real-valued functions on V. So an element of $A^1(V)$ is a continuous linear map $\mathscr{E}(V)\to V^*$, and more generally an element of $A^q(V)$ is a continuous multilinear map

$$\alpha\colon \underbrace{\mathscr{E}(V)\times\cdots\times\mathscr{E}(V)}_{\leftarrow q \rightarrow}\to(V^*)^{\otimes q}$$

which is alternating in the obvious sense. (Note that as we are using complex coefficients the dual V^* means $\mathrm{Hom}_{\mathbf{R}}(V;\mathbf{C})$.) In this notation $\mathrm{d}\alpha$ is the multilinear map obtained by making alternating the map

$$(f_1,\ldots,f_{q+1})\to\sum_i\alpha\left(\frac{\partial f_1}{\partial x_i}\cdot f_2,f_3,\ldots,f_{q+1}\right)\otimes e_i^*,$$

where $\{e_i^*\}$ is the standard basis of V^*.

Let P_q be the finite dimensional subspace of $\underbrace{\mathscr{E}(V)\otimes\cdots\otimes\mathscr{E}(V)}_{\leftarrow q \rightarrow}$ spanned by all elements $f_1\otimes\cdots\otimes f_q$, where $f_1,\ldots,f_q$ are polynomials such that $\Sigma\,\mathrm{degree}\,(f_i)\le q$. If $A_P{}^q(V)$ denotes the alternating linear maps $P_q\to(V^*)^{\otimes q}$ the preceding formula for d defines d: $A_P{}^q(V)\to A_P{}^{q+1}(V)$, making $A_P(V)$ a quotient cochain complex of $A(V)$. Let $B(V)$ be the kernel of $A(V)\to A_P(V)$.

LEMMA 2.3. *$B(V)$ is acyclic, and hence $A(V)\to A_P(V)$ is a cohomology equivalence.*

Remark. Because $A_P{}^q(V)$ is finite dimensional this implies that the cohomology groups of $A(V)$ are finite dimensional.

Proof. In the above notation the contraction $T_t\colon V\to V$ acts on $A(V)$ by $(T_t\alpha)(f_1,\ldots,f_q)=t^{-q}\alpha(T_t^*f_1,\ldots,T_t^*f_q)$. If $\alpha\in B(V)$ then $T_t\alpha\to 0$ as $t\to 0$, for if f vanishes to order k at the origin then T_t^*f is $O(t^k)$ as $t\to 0$, and any function f is the sum of a polynomial of degree $k-1$ and a function which vanishes to order k. (In fact $B(V)$ consists precisely of all $\alpha\in A(V)$ such that $T_t\alpha\to 0$ as $t\to 0$.)

Now because the group of diffeomorphisms $\{T_t\}$ is generated by a vector field ρ on V the derivative $T_t^{-1}\cdot t(\mathrm{d}/\mathrm{d}t)T_t$ is the Lie derivative $\theta(\rho)\colon A(V)\to A(V)$ along ρ. (One has $t(\mathrm{d}/\mathrm{d}t)$ rather than $\mathrm{d}/\mathrm{d}t$ because the group T_t is written multiplicatively rather than additively.) But for any Lie algebra

$$\theta(\rho)=i(\rho)\,\mathrm{d}+\mathrm{d}i(\rho),$$

where $i(\rho)\colon A(V)\to A(V)$ is the inner product with ρ. So

$$\frac{\mathrm{d}}{\mathrm{d}t}T_t\alpha=H_t\,\mathrm{d}\alpha+\mathrm{d}H_t\alpha,$$

where $H_t=t^{-1}T_ti(\rho)=i(\rho)t^{-1}T_t$. Then if $T_t\alpha$ is $O(t)$ as $t\to 0$ we have

$$\alpha=T_1\alpha-T_0\alpha=K\,\mathrm{d}\alpha+\mathrm{d}\,K\alpha,$$

where $K\alpha=i(\rho)\int_0^1t^{-1}T_t\alpha\cdot\mathrm{d}t$. Thus K is a contracting homotopy for $B(V)$.

Conclusion of proof of (2.1). $A_P(V)$ is a quotient of $A_0(V)$ as well as of $A(V)$. The kernel of $A_0(V)\to A_P(V)$ is $B_0(V)=B(V)\cap A_0(V)$. But the contracting homotopy K preserves $B_0(V)$, so $A_0(V)\to A_P(V)$ is a cohomology equivalence as well as $A(V)\to A_P(V)$.

Complement. The preceding argument can clearly be applied not only to $\mathbf{R}^n$ but to a finite disjoint union of copies of $\mathbf{R}^n$, and then will prove that the inclusion $A_0(\amalg^k \mathbf{R}^n) \to A(\amalg^k \mathbf{R}^n)$ is a cohomology equivalence, where $\amalg^k \mathbf{R}^n$ means $\underbrace{\mathbf{R}^n \amalg \ldots \amalg \mathbf{R}^n}_{k}$ and $A_0(\amalg^k \mathbf{R}^n)$ means the cochains whose supports involve only the origins in the various $\mathbf{R}^n$'s. This remark allows us to avoid using a Künneth theorem for topological vector spaces, for $A_0(\amalg^k \mathbf{R}^n)$ is just the algebraic tensor product $A_0(\mathbf{R}^n)^{\otimes k}$, and its cohomology is $H^*(A_0(\mathbf{R}^n))^{\otimes k}$ by the ordinary Künneth theorem.

§3. THE MANIFOLD F

Our next task is to construct a smooth manifold F whose cohomology is the same as that of $A(\mathbf{R}^n)$. More precisely we shall construct F and a cochain algebra A and maps of cochain algebras

$$A(\mathbf{R}^n) \leftarrow A \to \Omega(F)$$

which induce isomorphisms of cohomology. (Here $\Omega(F)$ denotes the differential forms on F). But we shall need to construct F so that the orthogonal group $G = O_n$ acts on it compatibly with its natural action on $A(\mathbf{R}^n)$. To explain this entails digressing to outline the theory of group actions on cochain algebras developed by Cartan in [3].

Definition (3.1). To say that a Lie group G acts on a topological cochain algebra A means that

(i) G acts smoothly on A by automorphisms, in the obvious sense, and

(ii) for each ξ in the Lie algebra $\mathfrak{g}$ of G there is given an antiderivation $i(\xi)\colon A \to A$ of degree -1 such that $\mathrm{d}\, i(\xi) + i(\xi)\,\mathrm{d} = \theta(\xi)$, where $\theta(\xi)$ is the derivation of degree 0 got by differentiating the G-action along the tangent vector ξ at the identity, and

(iii) $i(g \cdot \xi) = g \cdot i(\xi) \cdot g^{-1}$, where $g \cdot \xi$ denotes the adjoint action of $G \in G$ on $\xi \in \mathfrak{g}$.

A G-homomorphism of algebras with such G-action is one which commutes with the derivations $i(\xi)$ for all $\xi \in \mathfrak{g}$, as well as with the elements of G.

If G acts on a manifold X it acts on both $\Omega(X)$ and $A(X)$ in the sense of (3.1).

The G-cochain algebras we shall encounter resemble the algebra of differential forms on a smooth principal G-bundle P. In that case the G-action on P identifies $\mathfrak{g}$ with a subspace of the tangent space to P at each point, and one can choose an embedding $\omega\colon \mathfrak{g}^* \to \Omega^1(P)$ such that $i(\xi)\omega(\alpha) = \langle \alpha, \xi \rangle$ for $\xi \in \mathfrak{g}$, $\alpha \in \mathfrak{g}^*$. One can even make ω G-equivariant. To do so is exactly to choose a connection in P in the classical sense: one often thinks of ω as a G-invariant element of $\Omega^1(P;\mathfrak{g})$. It is explained in [3] that ω extends canonically to a G-homomorphism of cochain algebras $W \to \Omega(P)$, where W is the *Weil algebra* of G.

Definition (3.2). The Weil algebra W of G is $\Lambda\mathfrak{g}^* \otimes S\mathfrak{g}^*$ as algebra, where the exterior algebra $\Lambda\mathfrak{g}^*$ is generated by 1-forms $\alpha \in \mathfrak{g}^*$, and the symmetric algebra $S\mathfrak{g}^*$ by 2-forms Ω_α for $\alpha \in \mathfrak{g}^*$.

Its differential is defined by $\mathrm{d}\alpha = \mathrm{d}_1\alpha + \Omega_\alpha$, where $\mathrm{d}_1\alpha \in \Lambda^2\mathfrak{g}^*$ is the differential of α in the cochain complex of the Lie algebra $\mathfrak{g}$.

Its G-action is defined by making $i(\xi)$ (for $\xi \in \mathfrak{g}$) act as the obvious anti-derivation on $\Lambda\mathfrak{g}^*$, and trivially on $S\mathfrak{g}^*$.

The classical theory makes the following definition reasonable.

Definition (3.3). A *connection* in a G-cochain algebra A is a G-homomorphism $W \to A$, where W is the Weil algebra of G.

In the case of the forms on a principal bundle P choosing a connection permits one to identify $\Omega(P)$ as an algebra with $\Lambda\mathfrak{g}^* \otimes \Omega_{\mathrm{horiz}}(P)$, where $\Omega_{\mathrm{horiz}}(P)$ is the horizontal forms on P, i.e. those annihilated by $i(\xi)$ for all $\xi \in \mathfrak{g}$. (But notice that $\Omega_{\mathrm{horiz}}(P)$ is not closed under d.) In other words the connection $W \to \Omega(P)$ induces an isomorphism of algebras $W \otimes_{S\mathfrak{g}^*} \Omega_{\mathrm{horiz}}(P) \to \Omega(P)$. This motivates

Definition (3.4). A G-cochain algebra A with a connection $W \to A$ is *standard* if $W \otimes_{S\mathfrak{g}^*} A_{\mathrm{horiz}} \to A$ is an isomorphism.

If A is a G-cochain algebra its *basic* subalgebra A_{basic} consists of the G-invariant horizontal elements. Unlike A_{horiz} it is closed under d. If $A = \Omega(P)$, where P is a principal G-bundle, then A_{basic} can be identified with $\Omega(P/G)$. In the case of the Weil algebra W_{basic} is the invariant part of $S\mathfrak{g}^*$, which we shall call C. When $G = GL_n(\mathbf{R})$ it is well-known that C is the polynomial algebra on a sequence of elements c_i of dimension $2i (1 \le i \le n)$, called the universal Chern classes. The differential is zero.

A connection $W \to A$ in a G-cochain algebra induces a G-homomorphism of cochain algebras $W \otimes_C A_{\text{basic}} \to A$. The only technical result we need here is

PROPOSITION (3.5). *If G is connected and reductive, and A is a standard G-cochain algebra which as G-module is the sum of isotypical pieces corresponding to finite dimensional irreducible G-modules then $W \otimes_C A_{\text{basic}} \to A$ induces an isomorphism of cohomology. Furthermore a G-homomorphism $A \to A'$ of such algebras induces an isomorphism of cohomology if $A_{\text{basic}} \to A'_{\text{basic}}$ does.*

We shall leave the proof to the end of this section. One situation to which the proposition applies is when G is compact and $A = \Omega(P)$ is the forms on a principal G-bundle. Then it is an algorithm for calculating $H^*(P)$ from $\Omega(P/G)$ and the characteristic classes $C \to \Omega(P/G)$. Cartan[3] proved a slightly different result, which is much more useful for calculation, namely that there is a cohomology equivalence $E \otimes_C A_{\text{basic}} \to A$, where $E = \Lambda \otimes C$ as algebra, and Λ is the exterior algebra on elements $e_1, \ldots, e_n$ with $de_i = c_i$. Unfortunately that is of no use here as E is not a G-algebra.

The algebra $A_0(\mathbf{R}^n)$ of continuous cochains on the Lie algebra V_n of formal vector fields at the origin in $\mathbf{R}^n$ is a G-cochain algebra, where $G = GL_n(\mathbf{R})$, because the Lie algebra $\mathfrak{g}$ of G is a subalgebra of V_n. It has a natural connection, which makes it a standard G-algebra, because there is a G-stable splitting $V_n = \mathfrak{g} \oplus W$. In fact one can decompose V_n as $\prod_{k \ge -1} V_n^k$, where V_n^k is the part which is homogeneous of degree k under radial expansion, and $\mathfrak{g} = V_n^0$ and $W = \prod_{k \ne 0} V_n^k$. Gelfand and Fuchs[9] calculated $A_0(\mathbf{R}^n)_{\text{basic}}$, and proved it is generated by the characteristic classes, more precisely,

PROPOSITION (3.6) (Gelfand–Fuchs). *The connection $W \to A_0(\mathbf{R}^n)$ induces an isomorphism $\bar{C} \to A_0(\mathbf{R}^n)_{\text{basic}}$, where $\bar{C}$ is the quotient of the polynomial algebra $C = W_{\text{basic}}$ on $c_1, \ldots, c_n$ by the ideal of all elements of dimension greater than $2n$.*

We shall not prove this here (cf. [10]), but shall make some remarks about the proof. V_n has the topological basis $\{x_{j_0} \ldots x_{j_k}(\partial/\partial x_i)\}$, where $(i, j_0, \ldots, j_k)$ runs through the $(k+2)$-tuples such that $j_0 \le j_1 \le \cdots \le j_k$. Let $\{\theta^i_{j_0 \ldots j_k}\}$ be the dual basis of V_n^*. Then $A = A_0(\mathbf{R}^n)$ is the exterior algebra on the $\theta^i_{j_0 \ldots j_k}$, and A_{horiz} is generated by those with $k \ne 0$. The connection forms are the θ_j^i, and from the relation $d\theta_j^i = \Sigma \theta_k^i \wedge \theta_j^k + \Sigma \theta^k \wedge \theta^i_{jk}$ it follows that the curvature forms (the images of the generators of $S\mathfrak{g}^* \subset W$) are given by $\Omega_j^i = \Sigma \theta^k \wedge \theta^i_{jk}$. From invariant theory (cf. [2]) one knows that A_{basic} is generated by completely contracted expressions such as

$$\Omega_1 = \Sigma \Omega_i^i = \Sigma \theta^j \wedge \theta^i_{ij}$$

$$\Omega_2 = \Sigma \Omega_j^i \wedge \Omega_i^j = \Sigma \theta^i \wedge \theta^j \wedge \theta^j \wedge \theta^k_{il} \wedge \theta^l_{jk}$$

$$\Omega_n = \Sigma \Omega^{i_1}_{i_2} \wedge \cdots \wedge \Omega^{i_n}_{i_1} = \Sigma \theta^{j_1} \wedge \cdots \wedge \theta^{j_n} \wedge \theta^{i_1}_{i_2 j_1} \wedge \cdots \wedge \theta^{i_n}_{i_1 j_n}.$$

Because $\theta^i_{j_0 \ldots j_k}$ is symmetric in its lower indices it is easy to see that it cannot occur in any invariant expression if $k > 1$, and that any invariant is a polynomial in $\Omega_1, \ldots, \Omega_n$. These are essentially the Chern classes. The difficult step is to see that there are no relations among them except the obvious vanishing of expressions of dimension greater than $2n$.

In view of (3.5) if we could find a manifold B with a cohomology equivalence $\bar{C} \to \Omega(B)$ we could construct a bundle on B whose cohomology would be that of $A_0(\mathbf{R}^n)$. But is enough to find B and a cochain algebra M with equivalences $\bar{C} \leftarrow M \to \Omega(B)$. In fact *any* manifold B whose cohomology ring is $\bar{C}$ will do, for Sullivan's theory[16] shows that there is a unique minimal cochain algebra M (in his sense) whose cohomology is $\bar{C}$, and this M will automatically map to $\bar{C}$ and $\Omega(B)$. In other words the rational homotopy type of B is completely determined by its cohomology ring. There is in any case a natural candidate for B. The complex Grassmannian $G_{n,2n}$ of n-dimensional subspaces of $\mathbf{C}^{2n}$ has its cohomology ring generated by the Chern classes $c_1, \ldots, c_n$, and there are no relations between them up to dimension $2n$. It also has a natural cell decomposition given by its Schubert cells, which are complex manifolds and hence even-dimensional. The union B_0 of all the Schubert cells of $G_{n,2n}$ of dimension $\leq 2n$ has the desired cohomology ring $\bar{C}$. But as B_0 is an algebraic variety with singularities we take B to be an open neighborhood of B_0 in $G_{n,2n}$ to get a smooth manifold.

One further point needs to be mentioned. One cannot apply (3.5) to the forms on a principal G-bundle when G is not compact, as the condition about finite-dimensional irreducible G-modules does not hold. But *with complex coefficients* (which we are using) the Weil algebras of $GL_n(\mathbf{R})$ and U_n are identical, for invariant forms on either can be identified with holomorphic invariant forms on $GL_n(\mathbf{C})$. So we define F to be the natural principal U_n-bundle on the partial Grassmannian B, and then have cohomology equivalences

$$A_0(\mathbf{R}^n) \leftarrow W \otimes_C \bar{C} \leftarrow W \otimes_C M \rightarrow W \otimes_C \Omega(B) \rightarrow \Omega(F),$$

where W is the common Weil algebra of $GL_n(\mathbf{R})$ and U_n. The basic subalgebra of W is C with respect to any of the three groups $GL_n(\mathbf{R})$, $GL_n^+(\mathbf{R})$, U_n. ($GL_n^+(\mathbf{R})$ is the identity component of $GL_n(\mathbf{R})$.) And the basic subalgebra of $A_0(\mathbf{R}^n)$ is $\bar{C}$ for both $GL_n(\mathbf{R})$ and $GL_n^+(\mathbf{R})$. So the two maps on the right are U_n-equivalences by (3.5), and the two on the left are $GL_n^+(\mathbf{R})$-equivalences by (3.5), and hence $GL_n(\mathbf{R})$-equivalences. Thus all the maps are compatible with $O_n = GL_n(\mathbf{R}) \cap U_n$, and, taking $A = W \otimes_C M$ we have the desired O_n-equivalences

$$A_0(\mathbf{R}^n) \leftarrow A \rightarrow \Omega(F).$$

We conclude this section by giving the proof of Proposition (3.5).

We are given that $A \cong \Lambda\mathfrak{g}^* \otimes A_{\text{horiz}}$. Then A has a filtration $A = A_0 \supset A_1 \supset A_2 \supset \ldots$, where $A_p = \bigoplus_{q \geq p} \Lambda\mathfrak{g}^* \otimes A^q_{\text{horiz}}$, which is compatible with d. It is easy to check that the differential induced in the associated graded algebra gr A makes gr A into $C^*(\mathfrak{g}; A_{\text{horiz}})$, the cochain algebra of the Lie algebra $\mathfrak{g}$ acting on the graded $\mathfrak{g}$-module A_{horiz}. Thus there is a spectral sequence

$$H^*(\mathfrak{g}; A_{\text{horiz}}) \Rightarrow H^*(A).$$

On the other hand the horizontal part of $W \otimes_C A_{\text{basic}}$ is $S\mathfrak{g}^* \otimes_C A_{\text{basic}}$, so there is a similar spectral sequence

$$H^*(\mathfrak{g}; S\mathfrak{g}^* \otimes_C A_{\text{basic}}) \Rightarrow H^*(W \otimes_C A_{\text{basic}}),$$

and a map from the second to the first. But for a reductive algebra $\mathfrak{g}$ and a finite dimensional simple $\mathfrak{g}$-module M the cohomology. $H^*(\mathfrak{g}; M)$ is zero unless $\mathfrak{g}$ acts trivially on M. As the $\mathfrak{g}$-invariant part of $S\mathfrak{g}^* \otimes_C A_{\text{basic}}$ is $C \otimes_C A_{\text{basic}} = A_{\text{basic}}$, which coincides with the invariant part of A_{horiz}, the two spectral sequences have the same E_1-term $H^*(\mathfrak{g}; \mathbf{R}) \otimes A_{\text{basic}}$, and so $H^*(W \otimes_C A_{\text{basic}}) \xrightarrow{\cong} H^*(A)$. The last assertion of (3.5) follows from the fact that the E_2-term of the spectral sequence is $H^*(\mathfrak{g}; \mathbf{R}) \otimes H^*(A_{\text{basic}})$.

§4. THE FUNDAMENTAL MAP

Suppose that M and N are smooth manifolds of the same dimension n, and let P be a smooth family of immersions of M in N. That is, there is a smooth map $f: P \times M \rightarrow N$ such that $f_p = f|p \times M$ is an immersion for each p in P. The purpose of this section is to show that f induces a homomorphism of cochain algebras $\hat{f}: A(M) \rightarrow \Omega(P; A(N))$. The object on the right is the bigraded algebra of differential forms on P with coefficients in $A(N)$. Its differential is the sum of the de Rham differential d_P in $\Omega(P)$ and the Lie algebra differential d_A in $A(N)$.

Before constructing $\hat{f}$ we shall make some general remarks about it. First, if one restricts it to a point $p \in P$ the result is the map $A(M) \rightarrow A(N)$ induced by f_p. If one composes it with the augmentation $A(N) \rightarrow \mathbf{C}$ one obtains a map $A(M) \rightarrow \Omega(P)$ which is also familar. For recall that for any foliation of $P \times M$ of codimension n which is transversal to the fibres of the projection $P \times M \rightarrow P$ there is a "characteristic class" homomorphism $A(M) \rightarrow \Omega(P)$. Our map is associated to the foliation of $P \times M$ by the fibres of $f: P \times M \rightarrow N$.

If G is a Lie group the product $G \times G \to G$ induces a map $\Omega(G) \to \Omega(G \times G) = \Omega(G; \Omega(G))$. Restricting this to forms invariant under the right action of G it becomes $A(\mathfrak{g}) \to \Omega(G; A(\mathfrak{g}))$, for the right-invariant forms on G are the cochains of its Lie algebra $\mathfrak{g}$. The map we shall construct is an analogue of this. For $f: P \times M \to N$ induces a map $P \times \mathrm{Imm}(N; N) \to \mathrm{Imm}(M; N)$ which is equivariant with respect to the monoid $\mathrm{Imm}(N; N)$. ("Imm" denotes a space of immersions.) But $A(M)$ and $A(N)$ can be thought of as in some sense the differential forms on $\mathrm{Imm}(M; N)$ and $\mathrm{Imm}(N; N)$ which are left-invariant under $\mathrm{Imm}(N; N)$.

The construction is as follows. The tangent space to $\mathrm{Imm}(M; N)$ at any point can be identified with the vector fields on M. For if $f_t: M \to N$ is a one-parameter family of immersions then $m \to \{Df_t(m)^{-1}(d/dt)f_t(m)\}|_{t=0}$ is a vector field on M. Accordingly, given a family $f_p: M \to N$ parametrized by $p \in P$ one can associate to each tangent vector to P a vector field on M. I.e. one has a map $\omega_p: P \to \mathrm{Vect}(M)$ for each $p \in P$. On the other hand f_p induces $f_p^*: \mathrm{Vect}(N) \to \mathrm{Vect}(M)$. Now elements of $\Omega(P; A(N))$ can be thought of as functions which to each point p of P associate a multi-linear alternating form on $T_pP \oplus \mathrm{Vect}(N)$. Starting with $\alpha \in A(M)$ we get such a function by composing α with the family of maps $(-\omega_p) \oplus f_p^*: T_pP \oplus \mathrm{Vect}(N) \to \mathrm{Vect}(M)$. This defines a homomorphism of algebras $\hat{f}: A(M) \to \Omega(P; A(N))$. We have to prove that it is compatible with the differentials. It is enough to do so for elements of $A^1(M)$, for their products are dense in $A(M)$, and the differentials are antiderivations.

We shall think of the maps ω_p as constituting a 1-form ω on P with values in $\mathrm{Vect}(M)$, and of the f_p^* as a 0-form with values in $\mathrm{Hom}(A(M); A(N))$. Then the definition of ω can be re-expressed as $d_P f = -f \cdot \theta(\omega)$, where $\theta(\omega)$ is thought of as a 1-form with values in $\mathrm{End}(A(M))$. Applying d_P to this definition gives the "Maurer–Cartan" relation $d_P\theta(\omega) = -\theta(\omega) \wedge \theta(\omega)$, because f is injective at each point. Alternatively expressed, $d_P\omega = -\frac{1}{2}[\omega, \omega]$.

For $\alpha \in A^1(M)$ we have defined $\hat{f}\alpha = f \cdot \alpha - \alpha(\omega)$. So

$$
\begin{aligned}
d\hat{f}\alpha &= d_P f \cdot \alpha + d_A f\alpha - d_P\alpha(\omega) \quad (\text{for } d_A = 0 \text{ on } A^0(N)) \\
&= -f \cdot \theta(\omega) \cdot \alpha + f d_A\alpha + \frac{1}{2}\alpha([\omega, \omega]) \\
&= -f \cdot i(\omega) d_A\alpha + f d_A\alpha + \frac{1}{2} d_A\alpha(\omega, \omega) \\
&= f \cdot (1 - i(\omega) + \frac{1}{2} i(\omega)i(\omega)) \cdot dA\alpha \\
&= \hat{f} d_A\alpha,
\end{aligned}
$$

because for $\beta \in A^2(M)$ one has $\hat{f}\beta = f \cdot (1 - i(\omega) + \frac{1}{2}i(\omega)i(\omega)) \cdot \beta$. (The general formula, of course, is $\hat{f} = f \cdot \exp(-i(\omega))$.)

To conclude this section we need one more remark. If a Lie group G acts on P and N so that $f: P \times M \to N$ is equivariant then $\Omega(P; A(N))$ is a G-cochain-algebra in the sense of §3 and $\hat{f}: A(M) \to \Omega(P; A(N))$ has its image contained in $\Omega(P; A(N))_{\mathrm{basic}}$. Indeed the image consists of G-invariant elements by the naturality of the construction, and of horizontal elements because $\hat{f}$ is induced by maps $(-\omega_p) \oplus f_p^*: T_pP \otimes \mathrm{Vect}(N) \to \mathrm{Vect}(M)$ which vanish on the image of $\mathfrak{g}$.

§5. COCHAINS FOR MAPPING SPACES AND SPACES OF SECTIONS

In this section we shall describe how to construct a cohain algebra from which the cohomology of a mapping space, or more generally the space of cross-sections of a fibre bundle, can be calculated.

Let us begin for simplicity with the space of maps $X \to Y$, where X is a space which has an open covering $\mathfrak{u} = \{U_\alpha\}_{\alpha \in S}$ by contractible sets all of whose finite intersections are also contractible or empty. We shall call such a $\mathfrak{u}$ a *contractible covering.* We write $U_\sigma = U_{\alpha_0} \cap \cdots \cap U_{\alpha_q}$ when $\sigma = \{\alpha_0, \ldots, \alpha_q\}$ is a subset of S, and also $\sigma \leq \tau$ to denote $\sigma \supset \tau$, so that $\sigma \leq \tau \Rightarrow U_\sigma \subset U_\tau$.

The covering $\mathfrak{u}$ has a *nerve,* which is the simplicial set Σ whose set of p-simplexes Σ_p is the set of chains $\sigma_0 \leq \cdots \leq \sigma_p$ of finite subsets of S such that U_{σ_0} is non-empty. (Of course, this is the barycentric subdivision of what is usually called the nerve.)

We shall prove that in appropriate circumstances the cohomology of the mapping space Y^X is the total cohomology of a double complex

$$C(Y^{\Sigma}) = \{C(Y^{\Sigma_0}) \leftarrow C(Y^{\Sigma_1}) \leftarrow C(Y^{\Sigma_2}) \leftarrow \cdots\}$$

where Y^{Σ_p} denotes the product of a collection of copies of Y indexed by the set Σ_p of p-complexes of the nerve of $\mathfrak{u}$. The horizontal differential in $C(Y^{\Sigma})$ comes from the boundary operators in the nerve Σ. C can denote any reasonable kind of cochains.

To make this more precise it is best to divide the construction into two steps. Using the simplicial set $\{\Sigma_p\}$ we constructed a *covariant* functor $[p] \rightarrow Y^{\Sigma_p}$ from the category of finite ordered sets to spaces. Such a functor is called a *cosimplicial space.* To any cosimplicial space $[p] \rightarrow Z_p$ is associated a space $|Z|$, called its *realization.* This is defined as the subspace of the product $\prod_{p \geq 0} Z_p^{\Delta_p}$ consisting of all sequences $\alpha_p: \Delta^p \rightarrow Z_p$ such that the diagram

$$\begin{array}{ccc} \Delta^p & \xrightarrow{\alpha_p} & Z_p \\ {\scriptstyle \theta_*}\downarrow & & \downarrow{\scriptstyle \theta_*} \\ \Delta^q & \xrightarrow{\alpha_q} & Z_q \end{array}$$

commutes for all simplicial operations $\theta: [p] \rightarrow [q]$. (Here Δ^p denotes the standard p-simplex.) With this definition we have the following tautology.

PROPOSITION (5.1). *The realization of the cosimplicial space* $[p] \rightarrow Y^{\Sigma_p}$ *is exactly the space of maps from the realization of the simplicial set* $[p] \rightarrow \Sigma_p$ *to* Y. *That is* $|Y^{\Sigma}| = Y^{|\Sigma|}$.

For a contractible covering $\mathfrak{u}$ the nerve $|\Sigma|$ is homotopy equivalent to X (cf. [14, (4.1)] and [15, (A.1)]), so then $Y^X \simeq |Y^{\Sigma}|$.

If Z is a *simplicial* space, i.e. a contravariant functor $[p] \rightarrow Z_p$ it is well known that the cohomology of its realization is that of the double complex $\{C^q(Z_p)\}$. What we need here is an analogous statement for cosimplical spaces, but it is very much more delicate, and not true without restrictive hypotheses.

The first thing to notice is that taking the cochains on a cosimplical space Z gives one a simplicial cochain complex $[p] \rightarrow C(Z_p)$, i.e. an array of groups

$$\begin{array}{ccccc} \vdots & & \vdots & & \vdots \\ \uparrow & & \uparrow & & \uparrow \\ C_0^1 & \leftleftarrows & C_1^1 & \overset{\leftarrow}{\leftleftarrows} & C_2^1 \; \cdots \\ \uparrow & & \uparrow & & \uparrow \\ C_1^0 & \leftleftarrows & C_1^0 & \overset{\leftarrow}{\leftleftarrows} & C_1^0 \; \cdots, \end{array}$$

where $C_p^q = C^q(Z_p)$. In the total complex C_p^q has degree $q - p$, so in general there will be terms with both positive and negative total degree, whereas the cohomology of $|Z|$ is by definition all of positive degree. In any case we must be quite precise about the definition of the total complex.

Definition (5.2). The *realization* $|C|$ of a simplicial cochain complex $C = \{C_0 \leftleftarrows C_1 \overset{\leftarrow}{\leftleftarrows} \cdots\}$ is the total cochain complex of the normalized bicomplex $\{C_0 \leftarrow \bar{C}_1 \leftarrow \bar{C}_2 \leftarrow \cdots\}$, whose vertical differentials are those of the cochain complexes C_p and whose horizontal differentials are the alternating sums of the simplicial operations $\partial_i: C_p \rightarrow C_{p-1}$.

Normalized means that C_p has been replaced by $C_p = C_p/C_p^{\text{degen}}$, where C_p^{degen} is the sum of the images of the degeneracy operations $s_i: C_{p-1} \rightarrow C_p$. The *total complex* of the bicomplex has by definition in degree r the direct *sum* (not the product) of the $\bar{C}_p^q$ for $q - p = r$.

We want a theorem asserting that the realization $|C(Z)|$ of the cochains on a cosimplicial space Z has the same cohomology as the realization $|Z|$. Applying it to $Z = Y^{\Sigma}$ will give the desired model for the cochains of the mapping space $Y^{|\Sigma|}$.

Before stating the result we should mention multiplicative properties. We should like to consider not cochain complexes but cochain algebras, for in the light of the theories of

Quillen[13] and Sullivan[16] commutative cochain algebras over the real numbers or the rationals can be thought of as real or rational homotopy types. If C is simplicial cochain algebra its realization $|C|$ is also an algebra, the product being defined by "Eilenberg-Zilber shuffles". That is, the product of $x \in C^{q_1}_{p_1}$ and $y \in C^{q_2}_{p_2}$ is defined as $\sum_\theta \operatorname{sign}(\theta) \cdot \theta_1^* x \cdot \theta_2^* y$ in $C^{q_1+q_2}_{p_1+p_2}$, where $\theta = (\theta_1, \theta_2): [p_1 + p_2] \to [p_1] \times [p_2]$ runs through all (p_1, p_2)-shuffles. If each of the algebras C_p is commutative then the product in $|C|$ is also commutative. (This is of course in contrast with the non-commutative Alexander-Whitney product which applies to cosimplicial algebras.)

If X is a smooth manifold its de Rham complex $\Omega(X)$ is a convenient commutative cochain complex which represents its homotopy type over the real numbers. The simplest commutative cochain complex for a general space seems to be the singular de Rham complex, for which a q-cochain is a compatible collection of elements of $\Omega^q(\Delta^r)$, one for each singular simplex $\Delta^r \to X$. This is discussed in [17]. Proposition (5.9) below enables us to pass rather freely between different kinds of cochains.

We shall suppose at any rate that our cochains are such that there are natural integration maps $\int_{\Delta^p}: C^{p+q}(\Delta^p \times X) \to C^q(X)$ for any space X satisfying $(-1)^p\, d \int_{\Delta_p} \alpha = \int_{\Delta^p} d\alpha - \int_{\partial\Delta^p} \alpha$ and vanishing on elements pulled back from $\Delta^{p-1} \times X$ by degeneracy maps $\Delta^p \to \Delta^{p-1}$. Then for any cosimplicial space Z the natural maps $\Delta^p \times |Z| \to Z_p$ induce homomorphisms $C^{p+q}(Z_p) \to C^q(|Z|)$ which fit together to define a cochain map $|C(Z)| \to C(|Z|)$. It is even simple to check that this is a homomorphism of algebras when the cochains are multiplicative, providing $\int_{\Delta^p}: C(\Delta^p \times X) \to C(X)$ is a homomorphism of $C(X)$-modules.

For our main theorem we shall consider only a special class of cosimplical spaces. Suppose that $\{P_\alpha\}_{\alpha \in A}$ is an inverse system of spaces, i.e. A is a partially ordered set and there are given compatible maps $P_\beta \to P_\alpha$ whenever $\alpha \le \beta$. Then there is a cosimplicial space Z with $Z_p = \prod_{\alpha_0 \le \cdots \le \alpha_p} P_{\alpha_0}$. (Its realization is often called the "homotopical inverse limit" of the system.)

Proposition (5.3) (Anderson[1]). *In the foregoing situation if A is finite and each P_α is n-connected, where n is the dimension of the nerve of A, then the natural map $|C(Z)| \to C(|Z|)$ is a cohomology equivalence.*

(Of course the *nerve* of A means the simplicial set whose p-simplexes are the chains $\alpha_0 \le \cdots \le \alpha_p$ in A.)

Corollary (5.4). *If U is a finite open covering of X with nerve Σ, and $\pi_i(Y) = 0$ for $i \le \dim \Sigma$, then $|C(Y^\Sigma)| \to C(Y^{|\Sigma|})$ is a cohomology equivalence.*

The proof of (5.3) will be postponed to §7.

Without any more work we can now obtain a result for the space of sections of a bundle. First recall [14] that to an open covering $\{U_\alpha\}_{\alpha \in S}$ of a space X is associated a simplicial space X_Σ called its *thickened nerve*, defined by

$$(X_\Sigma)_p = \coprod_{\sigma_0 \le \cdots \le \sigma_p} U_{\sigma_0}.$$

There is a natural map $\pi: |X_\Sigma| \to X$, which is a homotopy equivalence if X is paracompact.

Now let $p: E \to X$ be a map, and let Γ_σ be the space of sections of p over U_σ. $\{\Gamma_\alpha\}$ is an inverse system of spaces, to which is associated the cosimplicial space Γ_Σ. We have the tautology.

Proposition (5.5). *The realization of Γ_Σ is the space of sections of $\pi^* E$, the pull-back of E by $\pi: |X_\Sigma| \to X$.*

Corollary (5.6). *If $p: E \to X$ is a Hurewicz fibration and X is paracompact then $|\Gamma_\Sigma| \simeq \Gamma(E)$, the space of sections of E.*

If we now choose the covering U contractible the spaces Γ_σ from which Γ_Σ is formed are all homotopy equivalent to fibres of $E \to X$, and (5.3) and (5.6) give a cochain complex for $\Gamma(E)$ providing the covering is finite and of dimension less than the connectivity of the fibres. But because the spaces Γ_σ are infinite-dimensional it will be convenient to replace Γ_Σ by an equivalent but very much smaller cosimplical space, constructed as follows.

For any subset σ of S such that U_σ is non-empty let U^σ denote $\bigcup_{\alpha \in \sigma} U_\alpha$, the *star* of U_σ. This is contractible if U is a contractible covering. Let $E^\sigma = p^{-1}(U^\sigma)$. Then $\{E^\sigma\}$ is an inverse system of spaces. It defines a cosimplicial space E^Σ which is the model we are seeking. It is comparable in size with the cosimplicial space Y^Σ for a mapping space. The essential property of E^Σ follows from

PROPOSITION (5.7). *If $p: E \to X$ is a Hurewicz fibration on a paracompact base, and U is a contractible covering of X then there is a transformation of inverse systems $\{E^\sigma\} \to \{\Gamma_\sigma\}$ such that $E^\sigma \to \Gamma_\sigma$ is a homotopy equivalence for each σ.*

Proof. Consider the relation $\sim$ on X defined by $x \sim y$ if and only if x and y both belong to a single set U_α of the covering U. A *slicing function* (cf. [12]) for the bundle E is a family of homotopy equivalences $T_{xy}: E_x \to E_y$ defined when $x \sim y$ and depending continuously on (x, y). (Here E_x means $p^{-1}(x)$.) It is easy to see that any Hurewicz fibration on a paracompact base admits a slicing function with respect to any contractible covering. Using it one defines $E^\sigma \to \Gamma_\sigma$ by associating to $e \in E_x$ the section whose value at y is $T_{xy}e$. That is possible because if $x \in U^\sigma$ and $y \in U_\sigma$ then $x \sim y$. The naturality with respect to σ is immediate.

COROLLARY (5.8). *If X has a finite contractible covering of dimension n, and $E \to X$ is a Hurewicz fibration whose fibres are n-connected, then there is a cohomology equivalence $|C(E^\Sigma)| \to C(\Gamma(E))$.*

To deduce the corollary from the preceding proposition one needs the following lemma, which we will use repeatedly. It is the analogue of a familiar property [15, (A.1)] of the realization of simplicial spaces.

PROPOSITION (5.9). *If $C \to C'$ is a morphism of simplicial cochain complexes such that $C_p \to C'_p$ is a cohomology equivalence for each p then $|C| \to |C'|$ is a cohomology equivalence.*

Proof. The realization $|C|$ of a simplicial cochain complex is the direct limit of its "skeletons" $|C|_n$, where $|C|_n$ is the total complex of the truncated bicomplex $(C_0 \leftarrow \bar{C}_1 \leftarrow \cdots \leftarrow \bar{C}_n)$. Because cohomology commutes with direct limits it is enough to show $|C|_n \to |C'|_n$ is an equivalence for each n. But $|C|_n$ fits into a short exact sequence $0 \to |C|_{n-1} \to |C|_n \to \bar{C}_n \to 0$ of cochain complexes, so by induction and the five lemma it is enough to show $\bar{C}_n \to \bar{C}'_n$ is an equivalence. This follows from the lemma.

LEMMA (5.10). *For any simplicial cochain complex C there is an exact sequence of cochain complexes*

$$0 \leftarrow \bar{C}_n \leftarrow C_n \leftarrow \bigoplus_i C^{(i)}_{n-1} \leftarrow \bigoplus_{i<j} C^{(ij)}_{n-2} \leftarrow \cdots \leftarrow C_0^{(12\ldots n)} \leftarrow 0,$$

where $C^{(i_1 \ldots i_k)}_{n-k}$ is a copy of C_{n-k}.

§6. THE MAIN THEOREM

Let M be a Riemannian manifold which has a finite covering U by geodesically convex open sets. If M is either compact or the interior of a compact manifold with boundary it has such a covering, as any covering by sufficiently small geodesic disks will do [4].

Let P be the principal bundle of M, with group O_n, and let E be the associated bundle whose fibre is the O_n-manifold F constructed in §3.

PROPOSITION (6.1). *The Gelfand–Fuchs algebra $A(M)$ has the homotopy type of the space of sections of E (over the complex numbers).*

This means of course that $A(M)$ is linked to the complex cochain algebra of $\Gamma(E)$ by a sequence of homomorphisms of cochain algebras (unfortunately not all going in the same direction) each of which is a cohomology equivalence.

In §5 we showed that a suitable cochain algebra for the space of sections is given by the realization of the simplicial algebra $\Omega(E^\Sigma)$, where E^Σ is the cosimplical space associated to the contractible covering U. Proposition (5.8) applies because the fibre F is $2n$-connected when M is n-dimensional. (Homotopically F is the fibre of the inclusion of the $2n$-skeleton of BU_n in BU_n.) One can use the differential forms on E^Σ as cochains by (5.9), for each term of E^Σ is a finite dimensional manifold.

The contractible covering U can also be used to decompose the Gelfand–Fuchs cochains $A(M)$. Because A is a covariant functor for immersions one can apply it to the thickened nerve M_Σ described in §5 to obtain a simplicial cochain algebra $A(M_\Sigma)$ with an augmentation to $A(M)$.

PROPOSITION (6.2). *The augmentation induces a cohomology equivalence*

$$|A(M_\Sigma)| \to A(M).$$

The proof of (6.2) will be given in §8. In fact a familiar argument with a partition of unity shows that the simplicial algebra $A(M_\Sigma)$ is a resolution of $A(M)$, and (6.2) would follow immediately if there were not considerable difficulties with convergence.

Granting (6.2) our task is to relate the simplicial algebras $A(M_\Sigma)$ and $\Omega(E^\Sigma)$. For each point p of the principal bundle P of M, i.e. each frame in the tangent bundle, there is an exponential map $\exp_p: \mathbf{R}^n \to M$ because U_α is geodesically convex this gives us an identification of U_α with an open set of $\mathbf{R}^n$ for each point of P_α, the part of the principal bundle over U_α. Thus P_α can be thought of as a family of embeddings of U_α in $\mathbf{R}^n$, and $P^\alpha = \bigcup_{\alpha \in \sigma} P_\alpha$ as a family of embeddings of U_σ in $\mathbf{R}^n$. The map $P^\sigma \times U_\sigma \to \mathbf{R}^n$ is equivariant with respect to the action of O_n on P^σ and $\mathbf{R}^n$. From §4 it induces a cochain homomorphism $A(U_\sigma) \to \Omega(P^\sigma; A(\mathbf{R}^n))_{\text{basic}}$.

On the other hand $E^\sigma = (P^\sigma \times F)/O_n$, so $\Omega(E^\sigma)$ can be identified with $\Omega(P^\sigma \times F)_{\text{basic}}$, i.e. with $\Omega(P^\sigma; \Omega(F))_{\text{basic}}$. Recalling from §3 the O_n-homomorphisms $A(\mathbf{R}^n) \leftarrow A \to \Omega(F)$, we have a chain

$$A(U_\sigma) \to \Omega(P^\sigma; A(\mathbf{R}^n))_{\text{basic}} \leftarrow \Omega(P^\sigma; A)_{\text{basic}} \to \Omega(P^\sigma; \Omega(F))_{\text{basic}} = \Omega(E^\sigma),$$

which is clearly natural with respect to σ.

All of these morphisms are cohomology equivalences. For $P^\sigma = O_n \times U^\sigma$ as O_n-space (non-canonically), so the chain can be rewritten

$$A(U_\sigma) \to \Omega(U^\sigma; A(\mathbf{R}^n)) \leftarrow \Omega(U^\sigma; A) \to \Omega(U^\sigma; \Omega(F)) = \Omega(E^\sigma).$$

Here the three terms of the form $\Omega(U^\sigma; B)$ are double complexes which (by de Rham's theorem) can be thought of as resolutions of B, so that $B \to \Omega(U^\sigma; B)$ is a cohomology equivalence, and the chain is equivalent to

$$A(U_\sigma) \to A(\mathbf{R}^n) \leftarrow A \to \Omega(F),$$

in which all the maps are equivalences—the first by (2.1).

Natural transformations linking $A(U_\sigma)$ and $\Omega(E^\sigma)$ are not exactly what is needed to link $A(M_\Sigma)$ and $\Omega(E^\Sigma)$. We need rather a relation between $A(\coprod_\sigma U_\sigma)$ and $\Omega(\prod_\sigma E^\sigma)$, where σ runs through the p-simplexes of the nerve of U. But $\prod_\sigma P^\sigma$ is a space of immersions of $\coprod_\sigma U_\sigma$ in $\coprod^k \mathbf{R}^n$ (where k is the the number of p-simplexes), and $\prod_\sigma E^\sigma = \{(\prod P^\sigma) \times F^k\}/G$, where $G = (O_n)^k$. The foregoing argument applies without change to this multiplied situation, providing one knows that the maps $A(\coprod^k \mathbf{R}^n) \leftarrow A^{\otimes k} \to \Omega(F^k)$ are cohomology equivalences. For the right-hand map this follows from the usual Künneth theorem, as F has finite dimensional cohomology. For the other, we saw in §2 that there is an equivalence $A_0(\coprod^k \mathbf{R}^n) \to A(\coprod^k \mathbf{R}^n)$. But $A_0(\coprod^k \mathbf{R}^n) = A_0(\mathbf{R}^n)^{\otimes k}$, which is equivalent to $A^{\otimes k}$ by the Künneth theorem.

Thus we have found equivalences linking $A(M_\Sigma)$ and $\Omega(E^\Sigma)$, and so completed the proof of (6.1).

§7. PROOF OF (5.3)

We have seen that for any cosimplicial space X there is a natural map $|H^*(X)| \to H^*(|X|)$. We shall say that X *converges* when it is an isomorphism.

The starting point of our proof is the "Eilenberg–Moore spectral sequence"[6]. This amount to the assertion that the cosimplicial space Π associated to the diagram $(X \to Z \leftarrow Y)$, i.e.

$$\Pi = \{X \times Y \rightrightarrows X \times Z \times Y \substack{\rightarrow\\[-1ex]\rightarrow\\[-1ex]\rightarrow} X \times Z \times Z \times Y \substack{\rightarrow\\[-1ex]\rightarrow\\[-1ex]\rightarrow\\[-1ex]\rightarrow} \cdots\},$$

converges providing the space Z is connected and simply connected. The realization of Π is the homotopical fibre product of X and Y over Z, which we shall denote by $X\pi_Z Y$. One should observe that the hypothesis on Z implies that the total complex $|H^*(\Pi)|$ involves only finitely many $H^{p+r}(\Pi_p)$ in each total dimension r, for $H^{p+r}(\Pi_p)$ will consist of degenerate elements when $p > r$.

If now $X \to Z \leftarrow Y$ is a diagram of cosimplicial spaces instead of spaces then the foregoing Π will be a bi-cosimplicial space. It has a total realization $|\Pi|$ which can be formed either by realizing the columns first, giving $|X|\pi_{|Z|}|Y|$, or else by realizing the rows first, giving $|X\pi_Z Y|$. That is to say, the realization of cosimplicial spaces commutes with homotopical fibre products.

LEMMA (7.1). *If the cosimplicial spaces X, Y, Z converge then $X\pi_Z Y$ converges, providing*
(i) *$|Z|$ and each Z_p is connected and simply connected, and*
(ii) *$|X|$, $|Y|$, $|Z|$ and each X_p, Y_p, Z_p are of finite type,*
i.e. *their cohomology is finitely generated in each dimension.*

Proof. One realizes in two ways the bi-simplicial graded abelian group $\{C_{pq}\}$ with $C_{pq} = H^*(X_q \times Z_q^{p-1} \times Y_q)$, and applies the Eilenberg–Moore theorem to each row. One needs to know that the product of convergent cosimplicial spaces is convergent. That is true because $|H^*(X \times Y)| \cong |H^*(X) \otimes H^*(Y)| \cong |H^*(X)| \otimes |H^*(Y)|$ by the Künneth theorem and the Eilenberg–Zilber theorem respectively. Using the Künneth theorem makes the hypotheses of finite type necessary, but they could be avoided by considering homology instead of cohomology. (One assumes the coefficients are a field to apply the Künneth theorem, but can then immediately remove the assumption.)

Now suppose that Σ is a simplicial set and $\sigma \mapsto Q_\sigma$ is a covariant functor from the simplexes of Σ to spaces. There is a cosimplicial space

$$Q^\Sigma = \left\{ \prod_{\sigma \in \Sigma_0} Q_\sigma \rightrightarrows \prod_{\sigma \in \Sigma_1} Q_\sigma \,\substack{\to\\\to\\\to}\, \cdots \right\}.$$

The situation of (5.3) is the particular case where Σ is the nerve of the partially ordered set A, and the functor is $[\alpha_0 \leq \cdots \leq \alpha_p] \to P_{\alpha 0}$. Using (7.1) we propose to prove Q^Σ is convergent by induction on the number of non-degenerate simplexes in Σ.

If Σ is the union of two simplicial sets Σ^1 and Σ^2 whose intersection is Σ^0 then the cosimplicial space Q^Σ is the fibre-product of Q^{Σ^1} and Q^{Σ^2} over Q^{Σ^0}. We should like to apply (7.1) to show that Q^Σ converges if the Q^{Σ^i} do. The fibre-product of Q^{Σ^1} and Q^{Σ^2} over Q^{Σ^0} maps to their homotopical fibre-product by a map which is obviously a homotopy equivalence at each level, so (7.1) can be applied providing the fibre-product and the homotopical fibre-product of $|Q^{\Sigma^1}|$ and $|Q^{\Sigma^2}|$ over $|Q^{\Sigma^0}|$ are equivalent. For this it suffices for $|Q^{\Sigma^1}| \to |Q^{\Sigma^0}|$ to be a fibration, which is easily seen to be the case. (Strictly speaking, one ought to say that a cosimplicial space of the type of Q^Σ is *good* (cf. [15] App. A) in the sense that its degeneracy maps are fibrations, and a map of good cosimplicial spaces induces an equivalence of their realizations of it is an equivalence at each level.)

The connectivity hypotheses in (7.1) will hold for Q^{Σ^i} providing each Q_σ is n-connected and $n > \dim \Sigma^i$. If so then the finiteness hypotheses will hold providing each Q_σ is of finite type, and Σ^i is finite, i.e. has only finitely many non-degenerate simplexes.

Thus for a finite simplicial set Σ to prove Q^Σ converges it is enough, by induction, to show that Q^Ω converges, where Ω runs through the irreducible sub-simplicial-sets of Σ, which consist of the faces of a single non-degenerate simplex. In proving (5.3) the Ω which arise are the nerves of totally ordered sets $A = \{\alpha_1 \leq \alpha_1 \leq \cdots \leq \alpha_p\}$. Then $|Q^\Omega| \simeq Q_{\alpha_0}$, and on the other hand $H^*(Q^\Omega)$ is a resolution of $H^*(Q_{\alpha_0})$, for the simplicial homotopy which contracts Ω induces a chain homotopy contracting $H^*(Q^\Omega)$ to $H^*(Q_{\alpha_0})$. The completes the proof of (5.3).

§8. PROOF OF (6.2)

If one has a double complex with augmentation

$$\begin{array}{ccccccccc}
\vdots & & \vdots & & \vdots & & \vdots & & \\
\uparrow & & \uparrow & & \uparrow & & \uparrow & & \\
A^2 & \leftarrow & C_0^2 & \leftarrow & C_1^2 & \leftarrow & C_2^2 & \leftarrow & \cdots \\
\uparrow & & \uparrow & & \uparrow & & \uparrow & & \\
A^1 & \leftarrow & C_0^1 & \leftarrow & C_1^1 & \leftarrow & C_2^1 & \leftarrow & \cdots \\
\uparrow & & \uparrow & & \uparrow & & \uparrow & & \\
A^0 & \leftarrow & C_0^0 & \leftarrow & C_1^0 & \leftarrow & C_2^0 & \leftarrow & \cdots
\end{array}$$

such that each row is an exact sequence, then A has the same cohomology as the "infinite total complex" whose cochains in dimension r are the product $\prod_{q\geq 0} C_r^{q+r}$. This is not the total complex in the sense we are using, for we take the direct sum rather than the product. Apart from this difficulty Proposition (6.2), which concerns the augmented double complex with $A^p = A^p(M)$ and $C_q^p = A^p(M_{\Sigma,q})$, where $M_{\Sigma,q}$ is the qth space of the simplical space M_Σ, would be almost trivial, for the rows form exact sequences by a standard argument using a partition of unity.

To deal with this "convergence" problem we shall use the Gelfand–Fuchs filtration on the cochain algebra $A(M)$. An element of $A^p(M)$ is a distribution on the product of p copies of M. We say it has filtration $\leq k$ if its support is contained in M_k^p, the subset of points $(x_1, \ldots, x_p)$ of M^p such that at most k of the x_i are distinct. If $A_k^p(M)$ denotes the cochains of filtration $\leq k$ then

$$0 = A_0^p(M) \subset A_1^p(M) \subset \cdots \subset A_p^p(M) = A^p(M).$$

Suppose that for some k we form an augmented double complex by applying the functor A_k instead of A to the augmented simplicial space $M_\Sigma \to M$. We shall prove

(a) each row is degenerate above dimension dk, where $d = \dim(\Sigma)$, and

(b) each row is an exact sequence.

The statements (a) and (b) imply that $|A_k(M_\Sigma)| \to A_k(M)$ is a cohomology equivalence, because the normalized double complex associated to $A_k(M_\Sigma)$ has then only $kd+1$ non-zero columns. But taking the union for all k we deduce that $|A(M_\Sigma)| \to A(M)$ is an equivalence, proving (6.2).

Proof of (*a*) (cf. [5, (4.18)]. Let F be a covariant functor from smooth manifolds and immersions to abelian groups. We shall say that F is of degree $\leq k$ if for any manifolds $X_1, \ldots, X_m$ (with $m \geq k$) the natural map

$$\bigoplus_{i_1 < \cdots < i_k} F(X_{i_1} \amalg \cdots \amalg X_{i_k}) \to F(\coprod_{1\leq i\leq m} X_i)$$

is surjective. The functor A_k^p is obviously of degree k.

If $U = \{U_\alpha\}_{\alpha\in S}$ is an open covering of M the simplicial space M_Σ is defined by $M_{\Sigma,q} = \coprod_{\sigma_0\leq\cdots\leq\sigma_l} U_{\sigma_0}$. The dimension of Σ is the largest d such that there is a chain $\sigma_0 < \sigma_1 < \cdots < \sigma_d$ of subsets of S with U_{σ_0} non-empty, i.e. such that M_Σ is degenerate above dimension d.

LEMMA (8.1). *If F is of degree k the simplicial vector space $F(M_\Sigma)$ is degenerate above dimension kd, where $d = \dim(\Sigma)$.*

Proof. Suppose $q > kd$. In dimension $qF(M_\Sigma)$ is $F(\coprod_c U_c)$, where c runs through the chains $c = (\sigma_0 \leq \cdots \leq \sigma_q)$ of Σ, and U_c denotes U_{σ_0}. So each element of it is a sum of elements coming from $F(U_{c_1} \amalg \ldots \amalg U_{c_k})$ for some chains $c_1, \ldots, c_k$. Bit each chain c_i has no more than d non-zero steps, so there must be at least one j such that the jth step of c_i is zero for each i. Then all of $F(U_{c_1} \amalg \cdots \amalg U_{c_k})$ is in the image of the jth degeneracy operation. This proves the lemma.

Proof of (*b*). We have to construct a filtration-preserving contracting homotopy of the simplicial vector space $A^p(M_\Sigma)$.

The simplicial space M_Σ comes from a finite open covering $\{U_\alpha\}_{\alpha\in S}$ of M. By choosing a shrinkage $\{U'_\alpha\}$ of $\{U_\alpha\}$, i.e., $\bar{U}'_\alpha \subset U_\alpha$, one obtains a sub-simplicial space M'_Σ of M_Σ. The complex $A^p(M_\Sigma)$ consists of distributions with compact support on the spaces of $(M_\Sigma)^p$, so it is the union of the $A^p(M'_\Sigma)$ for all shrinkages $\{U'_\alpha\}$ of $\{U_\alpha\}$. It will therefore suffice for us to give a filtration-preserving null-homotopy of the inclusion $A^p(M'_\Sigma) \to A^p(M_\Sigma)$.

Now for any manifold M $A^p(M)$ is the invariant subspace under an action of the pth symmetric group $\mathscr{S}_p$ in a larger space $B^p(M)$ obtained by omitting the alternating condition on cochains. Taking $\mathscr{S}_p$-invariants is an exact functor, so it will suffice for us to give a null-homotopy of $B^p(M'_\Sigma) \to B^p(M_\Sigma)$.

If $\pi = (\sigma_1, \ldots, \sigma_p)$ is a p-tuple of subsets of S let U_π denote $U_{\sigma_1} \times \cdots \times U_{\sigma_p}$. The set of such p-tuples π is partially ordered set Π such that $\pi \leq \pi' \Rightarrow U_\pi \subset U_{\pi'}$. The qth term of the simplicial space $(M_\Sigma)^p$ is

$$\coprod_{\pi_0\leq\cdots\leq\pi_q} U_{\pi_0},$$

and the action of $\mathscr{S}_p$ on $(M_\Sigma)^p$ comes from its action on Π. We define U'_π similarly, using the shrinkage $\{U'_\alpha\}$ of $\{U_\alpha\}$.

For each $\pi \in \Pi$ define $W_\pi = U_\pi - \bigcup_{p \geq \pi} \bar{U}_p$. Then $\{W_\pi\}$ is an open covering of M^p, for if $x \in M^p$ then $x \in W_\pi$, where $\pi = \inf\{p : x \in U_p\}$. Let $\{\varphi_\pi\}$ be a partition of unity subordinate to $\{W_\pi\}$. The vector space $B^p(M'_{\Sigma,q})$ is spanned by elements which can be denoted $[\pi_0\pi_1 \cdots \pi_q; f]$, where $\pi_0 \leq \cdots \leq \pi_q$ in Π, and f is a distribution with compact support in U'_{π_0}. Then

$$\partial: B^p(M'_{\Sigma,q}) \to B^p(M'_{\Sigma,q-1})$$

is given by

$$\partial[\pi_0 \cdots \pi_q; f] = \Sigma(-1)^i[\pi_0 \cdots \hat{\pi}_i \cdots \pi_q; f].$$

We define the desired chain-homotopy

$$h: B^p(M'_{\Sigma,q}) \to B^p(M'_{\Sigma,q+1})$$

by

$$h[\pi_0 \cdots \pi_q; f] = \sum_\pi [\pi\pi_0 \cdots \pi_q; \varphi_\pi f].$$

This makes sense because if $\pi_0 \not\geq \pi$ the sets W_π and U'_{π_0} are disjoint, and $\varphi_\pi f = 0$. It is trivial that $\partial h + h\partial$ is the inclusion; so h is a null-homotopy. The crucial point is to show that if $[\pi_0 \cdots \pi_q; f]$ has filtration k then so does $[\pi\pi_0 \cdots \pi_q; \varphi_\pi f]$, and for this to hold one must choose the partition $\{\varphi_\pi\}$ appropriately. The condition needed is that for every $x \in \operatorname{supp}(\varphi_\pi) \subset M^p$, and every $g \in \mathscr{S}_p$ such that $gx = x$, one has $g\pi = \pi$. That is, one needs

$$\operatorname{supp}(\varphi_\pi) \subset W'_\pi = W_\pi - \bigcup_{g\pi \neq \pi} (M^p)^g.$$

But $\{W'_\pi\}$ is an open covering of M^p, for if $x \in M^p$ and $\pi = \inf\{p : x \in U_p\}$ then $x \in W'_\pi$. So $\{\varphi_\pi\}$ can be chosen subordinate to $\{W'_\pi\}$, and the proof is complete.

REFERENCES

1. D. W. ANDERSON: A generalization of the Eilenberg–Moore spectral sequence, *Bull. Am. math. Soc.* **78** (1972), 784–786.
2. M. F. ATIYAH, R. BOTT and V. K. PATODI: On the heat equation and the index theorem, *Invent. Math.* **19** (1973), 279–330.
3. H. CARTAN: Notions d'algèbre différentielle. La transgression dans un groupe de Lie. *Colloque de Topologie*, pp. 15–27, 57–71. C.B.R.M., Bruxelles (1950).
4. G. DE RHAM: Complexes à automorphismes et homéomorphie différentiable. *Annls Inst. Fourier Univ. Grenoble*: **2** (1950), 51–67.
5. A. DOLD and D. PUPPE: Homologie nicht-additiver Funktoren, *Annl Inst. Fourier Univ. Grenoble* **11** (1961), 201–312.
6. S. EILENBERG and J. C. MOORE: Homology and Fibrations I, *Comment. math. Helvet.* **40** (1965), 199–236.
7. I. M. GEL'FAND and D. B. FUCHS: The cohomology of the Lie algebra of vector fields on the circle, *Funkt. Anal. i yevo prilozheniya* **2.4** (1968), 92–93.
8. I. M. GEL'FAND and D. B. FUCHS: The cohomology of the Lie algebra of tangent vector fields on a smooth manifold, *Funkt. Anal. i yevo prilozheniya* **3.3** (1969), 32–52, and **4.2** (1970), 23–31.
9. I. M. GEL'FAND and D. B. FUCHS: The cohomology of the Lie algebra of formal vector fields, *Izv. Akad. Nauk S.S.S.R., ser. matem* **34.2** (1970), 322–337.
10. C. GODBILLON: Cohomologie d'algèbres de Lie de champs de vecteurs formels, *Sém. Bourbaki* **421** (1972), 1–19.
11. A. HAEFLIGER: Sur la cohomologie de Gel'fand–Fuks. *Ann. scient. Éc. Norm. Sup.* **9** (1976), 503–532.
12. J. MILNOR: The construction of universal bundles I, *Ann. Math.* **63** (1956), 272–284.
13. D. QUILLEN: Rational homotopy theory, *Ann. Math.* **90** (1969), 205–295.
14. G. B. SEGAL: Classifying spaces and spectral sequences, *Publs math. Inst. ht. Étud. scient.* **34** (1968), 105–112.
15. G. B. SEGAL: Categonies and cohomology theories, *Topology* **13** (1974), 293–312.
16. D. SULLIVAN: Differential forms and the topology of manifolds. *Proceedings of an International Conference on Topology and related topics.* Univ. of Tokyo (1973).
17. R. G. SWAN: Thom's theory of differential forms on simplicial sets, *Topology* **14** (1975), 271–273.

Harvard University
Oxford University

Permissions — Volume 3

Birkhäuser Boston thanks the original publishers of the papers of Raoul Bott for granting permission to reprint specific papers in this collection.

Contemporary Mathematicians

Series Editor: Gian-Carlo Rota

Books in the Series:

Gabor Szego: Collected Papers
Edited by Richard Askey
0-8176-3063-5 (complete set)
0-8176-3056-2 (Volume 1: 1915-1927)
0-8176-3060-0 (Volume 2: 1927-1943)
0-8176-3061-9 (Volume 3: 1943-1972)

Lars Valerian Ahlfors: Collected Papers
0-8176-3077-5 (complete set)
0-8176-3075-9 (Volume 1: 1929-1955)
0-8176-3076-7 (Volume 2: 1955-1979)

The Selected Papers of Theodore S. Motzkin
Edited by D. Cantor, et al.
0-8176-3087-2

Fritz John: Collected Papers
Edited by Jurgen Moser
0-8176-3265-4 (complete set)
0-8176-3266-2 (Volume 1)
0-8176-3267-0 (Volume 2)

Kurt Otto Friedrichs: Collected Mathematical Papers
Edited by Cathleen Morawetz
0-8176-3270-0 (complete set)
0-8176-3268-9 (Volume 1)
0-8176-3269-7 (Volume 2)

Jakob Nielsen: Collected Papers
Edited by Vagn Lundsgaard Hansen
0-8176-3152-6 (complete set)
0-8176-3140-2 (Volume 1)
0-8176-3151-8 (Volume 2)

Selected Papers on Algebra and Topology
by Garrett Birkhoff
Edited by Gian-Carlo Rota *and* Joseph Oliveira
0-8176-3114-3

Shizuo Kakutani: Selected Papers
Edited by Robert Kallman
0-8176-3279-4 (complete set)
0-8176-3277-8 (Volume 1)
0-8176-3278-6 (Volume 2)

Alfred Tarski: Collected Papers
Edited by Steven R. Givant *and* Ralph McKenzie
0-8176-3284-0 (complete set)
0-8176-3280-8 (Volume 1)
0-8176-3281-6 (Volume 2)
0-8176-3282-4 (Volume 3)
0-8176-3283-2 (Volume 4)

Charles Loewner: Collected Papers
Edited by Lipman Bers
0-8176-3377-4

Collected Works of Arne Beurling
Edited by L. Carleson, et al.
0-8176-3412-6 (complete set)
0-8176-3415-0 (Volume 1)
0-8176-3416-9 (Volume 2)

The Dilworth Theorems:
Seleceted Papers of Robert P. Dilworth
Edited by K. Bogart, et al.
0-8176-3434-7

Nathan Jacobson: Collected Mathematical Papers
0-8176-3410-X (Volume 1: 1934-1946)
0-8176-3411-8 (Volume 2: 1947-1965)
0-8176-3446-0 (Volume 3: 1965-1985)

I.J. Schoenberg: Selected Papers
Edited by Carl de Boor
0-8176-3378-2 (complete set)
0-8176-3404-5 (Volume 1)
0-8176-3405-3 (Volume 2)

Hassler Whitney: Collected Papers
Edited by James Eells, et al.
0-8176-3560-2 (complete set)
0-8176-3558-0 (Volume 1)
0-8176-3559-9 (Volume 2)

Collected Papers of Raoul Bott
Edited by Robert MacPherson
0-8176-3613-7 (Volume 1)
0-8176-3646-3 (Volume 2)
0-8176-3647-1 (Volume 3)
0-8176-3648-X (Volume 4)

Collected Works of Emil L. Post
Edited by Martin Davis
0-8176-3579-3

MIX
Papier aus verantwortungsvollen Quellen
Paper from responsible sources
FSC® C105338

If you have any concerns about our products,
you can contact us on
ProductSafety@springernature.com

In case Publisher is established outside the EU,
the EU authorized representative is:
Springer Nature Customer Service Center GmbH
Europaplatz 3, 69115 Heidelberg, Germany

Printed by Libri Plureos GmbH
in Hamburg, Germany